Christian Constanda • Paul Harris
Editors

Integral Methods in Science and Engineering

Analytic Treatment and Numerical Approximations

 Birkhäuser

Editors
Christian Constanda
Department of Mathematics
The University of Tulsa
Tulsa, OK, USA

Paul Harris
Computing, Engineering, and Mathematics
University of Brighton
Brighton, UK

ISBN 978-3-030-16076-0   ISBN 978-3-030-16077-7   (eBook)
https://doi.org/10.1007/978-3-030-16077-7

Mathematics Subject Classification: 45Exx, 45E10, 65R20, 45D05

Preface

The international conferences on Integral Methods in Science and Engineering (IMSE) started in 1985 at the University of Texas–Arlington and, from 1996, continued biennially in a variety of venues around the world. These events bring together specialists who make use of integration techniques as essential tools in their research. These types of procedures are characterized by generality, elegance, and efficiency, which are essential ingredients in the work of a wide category of practitioners.

The time and place of the first 14 IMSE conferences are listed below:

1985, 1990:	University of Texas–Arlington, USA
1993:	Tohoku University, Sendai, Japan
1996:	University of Oulu, Finland
1998:	Michigan Technological University, Houghton, MI, USA
2000:	Banff, AB, Canada (organized by the University of Alberta, Edmonton)
2002:	University of Saint–Étienne, France
2004:	University of Central Florida, Orlando, FL, USA
2006:	Niagara Falls, ON, Canada (organized by the University of Waterloo)
2008:	University of Cantabria, Santander, Spain
2010:	University of Brighton, UK
2012:	Bento Gonçalves, Brazil (organized by the Federal University of Rio Grande do Sul)
2014:	Karlsruhe Institute of Technology, Germany
2016:	University of Padova, Italy

The 2018 event, the 15th in the series, was hosted by the University of Brighton, UK, July 16–20, gathering participants from 16 countries on 5 continents, whose high-quality presentations consolidated the well-deserved reputation of the IMSE conferences as a vehicle for scientists and engineers to communicate their most recent results and ideas and to forge contacts for future professional collaboration.

The Organizing Committee of the conference, assisted by Laura Williams and Richard Huck of Southwest Conferences, was comprised of:

Paul Harris (University of Brighton), Chairman
Jenny Venton (University of Brighton)
Dmitry Savostyanov (University of Brighton)

The accomplishments of the IMSE 2018 meeting were due to a large extent to the financial support received from the School of Computing, Engineering and Mathematics at the University of Brighton. The participants and the Organizing Committee wish to thank this academic body for its underwriting of the success of the conference.

IMSE 2018 included two minisymposia:

Asymptotic Analysis: Homogenization and Thin Structures; organizer: M.E. Pérez–Martínez (University of Cantabria)
Boundary: Domain Integral Equations; organizer: S. Mikhailov (Brunel University London)

The next IMSE conference will be held at the University of St. Petersburg, Russia, July 13–17, 2020. Further details will be posted in due course on the conference web site.

The peer-reviewed chapters of this volume, arranged alphabetically by first author's name, consist of 36 papers presented in Brighton. The editors would like to thank the reviewers for their valuable help and the staff at Birkhäuser–New York for their courteous and professional handling of the publication process.

Tulsa, OK, USA   Christian Constanda
Brighton, UK   Paul Harris
January 2019

The International Steering Committee of IMSE

Christian Constanda (The University of Tulsa), *Chairman*
Bardo E.J. Bodmann (Federal University of Rio Grande do Sul)
Haroldo F. de Campos Velho (INPE, Saõ José dos Campos)
Paul J. Harris (University of Brighton)
Andreas Kirsch (Karlsruhe Institute of Technology)
Mirela Kohr (Babes–Bolyai University of Cluj–Napoca)
Massimo Lanza de Cristoforis (University of Padova)
Sergey Mikhailov (Brunel University London)
Dorina Mitrea (University of Missouri–Columbia)
Marius Mitrea (University of Missouri–Columbia)
David Natroshvili (Georgian Technical University)
Maria Perel (St. Petersburg State University)
Maria Eugenia Pérez–Martínez (University of Cantabria)
Ovadia Shoham (The University of Tulsa)
Iain W. Stewart (University of Dundee)

Continuing a tradition started at the preceding conference, IMSE 2018 hosted an exhibition of digital art that consisted of five portraits of participants and a special conference poster, executed by artist Walid Ben Medjedel using different techniques. The exhibits were much appreciated by the participants, who, as before, commented on the relationship between digital art and mathematics.

The portraits and the poster have been reduced to scale and reproduced on the next page.

Digital Art by Walid Ben Medjedel

Christian Constanda: comics

Jenny Venton: watercolor

Paul Harris: caricature

Paul Harris: portrait

Marius Mitrea: retro vintage

IMSE2018 special poster

Contents

Contributors

Mario Ahues University of Lyon, Saint-Étienne, France

Heiko Andrä Fraunhofer-Institut für Techno- und Wirtschaftsmathematik, ITWM, Kaiserslautern, Germany

María Anguiano University of Seville, Sevilla, Spain

Amarisio S. Araújo Federal University of Viçosa (UFV), Viçosa, MG, Brazil

Tsegaye G. Ayele Addis Ababa University, Addis Ababa, Ethiopia

Laurent Baratchart INRIA Sophia Antipolis Méditerranée, Sophia Antipolis Cedex, France

Ricardo C. Barros Polytechnic Institute, University of the State of Rio de Janeiro, Nova Friburgo, RJ, Brazil

Solomon T. Bekele Addis Ababa University, Addis Ababa, Ethiopia

Bardo E. J. Bodmann Federal University of Rio Grande do Sul, Porto Alegre, Brazil

José Renato G. Braga National Institute for Space Research, São Joseé dos Campos, SP, Brazil

Renata Bunoiu University of Lorraine, Metz, France

Luana C. M. Cantagesso State University of Northern Rio de Janeiro, Macaé, RJ, Brazil

Leonardo D. Chiwiacowsky University of Caxias do Sul (UCS), Caxias do Sul, RS, Brazil

Christian Constanda The University of Tulsa, Tulsa, OK, USA

Jesús P. Curbelo Polytechnic Institute, University of the State of Rio de Janeiro, Nova Friburgo, RJ, Brazil

Filomena D. d'Almeida University of Porto, Porto, Portugal

Gerson da Penha Neto National Institute for Space Research, São Joseé dos Campos, SP, Brazil

Odair P. da Silva Polytechnic Institute, University of the State of Rio de Janeiro, Nova Friburgo, RJ, Brazil

Ramin Dabirian The University of Tulsa, Tulsa, OK, USA

Haroldo F. de Campos Velho National Institute for Space Research, São Joseé dos Campos, SP, Brazil

Sergey Dolgov University of Bath, Bath, UK

Dale Doty The University of Tulsa, Tulsa, OK, USA

Rosário Fernandes University of Minho, Braga, Portugal

Luiz F. Ferreira Federal University of Rio Grande do Sul, Porto Alegre, Brazil

Helaine C. M. Furtado Federal University of West Pará (UFOPA), Santarém, PA, Brazil

Delfina Gómez University of Cantabria, Santander, Spain

Evgeny A. Gorodnitskiy St. Petersburg State University, St. Peterburg, Russia

Anil Goyal University Institute of Technology, Bhopal, M.P., India

Layal Hakim University of Exeter, Exeter, UK

Paul J. Harris University of Brighton, Brighton, UK

Hiroshi Hirayama Kanagawa Institute of Technology, Atsugi, Kanagawa, Japan

Ronaldo Husemann Federal University of Rio Grande do Sul, Porto Alegre, Brazil

Andreas Kleefeld Forschungszentrum Jülich GmbH, Jülich, Germany

Cibele A. Ladeia Federal University of Rio Grande do Sul, Porto Alegre, Brazil

Juliette Leblond INRIA Sophia Antipolis Méditerranée, Sophia Antipolis Cedex, France

Paolo Luzzini University of Padova, Padova, Italy

Tamires A. Marotto State University of Northern Rio de Janeiro, Macaé, RJ, Brazil

Sergey E. Mikhailov Brunel University London, Uxbridge, UK

Mobina Mohammadikharkeshi The University of Tulsa, Tulsa, OK, USA

Ram S. Mohan The University of Tulsa, Tulsa, OK, USA

Asad Molayari The University of Tulsa, Tulsa, OK, USA

Paolo Musolino University of Padova, Padova, Italy

Santiago Navazo-Esteban University of Cantabria, Santander, Spain

Sergey A. Nazarov St. Petersburg State University, St. Petersburg, Russia

Hubert Nnang University of Yaoundé I and École Normale Supérieure de Yaoundé, Yaounde, Cameroon

Rafael Orive-Illera Autonomous University of Madrid, Ciudad Universitaria de Cantoblanco, Madrid, Spain

Julia Orlik Fraunhofer-Institut für Techno- und Wirtschaftsmathematik, ITWM, Kaiserslautern, Germany

Andreas Papoutsakis City University of London, London, UK

Aline R. Parigi Instituto Federal Farroupilha, Santa Maria, Brazil

Eduardo S. Pereira National Institute for Space Research, São Joseé dos Campos, SP, Brazil

Maria V. Perel St. Petersburg State University, St. Petersburg, Russia

Alvaro M. M. Peres State University of Northern Rio de Janeiro, Macaé, RJ, Brazil

María-Eugenia Pérez-Martínez University of Cantabria, Santander, Spain

Lukas Pieronek Forschungszentrum Jülich GmbH, Jülich, Germany

Adolfo Puime Pires State University of Northern Rio de Janeiro, Macaé, RJ, Brazil

Dmitry Ponomarev Vienna Institute of Technology, Vienna, Austria
St. Petersburg Department of Steklov Mathematical, Institute RAS, St. Petersburg, Russia

Diego Quiñones-Valles University of Brighton, Brighton, UK

Anna M. Radovanovic Universidade Estadual do Norte Fluminense, Macaé, RJ, Brazil

Sabrina B. M. Sambatti Climatempo, São José dos Campos, SP, Brazil

Pedro A. Santos Federal University of Rio Grande do Sul, Porto Alegre, Brazil

Dmitry Savostyanov University of Brighton, Brighton, UK

Juliana Schramm Federal University of Rio Grande do Sul, Porto Alegre, Brazil

Adalberto Schuck Jr. Federal University of Rio Grande do Sul, Porto Alegre, Brazil

Cynthia F. Segatto Federal University of Rio Grande do Sul, Porto Alegre, Brazil

Elcio H. Shiguemori Department of Science and Space Technology (DCTA), São Joseé dos Campos, SP, Brazil

Ovadia Shoham The University of Tulsa, Tulsa, OK, USA

Aditya Singh Indian Institute of Technology Indore, Indore, M.P., India

Deepak Singh National Institute of Technical Teachers' Training & Research, Bhopal, M.P., India

Luara K. S. Sousa State University of Northern Rio de Janeiro, Macaé, RJ, Brazil

Sarah Staub Fraunhofer-Institut für Techno- und Wirtschaftsmathematik, ITWM, Kaiserslautern, Germany

Joel Fotso Tachago University of Bamenda, Bambili, Bamenda, Cameroon

Paulo B. Vasconcelos University of Porto, Porto, Portugal

Marco T. Vilhena Federal University of Rio Grande do Sul, Porto Alegre, Brazil

Zenebe W. Woldemicheal Addis Ababa University, Addis Ababa, Ethiopia

Mudasir Younis University Institute of Technology, Bhopal, M.P., India

Elvira Zappale Universitá degli Studi di Salerno, Fisciano, SA, Italy

Barbara Zubik-Kowal Boise State University, Boise, ID, USA

Chapter 1
Singularity Subtraction for Nonlinear Weakly Singular Integral Equations of the Second Kind

Mario Ahues, Filomena D. d'Almeida, Rosário Fernandes, and Paulo B. Vasconcelos

1.1 Introduction

The reference Banach space is the set $X := C^0([a, b], \mathbb{R})$ with the supremum norm. We consider the operator K defined by

$$K(x)(s) := \int_a^b g(|s - t|) N(s, t, x(t)) \, dt, \quad x \in X, \ s \in [a, b],$$

where g is a weakly singular function in the following sense:

$$\lim_{s \to 0^+} g(s) = +\infty, \quad \text{and} \quad g \in C^0(]0, b - a], \mathbb{R}_+) \cap L^1([0, b - a], \mathbb{R}_+).$$

To be consistent with [An81] and [AhEtAl01], we assume that g is a decreasing function on $]0, b - a]$.

The factor N, containing the values $x(t) \in \mathbb{R}$ of the functional variable $x \in X$ for $t \in [a, b]$, is a continuous function

$$N : [a, b] \times [a, b] \times \mathbb{R} \to \mathbb{R}, \quad (s, t, u) \mapsto N(s, t, u),$$

with continuous partial derivative with respect to the third variable.

M. Ahues (✉)
Université de Lyon, Lyon, France
e-mail: mario.ahues@univ-st-etienne.fr

F. D. d'Almeida · P. B. Vasconcelos
Universidade do Porto, Porto, Portugal
e-mail: falmeida@fe.up.pt; pjv@fep.up.pt

R. Fernandes
Universidade do Minho, Braga, Portugal
e-mail: rosario@math.uminho.pt

© Springer Nature Switzerland AG 2019   1
C. Constanda, P. Harris (eds.), *Integral Methods in Science and Engineering*,
https://doi.org/10.1007/978-3-030-16077-7_1

Then the operator K maps X into itself, and it is Fréchet-differentiable over X.

When $N(s, t, x(t)) := \kappa(s, t) x(t)$ for some continuous function $\kappa :$ $[a, b] \times [a, b] \to \mathbb{R}$, then K is a linear bounded operator from X into itself.

In this paper, we are interested in the general, possibly nonlinear, case.

The main idea of the singularity subtraction method is to compensate the singularity of $g(|s - t|)$ along the diagonal $s = t$, by multiplying $g(|s - t|)$ by a factor which tends to 0 as $t \to s$. If K is linear, this factor is $\kappa(s, t)(x(t) - x(s))$. In the general case, the factor is $N(s, t, x(t)) - N(s, s, x(s))$.

This leads to rewrite K as

$$
K(x)(s) := \int_a^b g(|s - t|)[N(s, t, x(t)) - N(s, s, x(s))] \, dt
$$

$$
+ N(s, s, x(s)) \int_a^b g(|s - t|) \, dt. \tag{1.1}
$$

The singularity subtraction method builds an approximation of K as it is written in (1.1), and, as described in [An81] for the linear case, it is a double approximation scheme consisting of **truncation** and **numerical integration**.

The ideas worked out in [An81] and [AhEtAl01] for the linear case are extended here to the nonlinear case.

Truncation Given $\delta \in \,]0, b - a[$, we replace g with a truncated approximation g_δ in a δ-right-neighborhood of 0. This function coincides with g outside a small interval $[0, \delta]$, and is constantly equal to $g(\delta)$ in $[0, \delta]$. Hence g_δ is a continuous function. In the sequence of singularity subtraction approximations, the role of δ is played by a sequence $(a_n)_{n \geq 2}$ in $]0, b - a[$ leading to the function g_n defined by

$$
g_n(s) := \begin{cases} g(a_n) & \text{for } s \in [0, a_n], \\ g(s) & \text{for } s \in \,]a_n, b - a]. \end{cases}
$$

Numerical Integration To proceed with the singularity subtraction idea—like in the linear case—we define a general grid with $n \geq 2$ points on $[a, b]$:

$$
a \leq \tau_{n,1} < \tau_{n,2} < \ldots < \tau_{n,n} \leq b. \tag{1.2}
$$

This grid is called the basic grid, and it determines $n - 1$ subintervals of $[a, b]$.

The integrals in the first line of (1.1), after replacing g with g_n, are approximated by some quadrature rule Q_n with $p(n)$ nodes depending on the nodes of the basic grid. For instance, if Q_n is the composite trapezoidal rule, then the quadrature grid is the basic grid, so $p(n) = n$; if Q_n is the composite Simpson rule, then its nodes are the points of the basic grid and the mid-points of the corresponding subintervals, and hence $p(n) = 2n - 1$; for some other rules Q_n, the nodes are the so-called Gaussian points, which are obtained by shifting to each subinterval of the basic grid the zeros of a polynomial of a given degree m belonging to a complete sequence of orthogonal polynomials in some particular Hilbert space, and hence $p(n) = m(n - 1)$.

In this paper, we consider a sequence $(Q_n)_{n \geq 2}$ of quadrature rules built upon the basic grid. The nodes of Q_n are denoted by $t_{p(n),j}$, $j = 1, \ldots, p(n)$, and are numbered so that $a \leq t_{p(n),1} < \cdots < t_{p(n),p(n)} \leq b$. The weights of Q_n are denoted by $w_{p(n),j}$, $j = 1, \ldots, p(n)$. We suppose that they are all positive, and that there exists a constant $\gamma > 0$ such that

$$\sum_{t_{p(n),j} \in \mathscr{I}} w_{p(n),j} \leq \gamma \, (d - c) \quad \text{when} \quad a \leq c < d \leq b, \quad \text{and} \quad \mathscr{I} \text{ is }]c, d] \text{ or } [c, d[$$

(1.3)

(cf. hypothesis (H) in [AhEtAl01, page 225]). Almost all commonly used quadrature rules satisfy (1.3). The constant γ plays an active role in the proof of Theorem 1.

Ideally, $\int_a^b g(|s - t|) \, dt$ should be available in closed form. If not, a specially fine numerical quadrature formula should give an accurate value of this integral for any fixed value of $s \in [a, b]$.

1.2   Singularity Subtraction

Consider the basic grid (1.2) and define $h_{n,j} := \tau_{n,j+1} - \tau_{n,j}$ for $j = 1, \ldots, n - 1$, and $h_n := \max_{j=1,\ldots,n-1} h_{n,j}$. The singularity subtraction technique, as presented in [An81], relates truncation and numerical integration through the following condition on the sequences $(a_n)_{n \geq 2}$ and $(Q_n)_{n \geq 2}$: There exist constants $\alpha_1 > 0$ and $\beta_1 > 0$ such that

$$\alpha_1 h_n \leq a_n \leq \beta_1 h_n \quad \text{for all} \ \ n \geq 2,$$

i.e., the width of truncation must tend to zero at the same rate as the mesh sizes.

These considerations lead to approximate K, as written in (1.1), by the following operator K_n: For $x \in X$, and $s \in [a, b]$,

$$K_n(x)(s) := \sum_{j=1}^{p(n)} w_{p(n),j} g_n(|s - t_{p(n),j}|) \big[N(s, t_{p(n),j}, x(t_{p(n),j})) - N(s, s, x(s)) \big]$$

$$+ N(s, s, x(s)) \int_a^b g(|s - t|) \, dt.$$

The exact equation, to be solved numerically, is: For $y \in X$, find $\psi \in X$ such that

$$\psi = K(\psi) + y,$$

(1.4)

i.e., $\mathscr{F}(\psi) = 0$, if $\mathscr{F} : X \to X$ is the operator defined by $\mathscr{F}(x) := x - K(x) - y$ for all $x \in X$. We assume that 1 is not in the spectrum of the Fréchet-derivative of K at ψ, so ψ is an isolated solution of (1.4).

The approximate equation, to be solved exactly, is: Find $\psi_n \in X$ such that

$$\psi_n = K_n(\psi_n) + y, \qquad (1.5)$$

i.e., $\mathscr{F}_n(\psi_n) = 0$, if $\mathscr{F}_n : X \to X$ is the operator defined by $\mathscr{F}_n(x) := x - K_n(x) - y$ for all $x \in X$.

If we take the values of (1.5) at $t_{p(n),i}$, $i = 1, \ldots, p(n)$, we get the following, possibly nonlinear, system of order $p(n)$ with unknowns $x_n(i) := \psi_n(t_{p(n),i})$, $i = 1, \ldots, p(n)$:

$$x_n(i) = \sum_{j=1}^{p(n)} w_{p(n),j}\, g_n(|t_{p(n),i} - t_{p(n),j}|)[N(t_{p(n),i}, t_{p(n),j}, x_n(j))$$
$$- N(t_{p(n),i}, t_{p(n),i}, x_n(i))]$$
$$+ N(t_{p(n),i}, t_{p(n),i}, x_n(i))\int_a^b g(|t_{p(n),i} - t|)\, dt + y(t_{p(n),i}).$$

This $p(n)$-dimensional system can be written as

$$F_n(x_n) = 0, \qquad (1.6)$$

where, for all $x \in \mathbb{R}^{p(n)\times 1}$, and $i = 1, \ldots, p(n)$,

$$F_n(x)(i) := x(i) - \sum_{j=1}^{p(n)} w_{p(n),j}\, g_n(|t_{p(n),i} - t_{p(n),j}|)[N(t_{p(n),i}, t_{p(n),j}, x(j))$$
$$- N(t_{p(n),i}, t_{p(n),i}, x(i))]$$
$$- N(t_{p(n),i}, t_{p(n),i}, x(i))\int_a^b g(|t_{p(n),i} - t|)\, dt - y(t_{p(n),i}))$$
$$= x(i) - \sum_{j=1}^{p(n)} w_{p(n),j}\, g_n(|t_{p(n),i} - t_{p(n),j}|)N(t_{p(n),i}, t_{p(n),j}, x(j))$$
$$+ N(t_{p(n),i}, t_{p(n),i}, x(i))\Big[\sum_{j=1}^{p(n)} w_{p(n),j}\, g_n(|t_{p(n),i} - t_{p(n),j}|)$$
$$- \int_a^b g(|t_{p(n),i} - t|)\, dt\Big] - y(t_{p(n),i}).$$

The system (1.6) must be solved accurately by some numerical methods, for instance, Gauss' method in the linear case, and Newton's method—as described in the sequel—in the nonlinear case.

The Jacobian matrix of $\mathbf{F}_n : \mathbb{R}^{p(n) \times 1} \to \mathbb{R}^{p(n) \times 1}$ at $\mathbf{x} \in \mathbb{R}^{p(n) \times 1}$ is given by

$$\mathbf{F}'_n(\mathbf{x})(i, j) = \delta_{i,j} - w_{p(n),j}\, g_n(|t_{p(n),i} - t_{p(n),j}|)\frac{\partial N}{\partial u}(t_{p(n),i}, t_{p(n),j}, \mathbf{x}(j))$$

$$+ \delta_{i,j}\frac{\partial N}{\partial u}(t_{p(n),i}, t_{p(n),i}, \mathbf{x}(i))\left[\sum_{\ell=1}^{p(n)} w_{p(n),j}\, g_n(|t_{p(n),i} - t_{p(n),\ell}|)\right.$$

$$\left. - \int_a^b g(|t_{p(n),i} - t|)\, dt\right],$$

where $\delta_{i,j}$ is the Kronecker delta, and $i, j = 1, \ldots, p(n)$.

The Newton's sequence $\left(\mathbf{x}_n^{[k]}\right)_{k \geq 0}$ in $\mathbb{R}^{p(n) \times 1}$ is defined, for a given starting column $\mathbf{x}_n^{[0]} \in \mathbb{R}^{p(n) \times 1}$, by

$$\mathbf{F}'_n(\mathbf{x}_n^{[k]})\mathbf{c}_n^{[k]} = -\mathbf{F}_n(\mathbf{x}_n^{[k]}), \qquad \mathbf{x}_n^{[k+1]} := \mathbf{x}_n^{[k]} + \mathbf{c}_n^{[k]}, \qquad k \geq 0,$$

where $\mathbf{c}_n^{[k]}$ is the unknown. Equivalently,

$$\mathbf{F}'_n(\mathbf{x}_n^{[k]})\mathbf{x}_n^{[k+1]} = \mathbf{F}'(\mathbf{x}_n^{[k]})\mathbf{x}_n^{[k]} - \mathbf{F}_n(\mathbf{x}_n^{[k]}), \qquad k \geq 0,$$

where $\mathbf{x}_n^{[k+1]}$ is the unknown, i.e.

$$(\mathbf{I} - \mathbf{A}_n^{[k]} - \mathbf{D}_n^{[k]})\mathbf{x}_n^{[k+1]} = \mathbf{b}_n^{[k]},$$

where $\mathbf{I}$ is the identity matrix of order $p(n)$, and, for $i, j = 1, \ldots, p(n)$,

$$\mathbf{A}_n^{[k]}(i, j) := w_{p(n),j}\, g_n(|t_{p(n),i} - t_{p(n),j}|)\frac{\partial N}{\partial u}(t_{p(n),i}, t_{p(n),j}, \mathbf{x}_n^{[k]}(j)),$$

$$\mathbf{D}_n^{[k]}(i, j) := \delta_{i,j}\frac{\partial N}{\partial u}(t_{p(n),i}, t_{p(n),i}, \mathbf{x}_n^{[k]}(i))\left[\int_a^b g(|t_{p(n),i} - t|)\, dt\right.$$

$$\left. - \sum_{\ell=1}^{p(n)} w_{p(n),\ell}\, g_n(|t_{p(n),i} - t_{p(n),\ell}|)\right], \tag{1.7}$$

$$b_n^{[k]}(i) := -\sum_{j=1}^{p(n)} w_{p(n),j}\, g_n(|t_{p(n),i}-t_{p(n),j}|)\frac{\partial N}{\partial u}(t_{p(n),i}, t_{p(n),j}, \mathsf{x}_n^{[k]}(j))\mathsf{x}_n^{[k]}(j)$$

$$+\mathsf{x}_n^{[k]}(i)\frac{\partial N}{\partial u}(t_{p(n),i}, t_{p(n),i}, \mathsf{x}_n^{[k]}(i))\Big[\sum_{\ell=1}^{p(n)} w_{p(n),\ell}\, g_n(|t_{p(n),i}-t_{p(n),\ell}|)$$

$$-\int_a^b g(|t_{p(n),i}-t|)\, dt\Big]$$

$$+\sum_{j=1}^{p(n)} w_{p(n),j}\, g_n(|t_{p(n),i}-t_{p(n),j}|)\, N(t_{p(n),i}, t_{p(n),j}, \mathsf{x}_n^{[k]}(j))$$

$$-N(t_{p(n),i}, t_{p(n),i}, \mathsf{x}_n^{[k]}(i))\Big[\sum_{j=1}^{p(n)} w_{p(n),j}\, g_n(|t_{p(n),i}-t_{p(n),j}|)$$

$$-\int_a^b g(|t_{p(n),i}-t|)\, dt\Big] + y(t_{p(n),i}). \tag{1.8}$$

1.3 Convergence

Let $\overset{p}{\to}$ denote pointwise convergence, $\overset{n}{\to}$ norm convergence, $\overset{cc}{\to}$ collectively compact convergence (cf. [An71]), and $\overset{\nu}{\to}$ the ν-convergence (cf. [AhEtAl01]).

Theorem 1 *Let* $(Q_n)_{n\geq 2}$ *be a sequence of composite quadrature rules with nodes* $t_{p(n),j}$ *and weights* $w_{p(n),j}$, $j = 1, \ldots, p(n)$, *satisfying (1.3). Then* $(K_n)_{n\geq 2}$ *is pointwise convergent to* K, $(\mathscr{F}_n)_{n\geq 2}$ *is pointwise convergent to* $\mathscr{F}$, *and* $(\psi_n)_{n\geq 2}$ *is convergent with limit* ψ.

Proof The Fréchet-derivatives $T := K'$ and $T_n := (K_n)'$ at ψ are given by:

$$[T(\psi)f](s) = \int_a^b g(|s-t|)\frac{\partial N(s,t,\psi(t))}{\partial u} f(t)\, dt, \quad f \in X, s \in [a,b],$$

$$[T_n(\psi)f](s) = \sum_{j=1}^{p(n)} w_{p(n),j}\, g_n(|s-t_{p(n),j}|)\frac{\partial N(s,t_{p(n),j},\psi(t_{p(n),j}))}{\partial u} f(t_{p(n),j})$$

$$-\sum_{j=1}^{p(n)} w_{p(n),j}\, g_n(|s-t_{p(n),j}|)\frac{\partial N(s,s,\psi(s))}{\partial u} f(s)$$

$$+\frac{\partial N(s,s,\psi(s))}{\partial u} f(s)\int_a^b g(|s-t|)\, dt, \quad f \in X, s \in [a,b].$$

Let us consider the decomposition $T_n(\psi) = T_n^A(\psi) + T_n^B(\psi)$, where

$$[T_n^A(\psi)f](s) := \sum_{j=1}^{p(n)} w_{p(n),j}\, g_n(|s - t_{p(n),j}|) \frac{\partial N(s, t_{p(n),j}, \psi(t_{p(n),j}))}{\partial u} f(t_{p(n),j}),$$

$$[T_n^B(\psi)f](s) := -\sum_{j=1}^{p(n)} w_{p(n),j}\, g_n(|s - t_{p(n),j}|) \frac{\partial N(s, s, \psi(s))}{\partial u} f(s)$$

$$+ \frac{\partial N(s, s, \psi(s))}{\partial u} f(s) \int_a^b g(|s - t|)\, dt$$

for $f \in X$, and $s \in [a, b]$.

We define, for $x \in X$, and $s \in [a, b]$,

$$(Ux)(s) := \int_a^b g(|s - t|)x(t)\, dt \text{ and } (U_n x)(s)$$

$$:= \sum_{j=1}^{p(n)} w_{p(n),j} g_n(|s - t_{p(n),j}|)\, x(t_{p(n),j}).$$

By (1.3), $U_n \overset{cc}{\to} U$, so $U_n \overset{p}{\to} U$ (cf. Proposition 4.18 in [AhEtAl01, page 227]).

The proof is done in five steps:

1. We show that $T_n(\psi) \overset{p}{\to} T(\psi)$ and that $T_n(\psi) \overset{v}{\to} T(\psi)$ too:

As $T_n(\psi)$ and $T(\psi)$ are bounded linear operators, we use the results of [An81].

Since $(s, t, u) \mapsto \dfrac{\partial N(s, t, u)}{\partial u}$ is a continuous function, and since (1.3) holds, then $T_n(\psi)$ and $T_n^A(\psi)$ satisfy the hypotheses of Proposition 4.18 in [AhEtAl01, page 227], and $T_n^A(\psi) \overset{cc}{\to} T(\psi)$. Hence $T_n^A(\psi) \overset{p}{\to} T(\psi)$. Recall that $T(\psi)$ is compact because K is compact. This implies that $T_n^A(\psi) \overset{v}{\to} T(\psi)$.

For any $f \in X$ such that $\|f\| = 1$,

$$[T_n^B(\psi)f](s) = \frac{\partial N(s, s, \psi(s))}{\partial u} f(s) \left[\int_a^b g(|s - t|)\, dt \right.$$

$$- \sum_{j=1}^{p(n)} w_{p(n),j}\, g_n(|s - t_{p(n),j}|) \Bigg]$$

$$= \frac{\partial N(s, s, \psi(s))}{\partial u} f(s) \Big[(Ue)(s) - (U_n e)(s) \Big],$$

where $e(s) := 1$ for $s \in [a, b]$. Hence

$$\|T_n^B(\psi)f\| \leq \left\|\frac{\partial N(\cdot, \cdot, \psi(\cdot))}{\partial u}\right\| \, \|U_n e - U e\|,$$

which tends to 0 as $n \to \infty$, since $U_n \overset{p}{\to} U$. Hence $T_n^B(\psi) \overset{n}{\to} O$, so $T_n(\psi) \overset{p}{\to} T(\psi)$. Hence $T_n(\psi) \overset{\nu}{\to} T(\psi)$ (cf. Lemma 2.2 (b) (i) in [AhEtAl01, page 73]).

2. We show that $I - T_n(\psi)$ is invertible:

Since $(I - T(\psi))^{-1}$ exists, and $T_n(\psi) \overset{\nu}{\to} T(\psi)$, there exists $n_0 \geq 2$ such that, for $n \geq n_0$,

$$\|(I - T(\psi))^{-1}\| \, \|(T_n(\psi) - T(\psi)) T_n(\psi)\| < 1.$$

Hence $(I - T_n(\psi))^{-1}$ exists and is uniformly bounded (cf. [An81, page 413]). By continuity, the same holds for $((I - T_n(x))^{-1}$ for all x close enough to ψ.

3. We prove that $\mathscr{F}_n$ is locally invertible with continuous inverse in a neighborhood of 0:

$I - K_n$ is a continuously differentiable operator from the Banach space X into itself. By the Inverse Function Theorem, $I - T_n(\psi)$, being invertible, there is a neighborhood of ψ where $I - K_n$ is invertible with continuous inverse in some neighborhood of y. Hence $\mathscr{F}_n^{-1}$ exists and is continuous in some neighborhood of 0.

4. We prove that $(K_n)_{n \geq 2}$ is pointwise convergent to K, and $(\mathscr{F}_n)_{n \geq 2}$ is pointwise convergent to $\mathscr{F}$:

An auxiliary operator $\widehat{K}_n$ is used in the proof. For $x \in X$, and $s \in [a, b]$, define

$$\widehat{K}_n(x)(s) := \sum_{j=1}^{p(n)} w_{p(n), j} \, g_n(|s - t_{p(n), j}|) N(s, t_{p(n), j}, x(t_{p(n), j})).$$

K_n can be rewritten as

$$K_n(x)(s) = \widehat{K}_n(x)(s) + N(s, s, x(s))(U - U_n)e(s). \tag{1.9}$$

Define

$$\sigma(x) := \max_{a \leq s, t \leq b} |N(s, t, x(t))|,$$

which is finite because of the continuity of N in its three variables, and of x in its single one. In the linear case, $\sigma(x) = \rho\|x\|$ for some constant $\rho > 0$. Now,

$$|K_n(x)(s) - \widehat{K}_n(x)(s)| \leq \sigma(x)\|(U - U_n)e\| \to 0 \text{ and}$$

$$\|K_n(x) - \widehat{K}_n(x)\| \to 0 \text{ as } n \to \infty,$$

since $U_n \overset{p}{\to} U$ as stated in the beginning of this proof. Following the ideas of the proof of Proposition 4.18 in [AhEtAl01], we decompose

$$\widehat{K}_n(x) - K(x) = \lambda_\delta + \mu_n + \eta_n,$$

where λ_δ, μ_n and η_n are defined as follows. Let $\gamma > 0$ be the constant introduced in (1.3). Given $\epsilon > 0$, there exists $\delta \in]0, b - a]$ such that $\int_0^\delta g(u)\, du <$ $\frac{\epsilon}{18} \min\{1, \frac{1}{3\gamma}\}$. Set

$$\lambda_\delta(s) := \int_{\max\{a,s-\delta\}}^{\min\{b,s+\delta\}} [g(\delta) - g(|s - t|)]N(s, t, x(t))dt,$$

$$\mu_n(s) = \sum_{j=1}^{p(n)} w_{p(n),j}[g_n(|s - t_{p(n),j}|) - g_\delta(|s - t_{p(n),j}|)]N(s, t_{p(n),j}, x(t_{p(n),j})),$$

$$\eta_n(s) := \sum_{j=1}^{p(n)} w_{p(n),j}g_\delta(|s - t_{p(n),j}|)N(s, t_{p(n),j}, x(t_{p(n),j}))$$

$$- \int_a^b g_\delta(|s - t|)N(s, t, x(t))\, dt.$$

Then the following upper bounds hold for all n greater than some integer $n_0(x)$:

$$|\lambda_\delta(s)| \le 6\, \sigma(x) \int_0^\delta g(u)\, du \le \frac{\sigma(x)}{3}\epsilon,$$

$$|\mu_n(s)| \le \sigma(x)\Big[\sum_{|s-t_{p(n),j}|<\delta} w_{p(n),j}g_n(|s - t_{p(n),j}|) + 2\gamma\delta g(\delta)\Big] \le \frac{\sigma(x)}{3}\epsilon,$$

$$|\eta_n(s)| \le \frac{\sigma(x)}{3}\epsilon.$$

Since

$$|\widehat{K}_n(x)(s) - K(x)(s)| \le |\lambda_\delta(s)| + |\mu_n(s)| + |\eta_n(s)| \le \sigma(x)\epsilon,$$

we conclude that $\widehat{K}_n \overset{p}{\to} K$. $K_n \overset{p}{\to} K$, and $\mathscr{F}_n \overset{p}{\to} \mathscr{F}$.

5. We prove that $(\psi_n)_{n\ge2}$ is convergent with limit ψ:

Since $\mathscr{F}$ and $\mathscr{F}_n$ are invertible and Fréchet-differentiable, the derivative of their inverses at 0 is equal to the inverse of the derivative of the direct operators at the inverse image of 0, and the integral form of the Mean Value Theorem for

Derivatives gives:

$$\mathscr{F}_n^{-1}(0) - \mathscr{F}^{-1}(0) = \psi_n - \psi = \mathscr{F}_n^{-1}(\mathscr{F}(\psi)) - \mathscr{F}_n^{-1}(\mathscr{F}_n(\psi))$$

$$= \int_0^1 (\mathscr{F}_n^{-1})'(\mathscr{F}_n(\psi) + t(\mathscr{F}(\psi) - \mathscr{F}_n(\psi))\, dt \,\, (\mathscr{F}(\psi) - \mathscr{F}_n(\psi)).$$

Hence

$$\mathscr{F}_n(\psi_n) - \mathscr{F}_n(\psi) = \int_0^1 \mathscr{F}_n'(\psi + t(\psi_n - \psi))\, dt \,\, (\psi_n - \psi).$$

Since the sequence $(\mathscr{F}_n)_{n \geq 2}$ is pointwise convergent to $\mathscr{F}$ and $\mathscr{F}(\psi) = 0$, then $v_n(t) := \mathscr{F}_n(\psi) + t(\mathscr{F}(\psi) - \mathscr{F}_n(\psi))$ tends to 0 uniformly in $t \in [0, 1]$ as $n \to \infty$. On the other hand, $(\mathscr{F}_n^{-1})'(v_n(t)) = (I - T_n(u_n(t)))^{-1}$ is uniformly bounded for n large enough and $t \in [0, 1]$, where $u_n(t) := \mathscr{F}_n^{-1}(v_n(t))$. Finally, $\mathscr{F}_n'(x) = I - T_n(x)$ is bounded uniformly in x for x in any bounded set of X, and in $t \in [0, 1]$. Hence there exist constants $\alpha_2 > 0$ and $\beta_2 > 0$ such that

$$\alpha_2 \|\mathscr{F}_n(\psi)\| \leq \|\psi_n - \psi\| \leq \beta_2 \|\mathscr{F}_n(\psi)\|, \tag{1.10}$$

so the sequence $(\psi_n)_{n \geq 2}$ is convergent with limit ψ.

1.4   Numerics

Data The operator K is defined with $a = 0$, $b = 1$, $N(s, t, u) := u^3$, and with the weakly singular decreasing function g defined by $g(s) := -\log(s)$ for $s \in \,]0, 1]$.

The solution of (1.4) is chosen to be the function ψ defined by

$$\psi(s) := \left(s - \frac{1}{2}\right)^{2/3}, \quad s \in [0, 1].$$

Then the function $y = \psi - K(\psi)$ takes the values $y(0) = y(1) = 1/\sqrt[3]{4} - 1/9$, and

$$y(s) = \left(s - \frac{1}{2}\right)^{2/3} - \frac{1}{3}\left(s^2 - s + \frac{1}{3}\right) + s\log(s)\left(\frac{s^2}{2} - \frac{s}{2} + \frac{1}{4}\right)$$

$$+ \frac{(1-s)\log(1-s)}{3}\left(s^2 - \frac{s}{2} + \frac{1}{4}\right), \quad s \in \,]0, 1[.$$

Programming The numerical implementation was carried on MATLAB 2017b.

Newton Because of its fast convergence, we have performed seven iterations of Newton's method for each fixed value of n.

Table 1.1 The grid-valued
relative errors for nine values
of n

n	r_n
10	$1.8e - 03$
20	$2.9e - 04$
40	$4.9e - 05$
80	$8.5e - 06$
160	$1.5e - 06$
320	$2.6e - 07$
640	$4.6e - 08$
1280	$8.3e - 09$
2560	$1.6e - 09$

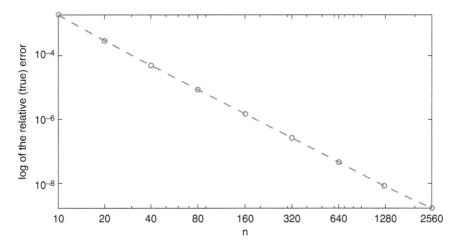

Fig. 1.1 The grid-valued relative errors in a loglog scale

Quadrature The basic grid was chosen to be uniform. We used a composite quadrature rule built with three Gauss points on each one of the $n - 1$ subintervals of the basic grid: the zeros of a Chebyshev polynomial of degree 3. Hence, we had been led to a nonlinear system of order $3(n - 1)$ for each value of n, and to a linear system of order $3(n-1)$ at each Newton's iteration. The numerical integration of the explicit integrals in (1.7) and (1.8) was performed by adaptive quadrature based on a Gauss-Kronrod method (cf. [Sh08]). The use of strict tolerances for the numerical adaptive quadrature approach is mandatory, otherwise finer discretizations would not deliver the predicted convergence results.

Results For some selected values of n, Table 1.1 shows the grid-valued relative errors

$$r_n := \frac{\max\limits_{j=1,\dots,p(n)} |\mathbf{x}_n^{[7]}(j) - \psi(t_{p(n),j})|}{\max\limits_{j=1,\dots,p(n)} |\psi(t_{p(n),j})|}.$$

Figure 1.1 shows the relative errors of Table 1.1 in a loglog scale.

1.5 Conclusions

In this paper, we have extended to nonlinear integral operators, the singularity subtraction technique for approaching linear weakly singular integral operators in the framework of real valued continuous functions. The singularity subtraction technique cannot be settled in Lebesgue spaces.

The equation to be solved numerically, written as $\mathcal{F}(\psi) = 0$, is discretized as $\mathcal{F}_n(\psi_n) = 0$, which leads to a finite-dimensional nonlinear problem which is solved by Newton's method.

We have proved that the sequence $(\mathcal{F}_n)_{n\geq 2}$ is pointwise convergent to $\mathcal{F}$, and that the sequence $(\psi_n)_{n\geq 2}$ is convergent with limit ψ. The double bound (1.10) shows that the convergence of the latter is neither slower nor faster than that of the sequence of $(\mathcal{F}_n(\psi))_{n\geq 2}$ to 0.

Since the values of ψ_n are approximated by Newton's method only at the $p(n)$ nodes of the quadrature grid, ψ_n could be globally approximated by interpolation, if needed. In fact, oppositely to the linear case, once the unknowns $x_n(i) = \psi_n(t_{p(n),i})$ are approximated by the values $x_n^{[7]}(i)$ issued from Newton's method, no natural interpolation formula is available for a closed formula of ψ_n since, for each $s \in [a, b]$, $\psi_n(s)$ is hidden implicitly in the nonlinear expression (1.5). This is a significant difference between the linear case and the nonlinear case.

The expression (1.9) shows that the order of the pointwise convergence to K of the sequence of approximations $(K_n)_{n\geq 2}$ is dominated by the order of convergence of the sequence of quadrature rules $(Q_n)_{n\geq 2}$. In the case of the numerical computations presented in this paper, the sequence of quadrature rules converges in theory to the exact integral at the same rate as n^{-2} tends to 0, when $n \to \infty$ (cf. [XiEtAl12]). This is confirmed in practice, as it is shown in the loglog plotting of Fig. 1.1, where we observe that the slope of the straight line is around -2.

A major survey on numerical approximation of nonlinear integral equations is [At92]. This paper studies numerical methods for calculating fixed points of nonlinear integral operators, i.e. equations of the form $\psi = K(\psi)$ with the notation of our paper. This corresponds to the case $y = 0$ and is less general than the work presented here since y cannot be incorporated as a part of the integral operator K. Methods treated in [At92] include a product integration type scheme for weakly singular Hammerstein operators, projection methods and Nyström methods. As in our paper, all those methods require the solution of finite-dimensional systems of nonlinear equations. An auxiliary numerical method is needed to solve these nonlinear finite-dimensional systems.

Acknowledgements The third author was partially supported by CMat (UID/MAT/00013/2013), and the second and fourth authors were partially supported by CMUP (UID/ MAT/ 00144/ 2013), which are funded by FCT (Portugal) with national funds (MCTES) and European structural funds (FEDER) under the partnership agreement PT2020.

References

[AhEtAl01] Ahues, M.; Largillier, A. and Limaye B.: *Spectral Computations for Bounded Operators*, Chapman & Hall/CRC, Boca Raton, FL (2001).

[An71] Anselone, P.: *Collectively compact operator approximation theory and applications to integral equations*, Prentice-Hall, Englewoodcliffs, NJ (1971).

[An81] Anselone, P.: Singularity subtraction in the numerical solution of integral equations, *J. Austral. Math. Soc. Ser. B*, **22**, 408–418 (1981).

[At92] Atkinson, K.: A survey of numerical methods for solving nonlinear integral equations, *Journal of Integral Equations*, **4**, 1, 15–46 (1992).

[Sh08] Shampine, L. F.: Vectorized Adaptive Quadrature in MATLAB, *J. Comput. Appl. Math.*, **211**, 131–140 (2008).

[XiEtAl12] Xiang S. and Bornemann, F.: On the Convergence Rates of Gauss and Clenshaw–Curtis Quadrature for Functions of Limited Regularity, *SIAM J. on Numer. Anal.*, **50**, 5, 2581–2587 (2012).

Chapter 2
On the Flow of a Viscoplastic Fluid in a Thin Periodic Domain

María Anguiano and Renata Bunoiu

2.1 Introduction

We study in this paper the steady incompressible nonlinear flow of a Bingham fluid in a thin periodic domain, which is a model of porous media. The model of thin porous media of thickness much smaller than the parameter of periodicity was introduced in [Zh08], where a stationary incompressible Navier-Stokes flow was studied. Recently, the model of the thin porous medium under consideration in this paper was introduced in [FaEtAl16], where the flow of an incompressible viscous fluid described by the stationary Navier-Stokes equations was studied by the multiscale asymptotic expansion method, which is a formal but powerful tool to analyze homogenization problems. These results were rigorously proved in [AS18] using an adaptation (introduced in [AS17]) of the unfolding method from [CiEtAl08]. This adaptation consists of a combination of the unfolding method with a rescaling in the height variable, in order to work with a domain of fixed height, and to use monotonicity arguments to pass to the limit. In [AS17], in particular, the flow of an incompressible stationary Stokes system with a nonlinear viscosity, being a power law, was studied. For nonstationary incompressible viscous fluid flow in a thin porous medium, we refer to [An17], where a nonstationary Stokes system is considered, and [An217], where a nonstationary non-Newtonian Stokes system, where the viscosity obeys the power law, is studied. For the unfolding method applied to the study of problems stated in other type of thin periodic domains, we

M. Anguiano
Universidad de Sevilla, Sevilla, Spain
e-mail: anguiano@us.es

R. Bunoiu (✉)
Université de Lorraine, Metz, France
e-mail: renata.bunoiu@univ-loraine.fr

© Springer Nature Switzerland AG 2019
C. Constanda, P. Harris (eds.), *Integral Methods in Science and Engineering*,
https://doi.org/10.1007/978-3-030-16077-7_2

refer, for instance, to [Gr04] for crane type structures and to [GrEtAl17] for thin layers with thin beam structures, where elasticity problems are studied.

Viscoplastic fluids are quite often encountered in real life. We mention oils, polymer solutions, volcanic lavas, muds and clays, avalanches, liquid chocolate. The theory of the fluid mechanics of such materials has several different applications, as for instance in the oil and gas industry, which can be found in the ground, which is a porous medium. The most commonly studied viscoplastic fluid is the Bingham fluid. In our thin porous medium, we consider the flow of a nonlinear viscoplastic Bingham flow, whose yield stress itself depends on the small parameter characterizing the geometry of the domain, denoted ε. The first study of this type of problem is due to [LiEtAl81], where the problem was studied in a classical porous medium, by using the multiscale asymptotic expansion method. A nonlinear Darcy law was obtained after the passage to the limit $\varepsilon \to 0$. The corresponding convergence result was proved in [BoEtAl93] with the two-scale convergence method and then recovered in [BuEtAl13] with the periodic unfolding method from [CiEtAl08]. For the study in a porous medium with a doubly periodic structure, we refer to [BuEtAl17], where a more involved nonlinear Darcy law was derived. The flow of a Bingham fluid was also studied in thin domains of small height, denoted ε. We refer the reader to [BuEtAl03, BuEtAl04] and [BuEtAl18] for these studies, where a lower dimensional Bingham-like law was exhibited from the limit problem, after the passage to the limit $\varepsilon \to 0$. This law was already used in the engineering (see [LiEtAl90]), but no rigorous mathematical derivation was previously known.

The paper is organized as follows. In Sect. 2.2 we state the problem: we define in (2.1) the thin porous medium (see also Fig. 2.3), in which we consider the flow of a viscoplastic Bingham fluid with velocity verifying the nonlinear variational inequality (2.3). In Sect. 2.3 we state and prove the main result of our paper, Theorem 1. We then give in Sect. 2.4 some conclusions and perspectives and we end the paper with a list of References.

2.2   Statement of the Problem

The Domain The periodic porous medium is defined by a domain ω and an associated microstructure, or periodic cell $Y' = [-1/2, 1/2]^2$, which is made of two complementary parts: the fluid part Y'_f, and the solid part Y'_s ($Y'_f \cup Y'_s = Y'$ and $Y'_f \cap Y'_s = \varnothing$). More precisely, we assume that ω is a smooth, bounded, connected set in $\mathbb{R}^2$, and that Y'_s is an open connected subset of Y' with a smooth boundary $\partial Y'_s$, such that $\overline{Y}'_s$ is strictly included in Y'.

The microscale of the porous medium is a small positive number ε. The domain ω is covered by a regular mesh of square of size ε: for $k' \in \mathbb{Z}^2$, each cell $Y'_{k',\varepsilon} = \varepsilon k' + \varepsilon Y'$ is divided in a fluid part $Y'_{f_{k'},\varepsilon}$ and a solid part $Y'_{s_{k'},\varepsilon}$, i.e. is similar to the unit cell Y' rescaled to size ε. We define $Y = Y' \times (0, 1) \subset \mathbb{R}^3$, which is divided in a fluid part $Y_f = Y'_f \times (0, 1)$ and a solid part $Y_s = Y'_s \times (0, 1)$, and consequently

Fig. 2.1 Views of the domain Λ_ε

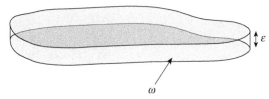

Fig. 2.2 Views of the domain ω_ε

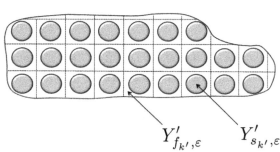

$$Y'_{f_{k'},\varepsilon} \qquad Y'_{s_{k'},\varepsilon}$$

$Y_{k',\varepsilon} = Y'_{k',\varepsilon} \times (0,1) \subset \mathbb{R}^3$, which is divided in a fluid part $Y_{f_{k'},\varepsilon}$ and a solid part $Y_{s_{k'},\varepsilon}$.

We define Λ_ε (see Fig. 2.1) by

$$\Lambda_\varepsilon = \omega \times (0,\varepsilon).$$

We denote by $\tau(\overline{Y}'_{s_{k'},\varepsilon})$ the set of all translated images of $\overline{Y}'_{s_{k'},\varepsilon}$. The set $\tau(\overline{Y}'_{s_{k'},\varepsilon})$ represents the solids in $\mathbb{R}^2$. The fluid part of the bottom $\omega_\varepsilon \subset \mathbb{R}^2$ of the porous medium is defined by $\omega_\varepsilon = \omega \backslash \bigcup_{k' \in \mathscr{K}_\varepsilon} \overline{Y}'_{s_{k'},\varepsilon}$ (see Fig. 2.2), where $\mathscr{K}_\varepsilon = \{k' \in \mathbb{Z}^2 : Y'_{k',\varepsilon} \cap \omega \neq \emptyset\}$. The whole fluid part $\Omega_\varepsilon \subset \mathbb{R}^3$ in the thin porous medium is defined by

$$\Omega_\varepsilon = \{(x_1, x_2, x_3) \in \omega_\varepsilon \times \mathbb{R} : 0 < x_3 < \varepsilon\}. \tag{2.1}$$

We make the assumption that the solids $\tau(\overline{Y}'_{s_{k'},\varepsilon})$ do not intersect the boundary $\partial \omega$ (see Fig. 2.2):

We define $Y^\varepsilon_{s_{k'},\varepsilon} = Y'_{s_{k'},\varepsilon} \times (0,\varepsilon)$. Denote by S_ε the set of the solids contained in Ω_ε (see Fig. 2.3). Then, S_ε is a finite union of solids, i.e.

$$S_\varepsilon = \bigcup_{k' \in \mathscr{K}_\varepsilon} \overline{Y}^\varepsilon_{s_{k'},\varepsilon}.$$

We define

$$\widetilde{\Omega}_\varepsilon = \omega_\varepsilon \times (0,1), \qquad \Omega = \omega \times (0,1). \tag{2.2}$$

Fig. 2.3 Views of the
domain Ω_ε

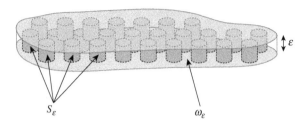

Fig. 2.3 Views of the domain Ω_ε

We observe that $\widetilde{\Omega}_\varepsilon = \Omega \setminus \bigcup_{k' \in \mathcal{K}_\varepsilon} \overline{Y}_{S_{k'},\varepsilon}$, and we define $T_\varepsilon = \bigcup_{k' \in \mathcal{K}_\varepsilon} \overline{Y}_{S_{k'},\varepsilon}$ as the set of the solids contained in $\widetilde{\Omega}_\varepsilon$.

The Problem In the domain Ω_ε defined in (2.1), we consider the stationary flow of an incompressible Bingham fluid. Following [DuEtAl72], the problem is formulated in terms of a nonlinear variational inequality.

For a vectorial function v, we define ($1 \le i, j \le 3$)

$$(D(v))_{i,j} = \frac{1}{2}\left(\partial_{x_j} v_i + \partial_{x_i} v_j\right), \quad |D(v)|^2 = D(v) : D(v),$$

where : denotes the full contraction of two matrices: for $A = (a_{i,j})_{1 \le i,j \le 3}$ and $B = (b_{i,j})_{1 \le i,j \le 3}$, we have $A : B = \sum_{i,j=1}^{3} a_{ij} b_{ij}$.

We consider the space

$$V(\Omega_\varepsilon) = \{v \in (H_0^1(\Omega_\varepsilon))^3 \mid \operatorname{div} v = 0 \text{ in } \Omega_\varepsilon\},$$

and for $u, v \in (H_0^1(\Omega_\varepsilon))^3$, we introduce

$$a(u, v) = 2\mu \int_{\Omega_\varepsilon} D(u) : D(v) dx,$$

$$j(v) = \sqrt{2} g \,\varepsilon \int_{\Omega_\varepsilon} |D(v)| dx,$$

$$(u, v)_{\Omega_\varepsilon} = \int_{\Omega_\varepsilon} u \cdot v \, dx,$$

where the positive real μ is the viscosity of the Bingham fluid and the positive real g is related to the yield stress of the Bingham fluid. More precisely, the yield stress of the Bingham fluid under consideration in this work is of the form $g\varepsilon$, where ε is the parameter related to the geometry of the domain. This yield stress is exactly the one considered by [LiEtAl81], where the flow of a Bingham fluid in a classical porous medium was studied.

Let $f \in (L^2(\Omega))^3$ be given such that $f = (f', 0)$ and $f_\varepsilon \in (L^2(\Omega_\varepsilon))^3$ be defined by

$$f_\varepsilon(x) = f(x', x_3/\varepsilon), \quad \text{a.e. } x \in \Omega_\varepsilon.$$

The model of the flow is described by the following nonlinear variational inequality: Find $u_\varepsilon \in V(\Omega_\varepsilon)$ such that

$$a(u_\varepsilon, v - u_\varepsilon) + j(v) - j(u_\varepsilon) \geq (f_\varepsilon, v - u_\varepsilon)_{\Omega_\varepsilon}, \quad \forall v \in V(\Omega_\varepsilon). \tag{2.3}$$

From [DuEtAl72], we know that for every fixed ε there exists a unique $u_\varepsilon \in V(\Omega_\varepsilon)$ solution of problem (2.3).

2.3   Main Convergence Result

Our aim is to study the asymptotic behavior of u_ε, solution of problem (2.3), when ε tends to zero. For this purpose, we first use the dilatation of the domain Ω_ε in the variable x_3, namely

$$y_3 = \frac{x_3}{\varepsilon}, \tag{2.4}$$

in order to have the functions defined in an open set with fixed height, denoted $\widetilde{\Omega}_\varepsilon$ and given by (2.2).

Namely, we define $\tilde{u}_\varepsilon \in (H_0^1(\widetilde{\Omega}_\varepsilon))^3$ by

$$\tilde{u}_\varepsilon(x', y_3) = u_\varepsilon(x', \varepsilon y_3) \ a.e. \ (x', y_3) \in \widetilde{\Omega}_\varepsilon.$$

Let us introduce some notation which will be useful in the following: for a vectorial function $v = (v', v_3)$ (and, analogously, for a scalar function w), associated with the change of variables (2.4), we introduce the operators D_ε, $\mathbb{D}_\varepsilon$, div_ε and ∇_ε, defined by

$$(D_\varepsilon v)_{i,j} = \partial_{x_j} v_i \text{ for } i = 1, 2, 3, \ j = 1, 2, \quad (D_\varepsilon v)_{i,3} = \frac{1}{\varepsilon} \partial_{y_3} v_i \text{ for } i = 1, 2, 3,$$

$$\mathbb{D}_\varepsilon [v] = \frac{1}{2} \left(D_\varepsilon v + D_\varepsilon^t v \right), \quad |\mathbb{D}_\varepsilon [v]|^2 = \mathbb{D}_\varepsilon [v] : \mathbb{D}_\varepsilon [v],$$

$$\text{div}_\varepsilon v = \text{div}_{x'} v' + \frac{1}{\varepsilon} \partial_{y_3} v_3, \quad \nabla_\varepsilon w = (\nabla_{x'} w, \frac{1}{\varepsilon} \partial_{y_3} w)^t.$$

We consider the space

$$V(\widetilde{\Omega}_\varepsilon) = \{\tilde{v} \in (H_0^1(\widetilde{\Omega}_\varepsilon))^3 \mid \mathrm{div}_\varepsilon \tilde{v} = 0 \text{ in } \widetilde{\Omega}_\varepsilon\},$$

and for $\tilde{u}, \tilde{v} \in V(\widetilde{\Omega}_\varepsilon)$, we introduce

$$a_\varepsilon(\tilde{u}, \tilde{v}) = 2\mu \int_{\widetilde{\Omega}_\varepsilon} \mathbb{D}_\varepsilon\left[\tilde{u}\right] : \mathbb{D}_\varepsilon\left[\tilde{v}\right] dx'dy_3,$$

$$j_\varepsilon(\tilde{v}) = \sqrt{2}g\,\varepsilon \int_{\widetilde{\Omega}_\varepsilon} |\mathbb{D}_\varepsilon[\tilde{v}]| dx'dy_3,$$

$$(\tilde{u}, \tilde{v})_{\widetilde{\Omega}_\varepsilon} = \int_{\widetilde{\Omega}_\varepsilon} \tilde{u} \cdot \tilde{v}\, dx'dy_3.$$

Using the transformation (2.4), the variational inequality (2.3) can be rewritten as:

Find $\tilde{u}_\varepsilon \in V(\widetilde{\Omega}_\varepsilon)$ such that

$$a_\varepsilon(\tilde{u}_\varepsilon, \tilde{v} - \tilde{u}_\varepsilon) + j_\varepsilon(\tilde{v}) - j_\varepsilon(\tilde{u}_\varepsilon) \geq (f, \tilde{v} - \tilde{u}_\varepsilon)_{\widetilde{\Omega}_\varepsilon}, \quad \forall \tilde{v} \in V(\widetilde{\Omega}_\varepsilon). \tag{2.5}$$

We start by obtaining some *a priori* estimates for $\tilde{u}_\varepsilon$, stated in the next lemma.

Lemma 1 *There exists a constant C independent of ε, such that if $\tilde{u}_\varepsilon \in (H_0^1(\widetilde{\Omega}_\varepsilon))^3$ is the solution of problem (2.5), one has*

$$\|\tilde{u}_\varepsilon\|_{(L^2(\widetilde{\Omega}_\varepsilon))^3} \leq C\varepsilon^2, \quad \left\|\mathbb{D}_\varepsilon\left[\tilde{u}_\varepsilon\right]\right\|_{(L^2(\widetilde{\Omega}_\varepsilon))^{3\times 3}} \leq C\varepsilon, \tag{2.6}$$

$$\|D_\varepsilon \tilde{u}_\varepsilon\|_{(L^2(\widetilde{\Omega}_\varepsilon))^{3\times 3}} \leq C\varepsilon. \tag{2.7}$$

We extend the velocity $\tilde{u}_\varepsilon$ by zero to the $\Omega \setminus \widetilde{\Omega}_\varepsilon$ and denote the extension by the same symbol. Obviously, estimates (2.6)–(2.7) remain valid for the extended function and the extension is divergence free too. According to this extension, problem (2.5) can be written as:

$$2\mu \int_\Omega \mathbb{D}_\varepsilon\left[\tilde{u}_\varepsilon\right] : \mathbb{D}_\varepsilon\left[\tilde{v} - \tilde{u}_\varepsilon\right] dx'dy_3 + \sqrt{2}g\,\varepsilon \int_\Omega |\mathbb{D}_\varepsilon[\tilde{v}]| dx'dy_3 \tag{2.8}$$

$$-\sqrt{2}g\,\varepsilon \int_\Omega |\mathbb{D}_\varepsilon[\tilde{u}_\varepsilon]| dx'dy_3 \geq \int_\Omega f \cdot (\tilde{v} - \tilde{u}_\varepsilon)\, dx'dy_3,$$

for every $\tilde{v}$ that is the extension by zero to the whole Ω of a function in $(H_0^1(\widetilde{\Omega}_\varepsilon))^3$.

Our main result is the following theorem:

Theorem 1 *Let $\tilde{u}_\varepsilon$ be the solution of problem (2.8). There exists $\hat{u} \in L^2(\omega; H^1_\sharp (Y)^3)$ (here "$\sharp$" denotes Y'-periodicity), such that*

$$\frac{\tilde{u}_\varepsilon}{\varepsilon^2} \rightharpoonup \int_{Y'} \hat{u} \, dy' \ in \ L^2(\Omega),$$

$\int_{Y'} \hat{u}_3 dy' = 0$, $\hat{u} = 0$ on $\omega \times Y_s$, $\hat{u} = 0$ on $y_3 = \{0, 1\}$, $\text{div}_y \hat{u} = 0$ in $\omega \times Y$,

$$\text{div}_{x'} \left(\int_Y \hat{u}'(x', y) dy \right) = 0 \ in \ \omega, \qquad \left(\int_Y \hat{u}'(x', y) dy \right) \cdot n = 0 \ on \ \partial\omega,$$

and $\hat{u}$ is the unique solution of the limit problem

$$2\mu \int_{\omega \times Y} \mathbb{D}_y [\hat{u}] : (\mathbb{D}_y [\tilde{v}] - \mathbb{D}_y [\hat{u}]) \, dx' dy + \sqrt{2}g \int_{\omega \times Y} |\mathbb{D}_y [\tilde{v}]| \, dx' dy$$

$$-\sqrt{2}g \int_{\omega \times Y} |\mathbb{D}_y [\hat{u}]| \, dx' dy \geq \int_{\omega \times Y} f' \cdot (\tilde{v}' - \hat{u}') \, dx' dy, \qquad (2.9)$$

for every $\tilde{v} \in L^2(\omega; H^1_\sharp (Y)^3)$ such that $\tilde{v}(x', y) = 0$ in $\omega \times Y_s$, $\text{div}_y \tilde{v} = 0$ in $\omega \times Y$,

$$\text{div}_{x'} \left(\int_Y \tilde{v}'(x', y) dy \right) = 0 \ in \ \omega, \qquad \left(\int_Y \tilde{v}'(x', y) dy \right) \cdot n = 0 \ on \ \partial\omega.$$

Proof

First Step   A priori estimates

The change of variable (2.4) does not provide the information we need about the behavior of $\tilde{u}_\varepsilon$ in the microstructure associated with $\tilde{\Omega}_\varepsilon$. To solve this difficulty, we use an adaptation of the unfolding method from [CiEtAl08]. In order to apply the unfolding method, we will need the following notation: for $k' \in \mathbb{Z}^2$, we define $\kappa : \mathbb{R}^2 \to \mathbb{Z}^2$ by

$$\kappa(x') = k' \iff x' \in Y'_{k',1}. \qquad (2.10)$$

Remark that κ is well defined up to a set of zero measure in $\mathbb{R}^2$ (the set $\cup_{k' \in \mathbb{Z}^2} \partial Y'_{k',1}$). Moreover, for every $\varepsilon > 0$, we have

$$\kappa \left(\frac{x'}{\varepsilon} \right) = k' \iff x' \in Y'_{k',\varepsilon}.$$

According to the adaptation introduced in [AS17] of the unfolding method, we divide the domain Ω in rectangular cuboids of lateral lengths ε and vertical length 1. For this purpose, given $\tilde{u}_\varepsilon \in (H_0^1(\Omega))^3$, we define $\hat{u}_\varepsilon$ by

$$\hat{u}_\varepsilon(x', y) = \tilde{u}_\varepsilon \left(\varepsilon\kappa \left(\frac{x'}{\varepsilon} \right) + \varepsilon y', y_3 \right), \quad \text{a.e. } (x', y) \in \omega \times Y, \quad (2.11)$$

where the function κ is defined in (2.10).

For $k' \in \mathcal{K}_\varepsilon$, the restriction of $\hat{u}_\varepsilon$ to $Y'_{k',\varepsilon} \times Y$ does not depend on x', whereas as a function of y it is obtained from $\tilde{u}_\varepsilon$ by using the change of variables

$$y' = \frac{x' - \varepsilon k'}{\varepsilon},$$

which transforms $Y_{k',\varepsilon}$ into Y.

We can obtain a priori estimates for the sequence $\hat{u}_\varepsilon$. There exists a constant C independent of ε, such that $\hat{u}_\varepsilon$ defined by (2.11) satisfies

$$\left\| \hat{u}_\varepsilon \right\|_{(L^2(\omega \times Y))^3} \leq C\varepsilon^2, \quad \left\| \mathbb{D}_y[\hat{u}_\varepsilon] \right\|_{(L^2(\omega \times Y))^{3 \times 3}} \leq C\varepsilon^2.$$

Second Step   Convergence results

As in [AS17], we can obtain compactness results for the sequences $\tilde{u}_\varepsilon$ and $\hat{u}_\varepsilon$ satisfying the a priori estimates given before.

For a subsequence of ε still denoted by ε, there exist $\tilde{u} \in H^1(0, 1; L^2(\omega)^3)$, where $\tilde{u}_3 = 0$ and $\tilde{u} = 0$ on $y_3 = \{0, 1\}$, $\hat{u} \in L^2(\omega; H_\#^1(Y)^3)$, with $\hat{u} = 0$ on $\omega \times Y_s$ and $\hat{u} = 0$ on $y_3 = \{0, 1\}$ such that $\int_Y \hat{u}(x', y) dy = \int_0^1 \tilde{u}(x', y_3) dy_3$ with $\int_Y \hat{u}_3 dy = 0$, such that

$$\frac{\tilde{u}_\varepsilon}{\varepsilon^2} \rightharpoonup (\tilde{u}', 0) \text{ in } H^1(0, 1; L^2(\omega)^3),$$

$$\frac{\hat{u}_\varepsilon}{\varepsilon^2} \rightharpoonup \hat{u} \text{ in } L^2(\omega; H^1(Y)^3), \quad (2.12)$$

$$\text{div}_y \hat{u} = 0 \text{ in } \omega \times Y, \quad (2.13)$$

$$\text{div}_{x'} \left(\int_0^1 \tilde{u}'(x', y_3) dy_3 \right) = 0 \text{ in } \omega, \quad \left(\int_0^1 \tilde{u}'(x', y_3) dy_3 \right) \cdot n = 0 \text{ on } \partial\omega,$$

$$\text{div}_{x'} \left(\int_Y \hat{u}'(x', y) dy \right) = 0 \text{ in } \omega, \quad \left(\int_Y \hat{u}'(x', y) dy \right) \cdot n = 0 \text{ on } \partial\omega.$$

$$(2.14)$$

Third Step   Passage to the limit

By using (2.11), we first transform the variational inequality (2.8) in a variational inequality stated in the domain $\omega \times Y$. Then, by choosing suitable test functions, we pass to the limit $\varepsilon \to 0$. By using convergences (2.12), (2.13), (2.14) we find the limit problem (2.9). The uniqueness of the solution $\hat{u}$ of problem (2.9) is proved by contradiction.

2.4   Conclusions and Perspectives

By using dimension reduction and homogenization techniques, we studied the limiting behavior of the velocity for a nonlinear viscoplastic Bingham flow with small yield stress $g\varepsilon$, in a thin porous medium of small height ε and for which the relative dimension of the pores is ε. We found in Theorem 1 the limit problem (2.9), in which both effects of a nonlinear Darcy law and a lower-dimensional *Bingham-like* law appear. Indeed, as in [LiEtAl81](see also [BuEtAl17]), problem (2.9) can be written as a nonlinear Darcy law set in the domain ω. The third component of the velocity of filtration appearing in the nonlinear Darcy law equals zero and this phenomenon corresponds precisely to a two-dimensional *Bingham-like* law (see [BuEtAl04]).

In the forthcoming work [AnEtAl] we study the cases of thin porous media, whose periodicity parameter is a_ε instead of ε. Different cases are analyzed, following the ratio between the height ε of the porous media and the relative dimension a_ε of the periodically distributed pores. Moreover, we consider the more involved case in which the convergence of the pressure of the flow is also studied.

Acknowledgements María Anguiano has been supported by Junta de Andalucía (Spain), Proyecto de Excelencia P12-FQM-2466.

References

[Zh08] Zhengan, Y., and Hongxing, Z.: Homogenization of a stationary Navier-Stokes flow in porous medium with thin film. *Acta Mathematica Scientia*, **28B**(4), 963–974 (2008).

[FaEtAl16] Fabricius, J., Hellström, J.G. I., Lundström, T.S., Miroshnikova E., and Wall, P.: Darcy's Law for Flow in a Periodic Thin Porous Medium Confined Between Two Parallel Plates. *Transp. Porous Med.*, **115**, 473–493 (2016).

[AS18] Anguiano, M., and Suárez-Grau, F.J.: The transition between the Navier-Stokes equations to the Darcy equation in a thin porous medium. *Mediterr. J. Math.*, **15:45** (2018).

[AS17] Anguiano, M., and Suárez-Grau, F.J.: Homogenization of an incompressible non-Newtonian flow through a thin porous medium. *Z. Angew. Math. Phys.*, **68:45** (2017).

[CiEtAl08] Cioranescu, D., Damlamian, A., and Griso, G.: The periodic Unfolding Method in Homogenization. *SIAM J. Math. Anal.*, **40**, No. 4, 1585–1620 (2008).

[An17] Anguiano, M.: Darcy's laws for non-stationary viscous fluid flow in a thin porous medium. *Math. Meth. Appl. Sci.*, **40**, No. 8, 2878–2895 (2017).

[An217] Anguiano, M.: On the non-stationary non-Newtonian flow through a thin porous medium. *ZAMM-Z. Angew. Math. Mech.*, **97**, No. 8, 895–915 (2017).

[Gr04] Griso, G.: Asymptotic behavior of a crane.*C.R.Acad.Sci. Paris, Ser. I*, **338**, No. 3, 261–266.

[GrEtAl17] Griso, G., Migunova, A., and Orlik, J.: Asymptotic analysis for domains separated by a thin layer made of periodic vertical beams. *Journal of Elasticity*, **128**, 291–331 (2017).

[LiEtAl81] Lions, J.L., and Sánchez-Palencia E.: Écoulement d'un fluide viscoplastique de Bingham dans un milieu poreux. *J.* Math. Pures Appl., **60**, 341–360 (1981).

[BoEtAl93] Bourgeat, A., and Mikelić, A.: A note on homogenization of Bingham flow through a porous medium. *J. Math. Pures Appl.*, **72**, 405–414 (1993).

[BuEtAl13] Bunoiu R., Cardone G., and Perugia C.: Unfolding Method for the Homogenization of Bingham flow. In *Modelling and Simulation in Fluid Dynamics in Porous Media*, Series: Springer Proceedings in Mathematics & Statistics, Vol. 28 (2013), pp. 109–123.

[BuEtAl17] Bunoiu, R., and Cardone, G.: Bingham Flow in Porous Media with Obstacles of Different Size. *Mathematical Methods in the Applied Sciences(MMAS)*, **40**, No. 12, 4514–4528 (2017).

[BuEtAl03] Bunoiu, R., and Kesavan, S.: Fluide de Bingham dans une couche mince, In *Annals of the University of Craiova*. Math. Comp. Sci. series 30, (2003), pp 1–9.

[BuEtAl04] Bunoiu, R., and Kesavan, S.: Asymptotic behaviour of a Bingham fluid in thin layers. *Journal of Mathematical Analysis and Applications*, **293**, No. 2, 405–418 (2004).

[BuEtAl18] Bunoiu, R., Gaudiello A., and Leopardi A.: Asymptotic Analysis of a Bingham Fluid in a Thin T-like Shaped Structure. *Journal de Mathématiques Pures et Appliquées*, **123**, 148–166 (2019).

[LiEtAl90] Liu K.F., and Mei, C.C.: Approximate equations for the slow spreading of a thin sheet of Bingham plastic fluid. *Phys. Fluids*, **A 2**, 30–36 (1990).

[DuEtAl72] Duvaut, G., and Lions, J.L.: *Les inéquations en mécanique et en physique*. Dunod, Paris (1972).

[AnEtAl] Anguiano, M., and Bunoiu, R.: Homogenization of Bingham flow in thin porous media. Submitted for publication.

Chapter 3
q-Calculus Formalism for Non-extensive Particle Filter

Amarisio S. Araújo, Helaine C. M. Furtado, and Haroldo F. de Campos Velho

3.1 Introduction

The q-calculus was developed by the English reverend Frank Hilton [Ja1908, Ja1910a, Ja1910b] defining q-derivative, q-integral, and q-functions. The quantum calculus (ordinary calculus without taking limits) is also based on the q-calculus [KaCh02]. Such formalism has been applied to different contexts: hypergeometric functions [Er03], statistical mechanics [UmEtAl08], optimization with q-gradient (q-steepest descent method) [SoEtAl11], and inverse vibration problem [ToEtAl15].

Campos Velho and Furtado have proposed a new approach for particle filter with an adaptive likelihood operator [CaFu11]. The proposed filter belongs to the Bayesian strategy for estimation theory, where a non-linear and non-Gaussian assumptions can be employed. For the proposed adaptive particle filter, the likelihood function is defined as the Tsallis' statistics distribution [Ts88, Ts99]. Such adaptive filter can also be applied to the probability density function with statistical moments not defined [CaFu11]. Tsallis' distribution[1] was derived to be associated with the non-extensive form for the entropy [Ts88], and the non-extensive parameter

---

[1] Here, the word *distribution* will have the same meaning as *probability density function* (PDF).

A. S. Araújo
Federal University of Viçosa (UFV), Viçosa, MG, Brazil
e-mail: amarisio@ufv.br

H. C. M. Furtado
Federal University of West Pará (UFOPA), Santarém, PA, Brazil
e-mail: helaine.furtado@ufopa.edu.br

H. F. de Campos Velho (✉)
National Institute for Space Research (INPE), São José dos Campos, SP, Brazil
e-mail: haroldo.camposvelho@inpe.br

© Springer Nature Switzerland AG 2019
C. Constanda, P. Harris (eds.), *Integral Methods in Science and Engineering*,
https://doi.org/10.1007/978-3-030-16077-7_3

q has a key role in the Tsallis' thermostatistics. When the non-extensive parameter $q \rightarrow 1$, the Boltzmann-Gibbs thermodynamics is recovered. Here, the worked particle filter will be named as *Non-Extensive Particle Filter* (NEx-PF). In the Adaptive NEx-PF, a secondary particle filter is employed to estimate the q-parameter.

Our goal is to apply the q-calculus formalism for deriving some properties of a framework to pave the road for enhancing the understanding and new developments to the NEx-PF. In particular, the q-stability is derived to the Tsallis' distribution, as a q-analogue to the standard definition in statistics [No05].

3.2   The Non-extensive Particle Filter

A class of sequential Monte Carlo estimation is usually called particle filter. This is one methodology available to the estimation theory—a branch of mathematics and statistics with focus on determining parameters and/or functions from measured or desired properties. A key issue to the particle filter is the likelihood operator [RiEtAl04] doing the comparison between the distribution of the observed/desired property and the distribution calculated from a mathematical model [KaSo04]. Particle filters are the most general and robust approaches dealing with non-linear problems and/or non-Gaussian distributions. The posterior density function is represented by a set of random samples (particles) with associated weights, computed from the product between the likelihood operator and the prior distribution.

The Tsallis' probability density function from the non-extensive thermo-statistics has used by Campos Velho and Furtado [CaFu11] to define a family of likelihood operators. As already mentioned, the value of the non-extensive parameter q for a particular application can be determined by using a secondary particle filter [CaFu11].

The algorithm for the standard and non-extensive particle filters is summarized in the steps below.

1. Compute the initial particle ensemble:

$$\left\{ w_{0|n-1}^{(i)} \right\}_{i=0}^{M} \sim p_{w_0}(w_0)$$

(initial PDF: Gaussian with zero mean and $\sigma = 5$, i.e., $p_{w_0}(w_0) = N(0, 5)$);
2. Compute:

$$r_n^{(i)} = p(y_n | w_{n|n-1}) = p_{\text{et}}(y_n - h(w_n, t_n))$$

where y_n denotes observations, $h(.)$ is the observation operator, and p_{et} is the likelihood operator:

$$p_{et}(z) = \begin{cases} N_z(0, 1) \text{ (regular assumption: Gaussian distribution)} \\ T_z(0, 1) \text{ (applying NEx-PF: Tsallis' distribution)} \end{cases}$$

being the innovation z expressed by:

$$z = y_n - h(w_n, t_n)$$

3. Normalize:

$$\tilde{r}_n^{(i)} = \frac{r_n^{(i)}}{\sum_{j=1}^{M} r_n^{(j)}} \; ;$$

4. Resampling: extract M particles, with substitution—see [Sc06]:

$$\Pr\{w_{n|n}^{(i)} = w_{n|n-1}^{(j)}\} = \tilde{q}_n^{(j)}, \quad i = 1, \dots, M \; ;$$

Resampling step:

– Generate M ordered numbers: $u_k = M^{-1}[(k-1) + \tilde{u}]$, with: $\tilde{u} \sim U(0, 1)$ (uniform distribution),
– Resampled particles are obtained by producing m_i copies of particle $w^{(i)}$:

$$m_i = \text{number of } u_k$$

$$u_k \in \left[\sum_{s=1}^{i-1} \tilde{r}_n^{(s)}, \sum_{s=1}^{i} \tilde{r}_n^{(s)} \right]$$

5. Time up-dating: compute the new particles:

$$w_{n+1|n}^{(i)} = f(w_{n|n}^{(i)}, t_n) + \mu_n \, , \quad \text{with: } \mu_n \in \mathcal{N}(0, 1)$$

and: $w_{n+1|n}^{(i)} \sim p(w_{n+1|n}^{(i)} | w_{n|n}^{(i)})$, being $i = 1, 2, \dots, M$;
6. Set: $t_{n+1} = t_n + \Delta t$, and go to step-2.

The kernel of the algorithm coming from the application of the Bayes' theorem and of the Markov property:

$$\begin{aligned} p(w_n | Y_n) &= p(w_n | y_n, Y_{n-1}) \\ &= \frac{p(y_n | w_n) \, p(w_n | Y_{n-1})}{p(y_n | Y_{n-1})} \\ &\propto p(y_n | w_n) \, p(w_n | Y_{n-1}) \end{aligned} \tag{3.1}$$

suggesting the following choice:

$$\underbrace{p(w_n|Y_n)}_{\text{posterior}_{(wn)}} \propto \underbrace{p(y_n|w_n)}_{\text{likelihood}_{(wn)}} \underbrace{p(w_n|Y_{n-1})}_{\text{prior}_{(wn)}} . \tag{3.2}$$

The equiprobability condition produces the maximum for the entropy. For the non-extensive form to the entropy, the latter condition leads to special distributions [Ts99]. Three conditions for the Tsallis' distribution are shown below.

q > 1 :

$$p_q(x) = \alpha_q^+ \left[1 - \frac{1-q}{3-q} \left(\frac{x}{\sigma} \right)^2 \right]^{-1/(q-1)} \tag{3.3}$$

q = 1 :

$$p_q(x) = \frac{1}{\sigma} \left[\frac{1}{2\pi} \right]^{1/2} e^{-(x/\sigma)^2/2} \tag{3.4}$$

q < 1 :

$$p_q(x) = \alpha_q^- \left[1 - \frac{1-q}{3-q} \left(\frac{x}{\sigma} \right)^2 \right]^{1/(q-1)} \tag{3.5}$$

where—here, $\Gamma(\cdot)$ denotes the gamma function:

$$\sigma^2 = \frac{\int_{-\infty}^{+\infty} x^2 [p_q(x)]^q dx}{\int_{-\infty}^{+\infty} [p_q(x)]^q dx} ,$$

$$\alpha_q^+ = \frac{1}{\sigma} \left[\frac{q-1}{\pi(3-q)} \right]^{1/2} \frac{\Gamma\left(\frac{1}{q-1} \right)}{\Gamma\left(\frac{3-q}{2(q-1)} \right)} ,$$

$$\alpha_q^- = \frac{1}{\sigma} \left[\frac{1-q}{\pi(3-q)} \right]^{1/2} \frac{\Gamma\left(\frac{5-3q}{2(1-q)} \right)}{\Gamma\left(\frac{2-q}{1-q} \right)} .$$

The distributions above applies if $|x| < \sigma[(3-q)/(1-q)]^{1/2}$, otherwise $p_q(x) = 0$. For distributions with $q < 5/3$, the standard central limit theorem applies, implying that if p_i is written as a sum of M random independent variables, in the limit case $M \to \infty$, the probability density function for p_i in the distribution space is the normal (Gaussian) distribution. However, for $5/3 < q < 3$ the Levy-Gnedenko's central limit theorem applies, resulting for $M \to \infty$ the Lévy distribution as the probability density function for the random variable p_i. The index in such Lévy distribution is $\alpha = (3-q)/(q-1)$ [Ts99, TsQu07].

3.3   Definition of Stable Probability Density Function

A probability density function (PDF) is defined as a *stable*-PDF if the distribution of the linear combination of two independent random copies of the same random variable belongs to the same class of the distribution, including location (parameter—scalar or vector—to determine the location or shift of the distribution: $p_z(x) = p(x - z)$) and scale parameters [No05]. The *scale* is a parameter where the cumulative distribution function satisfies: $F(x; s, \lambda) = F(xs^{-1}, 1, \lambda)$.

Examples of stable distributions:

$$\text{Gaussian:}\quad p_{\mu,\sigma}(x) = \frac{1}{\sigma}\left[\frac{1}{2\pi}\right]^{1/2} e^{-(x-\mu)/(2\sigma^2)}$$

$$\text{Cauchy:}\quad p_{\mu,\gamma}(x) = \frac{1}{\pi}\left[\frac{\gamma}{(x-\mu)^2 + \gamma^2}\right]$$

$$\text{Lévy:}\quad p_{\mu,c}(x) = \sqrt{\frac{c}{2\pi}}\frac{e^{c/[2(x-\mu)]}}{(x=\mu)^{3/2}} e^{-(x-\mu)/(2\sigma^2)}$$

For establishing the stability property for a given PDF, it is convenient to deal with the characteristic function of a distribution:

$$\varphi(w) = \int_{-\infty}^{+\infty} e^{iwx}\, p(x)\, dx \ . \tag{3.6}$$

For a Gaussian distribution it is easy to show the stability probability:

$$\varphi_{p_{\mu,\sigma}}(w) = \int_{-\infty}^{+\infty} e^{iwx}\frac{1}{\sqrt{2\pi\sigma^2}}e^{-(x-\mu)^2/(2\sigma^2)}\, dx = e^{[iw\mu - (\sigma^2 w^2)/2]} \tag{3.7}$$

Clearly, the distribution can be obtained from its characteristic function:

$$p(x) = \frac{1}{2\pi}\int_{-\infty}^{+\infty} \varphi(w)\, e^{-iwx}\, dx \ . \tag{3.8}$$

One important result is the Lévy-Khinchin theorem for the characteristic function of stable-PDF:

$$\log[\varphi(w)] = \begin{cases} i\mu w - c|w|^\alpha[1 - i\beta(w/|w|)\tan(\pi/1\alpha)] & \text{if } \alpha \neq 1 \\ i\mu w - c|w|^\alpha[1 + i\beta(2w/(\pi\,|w|))\log(|w|)] & \text{if } \alpha = 1 \end{cases} \tag{3.9}$$

where $0 \leq \alpha \leq 1$, $\beta \in [-1, 1]$, $c > 0$, and μ is the mean.

The *q*-calculus and *q*-algebra will help to define a *q*-stability for a distribution.

3.4 q-Calculus

Considering a differentiable function $\varphi(x)$, the q-derivative can be defined as [Er03]:

$$(D_q\varphi)(x) \equiv \begin{cases} \dfrac{\varphi(q\,x) - \varphi(x)}{q\,x - x}\,, & \text{if } x \neq 0 \text{ and } q \neq 1, \\ \dfrac{d\varphi(x)}{dx}\,, & \text{otherwise,} \end{cases} \tag{3.10}$$

In addition, at $x = 0$, the operator is defined as follows: $(D_q\varphi)(x) \equiv d\varphi(0)/dx$, for all q-value.

Considering the fundamental limits

$$\lim_{x \to \infty} \left(1 + \frac{x}{n}\right)^n = \lim_{x \to 0} (1 + n\,x)^{1/n} = e^x \tag{3.11}$$

$$\lim_{x \to 1} T(0, \sigma^2) = N(0, \sigma^2) \tag{3.12}$$

the q-exponential and q-logarithm can be defined as [UmEtAl08]:

$$e_q^x \equiv [1 + (1 - q)x]^{\frac{1}{1-q}}\,, \tag{3.13}$$

$$\log_q(x) \equiv \frac{x^{1-q} - 1}{1 - q}\,. \tag{3.14}$$

A q-**algebra** can also be defined [UmEtAl08] applying the generalized operation for sum and product:

$$x \boxplus y \equiv x + y + (1 - q)xy\,, \tag{3.15}$$

$$x \boxtimes y \equiv [x^{1-q} + y^{1-q} - 1]^{\frac{1}{1-q}}\,. \tag{3.16}$$

with the following neutral and inverse elements:

$$x \boxplus (-x)_q = 0\,, \quad \text{with: } (-x)_q = -x[1 + (1 - q)x]^{-1} \tag{3.17}$$

$$x \boxtimes (x^{-1})_q = 1\,. \quad \text{with: } (x^{-1})_q = [2 - x^{1-q}]^{\frac{1}{1-q}}\,. \tag{3.18}$$

Properties

1. $e_q^x \dot{e}_q^y = e_q^{x \boxplus y}$
2. $e_q^x \boxtimes e_q^y = e_q^{x+y}$
3. $\log_q(x\dot{y}) = \log_q(x) \boxplus \log_q(y)$

4. $\log_q(x \boxtimes y) = \log_q(x) + \log_q(y)$
5. *q*-sum and *q*-product ($\boxplus$, $\boxtimes$):

   a. Commutativity: **yes**
   b. Associativity: **yes**
   c. Distributivity: **no**

The real space vector with regular sum and product operations $\mathbb{R}(+, \times)$ is a *field*, and the $\mathbb{R}(\boxplus, \boxtimes)$ defines a *quasi-field*.

3.4.1   *q-Fourier Transform, q-Gaussian Function, and q-Stability*

From the definition of the *q*-exponential, the complex form to the *q*-exponential is given by

$$e_q^z = e_q^{x+iy} = e_q^x \boxtimes e_q^{iy} \tag{3.19}$$

with $i = \sqrt{-1}$ as the imaginary unit number.

Definition 1 (*q*-Fourier Transform)

$$\varphi_{[f]}^{[q]}(\omega) \equiv \int_{-\infty}^{+\infty} e_q^{ix\omega} \boxtimes f(x)dx \tag{3.20}$$

The expression in Eq. (3.20) can also be written as:

$$\varphi_{[f]}^{[q]}(\omega) = \int_{-\infty}^{+\infty} f(x)\, e_q^{ix\omega[f(x)]^{q-1}}\, dx \ . \tag{3.21}$$

The above expression is derived just applying the definitions:

$$e_q^{ix\omega} \boxtimes f(x) = [1 + (1-q)ix\omega]^{\frac{1}{1-q}} \boxtimes f(x)$$

$$= \{[1 + (1-q)ix\omega] + [f(x)]^{1-q} - 1\}^{\frac{1}{1-q}}$$

$$= \{[f(x)]^{1-q} + (1-q)ix\omega\}^{\frac{1}{1-q}}$$

$$= \{[f(x)]^{1-q}[1 + (1-q)ix\omega[f(x)]^{q-1}]\}^{\frac{1}{1-q}}$$

$$= f(x)[1 + (1-q)ix\omega[f(x)]^{q-1}]\}^{\frac{1}{1-q}}$$

$$= f(x)e_q^{ix\omega[f(x)]^{q-1}}$$

Definition 2 (q-Gaussian Function)

$$G_q(x) \equiv g_q \, e_q^{-\left(\frac{x}{\sigma}\right)^2} \tag{3.22}$$

the coefficient g_q is a parameter indicating a *family* of q-Gaussian functions.

Theorem 1 *The Tsallis' distribution is a q-Gaussian function.*

Proof Using a new notation for Eqs. (3.3) and (3.5)

$$h_q^{\pm} = \begin{cases} \alpha_q^+ \left[1 - \frac{1-q}{3-q}\left(\frac{x}{\sigma}\right)^2\right]^{\frac{1}{q-1}} , \text{ if } \quad 1 < q < 3 \\ \alpha_q^- \left[1 - \frac{1-q}{3-q}\left(\frac{x}{\sigma}\right)^2\right]^{\frac{1}{q-1}} , \text{ if } -\infty < q < 1 \end{cases} \tag{3.23}$$

doing some manipulation

$$\alpha_q^{\pm}\left[1 - \frac{1-q}{3-q}\left(\frac{x}{\sigma}\right)^2\right]^{\frac{1}{q-1}} = \alpha_q^{\pm}\left(\frac{1}{q-3}\right)\left[1 - (q-1)\left(\frac{x}{\sigma}\right)^2\right]^{\frac{1}{q-1}}$$

$$\equiv H_q^{\pm}\left[1 - (q-1)\left(\frac{x}{\sigma}\right)^2\right]^{\frac{1}{q-1}}$$

$$= H_q^{\pm} \, e_q^{\left(\frac{x}{\sigma}\right)^2} \diamond$$

where:

$$H_q^{\pm} = \begin{cases} \alpha_q^+\left(\frac{1}{q-3}\right) \text{ if } \quad 1 < q < 3 \\ \frac{1}{\sqrt{2\pi\sigma^2}} \quad \text{ if } \quad q = 1 \\ \alpha_q^-\left(\frac{1}{q-3}\right) \text{ if } -\infty < q < 1 \end{cases}$$

Definition 3 (q-Characteristic Function) Let $p(x)$ be a distribution. The q-characteristic function for the distribution $p(x)$ is expressed by

$$\varphi_p^q(w) = \int_{-\infty}^{+\infty} e_q^{ixw} \boxtimes p(x)\,dx \tag{3.24}$$

Definition 4 (q-Stable Distribution) A distribution is said q-stable if such distribution and its q-characteristic function are the same type, including location and scale parameters.

Theorem 2 *Tsallis' distribution is q-stable.*

Proof The Tsallis' distribution is a *q*-Gaussian function. Applying the definition of *q*-characteristic function to a *q*-Gaussian function, with $q \in [-1, 1)$, results (changing variable x):

$$\varphi^{[q]}_{[G_q]}(w) = \int_{-\infty}^{+\infty} e_q^{ixw} \boxtimes e_q^{\left(\frac{x}{\sigma}\right)^2} dx = \sigma H_q^+ \int_{-\infty}^{+\infty} e_q^{-y^2 - \frac{\sigma^2 [H_q^+]^{2(q-1)} \omega^2}{4}} dy .$$

Now, using the Cauchy theorem on integrals over closed curves

$$\varphi^{[q]}_{[G_q]}(w) = \exp_q \left\{ - \left(\frac{\omega^2 \sigma^2}{4 - \sigma^2 q [H_q^+]^{2(1-q)}} \right) \left(\frac{3-q}{2} \right) \right\} \diamond \qquad (3.25)$$

3.5   Final Remarks

The non-extensive particle filter obtained a better result for state estimation using an adaptive likelihood function based on Tsallis' distribution [CaFu11]. Here, the *q*-algebra defined the *quasi-field*. The quasi-field and the *q*-calculus are the framework to derive the *q*-Fourier transform and to define a *q*-Gaussian function.

The distribution associated with the Tsallis' non-extensive entropy is a *q*-Gaussian distribution [UmEtAl08]. From the framework established, the *q*-stability for *q*-Gaussian distribution was derived.

Appendix: Non-extensive Tsallis' Thermostatistics

A non-extensive form to the entropy can be expressed as [Ts88]:

$$S_q(p) \equiv \frac{k}{q-1} \left(1 - \sum_{i=1}^{N_p} p_i^q \right) . \qquad (3.26)$$

and the *q*-expectation of an observable is given by

$$O_q \equiv \langle O \rangle_q = \sum_{i=1}^{N_p} p_i^q O_i . \qquad (3.27)$$

Some properties are derived to the non-extensive entropy S_q.

1. If $q \to 1$:

$$S_1 = k \sum_{i=1}^{N_p} p_i \ln p_i , \tag{3.28}$$

$$O_1 = \sum_{i=1}^{N_p} p_i O_i . \tag{3.29}$$

2. The non-extensive entropy has a positive value: $S_q \geq 0$.
3. Non-extensivity:

$$S_q(A+B) = S_q(A) + S_q(B) + (1-q)S_q(A)S_q(B) \tag{3.30}$$

$$O_q(A+B) = O_q(A) + O_q(B) + (1-q)[O_q(A)S_q(B) + O_q S_q(A)] . \tag{3.31}$$

4. Maximum for the S_q under the constraint $O_q = \sum_i p_i^q \epsilon_i$ (canonical ensemble):

$$p_i = \frac{1}{Z_q}[1 - \beta(1-q)\epsilon_i]^{1/(1-q)} \tag{3.32}$$

where ϵ_i is the energy for the state-i, $O_q = U_q$ is the non-extensive form to the internal energy and the normalization factor Z_q (partition function), for $1 < q < 3$, is given by

$$Z_q = \left[\frac{\pi}{\beta(1-q)}\right]^{1/2} \frac{\Gamma[(3-q)/2(q-1)]}{\Gamma[1/(q-1)]} . \tag{3.33}$$

For $q = 1$, the normalization is written as

$$p_i = e^{\beta \epsilon_i}/Z_1 . \tag{3.34}$$

References

[CaFu11] Campos Velho, H. F., Furtado, H. C. M.: Adaptive particle filter for stable distribution. In: *Integral Methods in Science and Engineering*, C. Constanda and P. J. Harris (eds.), Birkhäuser, New York, NY (2011), pp. 47–57.
[Er03] Ernst, T. 2003. A Method for q-Calculus. *Journal of Nonlinear Mathematical Physics*, **10**,
[ToEtAl15] Hernández Torres, R., Scarabello, M., Campos Velho, H. F., Chiwiacowsky, L. D., Soterroni, A., Gouvea, E., Ramos, F. M.: A Hybrid Method using q-Gradient to Identify Structural Damage. In: *XXXVI Iberian Latin American Congress on Computational Methods in Engineering* (CILAMCE), 2015, Rio de Janeiro (RJ), Brazil (2015), pp. 1–14.

[Ja1908]   Jackson, F. H.: On *q*-functions and a certain difference operator. *Trans. Roy Soc. Edin.*, **46**, 253–281 (1908).

[Ja1910a]   Jackson, F. H.: On *q*-definite integrals. *Quart. J. Pure and Appl. Math.*, **41**, 193–203 (1910).

[Ja1910b]   Jackson, F. H. 1910b. *q*-difference equations. *American Journal of Mathematics*, **32**, 307–314 (1910).

[KaCh02]   Kac, V., Cheung, P.: *Quantum Calculus*. Springer-Verlag, New York (2002).

[KaSo04]   Kaipio, J., Somersalo, E.: *Statistical and Computational Inverse Problems*, In Applied Mathematical Sciences 160 – series, Springer-Verlag, (2004).

[No05]   Nolan, J. P.: *Stable Distributions: Models for Heavy Tailed Data*, Boston, MA: Birkhäuser (2005).

[RiEtAl04] Ristic, B., Arulampalam, S., Gordon, N.: *Beyond the Kalman Filter*, Boston, Artech House (2004).

[Sc06]   Schön, T. B.: *Estimation of Nonlinear Dynamic Systems: Theory and Applications*, Dissertations no. 998 – Linköping Studies in Science and Technology (2006).

[SoEtAl11] Soterroni, A. C., Galski, R. L., Ramos, F. M.: The *q*-gradient vector for unconstrained continuous optimization problems. In *Operations Research Proceedings*, B. Hu, K. Morasch, S. Pickl and M. Siegle (eds.), Springer, Berlin, Heidelberg (2011), pp. 365–370.

[Ts88]   Tsallis, C.: Possible Generalization of Boltzmann-Gibbs Statistics, *J. Statistical Phys.*, **52**, 479–487 (1988).

[Ts99]   Tsallis, C.: Nonextensive statistics: theoretical, experimental and computational evidences and connections, *Braz. J. Phys.*, **29**, 1–35 (1999).

[TsQu07]   C. Tsallis, S. M. D. Queiros, Nonextensive statistical mechanics and central limit theorems I - Convolution of independent random variables and q-product, in *Complexity, Metastability and Non-extensivity* (CTNEXT 07) (Editors: S. Abe, H. Herrmann, P. Quarati, A. Rapisarda, and C. Tsallis), American Institute of Physics, 2007.

[UmEtAl08] Umarov, S., Tsallis, C., Steinberg, S.: On a *q*-Central Limit Theorem Consistent with Nonextensive Statistical Mechanics. *Milan J. Math.*, **76**, 307–328 (2008).

Chapter 4
Two-Operator Boundary-Domain Integral Equations for Variable-Coefficient Dirichlet Problem with General Data

Tsegaye G. Ayele

4.1 Preliminaries

In this paper, the Dirichlet problem for linear second-order scalar elliptic PDE with variable coefficient is considered. The PDE on the right-hand side belongs to $H^{-1}(\Omega)$ and is extended to $\widetilde{H}^{-1}(\Omega)$ when necessary. Using the two-operator approach and an appropriate parametrix (Levi function) this problem is reduced to two different systems of boundary domain integral equations, briefly BDIEs. The equivalence of the original BVP to the two-operator BDIE systems is shown.

Let Ω be an open bounded three-dimensional region of $\mathbb{R}^3$. For simplicity, we assume that the boundary $\partial\Omega$ is simply connected, closed, infinitely smooth surface. Let $a \in C^\infty(\overline{\Omega})$, $a(x) > 0$ for $x \in \overline{\Omega}$. Let also $\partial_j = \partial_{x_j} := \partial/\partial x_j$ ($j = 1, 2, 3$), $\partial_x = (\partial_{x_1}, \partial_{x_2}, \partial_{x_3})$. We consider the scalar elliptic differential equation, which for sufficiently smooth u has the following strong form:

$$Au(x) := A(x, \partial_x)u(x) := \sum_{i=1}^{3} \frac{\partial}{\partial x_i}\left(a(x)\frac{\partial u(x)}{\partial x_i}\right) = f(x), \quad x \in \Omega, \qquad (4.1)$$

where u is the unknown function and f is a given function in Ω.

In what follows $\mathscr{D}(\Omega) = C_0^\infty(\Omega)$, $H^s(\Omega) = H_2^s(\Omega)$, $H^s(\partial\Omega) = H_2^s(\partial\Omega)$ are the Bessel potential spaces, where $s \in \mathbb{R}$ is an arbitrary real number (see, e.g., [Lio72, McL00]). We recall that H^s coincides with the Sobolev-Slobodetski spaces W_2^s for any non-negative s. We denote by $\widetilde{H}^s(\Omega)$ the subspace of $H^s(\mathbb{R}^3)$,

$$\widetilde{H}^s(\Omega) := \{g : g \in H^s(\mathbb{R}^3), \operatorname{supp}(g) \subset \overline{\Omega}\}$$

T. G. Ayele (✉)
Addis Ababa University, Addis Ababa, Ethiopia
e-mail: tsegaye.ayele@aau.edu.et

© Springer Nature Switzerland AG 2019
C. Constanda, P. Harris (eds.), *Integral Methods in Science and Engineering*,
https://doi.org/10.1007/978-3-030-16077-7_4

while $H^s(\Omega)$ denotes the space of restriction on Ω of distributions from $H(\mathbb{R}^3)$,

$$H^s(\Omega) = \{r_\Omega g : g \in H^s(\mathbb{R}^3)\}$$

where r_Ω denotes the restriction operator on Ω. We will also use the notation $g|_\Omega :=$ $r_\Omega g$. We denote by $H^s_{\partial\Omega}$ the following subspace of $H(\mathbb{R}^3)$ (and $\widetilde{H}(\Omega)$),

$$H^s_{\partial\Omega} := \{g : g \in H^s(\mathbb{R}^3), \ \mathrm{supp}(g) \subset \partial\Omega\}.$$

From the trace theorem (see, e.g., [Lio72, McL00]) for $u \in H^1(\Omega)$, it follows that $\gamma^+ u \in H^{\frac{1}{2}}(\partial\Omega)$, where $\gamma^+ = \gamma^+_{\partial\Omega}$ are the trace operators on $\partial\Omega$ from Ω. Let also $\gamma^{-1} : H^{\frac{1}{2}}(\partial\Omega) \to H^1(\Omega)$ denote a (non-unique) continuous right inverse to the trace operator γ^+, i.e., $\gamma^+_{\partial\Omega} \gamma^{-1}_{\partial\Omega} w = \gamma^+ \gamma^{-1} w = w$ for any $w \in H^{\frac{1}{2}}(\partial\Omega)$, and $(\gamma^{-1})^* : \widetilde{H}^{-1}(\Omega) \to H^{-\frac{1}{2}}(\partial\Omega)$ is continuous operator dual to $\gamma^{-1} : H^{\frac{1}{2}}(\partial\Omega) \to$ $H^1(\Omega)$, i.e., $\langle(\gamma^{-1})^* \tilde{f}, w\rangle_{\partial\Omega} := \langle \tilde{f}, \gamma^{-1} w\rangle_\Omega$ for any $\tilde{f} \in \widetilde{H}^{-1}(\Omega)$ and $w \in$ $H^{\frac{1}{2}}(\partial\Omega)$.

For $u \in H^2(\Omega)$, we denote by T_a^+ the canonical (strong) co-normal derivative operator on $\partial\Omega$ in the sense of traces,

$$T_a^+ u := \sum_{i=1}^{3} a(x) n_i(x) \gamma^+ \frac{\partial u(x)}{\partial x_i} = a(x) \gamma^+ \frac{\partial u(x)}{\partial n(x)}, \qquad (4.2)$$

where $n(x)$ is the outward (to Ω) unit normal vector at the point $x \in \partial\Omega$. However the classical conormal derivative operator in (4.2) is generally not well defined if $u \in H^1(\Omega)$ (see, e.g., [Cos88, Mik11] and [Mik15, Appendix Section A]).

For $u \in H^1(\Omega)$, the PDE Au in Eq. (4.1) is understood in the sense of distributions,

$$\langle Au, v\rangle_\Omega := -\mathscr{E}_a(u, v), \qquad \forall v \in \mathscr{D}(\Omega), \qquad (4.3)$$

where $\mathscr{E}_a(u, v) := \int_\Omega a(x) \nabla u(x).\nabla v(x) dx$ and the duality brackets $\langle g, \cdot\rangle_\Omega$ denote the value of a linear functional (distribution) g, extending the usual L_2 inner product.

Since the set $\mathscr{D}(\Omega)$ is dense in $\widetilde{H}^1(\Omega)$, Eq. (4.3) defines a continuous operator $A : H^1(\Omega) \to H^{-1}(\Omega) = [\widetilde{H}^1(\Omega)]^*$,

$$\langle Au, v\rangle_\Omega := -\mathscr{E}_a(u, v), \qquad \forall u \in H^1(\Omega), \ \forall v \in \widetilde{H}^1(\Omega).$$

Let us consider also the operator, $\check{A} : H^1(\Omega) \to \widetilde{H}^{-1}(\Omega) = [H^1(\Omega)]^*$,

$$
\begin{aligned}
\langle \check{A}u, v \rangle_\Omega := -\mathscr{E}_a(u, v) &= -\int_\Omega a(x)\nabla u(x).\nabla v(x)dx \\
&= -\int_{\mathbb{R}^3} \mathring{E}[a\nabla u](x).\nabla V(x)dx \\
&= \langle \nabla.\mathring{E}[a\nabla u], V \rangle_{\mathbb{R}^3} \\
&= \langle \nabla.\mathring{E}[a\nabla u], v \rangle_\Omega, \quad \forall u \in H^1(\Omega), \quad \forall v \in H^1(\Omega)
\end{aligned}
$$

which is evidently continuous and can be written as

$$
\check{A}u = \nabla.\mathring{E}[a\nabla u]. \tag{4.4}
$$

Here $V \in H^1(\mathbb{R}^3)$ is such that $r_\Omega V = v$ and $\mathring{E}$ denotes the operator of extension of the functions, defined in Ω, by zero outside Ω in $\mathbb{R}^3$. For any $u \in H^1(\Omega)$, the functional $\check{A}u$ belongs to $\widetilde{H}^{-1}(\Omega)$ and is the extension of the functional $Au \in H^{-1}(\Omega)$, which domain is thus extended from $\widetilde{H}^1(\Omega)$ to the domain $H^1(\Omega)$ for $\check{A}u$.

Inspired by the first Green identity for smooth functions, we can define *the generalised co-normal derivative* (cf. [McL00, Lemma 4.3], [Mik11, Definition 3.1]).

Definition 1 Let $u \in H^1(\Omega)$ and $Au = r_\Omega \tilde{f}$ in Ω for some $\tilde{f} \in \widetilde{H}^{-1}(\Omega)$. Then the generalised co-normal derivative $T_a^+(\tilde{f}, u) \in H^{-\frac{1}{2}}(\partial\Omega)$ is defined as

$$
\begin{aligned}
\langle T_a^+(\tilde{f}, u), w \rangle_{\partial\Omega} &:= \langle \tilde{f}, \gamma^{-1}w \rangle_\Omega + \mathscr{E}_a(u, \gamma^{-1}w) \\
&= \langle \tilde{f} - \check{A}u, \gamma^{-1}w \rangle_\Omega, \quad \forall w \in H^{\frac{1}{2}}(\Omega), \tag{4.5}
\end{aligned}
$$

that is, $T_a^+(\tilde{f}, u) := (\gamma^{-1})^*(\tilde{f} - \check{A}u)$.

Due to [McL00, Lemma 4.3] and [Mik11, Theorem 5.3], we have the estimate

$$
\|T_a^+(\tilde{f}, u)\|_{H^{-\frac{1}{2}}(\partial\Omega)} \le C_1 \|u\|_{H^1(\Omega)} + C_2 \|\tilde{f}\|_{\widetilde{H}^{-1}(\Omega)}, \tag{4.6}
$$

and for $u \in H^1(\Omega)$ such that $Au = r_\Omega \tilde{f}$ in Ω for some $\tilde{f} \in \widetilde{H}^{-1}(\Omega)$ the first Green identity holds in the following form:

$$
\langle T_a^+(\tilde{f}, u), \gamma^+ v \rangle_{\partial\Omega} := \langle \tilde{f}, v \rangle_\Omega + \mathscr{E}_a(u, v) = \langle \tilde{f} - \check{A}u, v \rangle_\Omega, \quad \forall v \in H^1(\Omega) \tag{4.7}
$$

As follows from Definition 1, the generalised co-normal derivative (4.5) is nonlinear with respect to u for a fixed $\tilde{f}$, but still linear with respect to the couple $(\tilde{f}, u)$, i.e.,

$$
\alpha_1 T_a^+(\tilde{f}_1, u_1) + \alpha_2 T_a^+(\tilde{f}_2, u_2) = T_a^+(\alpha_1 \tilde{f}_1, \alpha_1 u_1) + T_a^+(\alpha_2 \tilde{f}_2, \alpha_2 u_2)
$$
$$
= T_a^+(\alpha_1 \tilde{f}_1 + \alpha_2 \tilde{f}_2, \alpha_1 u_1 + \alpha_2 u_2)
$$

for any complex numbers α_1, α_2.

Let us also define some subspaces of $H^s(\Omega)$, cf. [Cos88, Gri85, Mik11, Mik13].

Definition 2 Let $s \in \mathbb{R}$ and $A_* : H^s(\Omega) \to \mathscr{D}^*(\Omega)$ be a linear operator. For $t \geq -\frac{1}{2}$ we introduce the space

$$
H^{s,t}(\Omega; A_*) := \{g : g \in H^s(\Omega) : A_* g|_\Omega = \tilde{f}_g|_\Omega, \quad \tilde{f}_g \in \tilde{H}^t(\Omega)\}
$$

endowed with the norm: $\|g\|_{H^{s,t}(\Omega; A_*)} := \left(\|g\|^2_{H^s(\Omega)} + \|\tilde{f}_g\|^2_{\tilde{H}^t(\Omega)} \right)^{\frac{1}{2}}$ and the inner product: $(g, h)_{H^{s,t}(\Omega; A_*)} = (g, h)_{H^s(\Omega)} + (\tilde{f}_g, \tilde{f}_h)_{\tilde{H}^t(\Omega)}$. We will mostly use the operators A, B or Δ as A_* in this definition.

Definition 3 For $u \in H^{1, -\frac{1}{2}}(\Omega; A)$, we define the canonical co-normal derivative $T_a^+ u \in H^{-\frac{1}{2}}(\partial\Omega)$ as

$$
\langle T_a^+ u, w \rangle_{\partial\Omega} := \langle \tilde{A}u, \gamma^{-1} w \rangle_\Omega + \mathscr{E}_a(u, \gamma^{-1} w)
$$
$$
= \langle \tilde{A}u - \check{A}u, \gamma^{-1} w \rangle_\Omega, \quad \forall w \in H^{\frac{1}{2}}(\Omega), \tag{4.8}
$$

that is, $T_a^+ u := (\gamma^{-1})^*(\tilde{A}u - \check{A}u)$.

The canonical co-normal derivative $T_a^+ u$ in (4.8) is independent of (non-unique) choice of the operator γ^{-1}, the operator $T_a^+ : H^{1, -\frac{1}{2}}(\Omega; A) \to H^{-\frac{1}{2}}(\partial\Omega)$ is continuous, and the first Green identity holds in the following form:

$$
\langle T_a^+ u, \gamma^+ v \rangle_{\partial\Omega} := \langle \tilde{A}u, v \rangle_\Omega + \mathscr{E}_a(u, v), \quad \forall v \in H^1(\Omega).
$$

Let $u \in H^{1, -\frac{1}{2}}(\Omega; A)$. Then Definitions 1 and 3 imply that the generalised co-normal derivative for arbitrary extension $\tilde{f} \in \tilde{H}^{-1}(\Omega)$ of the distribution Au can be expressed as

$$
\langle T_a^+(\tilde{f}, u), w \rangle_{\partial\Omega} := \langle T_a^+ u, w \rangle_{\partial\Omega} + \langle \tilde{f} - \check{A}u, \gamma^{-1} w \rangle_\Omega, \quad \forall w \in H^{\frac{1}{2}}(\Omega).
$$

For $b \in C^\infty(\overline{\Omega})$, $b(x) > 0$ for $x \in \overline{\Omega}$, consider the auxiliary linear elliptic partial differential operator B given by

$$
Bu(x) := B(x, \partial_x)u(x) := \sum_{i=1}^{3} \frac{\partial}{\partial x_i} \left(b(x) \frac{\partial u(x)}{\partial x_i} \right). \tag{4.9}
$$

Since $Au - a\Delta u = \nabla a \nabla u \in L_2(\Omega)$ and $Bu - b\Delta u = \nabla b \nabla u \in L_2(\Omega)$, for $u \in H^1(\Omega)$, we have $H^{1,0}(\Omega; A) = H^{1,0}(\Omega; \Delta) = H^{1,0}(\Omega; B)$. Then for $u \in H^1(\Omega)$ and $v \in H^{1,0}(\Omega; B)$, we write *the first Green identity* associated with operator B as

$$\mathcal{E}_b(u, v) + \int_\Omega u(x) Bv(x) dx = \langle T_b^+ v, \gamma^+ u \rangle_{\partial\Omega} \tag{4.10}$$

If $Au = \tilde{f}$ in Ω, where $\tilde{f} \in \tilde{H}^{-1}(\Omega)$, then subtracting (4.10) from (4.7), we obtain *the two-operator second Green identity* (cf. [Mik05b, AM11, AM10]),

$$\langle \tilde{f}, v \rangle_\Omega - \int_\Omega u(x) Bv(x) dx + \int_\Omega [a(x) - b(x)] \nabla u(x) \cdot \nabla v(x) dx$$

$$= \langle T_a^+ (\tilde{f}, u), \gamma^+ v \rangle_{\partial\Omega} - \langle T_b^+ v, \gamma^+ u \rangle_{\partial\Omega}. \tag{4.11}$$

If, moreover $u, v \in H^{1,0}(\Omega; A) = H^{1,0}(\Omega; B)$ then (4.11) becomes

$$\int_\Omega [v(x) Au(x) - u(x) Bv(x)] dx + \int_\Omega [a(x) - b(x)] \nabla u(x) \cdot \nabla v(x) dx$$

$$= \langle T_a^+ u, \gamma^+ v \rangle_{\partial\Omega} - \langle T_b^+ v, \gamma^+ u \rangle_{\partial\Omega}. \tag{4.12}$$

Note that if $a = b$ then the last domain integral in (4.11) and (4.12) disappears, and the generalised two-operator second Green identity (4.11) reduces to the one operator generalised second Green identity [Mik15, Eq.(2.17)] while (4.12) reduces to [Mik15, Eq.(2.18)].

4.2   Parametrix-Based Potential Operators

Definition 4 A function $P_b(x, y)$ of two variables $x, y \in \Omega$ is a parametrix (Levi function) for the operator $B(x; \partial_x)$ in $\mathbb{R}^3$ if

$$B(x, \partial_x) P_b(x, y) = \delta(x - y) + R_b(x, y),$$

where $\delta(.)$ is the Dirac distribution and $R_b(x, y)$ is the remainder with weak (integrable) singularity at $x = y$, i.e.,

$$R_b(x, y) = \mathcal{O}(|x - y|^{-\varkappa}) \quad \text{with } \varkappa < 3. \tag{4.13}$$

For the operator $B(x; \partial_x)$ defined by the right-hand side of (4.9) and $x, y \in \mathbb{R}^3$ the function

$$P_b(x, y) = \frac{1}{b(y)} P_\Delta(x, y) = \frac{-1}{4\pi b(y)|x - y|}, \tag{4.14}$$

is a parametrix and

$$R_b(x, y) = \nabla b(x).\nabla_x P_b(x, y) = -\frac{\nabla b(x).\nabla_y P_\Delta(x, y)}{b(y)} = \frac{(x - y).\nabla b(x)}{4\pi b(y)|x - y|^3},$$

(4.15)

is the corresponding remainder which satisfies estimate (4.13) with $\varkappa = 2$, due to smoothness of the function $b(x)$.

Evidently, the parametrix $P_b(x, y)$ given by (4.14) is the fundamental solution to the operator $B(y, \partial_x) := b(y)\Delta(\partial_x)$ with "frozen" coefficient $b(x) = b(y)$ and

$$B(y, \partial_x)P_b(x, y) = \delta(x - y).$$

Let $b \in C^\infty(\mathbb{R}^3)$ and $b(x) > 0$ a.e. in $\mathbb{R}^3$. Similar to [Mik02, CMN09b], we define the parametrix-based Newtonian and the remainder potential operators corresponding to the parametrix (4.14) and the remainder (4.15) by

$$\mathbf{P}_b g(y) := \int_{\mathbb{R}^3} P_b(x, y)g(x)dx$$

(4.16)

$$\mathscr{P}_b g(y) := \int_\Omega P_b(x, y)g(x)dx$$

(4.17)

$$\mathscr{R}_b g(y) := \int_\Omega R_b(x, y)g(x)dx.$$

(4.18)

For $g \in H^s(\Omega)$, $s \in \mathbb{R}$, (4.16) is understood as $\mathbf{P}_b g = \frac{1}{b}\mathbf{P}_\Delta g$, where the Newtonian potential operator $\mathbf{P}_\Delta$ for Laplacian Δ is well defined in terms of the Fourier transform (i.e., as pseudo-differential operator), on any space $H^s(\mathbb{R}^3)$. For $g \in \tilde{H}^s(\Omega)$, and any $s \in \mathbb{R}$, definitions in (4.17) and (4.18) can be understood as

$$\mathscr{P}_b g = \frac{1}{b}r_\Omega \mathbf{P}_\Delta g, \quad \mathscr{P}_b g = \frac{1}{b}r_\Omega \mathbf{P}_\Delta g \quad \text{and} \quad \mathscr{R}_b g = -\frac{1}{b}r_\Omega \nabla.\mathbf{P}_\Delta(g\nabla b),$$

(4.19)

while for $g \in H^s(\Omega), -\frac{1}{2} < s < \frac{1}{2}$, as (4.19) with g replaced by $\tilde{E}g$, where $\tilde{E} : H^s(\Omega) \to \tilde{H}^s(\Omega), -\frac{1}{2} < s < \frac{1}{2}$, is the unique extension operator related with the operator $\mathring{E}$ of extension by zero, cf. [Mik11, Theorem 16].

For $y \notin \partial\Omega$, the single layer and the double layer surface potential operators, corresponding to the parametrix (4.14) are defined as

$$V_b g(y) := -\int_{\partial\Omega} P_b(x, y)g(x)dS_x = \frac{1}{b}V_\Delta g(y),$$

(4.20)

$$W_b g(y) := -\int_{\partial\Omega} [T_b(x, n(x), \partial_x)P_b(x, y)]g(x)dS_x = \frac{1}{b}W_\Delta(bg)(y),$$

(4.21)

where g is some scalar density function, and the integrals are understood in the distributional sense if g is not integrable.

The corresponding boundary integral (pseudo-differential) operators of direct surface values of the single layer potential $\mathscr{V}_b$ and the double layer potentials $\mathscr{W}_b$ for $y \in \partial\Omega$ are

$$\mathscr{V}_b g(y) := -\int_{\partial\Omega} P_b(x, y)g(x)dS_x = \frac{1}{b}\mathscr{V}_\Delta g(y), \tag{4.22}$$

$$\mathscr{W}_b g(y) := -\int_{\partial\Omega} T_b(x, n(x), \partial_x)P_b(x, y)g(x)dS_x = \frac{1}{b}\mathscr{W}_\Delta (bg)(y) \tag{4.23}$$

We can also calculate at $y \in \partial\Omega$ the co-normal derivatives, associated with the operator A, of the single and of the double layer potentials corresponding to the parametrix (4.14).

$$T_a^\pm V_b g(y) = \frac{a(y)}{b(y)}T_b^\pm V_b g(y), \tag{4.24}$$

$$\mathscr{L}_{ab}^\pm g(y) := T_a^\pm W_b g(y) = \frac{a(y)}{b(y)}T_b^\pm W_b g(y) =: \frac{a(y)}{b(y)}\mathscr{L}_b^\pm g(y) \tag{4.25}$$

The direct value operators associated with (4.24) are

$$\mathscr{W}'_{ab} g(y) := -\int_{\partial\Omega} [T_a(y, n(y), \partial_y)P_b(x, y)]g(x)dS_x = \frac{a(y)}{b(y)}\mathscr{W}'_b g(y), \tag{4.26}$$

$$\mathscr{W}'_b g(y) := -\int_{\partial\Omega} [T_b(y, n(y), \partial_y)P_b(x, y)]g(x)dS_x. \tag{4.27}$$

From Eqs. (4.16)–(4.27) we deduce representations of the parametrix-based surface potential boundary operators in terms of their counterparts for $b = 1$, that is, associated with the fundamental solution $P_\Delta = -(4\pi|x - y|)^{-1}$ of the Laplace operator Δ.

$$\mathbf{P}_a g = \frac{1}{a}\mathbf{P}_\Delta g, \quad \mathbf{P}_b g = \frac{1}{b}\mathbf{P}_\Delta g, \quad \mathscr{P}_a g = \frac{1}{a}\mathscr{P}_\Delta g, \quad \mathscr{P}_b g = \frac{1}{b}\mathscr{P}_\Delta g. \tag{4.28}$$

$$\frac{a}{b}V_a g = V_b g = \frac{1}{b}V_\Delta g; \quad \frac{a}{b}W_a\left(\frac{bg}{a}\right) = W_b g = \frac{1}{b}W_\Delta (bg), \tag{4.29}$$

$$\frac{a}{b}\mathscr{V}_a g = \mathscr{V}_b g = \frac{1}{b}\mathscr{V}_\Delta g; \quad \frac{a}{b}\mathscr{W}_a\left(\frac{bg}{a}\right) = \mathscr{W}_b g = \frac{1}{b}\mathscr{W}_\Delta (bg), \tag{4.30}$$

$$\mathscr{W}'_{ab} g = \frac{a}{b}\mathscr{W}'_b g = \frac{a}{b}\left\{\mathscr{W}'_\Delta (bg) + \left[b\frac{\partial}{\partial n}\left(\frac{1}{b}\right)\right]\mathscr{V}_\Delta g\right\}, \tag{4.31}$$

$$\mathscr{L}_{ab}^\pm g = \frac{a}{b}\mathscr{L}_b^\pm g = \frac{a}{b}\left\{\mathscr{L}_\Delta (bg) + \left[b\frac{\partial}{\partial n}\left(\frac{1}{b}\right)\right]\gamma^\pm W_\Delta (bg)\right\}. \tag{4.32}$$

It is taken into account that b and its derivatives are continuous in $\mathbb{R}^3$ and

$$\mathscr{L}_\Delta(bg) := \mathscr{L}_\Delta^+(bg) = \mathscr{L}_\Delta^-(bg)$$

by the Lyapunov-Tauber theorem. Hence,

$$\Delta(bV_bg) = 0, \quad \Delta(bW_bg) = 0 \quad \text{in} \quad \Omega, \quad \forall g \in H^s(\partial\Omega) \quad (\forall s \in \mathbb{R}), \tag{4.33}$$

$$\Delta(b\mathscr{P}g) = g \quad \text{in } \Omega, \quad \forall g \in \tilde{H}^s(\Omega) \quad (\forall s \in \mathbb{R}). \tag{4.34}$$

For $g_1 \in H^{-\frac{1}{2}}(\partial\Omega)$, and $g_2 \in H^{\frac{1}{2}}(\partial\Omega)$. Then there hold the following relations on $\partial\Omega$,

$$\gamma^\pm V_bg_1 = \mathscr{V}_bg_1, \tag{4.35}$$

$$\gamma^\pm W_bg_2 = \mp\frac{1}{2}g_2 + \mathscr{W}_bg_2, \tag{4.36}$$

$$T_a^\pm V_bg_1 = \pm\frac{1}{2}\frac{a}{b}g_1 + \mathscr{W}'{}_{ab}g_1. \tag{4.37}$$

The jump relations (4.35)–(4.37) follow from the corresponding relations in [Mik15] and Eqs. (4.28)–(4.34).

4.3   Third Green Identities and Integral Relations

Applying some limiting procedures (see, e.g., [Mir70]), we obtain the parametrix based two-operator third Green identities.

Theorem 1

(i) *If $u \in H^1(\Omega)$, then the following third Green identity holds:*

$$u + \mathscr{Z}_bu + \mathscr{R}_bu + W_b\gamma^+u = \mathscr{P}_b\check{A}u \quad \text{in} \quad \Omega, \tag{4.38}$$

where the operator $\check{A}$ is defined in (4.4), and for $u \in C^1(\overline{\Omega})$,

$$\mathscr{P}_b\check{A}u(y) := \langle \check{A}u, P_b(., y) \rangle_\Omega$$

$$= -\mathscr{E}_a(u, P_b(., y)) = -\int_\Omega a(x)\nabla u(x).\nabla_x P_b(x, y)dx \tag{4.39}$$

(ii) If $Au = r_\Omega \tilde{f}$ in Ω, where $\tilde{f} \in \tilde{H}^{-1}(\Omega)$, then the generalised two-operator third Green identity has the form,

$$u + \mathscr{L}_b u + \mathscr{R}_b u - V_b T_a^+(\tilde{f}, u) + W_b \gamma^+ u = \mathscr{P}_b \tilde{f} \quad in \quad \Omega, \tag{4.40}$$

where it was taken into account that

$$\langle T_a^+(\tilde{f}, u), P_b(x, y) \rangle_{\partial\Omega} = -V_b T_a^+(\tilde{f}, u), \quad \langle \tilde{f}, P_b(x, y) \rangle_\Omega = \mathscr{P}_b \tilde{f}$$

and

$$\mathscr{L}_b u = -\int_\Omega [a(x) - b(x)] \nabla_x P_b(x, y) \cdot \nabla u(x) dx$$

$$= \frac{1}{b(y)} \sum_{j=1}^3 \partial_j \mathscr{P}_\Delta \big[(a - b) \partial_j u \big] \quad in \quad \Omega. \tag{4.41}$$

Proof

(i) Let first $u \in D(\overline{\Omega})$. Let $y \in \Omega$, $B_\epsilon(y) \subset \Omega$ be a ball centred at y with sufficiently small radius ϵ, and $\Omega_\epsilon := \Omega \setminus \overline{B}_\epsilon(y)$. For the fixed y, evidently, $P_b(., y) \in D(\overline{\Omega}_\epsilon) \subset H^{1,0}(A; \Omega_\epsilon)$ and has the coinciding classical and canonical co-normal derivatives on $\partial\Omega_\epsilon$. Then from (4.14) and the first Green identity (4.10) employed for Ω_ϵ with $v = P_b(., y)$ we obtain

$$-\int_{\partial B_\epsilon(y)} T_x^+ P_b(x, y) \gamma^+ u(x) ds_x - \int_{\partial\Omega} T_x P_b(x, y) \gamma^+ u(x) ds_x$$

$$= -\int_{\Omega_\epsilon} b(x) \nabla u(x) . \nabla_x P_b(x, y) dx,$$

which we rewrite as

$$-\int_{\partial B_\epsilon(y)} T_x^+ P_b(x, y) \gamma^+ u(x) ds_x - \int_{\partial\Omega} T_x P_b(x, y) \gamma^+ u(x) ds_x$$

$$- \int_{\Omega_\epsilon} [a(x) - b(x)] \nabla u(x) \nabla_x P_b(x, y) dx$$

$$= -\int_{\Omega_\epsilon} a(x) \nabla u(x) . \nabla_x P_b(x, y) dx. \tag{4.42}$$

Taking the limit as $\epsilon \to 0$, Eq. (4.42) reduces to the third Green identity (4.38)–(4.39) for any $u \in \mathscr{D}(\overline{\Omega})$. Taking into account the density of $\mathscr{D}(\overline{\Omega})$ in $H^1(\Omega)$, and the mapping properties of the integral potentials, see Appendix, we obtain that (4.38)–(4.39) hold true also for any $u \in H^1(\Omega)$.

(ii) Let $\{\tilde{f}_k\} \in \mathscr{D}(\Omega)$ be a sequence of covering to $\tilde{f}$ in $\tilde{H}^{-1}(\Omega)$ as $k \to \infty$. Then, according to [Mik15, Theorem B.1] there exists a sequence $\{u_k\} \in \mathscr{D}(\overline{\Omega})$ converging to u in $H^1(\Omega)$ such that $Au_k = r_\Omega \tilde{f}_k$ and $T_a^+(u_k) = T_a^+(\tilde{f}_k, u_k)$ converge to $T_a^+(\tilde{f}, u)$ in $H^{-\frac{1}{2}}(\partial\Omega)$. For such u_k by (4.39) and (4.7), we have

$$\mathscr{P}_b\check{A}u_k(y) = -\lim_{\epsilon\to 0} \int_{\Omega_\epsilon} a(x)\nabla u_k(x)P_\Delta(x,y)dx = -\lim_{\epsilon\to 0} \mathscr{E}_{\Omega_\epsilon}(u_k, P_b(.,y))$$

$$= -\lim_{\epsilon\to 0}\left[\int_{\Omega_\epsilon}\tilde{f}_k P_b(x,y)dx - \int_{\partial B_\epsilon(y)} P_b(x,y)T_a^+u_k(x)dS(x)\right]$$

$$+ \lim_{\epsilon\to 0}\int_{\partial\Omega} P_b(x,y)T_a^+u_k(x)dS(x) = \mathscr{P}_b\tilde{f}_k + V_bT_a^+u_k(y).$$

$$(4.43)$$

Taking the limits as $k \to \infty$, in (4.43), we obtain $\mathscr{P}_b\check{A}u(y) = \mathscr{P}_b\tilde{f} + V_bT_a^+\tilde{f}, u)$, which substitution to (4.38) gives (4.40). $\square$

Using the Gauss divergence theorem, we can rewrite Eq. (4.41) in the form that does not involve derivatives of u,

$$\mathscr{L}_bu(y) := \left[\frac{a(y)}{b(y)} - 1\right]u(y) + \widehat{\mathscr{L}_b}u(y),$$

$$\widehat{\mathscr{L}_b}u(y) := \frac{a(y)}{b(y)}W_a\gamma^+u(y) - W_b\gamma^+u(y) + \frac{a(y)}{b(y)}\mathscr{R}_au(y) - \mathscr{R}_bu(y),$$

which allows to call $\mathscr{L}_b$ integral operator in spite of its integro-differential representation (4.41).

For some functions $\tilde{f}$, Ψ, Φ let us consider a more general "indirect" integral relation, associated with (4.40).

$$u + \mathscr{L}_bu + \mathscr{R}_bu - V_b\Psi + W_b\Phi = \mathscr{P}_b\tilde{f} \quad \text{in} \quad \Omega. \tag{4.44}$$

Lemma 1 Let $u \in H^1(\Omega)$, $\Psi \in H^{-\frac{1}{2}}(\partial\Omega)$, $\Phi \in H^{\frac{1}{2}}(\partial\Omega)$ and $\tilde{f} \in \tilde{H}^{-1}(\Omega)$, satisfy (4.44). Then

$$Au = r_\Omega\tilde{f} \quad \text{in} \quad \Omega, \tag{4.45}$$

$$r_\Omega V_b(\Psi - T_a^+(\tilde{f}, u)) - r_\Omega W_b(\Phi - \gamma^+u) = 0 \quad \text{in} \quad \Omega, \tag{4.46}$$

$$\gamma^+u + \gamma^+\mathscr{L}_bu + \gamma^+\mathscr{R}_bu - \mathscr{V}_b\Psi - \frac{1}{2}\Phi + \mathscr{W}_b\Phi = \gamma^+\mathscr{P}_b\tilde{f} \quad \text{on} \quad \partial\Omega, \tag{4.47}$$

$$T_a^+(\tilde{f}, u) + T_a^+\mathscr{Z}_b u + T_a^+\mathscr{R}_b u - \frac{a}{2b}\Psi - \mathscr{W}'_{ab}\Psi + \mathscr{L}_{ab}^+\Phi$$
$$= T_a^+(\tilde{f} + \mathring{E}\mathscr{R}_*^b\tilde{f}, \mathscr{P}_b\tilde{f}) \quad \text{on} \quad \partial\Omega, \qquad (4.48)$$

where

$$\mathscr{R}_*^b\tilde{f}(y) := -\sum_{j=1}^{3} \partial_j[(\partial_j b)\mathscr{P}_b\tilde{f}]. \qquad (4.49)$$

Proof Subtracting (4.44) from identity (4.38), we obtain

$$V_b\Psi(y) - W_b(\Phi - \gamma^+ u)(y) = \mathscr{P}_b[\check{A}u(y) - \tilde{f}](y), \quad y \in \Omega. \qquad (4.50)$$

Multiplying equality (4.50) by $b(y)$, applying the Laplace operator Δ and taking into account (4.33), (4.34), we get $r_\Omega \tilde{f} = r_\Omega(\check{A}u) = Au$ in Ω. This means $\tilde{f}$ is an extension of the distribution $Au \in H^{-1}(\Omega)$ to $\tilde{H}^{-1}(\Omega)$, and u satisfies (4.45). Then (4.7) implies

$$\mathscr{P}_b[\check{A}u - \tilde{f}](y) = \langle \check{A}u - \tilde{f}, \mathscr{P}_b(., y)\rangle_\Omega$$
$$= -\langle T_a^+(\tilde{f}, u), \mathscr{P}_b(.y)\rangle_{\partial\Omega} = V_b T_a^+(\tilde{f}, u), \quad y \in \Omega. \quad (4.51)$$

Substituting (4.51) into (4.50) leads to (4.46). Equation (4.47) follows from (4.44) and jump relations in (4.35) and (4.36). To prove (4.48), let us first remark that for $u \in H^1(\Omega)$, we have $H^{1,0}(\Omega; A) = H^{1,0}(\Omega; \Delta) = H^{1,0}(\Omega; B)$ and similar to [Mik15, Eq. (4.13)], we have

$$B\mathscr{P}_b\tilde{f} = \tilde{f} + \mathscr{R}_*^b\tilde{f} \quad \text{in} \quad \Omega, \qquad (4.52)$$

due to (4.45) and (4.52), which implies $B(\mathscr{P}_b\tilde{f} - u) = \mathscr{R}_*^b\tilde{f}$ in Ω, where $\mathscr{R}_*^b\tilde{f}$ given by (4.49) and $\mathscr{R}_*^b\tilde{f} \in L_2(\Omega)$. Then $B(\mathscr{P}_b\tilde{f} - u)$ can be canonically extended (by zero) to $\tilde{B}(\mathscr{P}_b\tilde{f} - u) = \mathring{E}\mathscr{R}_*^b\tilde{f} \in \tilde{H}^0(\Omega) \subset \tilde{H}^{-1}(\Omega)$. Thus there exists a canonical co-normal derivative of $(\mathscr{P}_b\tilde{f} - u)$ associated with operator B, see, e.g., [Mik15, Eq. (4.14)] and written as

$$T_b^+(\mathscr{P}_b\tilde{f} - u) = T_b^+(\tilde{f} + \mathring{E}\mathscr{R}_*^b\tilde{f}, \mathscr{P}_b\tilde{f}) - T_b^+(\tilde{f}, u). \qquad (4.53)$$

Hence,

$$T_a^+\left(\mathscr{P}_b\tilde{f} - u\right) = \frac{a}{b}T_b^+(\mathscr{P}_b\tilde{f} - u) = T_a^+(\tilde{f} + \mathring{E}\mathscr{R}_*^b\tilde{f}, \mathscr{P}_b\tilde{f}) - T_a^+(\tilde{f}, u).$$
$$(4.54)$$

From (4.44) it follows that $\mathscr{P}_b \tilde{f} - u = \mathscr{L}_b u + \mathscr{R}_b u - V_b \Psi + W_b \Phi$ in Ω. Substituting this on the left-hand sides of (4.53) and (4.54) and taking into account the jump relation (4.37), we arrive at (4.48).   □

Remark 1 If $\tilde{f} \in \tilde{H}^{-\frac{1}{2}}(\Omega) \subset \tilde{H}^{-1}(\Omega)$, then $\tilde{f} + \mathring{E}\mathscr{R}_*^b \tilde{f} \in \tilde{H}^{-\frac{1}{2}}(\Omega)$ as well, which implies $\tilde{f} + \mathring{E}\mathscr{R}_*^b \tilde{f} = \tilde{B} \mathscr{P}_b \tilde{f}$. Then $T_b^+(\tilde{f} + \mathring{E}\mathscr{R}_*^b \tilde{f}, \mathscr{P}_b \tilde{f}) = T_b^+(\tilde{B} \mathscr{P}_b \tilde{f}, \mathscr{P}_b \tilde{f}) = T_b^+ \mathscr{P}_b \tilde{f}$, and

$$T_a^+(\tilde{f} + \mathring{E}\mathscr{R}_*^b \tilde{f}, \mathscr{P}_b \tilde{f}) = T_a^+(\tilde{B} \mathscr{P}_b \tilde{f}, \mathscr{P}_b \tilde{f}) = T_a^+ \mathscr{P}_b \tilde{f}. \tag{4.55}$$

Furthermore, if the hypotheses of Lemma 1 are satisfied, then (4.45) implies $u \in H^{1,-\frac{1}{2}}(\Omega; A)$ and $T_a^+(\tilde{f}, u) = T_a^+(\tilde{A}u, u) = T_a^+ u$. Henceforth (4.48) takes the familiar form, cf. [AM11, equation (3.23)],

$$T_a^+ u + T_a^+ \mathscr{L}_b u + T_a^+ \mathscr{R}_b u - \frac{a}{2b}\Psi - \mathscr{W}'_{ab}\Psi + \mathscr{L}_{ab}^+ \Phi = T_a^+ \mathscr{P}_b \tilde{f} \quad \text{on} \quad \partial\Omega.$$

Remark 2 Let $\tilde{f} \in \tilde{H}^{-1}(\Omega)$ and a sequence $\{\phi_i\} \in \tilde{H}^{-1}(\Omega)$ converge to $\tilde{f}$ in $\tilde{H}^{-1}(\Omega)$. By the continuity of operators [Mik15, C.1 and C.2], estimate (4.6) and relation (4.55) for ϕ_i, we obtain that

$$T_a^+(\tilde{f} + \mathring{E}\mathscr{R}_*^b \tilde{f}, \mathscr{P}_b \tilde{f}) = \lim_{i\to\infty} T_a^+(\phi_i + \mathring{E}\mathscr{R}_*^b \phi_i, \mathscr{P}_b \phi_i) = \lim_{i\to\infty} T_a^+ \mathscr{P}_b \phi_i.$$

in $H^{-\frac{1}{2}}(\partial\Omega)$, cf. also [Mik15, Theorem B.1].

Lemma 1 and the third Green identity (4.40) imply the following assertion.

Corollary 1 *If $u \in H^1(\Omega)$ and $\tilde{f} \in \tilde{H}^{-1}(\Omega)$ are such that $Au = r_\Omega \tilde{f}$ in Ω, then*

$$\gamma^+ u + \gamma^+ \mathscr{L}_b u + \gamma^+ \mathscr{R}_b u - \mathscr{V}_b(\tilde{f}, u) + \mathscr{W}_b \gamma^+ u = \gamma^+ \mathscr{P}_b \tilde{f} \quad \text{on} \quad \partial\Omega \tag{4.56}$$

$$\left(1 - \frac{a}{2b}\right) T_a^+(\tilde{f}, u) + T_a^+ \mathscr{L}_b u + T_a^+ \mathscr{R}_b u - \mathscr{W}'_{ab} T_a^+(\tilde{f}, u) + \mathscr{L}_{ab}^+ \gamma^+ u$$

$$= T_a^+(\tilde{f} + \mathring{E}\mathscr{R}_*^b \tilde{f}, \mathscr{P}_b \tilde{f}) \text{ on } \partial\Omega. \tag{4.57}$$

Similar to [Mik15, Lemma 4.6] and [Mik15, Theorem 4.7], we have

Lemma 2

(i) If $\Psi^* \in H^{-\frac{1}{2}}(\partial\Omega)$ and $r_\Omega V_b \Psi^* = 0$ in Ω, then $\Psi = 0$.
(ii) If $\Phi^* \in H^{\frac{1}{2}}(\partial\Omega)$ and $r_\Omega W_b \Phi^* = 0$ in Ω, then $\Phi = 0$.

Theorem 2 *Let $\tilde{f} \in \tilde{H}^{-1}(\Omega)$. A function $u \in H^1(\Omega)$ is a solution of PDE $Au = r_\Omega \tilde{f}$ in Ω if and only if it is a solution of BDIE (4.40).*

4.4   The Dirichlet Problem and Two-Operator BDIEs

We shall derive and investigate the *two-operator* BDIE systems for the following Dirichlet problem: *Find a function $u \in H^1(\Omega)$ satisfying equations*

$$Au = f \quad \text{in} \quad \Omega, \tag{4.58}$$

$$\gamma^+ u = \varphi_0 \quad \text{on} \quad \partial\Omega, \tag{4.59}$$

where $\varphi_0 \in H^{\frac{1}{2}}(\partial\Omega)$ and $f \in H^{-1}(\Omega)$.

Equation (4.58) is understood in the distributional sense (4.3) and the Dirichlet boundary condition (4.59) in the trace sense.

Theorem 3 *The Dirichlet problem* (4.58)–(4.59) *is uniquely solvable in $H^1(\Omega)$. The solution is $u = (\mathscr{A}^D)^{-1}(f, \varphi_0)^T$, where the inverse operator, $(\mathscr{A}^D)^{-1}$: $H^{\frac{1}{2}}(\partial\Omega) \times H^{-1}(\Omega) \to H^1(\Omega)$, to the left-hand side operator, $\mathscr{A}^D : H^1(\Omega) \to H^{\frac{1}{2}}(\partial\Omega) \times H^{-1}(\Omega)$, of the Dirichlet problem* (4.58)–(4.59)*, is continuous.*

For $u \in H^1(\Omega)$, we shall reduce the Dirichlet problem (4.58)–(4.59) with $f \in H^{-1}(\Omega)$ into two different systems of *segregated two-operator* BDIEs. Corresponding formulations for the mixed problem for $u \in H^{1,0}(\Omega, \Delta)$ with $f \in L_2(\Omega)$ were introduced and analysed in [Mik05b, AM11, AM10].

Let $\tilde{f} \in \tilde{H}^{-1}(\Omega)$ be an extension of $f \in H^{-1}(\Omega)$ (i.e., $f = r_\Omega \tilde{f}$), which always exists, see Lemma 2.15 and Theorem 2.16 in [Mik15]. We represent in (4.40), (4.56) and (4.57) the generalised co-normal derivative and the trace of the function u as

$$T^+(\tilde{f}, u) = \psi, \quad \gamma^+ u = \varphi_0$$

respectively, and will regard the new unknown function $\psi \in H^{-\frac{1}{2}}(\partial\Omega)$ as formally segregated of u. Thus we will look for the couple $(u, \psi) \in H^1(\Omega) \times H^{-\frac{1}{2}}(\partial\Omega)$.

BDIE System (D1) To reduce BVP (4.58)–(4.59) to one of the BDIE systems we will use Eq. (4.40) in Ω and Eq. (4.56) on $\partial\Omega$. Then we arrive at the system of BDIEs (D1),

$$u + \mathscr{L}_b u + \mathscr{R}_b u - V_b \psi = \mathscr{F}_1^{D1} \quad \text{in} \quad \Omega, \tag{4.60}$$

$$\gamma^+ \mathscr{L}_b u + \gamma^+ \mathscr{R}_b u - \mathscr{V}_b \psi = \mathscr{F}_2^{D1} \quad \text{on} \quad \partial\Omega, \tag{4.61}$$

where

$$\mathscr{F}^{D1} := \begin{bmatrix} \mathscr{F}_1^{D1} \\ \mathscr{F}_2^{D1} \end{bmatrix} = \begin{bmatrix} F_0^D \\ \gamma^+ F_0^D - \varphi_0 \end{bmatrix} \quad \text{and} \quad F_0^D := \mathscr{P}_b \tilde{f} - W_b \varphi_0. \tag{4.62}$$

For $\varphi_0 \in H^{\frac{1}{2}}(\partial\Omega)$, we have the inclusions $F_0^D \in H^1(\Omega)$ if $\tilde{f} \in \tilde{H}^{-1}(\Omega)$ and due to the mapping properties of operators involved in (4.62), we have the inclusion $\mathscr{F}^{D2} \in H^1(\Omega) \times H^{\frac{1}{2}}(\partial\Omega)$.

BDIE System (D2) To obtain a segregated BDIE system of *the second kind*, we will use Eq. (4.40) in Ω and Eq. (4.57) on $\partial\Omega$. Then we arrive at the system, (D2), of BDIEs,

$$u + \mathscr{Z}_b u + \mathscr{R}_b u - V_b \psi = \mathscr{P}_b \tilde{f} - W_b \varphi_0 \quad \text{in} \quad \Omega, \tag{4.63}$$

$$\left(1 - \frac{a}{2b}\right)\psi + T_a^+ \mathscr{Z}_b u + T_a^+ \mathscr{R}_b u - \mathscr{W}'_{ab}\psi$$
$$= T_a^+(\tilde{f} + \dot{E}\mathscr{R}_*^b \tilde{f}, \mathscr{P}_b \tilde{f}) - \mathscr{L}_{ab}^+ \varphi_0 \quad \text{on} \ \partial\Omega, \tag{4.64}$$

where

$$\mathscr{F}^{D2} := \begin{bmatrix} \mathscr{F}_1^{D2} \\ \mathscr{F}_2^{D2} \end{bmatrix} = \begin{bmatrix} \mathscr{P}_b \tilde{f} - W_b \varphi_0 \\ T_a^+(\tilde{f} + \dot{E}\mathscr{R}_*^b \tilde{f}, \mathscr{P}_b \tilde{f}) - \mathscr{L}_{ab}^+ \varphi_0 \end{bmatrix}. \tag{4.65}$$

Due to the mapping properties of operators involved in (4.65), we have the inclusion $\mathscr{F}^{D2} \in H^1(\Omega) \times H^{-\frac{1}{2}}(\partial\Omega)$.

4.5 Equivalence and Invertibility of BDIE Systems

Theorem 4 *Let $\varphi_0 \in H^{\frac{1}{2}}(\partial\Omega)$, $f \in H^{-1}(\Omega)$ and $\tilde{f} \in \tilde{H}^{-1}(\Omega)$ is such that $r_\Omega \tilde{f} = f$. Then*

(i) *If $u \in H^1(\Omega)$ solves the BVP (4.58)–(4.59), then the couple $(u, \psi) \in H^1(\Omega) \times H^{-\frac{1}{2}}(\Omega)$, where*

$$\psi = T_a^+(\tilde{f}, u), \qquad \text{on} \ \partial\Omega, \tag{4.66}$$

solves the BDIE systems (D1) and (D2).

(ii) *If a couple $(u, \psi) \in H^1(\Omega) \times H^{-\frac{1}{2}}(\partial\Omega)$ solves one of the BDIE systems, (D1) or (D2), then this solution is unique and solves the other system, while u solves the Dirichlet BVP, and ψ satisfies (4.66).*

Proof

(i) Let $u \in H^1(\Omega)$ be a solution to BVP (4.58)–(4.59). Due to Theorem 3 it is unique. Setting ψ by (4.66) evidently implies, $\psi \in H^{-\frac{1}{2}}(\partial\Omega)$. From Theorem 2 and relations (4.56)–(4.57) follows that the couple (u, ψ) satisfies the BDIE systems (D1) and (D2), with the right-hand sides (4.62) and (4.65) respectively, which completes the proof of item (i).

(ii) Let now a couple $(u, \psi) \in H^1(\Omega) \times H^{-\frac{1}{2}}(\partial\Omega)$ solve BDIE system (4.60)–(4.61). Taking trace of Eq. (4.60) on $\partial\Omega$ and subtracting Eq. (4.61) from it we obtain

$$\gamma^+ u = \varphi_0 \qquad \text{on } \partial\Omega, \tag{4.67}$$

i.e., u satisfies the Dirichlet condition (4.59).

Equation (4.60) and Lemma 1 with $\Psi = \psi$, $\Phi = \varphi_0$ imply that u is a solution of PDE (4.58) and $V_b \Psi^* - W_b \Phi^* = 0$, in Ω, where $\Psi^* = \psi - T_a^+(\tilde{f}, u)$ and $\Phi^* = \varphi_0 - \gamma^+ u$. Due to Eq. (4.67), $\Phi^* = 0$. Then Lemma (2)(i) implies $\Psi^* = 0$, which proves condition (4.66). Thus u obtained from the solution of BDIE system (D1) solves the Dirichlet problem and hence, by item (i) of the theorem, (u, ψ) solve also BDIE system (D2).

Due to (4.62), the BDIE system (4.60)–(4.61) with zero right-hand side can be considered as obtained for $\tilde{f} = 0$, $\varphi_0 = 0$, implying that its solution is given by a solution of the homogeneous problem (4.58)–(4.59), which is zero by Theorem 3. This implies uniqueness of the solution of the inhomogeneous BDIE system (4.60)–(4.61). Similar arguments work if we suppose that instead of the BDIE system (4.60)–(4.61), the couple $(u, \psi) \in H^1(\Omega) \times H^{-\frac{1}{2}}(\partial\Omega)$ solves BDIE system (4.63)–(4.64).   □

4.6   Conclusion

The Dirichlet problem for the linear second-order scalar elliptic differential equation with variable coefficient and extended right-hand side is considered. Using the two-operator approach and appropriate parametrix (Levi function) this problem is reduced to two different segregated systems of BDIEs. The equivalence of the original BVP to the two-operator BDIE systems is proved. The invertibility of the associated boundary integral operators in the appropriate Sobolev spaces can be also shown. In a similar way one can consider also the two-operator versions of the BDIEs for Dirichlet problem for non-smooth coefficient on Lipschitz domains, which were analysed for 3D case in [Mik18].

Acknowledgements The work on this paper of the author was supported by EMS-Simmons for Africa and ISP Sweden. He also thanks the Department of Mathematics at Brunel University for hosting his research visit and the Department of Mathematics at Addis Ababa University for granting leave of absence.

References

[AM11] Ayele, T.G. and Mikhailov, S.E.: Analysis of two-operator boundary-domain integral equations for a variable-coefficient BVP, *Eurasian Math. J., 2:3 (2011), 20–41*.

[AM10] Ayele, T.G. and Mikhailov, S.E.: *Two-operator boundary-domain integral equations for a variable-coefficient BVP*, in Integral Methods in Science and Engineering, C. Constanda and M. Pérez, eds.), Vol. 1: Analytic Methods, Birkhäuser, Boston-Basel-Berlin (2010), ISBN 978-08176-4898-5, 29–39.

[CMN09b] Chakuda, O., Mikhailov, S.E. and Natroshvili, D.: Analysis of direct boundary-domain integral equations for a mixed BVP with variable coefficient. I: Equivalence and Invertibility. *J. Integral Equations Appl.* 21 (2009), 499–543.

[Cos88] Costabel, M.: Boundary integral operators on Lipschitz domains: elementary results. *SIAM J. Math. Anal.* 19 (1988), 613–626.

[Gri85] Grisvard, P.: *Elliptic Problems in Nonsmooth Domains*, Pitman, Boston–London–Melbourne, 1985.

[Lio72] Lions, J.-L. and Magenes, E.: *Non-Homogeneous Boundary Value Problems and Applications*, Vol. 1. Springer, Berlin, Heidelberg, New York 1972.

[McL00] McLean, W.: *Strongly Elliptic Systems and Boundary Integral Equations.* Cambridge University Press, Cambridge, 2000.

[Mik02] Mikhailov, S.E.: Localized boundary–domain integral formulations for problems with variable coefficients. *Int. J. Engng. Anal. Boundary Elements* 26 (2002), 681–690.

[Mik05b] Mikhailov, S.E.: Localized formulations for scalar nonlinear BVPs with variable coefficients. *J. Engng. Math.*, 51 (2005), 283–302.

[Mik11] Mikhailov, S.E.: Traces, extensions and co-normal derivatives for elliptic systems on Lipschitz domains. *J. Math. Anal. Appl.* 378 (2011), 324–342.

[Mik13] Mikhailov, S.E.: Solution regularity and co-normal derivatives for elliptic systems with non-smooth coefficients on Lipschitz domains. *J. Math. Anal. Appl.*, 400(1):48–67, 2013.

[Mik15] Mikhailov, S.E.: Analysis of Segregated BDIEs for Variable-Coefficient Dirichlet and Neumann Problems with General Data. *ArXiv:1509.03501*,1–32(2015).

[Mik18] Mikhailov S.E.: Analysis of segregated boundary-domain integral equations for BVPs with non-smooth coefficient on Lipschitz domains, *Boundary Value Problems*, 87 (2018), 1–52, DOI: 10.1186/s13661-018-0992-0.

[Mir70] Miranda, C.: *Partial Differential Equations of Elliptic Type.*, 2nd ed. Springer, Berlin–Heidelberg–New York, 1970.

Chapter 5
Two-Operator Boundary-Domain Integral Equations for Variable Coefficient Dirichlet Problem in 2D

Tsegaye G. Ayele and Solomon T. Bekele

5.1 Preliminaries

In this paper we will consider the Dirichlet problem for the second order elliptic PDE with variable coefficient in two-dimensional bounded domain. Using an appropriate parametrix (Levi function) and applying the two-operator approach the problem is reduced to two systems of boundary-domain integral equations (BDIEs). Although the theory of BDIEs in 3D is well developed, cf. [Mi02, ChEtAl09b, Mi05b, AyMi11], the BDIEs in 2D need a special consideration due to their different equivalence properties. As a result, we need to set conditions on the domain or on the associated Sobolev spaces to ensure the invertibility of corresponding parametrix-based integral layer potentials and hence the unique solvability of BDIEs. The properties of corresponding potential operators are investigated. The equivalence of the original BVP and the obtained BDIEs is analysed and the invertibility of the BDIE operators is proved in appropriate Sobolev-Slobodecki (Bessel potential) spaces.

Let Ω be a domain in $\mathbb{R}^2$ bounded by simple closed infinitely smooth curve $\partial\Omega$, the set of all infinitely differentiable function on Ω with compact support is denoted by $\mathscr{D}(\Omega)$. The function space $\mathscr{D}'(\Omega)$ consists of all continuous linear functionals over $\mathscr{D}(\Omega)$. For $s \in \mathbb{R}$, we denote by $H^s(\mathbb{R}^2)$ the Bessel potential space. For any non-empty open set $\Omega \subset \mathbb{R}^2$, we define $H^s(\Omega) = \{u \in \mathscr{D}'(\Omega) : u = U|_{\Omega}$ for some $U \in H^s(\mathbb{R}^2)\}$. The space $\widetilde{H}^s(\Omega)$ is defined to be the closure of $\mathscr{D}(\Omega)$ with respect to the norm of $H^s(\mathbb{R}^2)$. Note that the space $H^1(\mathbb{R}^2)$ coincides with the Sobolev space $W_2^1(\mathbb{R}^2)$ with equivalent norm and $H^{-s}(\mathbb{R}^2)$ is the dual space to $H^s(\mathbb{R}^2)$ (see, e.g., [Mc00]).

T. G. Ayele · S. T. Bekele (✉)
Addis Ababa University, Addis Ababa, Ethiopia
e-mail: tsegaye.ayele@aau.edu.et

© Springer Nature Switzerland AG 2019
C. Constanda, P. Harris (eds.), *Integral Methods in Science and Engineering*,
https://doi.org/10.1007/978-3-030-16077-7_5

We shall consider the scalar elliptic differential equation

$$Au(x) = \sum_{i=1}^{2} \frac{\partial}{\partial x_i}\left[a(x)\frac{\partial u(x)}{\partial x_i}\right] = f(x) \qquad \text{in } \Omega,$$

where u is the unknown function and f is a given function in Ω. We assume that $a \in C^{\infty}(\mathbb{R}^2)$, $0 < a_{\min} \le a(x) \le a_{\max} < \infty$, $\forall x \in \mathbb{R}^2$. For $u \in H^2(\Omega)$ and $v \in H^1(\Omega)$, if we put $h(x) = a(x)\frac{\partial u(x)}{\partial x_j}v(x)$ and applying the Gauss-Ostrogradsky Theorem, we obtain the following *Green's first identity*:

$$\mathcal{E}_a(u, v) = -\int_{\Omega}(Au)(x)v(x)dx + \int_{\partial\Omega} T^{c+}u(x)\gamma^+v(x)ds_x, \tag{5.1}$$

where $\mathcal{E}_a(u, v) := \sum_{i=1}^{2}\int_{\Omega} a(x)\frac{\partial u(x)}{\partial x_i}\frac{\partial v(x)}{\partial x_i}dx$ is the symmetric bilinear form, γ^+ is the trace operator and

$$T^{c+}u(x) := \sum_{i=1}^{2} n_i(x)\gamma^+\left[a(x)\frac{\partial}{\partial x_i}u(x)\right] \quad \text{for } x \in \partial\Omega, \tag{5.2}$$

is the *classical co-normal derivative*.

Remark 1 For $v \in \mathscr{D}(\Omega)$, $\gamma^+v = 0$. If $u \in H^1(\Omega)$, then we can define Au as a distribution on Ω by $(Au, v) = -\mathcal{E}_a(u, v)$ for $v \in \mathscr{D}(\Omega)$.

The subspace $H^{1,0}(\Omega; A)$ is defined as in [Co88] (see also, [Mi11])

$$H^{1,0}(\Omega; A) := \{g \in H^1(\Omega) : Ag \in L_2(\Omega)\},$$

with the norm $\|g\|^2_{H^{1,0}(\Omega;A)} := \|g\|^2_{H^1(\Omega)} + \|Ag\|^2_{L_2(\Omega)}$.

For $u \in H^1(\Omega)$ the classical co-normal derivative (5.2) is not well defined, but for $u \in H^{1,0}(\Omega; A)$, there exists the following continuous extension of this definition hinted by the first Green identity (5.1) (see, e.g., [Co88, Mi11] and the references therein).

Definition 1 For $u \in H^{1,0}(\Omega; A)$ the co-normal derivative $T_a^+u \in H^{-\frac{1}{2}}(\partial\Omega)$ is defined in the following weak form,

$$\langle T_a^+u, w\rangle_{\partial\Omega} := \mathcal{E}_a(u, \gamma^+_{-1}w) + \int_{\Omega}(Au)\gamma^+_{-1}w dx \quad \text{for all } w \in H^{\frac{1}{2}}(\partial\Omega) \tag{5.3}$$

where $\gamma_{-1}^+ : H^{\frac{1}{2}}(\partial\Omega) \to H^1(\Omega)$ is a continuous right inverse of the trace operator γ^+, which maps $H^1(\Omega) \to H^{\frac{1}{2}}(\partial\Omega)$, while $\langle \cdot, \cdot \rangle_{\partial\Omega}$ denote the duality brackets between the spaces $H^{-\frac{1}{2}}(\partial\Omega)$ and $H^{\frac{1}{2}}(\partial\Omega)$, which extend the usual $L_2(\partial\Omega)$ inner product.

Remark 2 The first Green identity (5.1) also holds for $u \in H^{1,0}(\Omega; A)$ and $v \in H^1(\Omega)$, cf. [Co88, Mi11].

$$\mathcal{E}_a(u, v) = -\int_\Omega (Au)(x)v(x)dx + \int_{\partial\Omega} T_a^+ u(x)\gamma^+ v(x)ds_x, \qquad (5.4)$$

By interchanging the role of u and v in the first Green identity and subtracting the result, we obtain *the second Green identity* for $u, v \in H^{1,0}(\Omega; A)$,

$$\int_\Omega (vAu - uAv)dx = \langle T_a^+ u, \gamma^+ v \rangle_{\partial\Omega} - \langle T_a^+ v, \gamma^+ u \rangle_{\partial\Omega}.$$

For $b \in C^\infty(\mathbb{R}^2)$, $\quad 0 < b_{\min} \le b(x) \le b_{\max} < \infty$, $\quad \forall x \in \mathbb{R}^2$, let us consider the auxiliary linear elliptic partial differential operator B defined by

$$Bu(x) := B(x, \partial_x)u(x) := \sum_{i=1}^2 \frac{\partial}{\partial x_i}\left[b(x)\frac{\partial u(x)}{\partial x_i}\right]. \qquad (5.5)$$

Then for $u \in H^{1,0}(\Omega, \Delta) = H^{1,0}(\Omega; B)$ the associate co-normal derivative operator T_b^+ is defined by (5.3) (and for $u \in H^2(\Omega)$ by (5.2)) with a replaced by b. For $v \in H^{1,0}(\Omega; B)$ and $u \in H^1(\Omega)$ the first Green identity associated with operator B in (5.5) is

$$\mathcal{E}_b(u, v) = -\int_\Omega (Bv)(x)u(x)dx + \int_{\partial\Omega} T_b^+ v(x)\gamma^+ u(x)ds_x, \qquad (5.6)$$

If $u, v \in H^{1,0}(\Omega, \Delta)$, then subtracting (5.6) from (5.4) we obtain *the two-operator second Green identity* (cf. [Mi05b, AyMi11]),

$$\int_\Omega [u(x)Bv(x) - v(x)Au(x)] dx$$

$$= \int_{\partial\Omega} [\gamma^+ u(x)T_b^+ v(x) - \gamma^+ v(x)T_a^+ u(x)] dS(x)$$

$$- \int_\Omega [a(x) - b(x)]\nabla v(x) \cdot \nabla u(x)dx. \qquad (5.7)$$

Note that if $a = b$ in then the last domain integral in (5.7) disappears, and the two-operator second Green identity (5.7) reduces to the classical second Green identity.

5.2   Parametrix and Potential Type Operators

Definition 2 A function $P_b(x, y)$ of two variables $x, y \in \Omega$ is a parametrix (Levi function) for the operator $B(x; \partial_x)$ in $\mathbb{R}^2$ if

$$B(x, \partial_x) P_b(x, y) = \delta(x - y) + R_b(x, y),$$

where $\delta(.)$ is the Dirac-delta distribution, while $R_b(x, y)$ is the remainder with weak (integrable) singularity at $x = y$.

For 2D, the parametrix and hence the corresponding remainder can be chosen as

$$P_b(x, y) = \frac{1}{2\pi b(y)} \log \left(\frac{|x - y|}{r_0} \right),$$

$$R_b(x, y) = \sum_{i=1}^{2} \frac{x_i - y_i}{2\pi b(y)|x - y|^2} \frac{\partial b(x)}{\partial x_i}, \quad x, y \in \mathbb{R}^2.$$

Similar to [Mi02, ChEtAl09b], we define the parametrix-based Newtonian and remainder potential operators as

$$\mathscr{P}_b g(y) := \int_\Omega P_b(x, y) g(x) dx, \quad \mathscr{R}_b g(y) := \int_\Omega R_b(x, y) g(x) dx \qquad (5.8)$$

and the single layer and double layer potential operators as

$$V_b g(y) := - \int_{\partial\Omega} P_b(x, y) g(x) dS_x, \qquad (5.9)$$

$$W_b g(y) := - \int_{\partial\Omega} [T_b^+(x, n(x), \partial_x) P_b(x, y)] g(x) dS_x \qquad (5.10)$$

where g is some scalar function and the integrals are understood in the distributional sense if g is not integrable. For $y \in \partial\Omega$, the corresponding boundary integral (pseudo-differential) operators of direct surface values of the single layer potential $\mathscr{V}_b$ and of the double layer potential $\mathscr{W}_b$ are

$$\mathscr{V}_b g(y) := - \int_{\partial\Omega} P_b(x, y) g(x) dS_x, \qquad (5.11)$$

$$\mathscr{W}_b g(y) := - \int_{\partial\Omega} [T_b^+(x, n(x), \partial_x) P_b(x, y)] g(x) dS_x. \qquad (5.12)$$

We can also calculate at $y \in \partial\Omega$, the co-normal derivatives associated with the operator A of the single layer potential and of the double layer potential:

$$T_a^+ V_b g(y) := \frac{a(y)}{b(y)} T_b^+ V_b g(y), \tag{5.13}$$

$$\mathcal{L}_{ab}^+ g(y) := T_a^+ W_b g(y) = \frac{a(y)}{b(y)} T_b^+ W_b g(y) = \frac{a(y)}{b(y)} \mathcal{L}_b^+ g(y), \tag{5.14}$$

The direct value operators associated with (5.13) are

$$\mathcal{W}_{ab}' g(y) := -\int_{\partial\Omega} [T_a^+(y, n(y), \partial_y) P_b(x, y)] g(x) dS_x = \frac{a(y)}{b(y)} \mathcal{W}_b' g(y), \tag{5.15}$$

$$\mathcal{W}_b' g(y) := -\int_{\partial\Omega} [T_b^+(y, n(y), \partial_y) P_b(x, y)] g(x) dS_x. \tag{5.16}$$

From Eqs. (5.8)–(5.16), we deduce representations of the parametrix-based surface potential boundary operators in terms of their counterparts for $b = 1$, that is, associated with the fundamental solution $P_\Delta = (2\pi)^{-1} \log\left(\frac{|x-y|}{r_0}\right)$ of the Laplace operator Δ.

$$\mathscr{P}_b g = \frac{1}{b} \mathscr{P}_\Delta g, \quad \mathscr{R}_b g = -\frac{1}{b} \sum_{j=1}^{3} \partial_j \mathscr{P}_\Delta [g(\partial_j b)], \tag{5.17}$$

$$\frac{a}{b} V_a g = V_b g = \frac{1}{b} V_\Delta g; \quad \frac{a}{b} W_a\left(\frac{bg}{a}\right) = W_b g = \frac{1}{b} W_\Delta (bg), \tag{5.18}$$

$$\frac{a}{b} \mathcal{V}_a g = \mathcal{V}_b g = \frac{1}{b} \mathcal{V}_\Delta g; \quad \frac{a}{b} \mathcal{W}_a\left(\frac{bg}{a}\right) = \mathcal{W}_b g = \frac{1}{b} \mathcal{W}_\Delta (bg), \tag{5.19}$$

$$\mathcal{W}_{ab}' g = \frac{a}{b} \mathcal{W}_b' g = \frac{a}{b} \left\{ \mathcal{W}_\Delta' (g) + \left[b\frac{\partial}{\partial n}\left(\frac{1}{b}\right)\right] \mathcal{V}_\Delta g \right\}, \tag{5.20}$$

$$\mathcal{L}_{ab}^\pm g = \frac{a}{b} \mathcal{L}_b^\pm g = \frac{a}{b} \left\{ \mathcal{L}_\Delta(bg) + \left[b\frac{\partial}{\partial n}\left(\frac{1}{b}\right)\right] \gamma^\pm W_\Delta(bg) \right\}. \tag{5.21}$$

It is taken into account that b and its derivatives are continuous in $\mathbb{R}^2$ and $\mathcal{L}_\Delta(bg) := \mathcal{L}_\Delta^+(bg) = \mathcal{L}_\Delta^-(bg)$ by the Lyapunov-Tauber theorem.

Let $g_1 \in H^{-\frac{1}{2}}(\partial\Omega)$, and $g_2 \in H^{\frac{1}{2}}(\partial\Omega)$. Then there hold the following relations on $\partial\Omega$,

$$\gamma^\pm V_b g_1 = \mathcal{V}_b g_1, \tag{5.22}$$

$$\gamma^\pm W_b g_2 = \mp\frac{1}{2} g_2 + \mathcal{W}_b g_2, \tag{5.23}$$

$$T_a^\pm V_b g_1 = \pm\frac{1}{2}\frac{a}{b} g_1 + \mathcal{W}_{ab}' g_1. \tag{5.24}$$

The jump relations (5.22)–(5.24) follow from the corresponding relations in [DuMi15, Theorem 2] and relations (5.17)–(5.21), see also, [AyMi11, Theorem A.6].

5.3 Invertibility of the Single Layer Potential Operator in 2D

Due to [Mc00, Theorem 7.6] and the first relation in (5.19), the boundary integral operator $\mathcal{V}_b : H^{-\frac{1}{2}}(\partial\Omega) \to H^{\frac{1}{2}}(\partial\Omega)$ is a Fredholm operator of index zero, and by [ChEtAl09b, Lemma 4.2(i)] is also injective implying its invertibility in 3D. But this is not the case in 2D. Remark 1.42(ii) in [Co00] and the proof of Theorem 6.22 in [Se08] assert that for some 2D domains the Kernel of the operator $\mathcal{V}_\Delta$ is non-zero. Due to the first relation in (5.19) this shows the Kernel of the operator $\mathcal{V}_b$ is also non-zero for some domains.

To ensure the invertibility of the single layer potential operator in 2D, we define (see, e.g., [Se08, p. 147]) the subspaces $H^s_{**}(\partial\Omega)$ of $H^s(\partial\Omega)$ by:

$$H^s_{**}(\partial\Omega) := \{g \in H^s(\partial\Omega) : \langle g, 1\rangle_{\partial\Omega} = 0\}$$

where the norm in $H^s_{**}(\partial\Omega)$ is the induced norm of $H^s(\partial\Omega)$.

The following assertions extend the result in Theorems 4 and 5 in [DuMi15].

Theorem 1 *If* $\Psi \in H^{-\frac{1}{2}}_{**}(\partial\Omega)$ *satisfies* $\mathcal{V}_b\Psi = 0$ *on* $\partial\Omega$, *then* $\Psi = 0$.

Proof For the case $a = b$, the proof follows from [Mc00, Corollary 8.11(ii)] and [DuMi15, Theorem 4]. The case $a \neq b$ then follows from the relations in (5.18) and [DuMi15, Theorem 4].

Theorem 2 *Let* $\Omega \subset \mathbb{R}^2$ *with* $diam(\Omega) < r_0$. *Then the single layer potential operator* $\mathcal{V}_b : H^{-\frac{1}{2}}(\partial\Omega) \to H^{\frac{1}{2}}(\partial\Omega)$ *is invertible.*

Proof For the case $a = b$ the proof follows from [Se08, Theorem 6.23] and [DuMi15, Theorem 5]. The case $a \neq b$ then follows from the relations in (5.18) and [DuMi15, Theorem 5].

5.4 Dirichlet Problem and Two-Operator Third Green Identity

We shall derive and investigate the *two-operator* BDIE systems for the following Dirichlet boundary value problem: *Find a function* $u \in H^1(\Omega)$ *satisfying equations*

$$Au = f \quad in \quad \Omega, \tag{5.25}$$

$$\gamma^+ u = \varphi_0 \quad on \quad \partial\Omega. \tag{5.26}$$

where $\varphi_0 \in H^{\frac{1}{2}}(\partial\Omega)$ *and* $f \in L_2(\Omega)$.

Equation (5.25) is understood in the distributional sense as in Remark 1 and the Dirichlet boundary condition (5.26) in the trace sense.

The following assertion is well-known and can be proved by using variational settings and the Lax-Milgram lemma (see, e.g., [Mi05a, Corollary and Theorem 5], [Mi15, Theorem 5.1])

Theorem 3 *The Dirichlet problem* (5.25)–(5.26) *is uniquely solvable in* $H^{1,0}(\Omega; A)$. *The solution is* $u = (\mathscr{A}^D)^{-1}(f, \varphi_0)^T$, *where the inverse operator,* $(\mathscr{A}^D)^{-1} : L_2(\Omega) \times H^{\frac{1}{2}}(\partial\Omega) \to H^{1,0}(\Omega; A)$, *to the left-hand side operator,* $\mathscr{A}^D : H^{1,0}(\Omega; A) \to L_2(\Omega) \times H^{\frac{1}{2}}(\partial\Omega)$, *of the Dirichlet problem* (5.25)–(5.26), *is continuous.*

If $u \in H^{1,0}(\Omega; A)$, then substituting $v(x)$ by $P_b(x, y)$ in the two-operator second Green identity (5.7) for $\Omega \setminus B(y, \varepsilon)$, where $B(y, \varepsilon)$ is a disc of radius ε centred at y, and taking the limit $\varepsilon \to 0$, we arrive at the following parametrix-based two-operator third Green identity (cf. e.g.,[Mi02]),

$$u + \mathscr{Z}_b u + \mathscr{R}_b u - V_b T_a^+ u + W_b \gamma^+ u = \mathscr{P}_b A u \quad \text{in} \quad \Omega, \tag{5.27}$$

$$\text{where} \quad \mathscr{Z}_b u(y) = -\int_\Omega (a(x) - b(x)) \nabla_x P_b(x, y) \cdot \nabla u(x) dx. \tag{5.28}$$

Using the Gauss Divergence Theorem, we can rewrite $\mathscr{Z}_b u$ in (5.27) in the form that does not involve derivative of u, i.e.,

$$\mathscr{Z}_b u = \left[\frac{a(y)}{b(y)} - 1\right] u(y) + \widehat{\mathscr{Z}_b} u(y), \tag{5.29}$$

$$\text{where} \quad \widehat{\mathscr{Z}_b} u(y) = \frac{a(y)}{b(y)} W_a \gamma^+ u(y) - W_b \gamma^+ u(y) + \frac{a(y)}{b(y)} \mathscr{R}_a u(y) - \mathscr{R}_b u(y), \tag{5.30}$$

which allows to call $\mathscr{Z}_b$ integral operator in spite of its integro-differential representation (5.28).

Note that substituting (5.29) and (5.30) in (5.27) and multiplying by $b(y)/a(y)$ one reduces (5.27) to the one-operator parametrix-based third Green identity obtained in [ChEtAl09b],

$$u + \mathscr{R}_a u - V_a T_a^+ u + W_a \gamma^+ u = \mathscr{P}_a A u \quad \text{in} \quad \Omega.$$

Relations (5.28)–(5.30) and the mapping properties of $\mathscr{P}_a, \mathscr{R}_a, \mathscr{R}_b, W_a$ and W_b, imply the following assertion (see, e.g., [AyMi11, Appendix A]).

Theorem 4 *The operators*

$$\mathcal{Z}_b : H^s(\Omega) \to H^s(\Omega), \quad s > \frac{1}{2},$$

$$\widehat{\mathcal{Z}_b} : H^s(\Omega) \to H^{s,0}(\Omega; \Delta), \quad s \geq 1,$$

are continuous.

If $u \in H^{1,0}(\Omega; \Delta)$ is a solution to Eq. (5.25) with $f \in L_2(\Omega)$, then (5.27) gives

$$u + \mathcal{Z}_b u + \mathcal{R}_b u - V_b T_a^+ u + W_b \gamma^+ u = \mathcal{P}_b f \quad \text{in} \quad \Omega, \tag{5.31}$$

Applying *the trace operator* to Eq. (5.31) and using the jump relations (5.22) and (5.23), we have

$$\frac{1}{2}\gamma^+ u + \gamma^+ \mathcal{Z}_b u + \gamma^+ \mathcal{R}_b u - \mathcal{V}_b T_a^+ u + \mathcal{W}_b \gamma^+ u = \gamma^+ \mathcal{P}_b f \quad \text{on} \quad \partial\Omega, \tag{5.32}$$

Similarly, applying *the co-normal derivative operator* to Eq. (5.31), and using again the jump relation (5.24) and relations (5.13) and (5.14), we obtain

$$\left(1 - \frac{a}{2b}\right) T_a^+ u + T_a^+ \mathcal{Z}_b u + T_a^+ \mathcal{R}_b u - \mathcal{W}'_{ab} T_a^+ u + \mathcal{L}^+_{ab} \gamma^+ u = T_a^+ \mathcal{P}_b f \quad \text{on} \quad \partial\Omega. \tag{5.33}$$

Note that if $\mathcal{P}_b$ is not only the parametrix but also the fundamental solution of the operator B, then the remainder operator $\mathcal{R}_b$ vanishes in (5.31)–(5.33) (and everywhere in the paper), while the operator $\mathcal{Z}_b$ does not unless operators A and B are equal.

For some functions f, Ψ, Φ, let us consider a more general "indirect" integral relation, associated with (5.31),

$$u + \mathcal{Z}_b u + \mathcal{R}_b u - V_b \Psi + W_b \Phi = \mathcal{P}_b f, \quad \text{in} \quad \Omega \tag{5.34}$$

Lemma 1 *Let $f \in L_2(\Omega)$, $\Psi \in H^{-\frac{1}{2}}(\partial\Omega)$, $\Phi \in H^{\frac{1}{2}}(\partial\Omega)$, and $u \in H^1(\Omega)$ satisfy (5.34). Then $u \in H^{1,0}(\Omega; \Delta)$ and is a solution of PDE $Au = f$ in Ω and*

$$V_b \left(\Psi - T_a^+ u\right) - W_b \left(\Phi - \gamma^+ u\right) = 0 \quad \text{in} \quad \Omega. \tag{5.35}$$

Proof The proof is similar to the one in 3D case in [AyMi11, Lemma 3.1].

Lemma 2

i) *Let either* $\Psi^* \in H_{**}^{-\frac{1}{2}}(\partial\Omega)$ *or* $\Psi^* \in H^{-\frac{1}{2}}(\partial\Omega)$ *but* $\mathrm{diam}(\Omega) < r_0$. *If* $V_b\Psi^* = 0$
   in Ω, *then* $\Psi^* = 0$.

ii) *Let* $\Phi^* \in H^{\frac{1}{2}}(\partial\Omega)$. *If* $W_b\Phi^* = 0$ *in* Ω, *then* $\Phi^* = 0$.

Proof For the case $a = b$ the proof follows from [DuMi15, Lemma 2]. The case $a \neq b$ then follows from the relation (5.18) and [DuMi15, Lemma 2].

5.5 Two-Operator BDIEs for Dirichlet BVP

To reduce the variable-coefficient Dirichlet BVP (5.25)–(5.26) to a *segregated* boundary-domain integral equation system, let us denote the unknown conormal derivative as $\psi = T^+u \in H^{-\frac{1}{2}}(\partial\Omega)$ and will further consider ψ as formally independent on u.

Assuming that the function u satisfies (5.25), substituting the Dirichlet condition into the third Green identity (5.31) and either into its trace (5.32) or into its co-normal derivative (5.33) on $\partial\Omega$, we can reduce the BVP (5.25)–(5.26) to two different *two-operator BDIE systems* for the unknown functions $u \in H^{1,0}(\Omega; A)$ and $\psi = T^+u \in H^{-\frac{1}{2}}(\partial\Omega)$.

BDIE system (D1) obtained from Eq. (5.31) in Ω, and Eq. (5.32) on the whole boundary $\partial\Omega$ is:

$$u + \mathcal{Z}_b u + \mathcal{R}_b u - V_b\psi = F_0 \quad \text{in} \quad \Omega, \tag{5.36}$$

$$\gamma^+ \mathcal{Z}_b u + \gamma^+ \mathcal{R}_b u - \mathcal{V}_b\psi = \gamma^+ F_0 - \varphi_0 \quad \text{in} \quad \partial\Omega, \tag{5.37}$$

where $F_0 := \mathcal{P}_b f - W_b\varphi_0$.

BDIE system (D2) obtained from Eq. (5.31) in Ω, and Eq. (5.33) on the whole boundary $\partial\Omega$ is:

$$u + \mathcal{Z}_b u + \mathcal{R}_b u - V_b\psi = F_0 \quad \text{in} \quad \Omega, \tag{5.38}$$

$$\left(1 - \frac{a}{2b}\right)\psi + T_a^+ \mathcal{Z}_b u + T_a^+ \mathcal{R}_b u - \mathcal{W}_{ab}'\psi = T_a^+ F_0 \quad \text{on} \quad \partial\Omega. \tag{5.39}$$

where $F_0 := \mathcal{P}_b f - W_b\varphi_0$.

5.6   Equivalence and Invertibility Theorems

Now let us prove the equivalence of the original Dirichlet BVP (5.25)–(5.26) with
the two-operator BDIE systems (D1) and (D2).

Theorem 5 *Let $f \in L_2(\Omega)$ and $\varphi_0 \in H^{\frac{1}{2}}(\partial\Omega)$.*

i. *If some $u \in H^1(\Omega)$ solves the Dirichlet BVP (5.25)–(5.26) in Ω, then the pair*
   (u, ψ) *where*

$$\psi = T_a^+ u \in H^{-\frac{1}{2}}(\partial\Omega) \tag{5.40}$$

   solves BDIEs (D1) and (D2).
ii. *If a pair $(u, \psi) \in H^1(\Omega) \times H^{-\frac{1}{2}}(\partial\Omega)$ solves BDIE system (D1) and*
    $diam(\Omega) < r_0$, *then u solves BDIEs (D2) and the BVP (5.25)–(5.26), this*
    solution is unique, and ψ satisfies (5.40).
iii. *If a pair $(u, \psi) \in H^1(\Omega) \times H^{-\frac{1}{2}}(\partial\Omega)$ solves BDIE system (D2), then u*
     solves BDIEs (D1) and the BVP (5.25)–(5.26), this solution is unique, and ψ
     satisfies (5.40).

Proof

(i) Let $u \in H^1(\Omega)$ be a solution of the BVP (5.25)–(5.26) then since $f \in L_2(\Omega)$,
    we have that $u \in H^{1,0}(\Omega; \Delta)$. Setting ψ by (5.40) and recalling how the BDIE
    systems (D1) and (D2) were constructed, we obtain that (u, ψ) solves them.
    This completes the proof of item (i).

    Let the pair $(u, \psi) \in H^1(\Omega) \times H^{-\frac{1}{2}}(\partial\Omega)$ solve system (D1) or (D2). Due
    to the first equations in BDIE systems, (5.36) and (5.38), the hypotheses of
    Lemma 1 are satisfied implying that u belongs to $H^{1,0}(\Omega; A)$ and solves
    PDE (5.25) in Ω, while Eq. (5.35) also holds.

(ii) Let $(u, \psi) \in H^1(\Omega) \times H^{-\frac{1}{2}}(\partial\Omega)$ solve system (D1). Taking the trace
     of (5.36) and subtracting (5.37) from it, we get $\gamma^+ u = \varphi_0$ on $\partial\Omega$. Thus,
     the Dirichlet boundary condition is satisfied, and using it in (5.35), we have
     $V_b(\psi - T^+ u)(y) = 0$ in Ω. Item (i) in Lemma 2 then implies $\psi = T_a^+ u$.

(iii) Let now the pair $(u, \psi) \in H^1(\Omega) \times H^{-\frac{1}{2}}(\partial\Omega)$ solve system (D2). Taking the
      co-normal derivative of (5.38) and subtracting the second equation from it, we
      get $\psi = T_a^+ u$ on $\partial\Omega$. Then substituting this in (5.35) gives $W(\varphi_0 - \gamma^+ u)(y) =$
      0 in Ω and item (ii) in Lemma 2 then implies $\varphi_0 = \gamma^+ u$ on $\partial\Omega$.

The uniqueness of the BDIE system solutions follows form the fact that the corresponding homogeneous BDIE systems can be associated with the homogeneous
Dirichlet problem, which has only the trivial solution. Then paragraphs (ii) and
(iii) above imply that the homogeneous BDIE systems also have only the trivial
solutions.

BDIE systems (D1) and (D2) with the right-hand sides (5.36) and (5.37) and (5.38) and (5.39) can be written as $\mathfrak{D}^1 \mathcal{U}^D = \mathcal{F}^{D1}$ and $\mathfrak{D}^2 \mathcal{U}^D = \mathcal{F}^{D2}$,, respectively. Here $\mathcal{U}^D := (u, \psi)^T \in H^1(\Omega) \times H^{-\frac{1}{2}}(\partial\Omega)$, and

$$\mathfrak{D}^1 := \begin{bmatrix} I + \mathcal{Z}_b + \mathcal{R}_b & -V_b \\ \gamma^+ \mathcal{Z}_b + \gamma^+ \mathcal{R}_b & \mathcal{V}_b \end{bmatrix} ;$$

$$\mathfrak{D}^2 := \begin{bmatrix} I + \mathcal{Z}_b + \mathcal{R}_b & -V_b \\ T_a^+ \mathcal{Z}_b + T_a^+ \mathcal{R}_b & \left(1 - \frac{a}{2b}\right) I - \mathcal{W}_{ab}' \end{bmatrix}$$

while $\mathcal{F}^{D1}$ and $\mathcal{F}^{D2}$ are given by:

$$\mathcal{F}^{D1} := \begin{bmatrix} \mathcal{F}_1^{D1} \\ \mathcal{F}_2^{D1} \end{bmatrix} = \begin{bmatrix} F_0^D \\ \gamma^+ F_0^D - \varphi_0 \end{bmatrix} ; \qquad \mathcal{F}^{D2} := \begin{bmatrix} \mathcal{F}_1^{D2} \\ \mathcal{F}_2^{D2} \end{bmatrix} = \begin{bmatrix} F_0 \\ T_a^+ F_0 \end{bmatrix} .$$

Due to the mapping properties of the operators participating in $\mathfrak{D}^1$ and $\mathfrak{D}^2$ as well as the right-hand sides $\mathcal{F}^{D1}$ and $\mathcal{F}^{D2}$ (see, e.g., [AyMi11, Appendix A]), we have $\mathcal{F}^{D1} \in H^{1,0}(\Omega; A) \times H^{\frac{1}{2}}(\partial\Omega)$, $\mathcal{F}^{D2} \in H^{1,0}(\Omega; A) \times H^{-\frac{1}{2}}(\partial\Omega)$, while the operators

$$\mathfrak{D}^1 : H^{1,0}(\Omega; A) \times H^{-\frac{1}{2}}(\partial\Omega) \to H^1(\Omega) \times H^{\frac{1}{2}}(\partial\Omega) \tag{5.41}$$

$$\mathfrak{D}^2 : H^{1,0}(\Omega; A) \times H^{-\frac{1}{2}}(\partial\Omega) \to H^1(\Omega) \times H^{-\frac{1}{2}}(\partial\Omega) \tag{5.42}$$

are continuous. Due to Theorem 5 item (ii) and (iii), operators (5.41) and (5.42) are injective.

Theorem 6 *If $diam(\Omega) < r_0$, then operator (5.41) is continuous and continuously invertible.*

Proof The continuity of operator (5.41) is proved above. Theorem 5 (ii) implies that operator (5.41) is injective. To prove the invertibility of operator (5.41), let us consider the BDIE system (D1) with arbitrary right-hand side

$$\mathcal{F}_*^{D1} = (\mathcal{F}_{*1}^{D1}, \mathcal{F}_{*2}^{D1})^T \in H^{1,0}(\Omega; A) \times H^{\frac{1}{2}}(\partial\Omega).$$

Taking $\mathcal{F}_1 = \mathcal{F}_{*1}^{D1}$ and $\Phi_* = \gamma^+ \mathcal{F}_{*1}^{D1} - \mathcal{F}_{*2}^{D1}$ in [AyMi11, Lemma B.3], we obtain the representation of $\mathcal{F}_*^{D1}$ as: $\mathcal{F}_*^{D1} = (\mathcal{F}_{*1}^{D1}, \mathcal{F}_{*2}^{D1})^T = (\mathcal{F}_1, \gamma^+ \mathcal{F}_1 - \Phi_*)^T$ where the couple

$$(f_*, \Phi_*) = \mathcal{C}_\Phi \mathcal{F}_1 \in L_2(\Omega) \times H^{\frac{1}{2}}(\partial\Omega) \tag{5.43}$$

is unique and the operator

$$\mathscr{C}_{\mathscr{F}_1} : H^{1,0}(\Omega; A) \to L_2(\Omega) \times H^{\frac{1}{2}}(\partial\Omega) \tag{5.44}$$

is linear and continuous. If $\mathrm{diam}(\Omega) < r_0$, then applying Theorem 5 with $f = f_*$, $\Phi_* = \varphi_0$, we obtain that BDIE system (D1) is uniquely solvable and its solution is of the form: $\mathscr{U}_1 = (\mathscr{A}^D)^{-1}(f, \varphi_0)^T$ and $\mathscr{U}_2 = \gamma^+ \mathscr{U}_1 - \varphi_0$, where the inverse operator, $(\mathscr{A}^D)^{-1} : L_2(\Omega) \times H^{\frac{1}{2}}(\partial\Omega) \to H^{1,0}(\Omega; A)$, to the left-hand side operator, $\mathscr{A}^D : H^{1,0}(\Omega; A) \to L_2(\Omega) \times H^{\frac{1}{2}}(\partial\Omega)$, of the Dirichlet problem (5.25)–(5.26), is continuous. Representation (5.43) and continuity of the operator (5.44) imply invertibility of (5.41).

The following assertion is [Mi15, Lemma 6.6] redone for a more narrow space.

Lemma 3 *For any couple* $(\mathscr{F}_1, \mathscr{F}_2) \in H^1(\Omega) \times H^{-\frac{1}{2}}(\partial\Omega)$ *there exists a unique couple* $(f_{**}, \Phi_*) \in L_2(\Omega) \times H^{\frac{1}{2}}(\partial\Omega)$ *such that*

$$\mathscr{F}_1 = \mathscr{P}_b f_{**} - W_b \Phi_* \qquad \text{in} \quad \Omega, \tag{5.45}$$

$$\mathscr{F}_2 = T_a^+ (\mathscr{P}_b f_{**} - W_b \Phi_*) \qquad \text{on} \quad \partial\Omega. \tag{5.46}$$

Moreover, $(f_{**}, \Phi_*) = C_\Phi(\mathscr{F}_1, \mathscr{F}_2)$
and $C_\Phi : H^1(\Omega) \times H^{-\frac{1}{2}}(\partial\Omega) \to L_2(\Omega) \times H^{\frac{1}{2}}(\partial\Omega)$ *is a continuous linear operator given by*

$$f_{**} = \Delta(b\mathscr{F}_1), \tag{5.47}$$

$$\Phi_* = \frac{1}{b} \left(-\frac{1}{2}I + \mathscr{W}_\Delta \right)^{-1} \gamma^+ \{ -b\mathscr{F}_1 + \mathscr{P}_\Delta[\Delta(b\mathscr{F}_1)] \}. \tag{5.48}$$

The following similar assertion for the operator $\mathfrak{D}^2$ holds without limitations on the diameter of the domain Ω.

Theorem 7 *The operator* (5.42) *is continuous and continuously invertible.*

The continuity of operator (5.42) is proved above. Theorem 5 (iii) implies that operator (5.42) is injective. To prove the invertibility of operator (5.42), let us consider the BDIE system (D2) with arbitrary right-hand side $\mathscr{F}_*^{D2} = (\mathscr{F}_{*1}^{D2}, \mathscr{F}_{*2}^{D2})^T \in H^{1,0}(\Omega; A) \times H^{-\frac{1}{2}}(\partial\Omega)$. Take $\mathscr{F}_1 = \mathscr{F}_{*1}^{D2}$ and $\mathscr{F}_2 = T_a^+ \mathscr{F}_1 = \mathscr{F}_{*2}^{D2}$ in Lemma 3 to represent $\mathscr{F}_*^{D2}$ as: $\mathscr{F}_*^{D2} = (\mathscr{F}_{*1}^{D2}, \mathscr{F}_{*2}^{D2})^T = (\mathscr{F}_1, T_a^+ \mathscr{F}_1)^T$ where $\mathscr{F}_1$ and $\mathscr{F}_2$ are given, respectively, by (5.45) and (5.46), the couple

$$(f_{**}, \Phi_*) = \mathscr{C}_{**}(\mathscr{F}_1, \mathscr{F}_2) \in L_2(\Omega) \times H^{-\frac{1}{2}}(\partial\Omega) \tag{5.49}$$

given by (5.47) and (5.48) is unique and the operator

$$\mathscr{C}_{**} : H^{1,0}(\Omega; A) \times H^{-\frac{1}{2}}(\partial\Omega) \to L_2(\Omega) \times H^{-\frac{1}{2}}(\partial\Omega) \tag{5.50}$$

is linear and continuous.

Applying Theorem 5 with $f_{**} = f$, $\Phi_* = \varphi_0$, we obtain that BDIE system (D2) is uniquely solvable and its solution is of the form: $\mathscr{U}_1 = (\mathscr{A}^D)^{-1}(f, \varphi_0)^T$ and $\mathscr{U}_2 = T_a^+(f, \mathscr{U}_1)$, for the same continuous inverse operator, $(\mathscr{A}^D)^{-1}$. Representation (5.49) and continuity of the operator (5.50) imply invertibility of (5.42).

5.7 Conclusion

For a variable coefficient PDE in a two-dimensional domain, the two-operator boundary-domain integral equations associated with the Dirichlet boundary condition in the interior domain have been formulated and analysed in this paper. Equivalence of the BDIEs to the original BVP was shown in the case when the PDE right-hand side is from $L_2(\Omega)$ and the Dirichlet data from the space $H^{\frac{1}{2}}(\partial\Omega)$. The invertibility of the associated boundary domain integral operators in the corresponding Sobolev spaces was also proved. In a similar way one can consider also the 2D versions of the two-operator BDIEs for the Neumann problem, mixed problem in interior and exterior domains formulated and analysed in [AyMi11].

Acknowledgements The work on this paper of the first author was supported by EMS-Simmons for Africa and ISP Sweden and of the second author was supported by ISP, Sweden. They also thank the Department of Mathematics at Brunel University for hosting their research visit.

Reference

[AyMi11] Tsegaye G. Ayele, Sergey E. Mikhailov (2011), Analysis of two-operator boundary-domain integral equations for a variable-coefficient BVP, in: *Eurasian Math. J., 2:3 (2011), 20–41.*

[ChEtAl09b] O. Chakuda, S.E. Mikhailov, D. Natroshvili, *Analysis of direct boundary-domain integral equations for a mixed BVP with variable coefficient. I: Equivalence and Invertibility.* J. Integral Equat. and Appl. 21 (2009), 499–543.

[Co00] Constanda, C.: *Direct and Indirect Boundary Integral Equation Methods.* Chapman & Hall/CRC (2000).

[Mi15] Mikhailov, S.E.: Analysis of Segregated Boundary-Domain Integral Equations for variable coefficient Dirichlet and Neumann problems with general date. *ArXiv:1509.03501*, 1–32 (2015).

[Co88] M. Costabel, *Boundary integral operators on Lipchitz domains: elementary results.* SIAM journal on Mathematical Analysis 19 (1988), 613–626.

[DuMi15] Dufera, T.T. and Mikhailov, S.E.: Analysis of Boundary-Domain Integral Equations for Variable-Coefficient Dirichlet BVP in 2D. In: *Integral Methods in Science and Engineering: Computational and Analytic Aspects, C. Constanda and A. Kirsch (eds.),Birkhäuser, Boston (2015)*, pp. 163–175.

[Mc00] W. McLean, *Strongly Elliptic Systems and Boundary Integral Equations.* Cambridge University Press, Cambridge, 2000.

[Mi05a] Mikhailov S.E.*Analysis of boundary-domain integral and integro-differential equations for a Dirichlet problem with variable coefficient.* In: Integral Methods in Science and Engineering: Theoretical and Practical Aspects (Edited by C.Constanda, Z.Nashed, D.Rolins), Boston-Basel-Berlin: Birkhäuser, ISBN 0-8176-4377-X, 161–176.

[Mi02] S.E. Mikhailov, *Localized boundary-domain integral formulations for problems with variable coefficients*, Int. J. Engineering Analysis with Boundary Elements 26 (2002), 681–690.

[Mi05b] S.E. Mikhailov, *Localized formulations for scalar nonlinear BVPs with variable coefficients.* J. Eng. Math., 51 (2005), 283–302.

[Mi11] S.E. Mikhailov, *Traces, extensions and co-normal derivatives for elliptic systems on Lipschitz domains.* J. Math. Analysis and Appl. 378 (2011), 324–342.

[Se08] Steinbach, O.: *Numerical Approximation Methods for Elliptic Boundary Value Problems: Finite and Boundary Elements*, Springer New York (2008).

Chapter 6
Solution of a Homogeneous Version of Love Type Integral Equation in Different Asymptotic Regimes

Laurent Baratchart, Juliette Leblond, and Dmitry Ponomarev

6.1 Introduction

For $h, a > 0$, we consider the following homogeneous Fredholm integral equation of the second kind

$$\frac{h}{\pi} \int_{-a}^{a} \frac{f(t)}{(x-t)^2 + h^2} dt = \lambda f(x), \quad x \in (-a, a), \tag{6.1}$$

which can be viewed as a problem of finding eigenfunctions of the integral operator $P_h \chi_A : L^2(A) \to L^2(A)$ with

$$P_h[f](x) := (p_h \star f)(x) = \frac{h}{\pi} \int_{-\infty}^{\infty} \frac{f(t)}{(x-t)^2 + h^2} dt, \tag{6.2}$$

$$p_h(x) := \frac{h}{\pi} \frac{1}{x^2 + h^2}, \tag{6.3}$$

and χ_A being the characteristic function of the interval $A := (-a, a)$.

Integral equations with kernel function (6.3) have a long history (starting with [Sn23] as the earliest mention we could trace and up to recent papers [TrWi16a, TrWi16b, TrWi18, Pr17]). It most commonly arises in rotationally symmetric

L. Baratchart · J. Leblond
INRIA Sophia Antipolis Méditerranée, Valbonne, France
e-mail: laurent.baratchart@inria.fr; juliette.leblond@inria.fr

D. Ponomarev (✉)
Vienna University of Technology, Institute of Analysis and Scientific Computing, Vienna, Austria

St. Petersburg Department of Steklov Mathematical Institute RAS, St. Petersburg, Russia
e-mail: dmitry.ponomarev@asc.tuwien.ac.at

© Springer Nature Switzerland AG 2019
C. Constanda, P. Harris (eds.), *Integral Methods in Science and Engineering*,
https://doi.org/10.1007/978-3-030-16077-7_6

electrostatic [Lo49] or fluid dynamics problems [Hu64a] (in such contexts it is most famously known as Love equation), quantum-mechanical statistics of Fermi/Bose gases (known there as Gaudin/Lieb-Liniger equation, respectively) [Ga71, LiLi63], antiferromagnetic one-dimensional Heisenberg chains [Gr64], and is relevant as well in other contexts such as probability theory [KaPo50] and radiative transfer [Tr69]. Since p_h is the two-dimensional Poisson kernel for the upper half-plane, the integral equation (6.1) has also applications to problems of approximation by harmonic functions [LePo17] and it is instrumental in some inverse source problems for the Poisson PDE (e.g., the so-called inverse magnetization problems, see [BaEtAl13]).

The class of exactly solvable convolution integral equations on interval is extremely narrow and rarely exceeds the class of equations with kernels whose Fourier transform is a rational function. Such approaches usually hinge on matrix Wiener-Hopf factorization which are inapplicable due to non-smooth and non-algebraic behavior at infinity of the Fourier transform of the kernel function (6.3): $\hat{p}_h(k) = e^{-2\pi h |k|}$. Therefore, the main hope for an analytical solution is a structural approach (i.e., when exact solutions are determined up to constants that cannot be expressed in a closed-form) or an asymptotic one. Despite seeming simplicity of the kernel function p_h, the integral equation (6.1) evades applicability of relevant constructive techniques: both for exact structural [LeMu65] and asymptotical solutions [KnKe91, Hu64b]. The problem of failure of asymptotic approaches (when the length of the interval is large) in [KnKe91, Hu64b] is the lack of sufficient decay of the kernel function at infinity (alternatively, the lack of existence of second-order derivative at the origin of the Fourier transform of the kernel function). The powerful approach of Leonard and Mullikin [LeMu65] aiming to obtain essentially exact sine/cosine solutions (with frequencies to be determined from unsolvable explicitly auxiliary equations) breaks down since the inverse Laplace transform of the kernel function is not of constant sign which the authors claim to be merely a technical problem (according to them, the assumption of constant sign is made to "simplify the discussion"). However, from results of our approach we will see that change of the sign of this function that occurs infinite number of times results in a qualitatively different form of solutions which are beyond simple trigonometric functions shifted by constants (see the results for the small interval and eigenfunctions of a higher order in case of the large interval).

To the best of our knowledge, the only available result in the literature regarding Eq. (6.1) (except for its non-homogeneous version with $\lambda = \pm 1$) is the exact exponential decay law of eigenvalues [Wi64] and a relevant reduction to a hypersingular equation which "appears too difficult to solve explicitly" [KnKe91] (see also Sect. 6.6).

Consideration of the homogeneous version of the equation with the kernel (6.3) is the most general in a sense that the obtained solutions (eigenfunctions) permit construction of the resolvent kernel (in a form of a uniformly and absolutely converging series, as a consequence of Mercer theorem for positive definite kernels) for a general non-homogeneous equation or, as a more practical alternative (due to completeness and orthogonality of eigenfunctions and the fast decay of eigenvalues), provide solution in a form of expansion over first few eigenfunctions.

After discussing general spectral properties of the integral operator (Sect. 6.2), we propose original constructive techniques for obtaining asymptotical solution for the case of small (Sect. 6.3) and large size of interval (Sect. 6.4). When the interval is small, the integral equation can be approximated by another one which admits a commuting differential operator. This fact allows reduction of the problem to solving a second-order boundary-value problem whose solutions upon further approximations are prolate spheroidal harmonics (Slepian functions). When the interval is large, the problem can be transformed into an integro-differential equation on a shifted half-line. Integral kernel function of such problem consists of two terms: one depends on the difference of the arguments, the other—on their sum. The latter turns out to be small for large interval and hence we end up with approximation by a convolution integro-differential equation which we solve by an extended Wiener-Hopf method. Connection of this half-line problem solution to the solution of the original equation inside the original interval is provided by analytic continuation that can be performed by means of solution of an elementary non-homogeneous ODE. Finally, we illustrate the obtained asymptotical results, compare them with numerical solution (Sect. 6.5), and outline potential further work (Sect. 6.6).

6.2   General Properties

Since the kernel $p_h(x)$ is an even and real-valued function, the operator $P_h \chi_A$ is self-adjoint, and because of the regularity of $p_h(x)$, the operator is also compact (e.g., as a Hilbert-Schmidt operator), and, by its analytic properties, has a dense range in $L^2(A)$. Hence the standard spectral theorem [NaSe00] reformulates as

Theorem 1 *There exists $(\lambda_n)_{n=1}^{\infty} \in \mathbb{R}$, $\lambda_n \to 0$ as $n \to \infty$ and $(f_n)_{n=1}^{\infty}$ is a complete set in $L^2(A)$.*

Basic properties of eigenfunctions and eigenvalues can be outlined in the following proposition (see [Po16]).

Proposition 1 *For λ, f satisfying (6.1), the following statements hold true:*
(a) All $(\lambda_n)_{n=1}^{\infty}$ are simple, and $\lambda_n \in (0, 1)$,
(b) Each f_n is either even or odd, real-valued (up to a constant multiplicator), and $f_n \in C^{\infty}(\bar{A})$. Moreover, $f_n(\pm a) \neq 0$.

The key result here is non-vanishing behavior of eigenfunctions at the endpoints implying the multiplicity (simplicity) of the spectrum which, in particular, along with the evenness of p_h, entails further the real-valuedness and a certain parity of each solution f_n, a fact that will be used constructively in Sect. 6.4.

The upper bound for the eigenvalues in part (a) of Proposition 1 can be improved to

$$\lambda_n \leq \frac{2}{\pi} \arctan \frac{a}{h}, \quad n \in \mathbb{N}_+,$$

and asymptotically exponential decay of higher-order eigenvalues is given by

$$\log \lambda_n \simeq -n\pi \frac{K\left(\operatorname{sech}\left(\pi a / h\right)\right)}{K\left(\tanh\left(\pi a / h\right)\right)}, \quad n \gg 1 \tag{6.4}$$

where $K(x) := \int_0^{\pi/2} \frac{d\theta}{\sqrt{1-x^2 \sin^2 \theta}}$ the complete elliptic integral of the first kind.

Note that, since the spectrum is simple, we can uniquely order eigenvalues as

$$0 < \cdots < \lambda_3 < \lambda_2 < \lambda_1 < 1,$$

and denote f_n the eigenfunction corresponding to λ_n, $n \in \mathbb{N}_+$. In what follows, when no comparison between different eigenvalues/eigenfunctions is made, we will continue writing simply f, λ instead of f_n, λ_n.

Finally, observe that a scaling argument (with a change of variable; see further) implies that the spectrum actually depends only on one parameter $\beta := h/a$. The main results will be formulated in terms of the magnitude of this parameter.

6.3 Small Interval ($\beta \gg 1$)

Setting $\phi(x) := f(ax)$ for $x \in (-1, 1)$, Eq. (6.1) rewrites as

$$\frac{\beta}{\pi} \int_{-1}^{1} \frac{\phi(t)}{(x-t)^2 + \beta^2} dt = \lambda \phi(x), \quad x \in (-1, 1), \tag{6.5}$$

Since eigenfunctions are defined up to a multiplicatory constant, for the sake of determinicity, let us choose this constant to be real and so that $\|\phi\|_{L^2(-1,1)} = 1$.

Observe that the kernel function essentially coincides with [0/2] Padé approximant of hyperbolic secant function

$$\operatorname{sech}(x) = \frac{1}{1 + x^2/2} + \mathscr{O}\left(x^4\right), \quad |x| \ll 1,$$

hence the formulation (6.5) can be approximated by

$$\int_{-1}^{1} \operatorname{sech}\left((x-t)\sqrt{2}/\beta\right) \phi(t)\, dt = \pi\beta\lambda\phi(x) + \mathscr{O}\left(1/\beta^4\right), \quad x \in (-1, 1),$$
$$\tag{6.6}$$

and we therefore expect its solutions to be close to those of (6.5) for large β.

We drop the error term, postponing precise approximation error analysis to a further paper, and now focus on an eigenvalue problem for the integral operator on the left of (6.6), which turns out to be again a positive compact self-adjoint operator on $L^2(-1, 1)$ with a simple spectrum and the same law of decay of eigenvalues (6.4). However, this seemingly more complicated integral operator has an advantage over the original one since it belongs to a rather unique family of convolution integral operators that admit a commuting differential operator [Gr83, Wi64]: eigenfunctions of an integral operator with the kernel $\frac{b \sin cx}{c \sinh bx}$ (with constants $b, c \in \mathbb{R} \cup i\mathbb{R}$) are also eigenfunctions of the differential operator $-\frac{d}{dx}\left(1 - \frac{\sinh^2(bx)}{\sinh^2 b}\right)\frac{d}{dx} + (b^2 + c^2)\frac{\sinh^2(bx)}{\sinh^2 b}$ with condition of finiteness at $x = \pm 1$. Therefore, taking $c = i\sqrt{2}/\beta$, $b = 2\sqrt{2}/\beta$, and denoting $\dfrac{\mu}{2\sinh^2\left(2\sqrt{2}/\beta\right)}$ an eigenvalue of the differential operator, we reduce (6.6) to solving a boundary-value problem for ODE, for $x \in (-1, 1)$,

$$\left(\left(\cosh\frac{4\sqrt{2}}{\beta} - \cosh\frac{4\sqrt{2}x}{\beta}\right)\phi'(x)\right)' + \left(\mu - \frac{6}{\beta^2}\left(\cosh\frac{4\sqrt{2}x}{\beta} - 1\right)\right)\phi(x) = 0$$

(6.7)

with boundary conditions

$$\phi'(\pm 1) = \mp\frac{\mu + 6/\beta^2\left(1 - \cosh\left(4\sqrt{2}/\beta\right)\right)}{4\sqrt{2}\beta\sinh\left(4\sqrt{2}/\beta\right)}\phi(\pm 1).$$

(6.8)

Alternatively, introducing

$$\psi(s) := \frac{\phi\left(\frac{\beta}{2\sqrt{2}}\log\left[\left(e^{-2\sqrt{2}/\beta} - e^{2\sqrt{2}/\beta}\right)s + e^{-2\sqrt{2}/\beta}\right]\right)}{\left[\left(e^{-2\sqrt{2}/\beta} - e^{2\sqrt{2}/\beta}\right)s + e^{-2\sqrt{2}/\beta}\right]^{1/2}},$$

(6.9)

Equation (6.6) can be brought into a simpler integral equation arising in the context of singular-value analysis of the finite Laplace transform [BeGt85]

$$\int_0^1 \frac{\psi(t)}{s + t + \gamma}dt = -\pi\sqrt{2}\lambda\psi(s), \quad s \in (0, 1),$$

(6.10)

with $\gamma := 2e^{-2\sqrt{2}/\beta}$. The operator in the left-hand side of (6.10) is a truncated Stieltjes transform which again, by commutation with a differential operator, can be reduced to solving an ODE, for $s \in (0, 1)$,

$$\left(s\left(1-s\right)\left(\gamma+s\right)\left(\gamma+1+s\right)\psi'\left(s\right)\right)' - \left(2s\left(s+\gamma\right)+\mu\right)\psi\left(s\right) = 0 \qquad (6.11)$$

with boundary conditions enforcing regularity of solutions at the endpoints

$$\psi'\left(0\right) = \frac{\mu}{\gamma\left(\gamma+1\right)}\psi\left(0\right), \quad \psi'\left(1\right) = -\frac{2\left(\gamma+1\right)+\mu}{\left(\gamma+1\right)\left(\gamma+2\right)}\psi\left(1\right). \qquad (6.12)$$

Finally, it is remarkable that if we get back to (6.7) and Taylor-expand hyperbolic cosine functions due to smallness of $1/\beta$, we obtain

$$\left(\left(1-x^2\right)\phi'\left(x\right)\right)' + \left(\mu - \frac{6}{\beta^2}x^2\right)\phi\left(x\right) = 0, \quad x \in \left(-1, 1\right), \qquad (6.13)$$

an ODE that coincides with the well-studied equation [OsEtAl13, SlPo61] whose solutions are bounded on $[-1, 1]$ only for special values $\mu_n = \chi_n\left(\frac{\sqrt{6}}{\beta}\right)$, $n = \mathbb{N}_0$, and termed as prolate spheroidal (Slepian) wave functions $S_{0n}\left(\frac{\sqrt{6}}{\beta}, x\right)$ (with notation as in [SlPo61]).

Note that even though differential operators presented here have the same eigenfunctions as integral ones, eigenvalues are different. Once an eigenfunction ϕ_n is obtained, the corresponding eigenvalue of the original integral operator can be computed as $\lambda_n = \left\langle P_\beta\left[\chi_{(-1,1)}\phi_n\right], \phi_n\right\rangle_{L^2(-1,1)}$.

6.4 Large Interval ($\beta \ll 1$)

Let us set $\varphi\left(x\right) := f\left(xh\right)$ for $x \in \left(-a/h, a/h\right)$ and, by a change of variable, rewrite (6.1) as

$$\frac{1}{\pi}\int_{-1/\beta}^{1/\beta}\frac{\varphi\left(t\right)}{\left(x-t\right)^2+1}dt = \lambda\varphi\left(x\right), \quad x \in \left(-1/\beta, 1/\beta\right), \qquad (6.14)$$

Denote $B := \left(-1/\beta, 1/\beta\right)$, choose normalization $\|\varphi\|_{L^2(B)} = 1$, and define the analytic continuation to $\mathbb{R}$ of the solution of (6.14) as

$$\varphi\left(x\right) = \frac{1}{\lambda\pi}\int_{-1/\beta}^{1/\beta}\frac{\varphi\left(t\right)}{\left(x-t\right)^2+1}dt. \qquad (6.15)$$

Then, building up on a transformation introduced in [Gr64], we can prove a non-evident yet very important result [Po16]

Lemma 1 *The analytic continuation of solution of (6.14) given by (6.15) satisfies*

$$\int_{\mathbb{R}\backslash B} R_0\left(x-t\right)\varphi\left(t\right)dt = \varphi\left(x\right), \quad x \in \mathbb{R}, \qquad (6.16)$$

with

$$R_0(x) := -\frac{\sin(x \log \lambda)}{\tanh(\pi x)} - \frac{1}{\pi} \sum_{n=1}^{\infty} \frac{n\lambda^n}{n^2 + x^2}. \tag{6.17}$$

The parity of solutions (part (b) of Proposition 1) allows reducing an integration to only one half-line.

Theorem 2 *The analytic continuations $\varphi_{ext}(x) := \varphi(x + 1/\beta)$ of even/odd solutions of (6.14) satisfy, for $x \in \mathbb{R}$,*

$$\int_0^\infty \left[R_0(x - t) \pm R_0\left(x + t + \frac{2}{\beta}\right) \right] \varphi_{ext}(t)\, dt = \varphi_{ext}(x), \tag{6.18}$$

as well as an integro-differential equation

$$\int_0^\infty \left[\mathscr{K}(x - t) \pm \mathscr{K}\left(x + t + \frac{2}{\beta}\right) \right] \varphi_{ext}(t)\, dt = \varphi_{ext}''(x) + \log^2 \lambda\, \varphi_{ext}(x) \tag{6.19}$$

with the kernel function

$$\mathscr{K}(x) := -\left(\frac{d^2}{dx^2} + \log^2 \lambda \right) \left(\frac{\sin(x \log \lambda)}{\tanh(\pi x)} + \frac{1}{\pi} \sum_{n=1}^{\infty} \frac{n\lambda^n}{n^2 + x^2} \right). \tag{6.20}$$

Here and onwards the upper sign corresponds to even solutions, the lower to odd ones.

We note that the first equation, which is a direct rephrasing of Lemma 1, even though simpler than the integro-differential equation, has a kernel (6.17) with an oscillatory behavior at infinity whereas $\mathscr{K}(x)$ decays. Indeed, it is easy to see that

$$R_0(x) \underset{|x| \gg 1}{\approx} \frac{\sin(x \log \lambda)}{\tanh(\pi x)} \approx \sin(|x| \log \lambda), \qquad \mathscr{K}(x) \underset{|x| \gg 1}{=} \mathscr{O}\left(\frac{1}{x^2} \right).$$

This decaying property of the kernel function of (6.19) is crucial for construction of approximation on the right half-line region since the sum part of the kernel in (6.19) is uniformly small for $\beta \ll 1$ and $x, t > 0$: $\mathscr{K}(x + t + 2/\beta) = \mathscr{O}(\beta^2)$. Neglecting this small term (and thus again postponing tedious error analysis to a further paper), we end up with an equation of Wiener-Hopf type. Even though the presence of the derivative prohibits application of a standard Wiener-Hopf method, this difficulty can still be overcome by means of additional transformation leading to an explicitly solvable scalar Riemann-Hilbert problem giving thus an exact solution of the approximate equation. These results presented in greater detail in [Po16] are summarized here in Theorem 3 below. First of all, however, we should set up notations and define few auxiliary quantities, for $k \in \mathbb{R}$,

$$k_0 := -\frac{\log \lambda}{2\pi}, \quad \kappa := -\frac{\pi}{6} + 2k_0 \log \left(e^{2\pi k_0} - 1\right) + \frac{1}{\pi} \mathrm{Li}_2 \left(1 - e^{2\pi k_0}\right),$$

$$\hat{\mathscr{K}}(k) = \frac{2\pi^2 \left(k_0^2 - k^2\right) e^{\pi (k_0 - |k|)}}{\sinh (\pi (k_0 - |k|))},$$

$$G(k) := \frac{k^2 - k_0^2}{2 \left(k^2 + 1\right)} \left[1 + \coth (\pi (|k| - k_0))\right],$$

$$X_\pm(k) := \exp \left(P_\pm \left[\log G\right](k)\right) = G^{1/2}(k) \exp \left[\pm \frac{1}{2\pi i} \mathrm{p.v.} \int_{\mathbb{R}} \frac{\log G(\tau)}{\tau - k} d\tau\right],$$

$$\mathscr{C}(k) := \frac{(1 + \kappa) \left(1 + 4\pi^2 k_0^2\right)}{(1 - 2\pi i k)^2} - \frac{1 - 4\pi^2 k_0^2 + 2\kappa}{1 - 2\pi i k}$$

$$- P_+ \left[\frac{2 (1 - \pi i \cdot) + \kappa}{(1 - 2\pi i \cdot)^2} \hat{\mathscr{K}}(\cdot)\right](k),$$

where we defined the Euler dilogarithm/Spence's function, Fourier transform, and projection operators on spaces of analytic functions of upper and lower half-planes as follows:

$$\mathrm{Li}_2(x) := -\int_0^x \frac{\log (1 - t)}{t} dt = \sum_{n=1}^{\infty} \frac{x^n}{n^2},$$

$$\hat{F}(k) := \mathscr{F}[F](x) = \int_{\mathbb{R}} F(x) e^{2\pi i k x} dx,$$

$$P_\pm[F](k) := \mathscr{F} \chi_{\mathbb{R}_\pm} \mathscr{F}^{-1}[F](k) = \frac{1}{2} F(k) \pm \frac{1}{2\pi i} \mathrm{p.v.} \int_{\mathbb{R}} \frac{F(t)}{t - k} dt.$$

Now we are ready to state the following

Theorem 3 *The integro-differential equation*

$$\int_0^{\infty} \mathscr{K}(x - t) \varphi_{ext}(t) dt = \varphi_{ext}''(x) + \log^2 \lambda \, \varphi_{ext}(x), \qquad x > 0, \qquad (6.21)$$

possesses the unique solution given by

$$\varphi_{ext}(x) = \varphi \left(\frac{1}{\beta}\right) \left[e^{-x} (1 + (1 + \kappa) x) + \int_{\mathbb{R}} e^{-2\pi i k x} \frac{P_+ [\mathscr{C} / X_-](k)}{4\pi^2 (k^2 + 1) X_+(k)} dk\right].$$

$$(6.22)$$

Moreover, this solution satisfies the endpoint condition $\varphi_{ext}'(0) = \kappa \varphi_{ext}(0)$.

Now we reuse Theorem 2 to recover solutions $\varphi(x) = \varphi_{ext}(x - 1/\beta)$ inside the interval B due to the fact that the left-hand side of (6.19) is now computable from (6.22). This non-homogeneous ODE is easily solvable and depending on a choice of the sign in the integral term of (6.19) we obtain either even or odd family of solutions.

We conclude that even eigenfunctions are given by

$$\varphi(x)/\varphi\left(\frac{1}{\beta}\right) = C_1(\lambda, \beta)\cos(x \log \lambda) - \int_0^x N_0^+(t, \lambda, \beta)\sin((x-t)\log\lambda)\,dt,$$

(6.23)

and odd ones by

$$\varphi(x)/\varphi\left(\frac{1}{\beta}\right) = C_2(\lambda, \beta)\sin(x \log \lambda) - \int_0^x N_0^-(t, \lambda, \beta)\sin((x-t)\log\lambda)\,dt,$$

(6.24)

where

$$C_1(\lambda, \beta) := \frac{1}{\cos\left(\frac{1}{\beta}\log\lambda\right)}\left[1 + \int_0^{\frac{1}{\beta}} N_0^+(t, \lambda, \beta)\sin\left(\left(\frac{1}{\beta}-t\right)\log\lambda\right)dt\right],$$

$$C_2(\lambda, \beta) := -\frac{1}{\sin\left(\frac{1}{\beta}\log\lambda\right)}\left[1 + \int_0^{\frac{1}{\beta}} N_0^-(t, \lambda, \beta)\sin\left(\left(\frac{1}{\beta}-t\right)\log\lambda\right)dt\right],$$

$$N_0^\pm(x, \lambda, \beta) := \frac{1}{2\pi k_0}\int_0^\infty \left(\mathscr{K}\left(x-t-\frac{1}{\beta}\right) \pm \mathscr{K}\left(x+t+\frac{1}{\beta}\right)\right)\cdot$$
$$\left[e^{-t}(1+(1+\kappa)t) + \int_\mathbb{R} e^{-2\pi i k t}\frac{P_+[\mathscr{C}/X_-](k)}{4\pi^2(k^2+1)X_+(k)}dk\right]dt.$$

Evaluation of derivatives and use of the boundary condition obtained in Theorem 3 lead to characteristic equations for even and odd eigenvalues, respectively,

$$\frac{\kappa}{\log\lambda}\cos\left(\frac{1}{\beta}\log\lambda\right) + \sin\left(\frac{1}{\beta}\log\lambda\right) = -\int_0^{\frac{1}{\beta}} N_0^+(t, \lambda, \beta)\cos(t\log\lambda)\,dt,$$

(6.25)

$$\frac{\kappa}{\log\lambda}\sin\left(\frac{1}{\beta}\log\lambda\right) - \cos\left(\frac{1}{\beta}\log\lambda\right) = -\int_0^{\frac{1}{\beta}} N_0^-(t, \lambda, \beta)\sin(t\log\lambda)\,dt.$$

(6.26)

6.5 Numerical Illustrations

We verify our results of both Sects. 6.3 and 6.4 by comparing them to a numerical (Nyström) method applied to a rescaled formulation (6.5). We use $N = 100$ points Gauss-Legendre quadrature rule to approximate the integral operator

$$\sum_{j=1}^{N} \omega_j p_\beta \left(x - t_j \right) \phi_j = \lambda \phi \left(x \right), \quad x \in (-1, 1) \tag{6.27}$$

with $\omega_j := \dfrac{2\left(1 - t_j^2\right)}{N^2 P_{N-1}^2(t_j)}$, P_{N-1} being a $(N-1)$-th Legendre polynomial, and solve for $\phi_j := \phi \left(t_j \right)$, $j = 1, \ldots, N$, the following linear system

$$\sum_{j=1}^{N} p_\beta \left(t_i - t_j \right) \omega_j \phi_j = \lambda \phi_i, \quad i = 1, \ldots, N. \tag{6.28}$$

Eigenvalues are found from equating determinant of the system to zero, and continuous eigenfunctions are then reconstructed from (6.27) as

$$\phi \left(x \right) = \frac{1}{\lambda} \sum_{j=1}^{N} \omega_j p_\beta \left(x - t_j \right) \phi_j, \quad x \in (-1, 1). \tag{6.29}$$

Numerical solutions demonstrate properties of a Sturm-Liouville sequence: even and odd eigenfunctions interlace and each ϕ_n, $n \in \mathbb{N}_+$, has exactly $n - 1$ zeros.

In the case $\beta \gg 1$, we compare numerical results with prolate spheroidal wave functions which were computed using a Fortran code provided in [ZhJi96] and converted into a MATLAB program with the software f2matlab. We see in Fig. 6.1 that even double approximation (first, by a cumbersome boundary-value

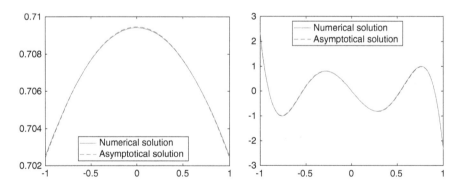

Fig. 6.1 Eigenfunctions ϕ_1 (left plot) and ϕ_6 (right plot). $\beta = 10$

problem (6.7) and (6.8) and then, proceeding further, by the one with ODE (6.13) for standard special functions) already furnishes excellent results.

In the case $\beta \ll 1$, we first solve characteristic Eqs. (6.25) and (6.26) by finding intersection of curves in left- and right-hand sides as function of $k_0 = -\frac{\log \lambda}{2\pi}$. They are plotted in Fig. 6.2 along with vertical lines which correspond to eigenvalues obtained from the numerical solution described above. Plugging found eigenvalues back in (6.23) and (6.24), we obtain even and odd family of solutions, respectively. We plot a couple of eigenfunctions in Fig. 6.3, namely, the third even eigenfunction and the tenth odd. As in Fig. 6.1, asymptotic solutions are almost indistinguishable from the numerical, however, Fig. 6.4 shows a breakdown of the asymptotic approximation for higher-order eigenfunctions (note also the discrepancies between abscisses of circled intersection points and vertical lines in Fig. 6.2).

More plots of eigenfunctions and approximation errors are available in [Po16].

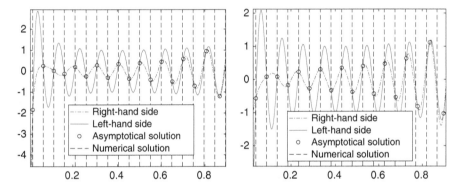

Fig. 6.2 Solving characteristic Eqs. (6.25) (left plot) and (6.26) (right plot). $\beta = 0.1$

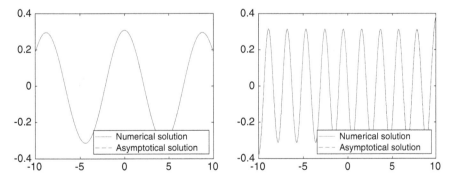

Fig. 6.3 Eigenfunctions φ_5 (left plot) and φ_{20} (right plot). $\beta = 0.1$

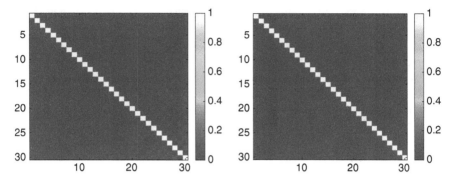

Fig. 6.4 Inner product matrices for solutions: asymptotic (left plot) and numerical (right plot). $\beta = 0.1$

6.6 Conclusion

We have presented two different methods to construct asymptotic solutions in cases when the interval is small and large. In the first case we have exploited a rather specific property of asymptotical closedness of the problem to an integral equation with an admissible commuting differential operator and concluded that solutions (eigenfunctions) can be approximated by those coming from either of two auxiliary Sturm-Liouville problems and, if further approximation is pursued, they coincide with scaled versions of prolate spheroidal wave functions. In the second case, when the interval is large, the developed approach is rather general and should, in principle, be applicable to a wide class of integral equations with even kernels. Computational details (and simplicity of the form of the kernel for the integral equation on the half-line), however, will depend on analytical structure of the Fourier transform of the kernel. This is a natural topic for further investigation. Also, in the case of the large interval, it is interesting to attempt to extend the results for $\lambda = -1$ (and a non-homogeneous term) recently obtained by Tracy and Widom [TrWi16b, TrWi18] or those given by a boundary-layer type of asymptotic constructions in [AtLe83], and compare these results with ours. Moreover, in the same large interval case, it was proven in [Po16] that Eq. (6.1) can be approximately reduced to a known non-homogeneous hypersingular equation known in air-foil theory p.v. $\int_{-a}^{a} \frac{f'(t)}{x-t} dt = \mu f(x) + g(x)$ which so far has been efficiently solved only numerically [KaPo50, Tr69]. It seems worthy exploring this connection deeper on a constructive level. Nevertheless, of the primary importance is to provide rigorous justification of the obtained results (initiated in [Po16]) which were presented here heuristically and verified only numerically. This work in progress will soon be published in a forthcoming paper.

Acknowledgements D. Ponomarev is grateful to Austrian Science Fund for its current support (FWF project I3538-N32)

References

[AtLe83] Atkinson, C., Leppington, F. G.: The asymptotic solution of some integral equations. *IMA J. App. Math.*, **31 (3)**, 169–182 (1983).

[BeGt85] Bertero, M., Grunbaum, F. A.: Commuting differential operators for the finite Laplace transform. *Inverse Problems*, **1**, 181–192 (1985).

[BaEtAl13] Baratchart, L., Hardin, D. P., Lima, E. A., Saff, E. B., Weiss, B. P.: Characterizing kernels of operators related to thin-plate magnetizations via generalizations of Hodge decompositions. *Inverse Problems*, **29 (1)**, 29 pp. (2013).

[Ga71] Gaudin, M.: Boundary energy of a Bose gas in one dimension. *Phys. Rev. A*, **4 (1)**, 386–394 (1971).

[Gr64] Griffiths, R. B.: Magnetization curve at zero temperature for the antiferromagnetic Heisenberg linear chain. *Phys. Rev.*, **133 (3A)**, A768–A775 (1964).

[Gr83] Grunbaum, F. A.: Differential operators commuting with convolution integral operators. *J. Math. Anal. Appl.*, **91**, 80–93 (1983).

[Hu64a] Hutson, V.: The coaxial disc viscometer. *ZAMM*, **44**, 365–370 (1964).

[Hu64b] Hutson, V.: Asymptotic solutions of integral equations with convolution kernels. *Proc. Edinburgh Math. Soc.*, **14**, 5–19 (1964).

[KaPo50] Kac, M., Pollard, H.: The distribution of the maximum of partial sums of independent random variables. *Canad. J. Math.*, **2**, 375–384 (1950).

[KnKe91] Knessl, C., Keller, J. B.: Asymptotic properties of eigenvalues of integral equations. *SIAM J. Appl. Math.*, **51 (1)**, 214–232 (1991).

[LePo17] Leblond, J., Ponomarev, D.: On some extremal problems for analytic functions with constraints on real or imaginary parts. *Advances in Complex Analysis and Operator Theory*, 219–236 (2017).

[LeMu65] Leonard, A., Mullikin, T. W.: Integral equations with difference kernels on finite intervals. *Trans. Amer. Math. Soc.*, **116**, 465–473 (1965).

[LiLi63] Lieb, E. H., Liniger, W.: Exact analysis of an interacting Bose gas. I. The general solution and the ground state. *Phys. Rev.*, **130 (4)**, 1605–1616 (1963).

[Lo49] Love, E. R.: The electrostatic field of two equal circular co-axial conducting disks. *Quat. Journ. Mech. & Appl. Math.*, **2 (4)**, 428–451 (1949).

[NaSe00] Naylor, A. W., Sell, G. R.: Linear operator theory in engineering and science. Springer (2000).

[OsEtAl13] Osipov, A., Rokhlin, V., Xiao, H.: Prolate spheroidal wave functions of order zero. Springer (2013).

[Po16] Ponomarev, D.: On some inverse problems with partial data. *Doctoral thesis, Université Nice - Sophia Antipolis*, 167 pp. (2016).

[Pr17] Prolhac, S.: Ground state energy of the δ-Bose and Fermi gas at weak coupling from double extrapolation. *J. Phys. A: Math. Theor.*, **50**, 10 pp. (2017).

[SlPo61] Slepian, D., Pollack, H. O.: Prolate spheroidal wave functions, Fourier analysis and uncertainty - 1. *B. S. T. J.*, **40**, 43–64 (1961).

[Sn23] Snow, C.: Spectroradiometric analysis of radio signals. *Scientific papers of the Bureau of Standards*, **477**, 231–261 (1923).

[TrWi16a] Tracy, C. A., Widom, H.: On the ground state energy of the δ-function Bose gas. *J. Phys. A.: Math. Theor.*, **49**, 19 pp. (2016).

[TrWi16b] Tracy, C. A., Widom, H.: On the ground state energy of the δ-function Fermi gas. *J. Math. Phys.*, **57**, 14 pp. (2016).

[TrWi18] Tracy, C. A., Widom, H.: On the ground state energy of the δ-function Fermi gas II: Further asymptotics *Geom. Meth. in Phys. XXXV Workshop*, 201–212, (2018).

[Tr69] Trigt, C., van: Analytically solvable problems in radiative transfer - 1. *Phys. Rev.*, **181 (1)**, 97–114 (1969).

[Wi64] Widom, H.: Asymptotic behaviour of the eigenvalues of certain integral equations II. *Arch. Rat. Mech. Anal.*, **17**, 215–229 (1964).

[ZhJi96] Zhang, S., Jin, J.: Computations of special functions. Wiley-Interscience (1996).

Chapter 7
A Semi-analytical Solution for One-Dimensional Oil Displacement by Miscible Gas in a Homogeneous Porous Medium

Luana C. M. Cantagesso, Luara K. S. Sousa, Tamires A. Marotto, Anna M. Radovanovic, Adolfo Puime Pires, and Alvaro M. M. Peres

7.1 Introduction

Enhanced oil recovery (EOR) methods are characterized by the injection of different fluids into a hydrocarbon reservoir to increase the recovery. According to the main physical–chemical mechanism that governs the oil displacement, EOR techniques can be classified into three major categories: thermal (hot water flooding, steam drive, or in situ combustion), chemical (alkaline flooding, surfactant flooding, or micellar polymer flooding), or solvent (carbon dioxide, hydrocarbon, nitrogen, or natural gas injection) [La89].

Gas flooding is the second most applied EOR method to date. In recent years, CO_2 injection has become even more attractive for those projects that combine CO_2 sequestration and EOR objectives [Ko14]. Over the last decades, a considerable number of gas injection projects have been undertaken [Ma00, Mc95, Mi92, Or84, Sh02, Ta92, Va86]. In all of them, mass transfer between displacing and displaced phases strongly affects the efficiency of the process. Miscible methods decrease capillary and interfacial forces through mass transfer [Pi05].

We consider oil displacement by miscible gas injection at constant rate through a homogeneous porous medium. The mathematical model consists of a three-component, two-phase one-dimensional incompressible and isothermal flow. This fluid flow problem is governed by a system of two hyperbolic equations, which is solved by the method of characteristics for saturation and concentrations. The pressure profile is then obtained by integrating Darcy's law over the spatial domain.

The next section describes the mathematical model, followed by its solution for a typical set of rock and fluid properties illustrating a miscible gas flood calculation.

L. C. M. Cantagesso · L. K. S. Sousa · T. A. Marotto · A. M. Radovanovic
A. P. Pires (✉) · A. M. M. Peres
Universidade Estadual do Norte Fluminense, Macaé, RJ, Brazil

© Springer Nature Switzerland AG 2019   81
C. Constanda, P. Harris (eds.), *Integral Methods in Science and Engineering*,
https://doi.org/10.1007/978-3-030-16077-7_7

7.2   Physical and Mathematical Model

In this section, we present the physical and mathematical model for one-dimensional oil displacement by miscible gas injection at constant rate. The porous medium is homogeneous with uniform cross-sectional area, initially filled with a liquid phase. After the beginning of the gas injection, three different fluid regions may appear: a single-phase gas region beginning at the inlet point, followed by a two-phase region where mass transfer takes place, and a single-phase oil region up to the porous medium outlet. A schematic representation of these regions is pictured in Fig. 7.1.

The basic model assumptions are as follows:

- Isothermal fluid flow through a one-dimensional homogeneous porous medium.
- Incompressible rock and fluid system.
- Two-phase three-component flow.
- Instantaneous phase equilibrium.
- No adsorption, no chemical reactions.
- Gravity, dispersion, and capillary effects are negligible.
- Density of the component is independent of the phase.
- Amagat and Darcy laws are valid.

Under these assumptions, mass conservation for an n-component system is given by

$$\frac{\partial}{\partial t}\left(\phi \sum_{j=1}^{N_p} \rho_j s_j w_{ij}\right) + \frac{\partial}{\partial x}\left(\sum_{j=1}^{N_p} \rho_j w_{ij} u_j\right) = 0, \tag{7.1}$$

where ϕ is the porosity, ρ_j is the density of phase j, w_{ij} is the mass fraction of component i in phase j, s_j is the saturation of phase j, u_j is the velocity of phase j, t is time, x is the spatial variable, and N_p is the number of phases.

If the pure component density is independent of the phase, we can rewrite the mass conservation equation for component i in terms of the volume fraction in phase j by using the relationship

$$\rho_j w_{ij} = \rho_i c_{ij}, \tag{7.2}$$

where c_{ij} is the volume fraction of the component i in phase j and ρ_i is the pure component density at pressure (P) and temperature (T) of the system.

Fig. 7.1 Representation of the three regions

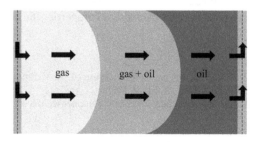

gas   gas + oil   oil

The macroscopic phase velocity can be expressed in terms of the fractional flow function (f_j) by

$$f_j = \frac{u_j}{u_T} \Leftrightarrow u_j = f_j u_T, \tag{7.3}$$

where the total velocity u_T is given by

$$u_T = -k \left[\sum_{j=1}^{N_p} \frac{k_{rj}(s_o)}{\mu_j(\vec{C})} \right] \frac{\partial p}{\partial x}, \tag{7.4}$$

in which p is the pressure, k denotes the absolute permeability, k_{rj} is the relative permeability of the phase j, μ_j is the viscosity of phase j, s_o is the oil saturation, and $\vec{C}$ is the concentration vector.

The term inside the brackets in Eq. (7.4) represents the total mobility (λ_T), so this equation can be written as

$$u_T = -\lambda_T(s_o, \vec{C}) k \frac{\partial p}{\partial x}. \tag{7.5}$$

The total concentration (C_i) and total flow (F_i) variables for component i are defined by

$$C_i = \sum_{j=1}^{N_p} s_j c_{ij} \tag{7.6}$$

and

$$F_i = \sum_{j=1}^{N_p} f_j c_{ij}. \tag{7.7}$$

The dimensionless time (t_D) and spatial coordinate (x_D) variables are defined by

$$x_D = \frac{x}{L}, \quad t_D = \frac{u_T t}{\phi L}, \tag{7.8}$$

where L is the porous medium length.

Using Amagat's law [PrEtAl86] and applying Eqs. (7.2), (7.3), (7.6), (7.7), and (7.8) in Eq. (7.1), we arrive at the following hyperbolic system of equations for a three-component system:

$$\begin{cases} \dfrac{\partial C_2}{\partial t_D} + \dfrac{\partial F_2}{\partial x_D} = 0 \\ \dfrac{\partial C_3}{\partial t_D} + \dfrac{\partial F_3}{\partial x_D} = 0. \end{cases} \tag{7.9}$$

The solution of the hyperbolic system (7.9) depends on the phase equilibrium conditions at system pressure and temperature, in this case, a ternary phase diagram [Pi05]. The lines connecting the bubble and dew points in a ternary diagram define the vapor and liquid phases composition at equilibrium. Those lines are known as tie lines [Or07], and can be parameterized by two thermodynamic geometric variables α and β [Be93], given by

$$\alpha = \frac{c_{2o} - c_{2g}}{c_{3o} - c_{3g}} \tag{7.10}$$

and

$$\beta = c_{2g} - \alpha c_{3g}. \tag{7.11}$$

The subscripts o and g denote the oil phase and the gas phase, respectively. The variable α represents the tie-line slope, and the β values represent the intercept of the tie line with the vertical axis. From a collection of phase diagram tie lines, a specific phase equilibrium relationship of α as a function of β can be constructed.

Using the above defined geometric variables, System (7.9) becomes

$$\begin{cases} \dfrac{\partial C}{\partial t_D} + \dfrac{\partial F}{\partial x_D} = 0 \\[2mm] \dfrac{\partial (\alpha C + \beta)}{\partial t_D} + \dfrac{\partial (\alpha F + \beta)}{\partial x_D} = 0, \end{cases} \tag{7.12}$$

where $C = C_3$ and $F = F_3$.

The initial and boundary conditions for this problem are

$$\begin{cases} C(x_D, t_D = 0) = C^{(I)}, & \beta(x_D, t_D = 0) = \beta^{(I)} \\ C(x_D = 0, t_D) = C^{(J)}, & \beta(x_D = 0, t_D) = \beta^{(J)}, \end{cases}$$

where $C^{(I)}$ and $\beta^{(I)}$ denote, respectively, the total concentration of the third component and the tie-line intercept of fluid that initially saturates the porous media. The third component total concentration and the tie-line intercept of the injected gas at the inlet are represented by $C^{(J)}$ and $\beta^{(J)}$, respectively.

The hyperbolic system can be recast in conservative form as

$$u_{t_D} + A u_{x_D} = 0, \tag{7.13}$$

where the vector-column u and the 2×2 matrix A are given, respectively, by

$$u = \begin{pmatrix} C \\ \beta \end{pmatrix}$$

and

$$A = \begin{bmatrix} \dfrac{\partial F}{\partial C} & \dfrac{\partial F}{\partial \beta} \\ & F\dfrac{\partial \alpha}{\partial \beta} + 1 \\ 0 & \dfrac{}{C\dfrac{\partial \alpha}{\partial \beta} + 1} \end{bmatrix}.$$

In Eq. (7.13), subscripts t_D and x_D denote partial derivatives taken with respect to dimensionless time and dimensionless distance.

The main diagonal elements of the upper-triangular matrix A are the eigenvalues of the hyperbolic system, with eigenpairs given by

$$\lambda^{(I)} = \frac{\partial F}{\partial C}, \quad r^{(I)} = \begin{pmatrix} 1 \\ 0 \end{pmatrix}$$

and

$$\lambda^{(II)} = \frac{F\dfrac{\partial \alpha}{\partial \beta} + 1}{C\dfrac{\partial \alpha}{\partial \beta} + 1}, \quad r^{(II)} = \begin{pmatrix} -\dfrac{\partial F}{\partial \beta} \\ \dfrac{\partial F}{\partial C} - \dfrac{F\dfrac{\partial \alpha}{\partial \beta} + 1}{C\dfrac{\partial \alpha}{\partial \beta} + 1} \end{pmatrix}.$$

The Riemann invariants of the problem are

$$R^{(I)} = F - \int \left(\frac{F\dfrac{\partial \alpha}{\partial \beta} + 1}{C\dfrac{\partial \alpha}{\partial \beta} + 1} - \frac{\partial F}{\partial C} \right) dC, \quad R^{(II)} = \beta.$$

The rarefaction waves calculated from the right eigenvectors are

$$\beta = R^{(II)}$$

and

$$F = \int \left(\frac{F\dfrac{\partial \alpha}{\partial \beta} + 1}{C\dfrac{\partial \alpha}{\partial \beta} + 1} - \frac{\partial F}{\partial C} \right) dC + R^{(I)}.$$

The total concentration C changes, whereas the tie-line geometric variable β remains constant along the first rarefaction. In the second rarefaction family, both β and C vary.

The shock wave speeds are obtained from the Rankine–Hugoniot conditions; they are

$$D^{(I)} = \frac{[F]}{[C]}$$

and

$$D^{(II)} = \frac{F^+ + \dfrac{[\beta]}{[C]}}{C^+ + \dfrac{[\beta]}{[C]}} = \frac{F^- + \dfrac{[\beta]}{[C]}}{C^- + \dfrac{[\beta]}{[C]}},$$

where $[A] = A^+ - A^-$ is the jump between the right and left conditions. Thus, $[A]$ could represent the jump of C, F, or β variables.

The solution of the hyperbolic system for a constant injected and original fluid composition together with the fractional flow curves yields the saturation and phase composition profile along the porous medium. From this solution, the total mobility λ_T spatial profile can be calculated for any dimensionless time. Then the pressure drop across the porous medium can be obtained by a straightforward integration of Darcy's law over the spatial coordinate [Pe03].

For constant gas injection rate at the inlet ($x = 0$), the inner boundary condition in dimensional variables is

$$A\frac{kk_{rg}(s_{or})}{\mu_{g,i}^{(J)}}\frac{\partial p}{\partial x}\bigg|_{x=0} = -q_{g,sc}^{(J)}B_{g,i},$$

where A is the porous medium cross-sectional area, s_{or} is the residual oil saturation, $\mu_{g,i}^{(J)}$ and $B_{g,i}$ represent the gas viscosity and the gas formation volume factor, respectively, both evaluated for constant injection fluid composition at the system initial pressure. The term $q_{g,sc}^{(J)}$ is a constant volumetric gas injection rate at standard conditions.

At the outlet ($x = L$), fluid flows out of the porous medium under constant pressure (initial system pressure (p_i)). Thus, the external boundary condition is

$$p(x = L, t) = p_i.$$

The dimensionless pressure variable is defined by

$$p_D = \frac{kk_{rg}(s_{or})A}{q_{g,sc}^{(J)}B_{g,i}\mu_{g,i}^{(J)}L}\Delta p.$$

Using the dimensionless variables definitions, Eq. (7.5) becomes

$$\frac{u_T(x,t)A}{q_{g,sc}^{(J)} B_{g,i}(x_D,t_D)} = -\lambda_{TD}(s_o,\vec{C})\frac{\partial p_D(x_D,t_D)}{\partial x_D}, \tag{7.14}$$

where the dimensionless total mobility is defined by

$$\lambda_{TD}(s_o,\vec{C}) = \left(\frac{k_{ro}(s_o)}{\mu_o(\vec{C})} + \frac{k_{rg}(s_g)}{\mu_g(\vec{C})}\right)\frac{\mu_{g,i}^{(J)}}{k_{rg}(s_{or})}.$$

The dimensionless inner and external boundary conditions are

$$\left.\frac{\partial p_D}{\partial x_D}\right|_{x_D=0} = -1$$

and

$$p_D(x_D = 1, t_D) = 0. \tag{7.15}$$

In Eq. (7.14), the product $u_t(x,t)A$ represents the total volumetric flow rate at (x,t). As the fluids are considered incompressible and the injection rate is constant, the total volumetric flow rate along the porous medium is also a constant at any time. Thus, we must have $u_t(x,t)A = q_{g,sc}^{(J)} B_{g,i}$, and Eq. (7.14) becomes

$$\frac{\partial p_D(x_D,t_D)}{\partial x_D} = -\frac{1}{\lambda_{TD}(x_D,t_D)}.$$

Integrating the above equation from a given position x_D to the outlet $x_D = 1$ and using Eq. (7.15), we calculate the dimensionless pressure drop from x_D to $x_D = 1$ for any dimensionless time, that is,

$$p_D(x_D,t_D) = \int_{x_D}^{1} \frac{1}{\lambda_{TD}(x_D',t_D)} dx_D'. \tag{7.16}$$

7.3 Example

This section presents an application of the solution presented in the previous section for a specific set of rock and fluid properties. The complete fluid data used here can be found in Chap. 3, page 47 of Pedersen and Christensen's book [Pe06]. As our model has three components, it is necessary to lump the original fluid into three pseudo-components. Table 7.1 shows the lumping of the original fluid obtained by a genetic algorithm [Sc17]. The pseudo-components critical properties, acentric

Table 7.1 Component
lumping of the original fluid

Pseudo-components	Lumping
Pseudo 1	$N_2, CO_2, C_1 - C_3$
Pseudo 2	$iC_4 - C_{13}$
Pseudo 3	$C_{14} - C_{20+}$

Table 7.2 The pseudo-component critical properties, acentric factor, molar weight, and global composition

Pseudo-components	Pc (psia)	Tc (R)	w	M (g/mol)	Z (gmol/gmol)
Pseudo 1	6.6614E + 2	4.1659E + 2	4.4540E − 2	2.2116E + 1	0.5645
Pseudo 2	4.3762E + 2	9.8297E + 2	3.3989E − 1	1.0361E + 2	0.3093
Pseudo 3	1.9817E + 2	1.4952E + 3	1.0164E + 0	3.4442E + 2	0.1262

Table 7.3 The
pseudo-component
concentration for the initial
and injected fluid

	Pseudo 1	Pseudo 2	Pseudo 3
Injected	0.9899	0.0101	–
Initial	–	0.0982	0.9018

Fig. 7.2 Pseudo-fluid ternary
diagram showing the binodal
curve, the injected, and initial
tie lines

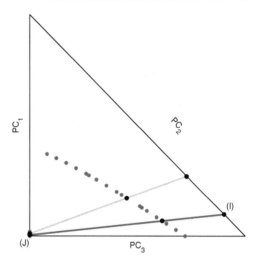

factor, molar mass, and global composition are presented in Table 7.2. The initial and injected concentration for each pseudo-component is shown in Table 7.3.

The thermodynamic equilibrium for the lumped fluid was calculated by means of the Peng–Robinson equation of state [PR76]. Figure 7.2 shows the binodal curve for 200 Bar and 338.75 K. The initial and injected fluids tie lines are indicated by the red and green curves, respectively. Several tie lines were calculated to obtain a set of geometric variables pairs (α, β) for this pseudo-fluid. A polynomial fit to these variable pair was then implemented:

$$\alpha = 4517.9\beta^2 - 7.1254\beta.$$

Fig. 7.3 Relative permeability curves

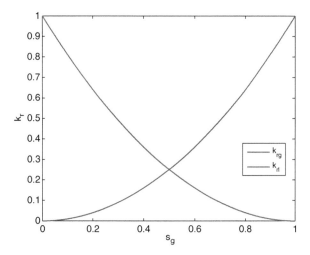

Table 7.4 Physical properties

	Symbol	Value	Unit
System pressure	P	200	Bar
System temperature	T	338.75	K
Oil viscosity (initial condition)	$\mu_o^{(I)}$	2.7497	Pa s
Gas viscosity (injection condition)	$\mu_g^{(J)}$	0.0366	Pa s
Gas residual saturation	S_{gr}	0	–
Oil residual saturation	S_{or}	0	–
End-point relative permeability of gas phase	k_{rg}^0	1	–
End-point relative permeability of oil phase	k_{ro}^0	1	–
Corey's exponent of gas phase	n_g	2	–
Corey's exponent of oil phase	n_o	2	–

Here, the relative permeability curves are given by Corey's model [CoEtAl56]:

$$k_{rj}(s_o) = k_{rj}^0 \left(\frac{s_j - s_{rj}}{1 - \sum\limits_{j=1}^{N_p} s_{rj}} \right)^{n_j},$$

where k_{rj} is the relative permeability of phase j, k_{rj}^0 is the end-point relative permeability of phase j, s_j is the saturation of phase j, s_{jr} is the residual saturation of phase j, n_j is Corey's exponent of phase j, and N_p is the number of phases.

Figure 7.3 shows the gas phase and the oil phase relative permeability curves obtained with Corey's parameters shown in Table 7.4.

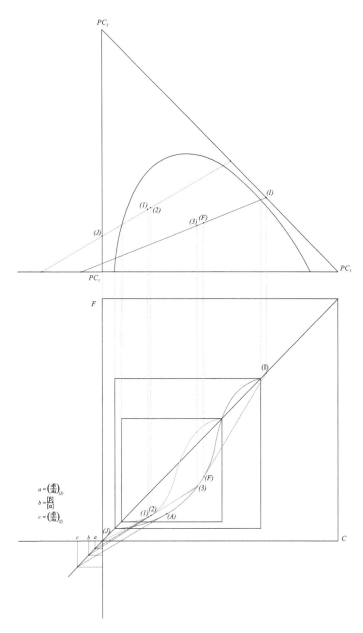

Fig. 7.4 Ternary diagram and solution path

System (7.12) was solved by the method of characteristics for saturation and concentration. The structure of the solution path (Fig. 7.4) is given by (J) → (1) − (2) → (3) − (F) → (I), where → denotes a shock wave and − indicates a rarefaction wave. The solution begins at injection conditions (J), which corresponds to single phase gas (region 1), connected to point (1) in the two-phase region (region 2) through a concentration shock. The transition from single phase to two-phase or from two-phase to single phase is always a concentration shock [Be93]. From point (1), there is a concentration rarefaction up to point (2). Next, there is a concentration and β shock linking points (2) and (3). From (3), there is another concentration rarefaction wave up to (F), which is connected to initial conditions (I) through a concentration shock. The solution is presented in Eq. (7.17) and the solution path in Fig. 7.4.

$$
\begin{matrix}
C(x_D, t_D) \\
\beta(x_D, t_D)
\end{matrix}
=
\begin{cases}
C^{(J)}, \beta^{(J)} & 0 < \frac{x_D}{t_D} < D^{(I)} = 7.71 \times 10^{-4} \\
C_{(1R)}, \beta^{(J)} & D^{(I)} < \frac{x_D}{t_D} < D^{(II)} = 0.13 \\
C_1 = 0.42, \beta^{(I)} & D^{(II)} < \frac{x_D}{t_D} < \frac{\partial F}{\partial C} = 1.80 \\
C_{(2R)}, \beta^{(I)} & \frac{\partial F}{\partial C} = 1.80 < \frac{x_D}{t_D} < D^{(I)} = 1.94 \\
C^{(I)}, \beta^{(I)} & D^{(I)} < \frac{x_D}{t_D} < +\infty,
\end{cases}
\tag{7.17}
$$

where $C_{(iR)}$ represents the concentration change along the i-th concentration rarefaction wave.

The miscible solution described is compared to an equivalent immiscible solution given (with the same porous medium and fluid properties) by

$$
s_g(x_D, t_D) =
\begin{cases}
s_g^{(J)} & \frac{x_D}{t_D} = 0 \\
s_{g(R)} & 0 < \frac{x_D}{t_D} < D_{BL} = 4.66 \\
s_g^{(I)} & D_{BL} < \frac{x_D}{t_D} < +\infty,
\end{cases}
$$

where $s_{g(R)}$ represents the gas saturation along the rarefaction wave and D_{BL} is the Buckley–Leverett shock [Bu42].

The gas saturation for both solutions is presented in Fig. 7.5. It is important to point out that the gas saturation for the miscible case is calculated with Eqs. (7.6), (7.7), (7.10), and (7.11). An important technical and economic factor of every EOR design project is the recovery factor, which is defined as the relationship between the accumulated produced oil and the total original volume of oil in the reservoir. Figure 7.6 shows that miscible displacement is clearly a much more efficient process than the immiscible one for the fluid and rock data considered in this sample calculation.

Equation (7.16) can be used to compute the dimensionless pressure at the injection point ($x_D = 0$) for any dimensionless time once the saturation and concentration profiles are solved. As the single-phase gas region 1 is very small and can be neglected, Eq. (7.16) is rewritten as

Fig. 7.5 Saturation profile comparison: miscible versus immiscible cases

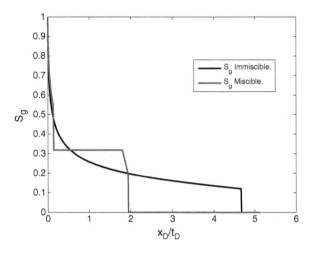

Fig. 7.6 Recovery factor comparison: miscible versus immiscible cases

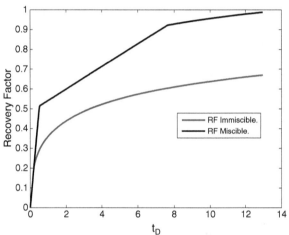

$$p_D(x_D = 0, t_D) = p_{wD}(t_D)$$

$$= \int_0^{D^{(II)}t_D} \frac{1}{\lambda_T^{(2a)}(x_D', t_D)} dx_D' + \frac{1}{\lambda_T^{(2b)}} \left(x_{DF} - D^{(II)}t_D \right)$$

$$+ \int_{x_{DF}}^{D^{(I)}t_D} \frac{1}{\lambda_T^{(2c)}(x_D', t_D)} dx_D' + \frac{1}{\lambda_T^{(3)}} \left(1 - D^{(I)}t_D \right),$$

$$\tag{7.18}$$

where $x_{DF} = \dfrac{\partial F}{\partial C} t_D$ and the superscript in the dimensionless total mobility variable denotes the saturation regions. Note that the integration is split according to the rarefaction and constant states. Furthermore, the integration along the

Fig. 7.7 Pressure behavior at the injection point: miscible versus immiscible displacement

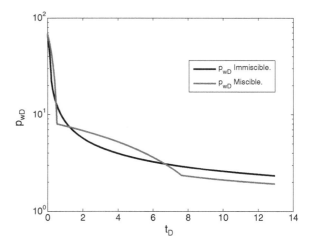

two-phase region 2 $\left(0 < x_D < D^{(I)}t_D\right)$ is subdivided into a rarefaction region 2a $\left(0 < x_D < D^{(II)}t_D\right)$, a constant state region 2b $\left(D^{(II)}t_D < x_D < x_{DF}\right)$, as well as a rarefaction region 2c $\left(x_{DF} < x_D < D^{(I)}t_D\right)$. The last term in Eq. (7.18) represents the pressure change due to the single-phase oil region 3. Equation (7.18) holds up to $t_D < \dfrac{1}{D^{(II)}}$, which is the time required to completely displace the single-phase oil region out the porous medium.

The pressure evolution with time at the inlet point is shown in Fig. 7.7. This plot shows that until the time when the injected gas reaches the outlet, there is a sharp decline in injection pressure because a low mobility oil (original porous medium liquid) is displaced by a higher mobility injected gas. After that time, the pressure curve flattens out as gas saturation increases at the porous medium exit point. Figure 7.7 shows that the pressure curve for an immiscible displacement has a similar behavior.

7.4 Summary and Conclusions

We have presented an analytical solution for one-dimensional oil displacement by miscible gas injection at constant rate for a three-component fluid. This problem is described by a system of two hyperbolic equations which is solved by the method of characteristics for saturation and concentrations. The solution path is composed of rarefaction and shock waves and constant states. The pressure profile is obtained by integrating Darcy's law over the spatial domain once the saturation, composition, and total mobility are calculated. The results for a specific set of rock and fluid properties are presented to illustrate this miscible gas flood calculation method. For this data set, the solution shows a two-phase region displacing the original single oil phase. The two-phase region consists of two rarefaction waves separated by a constant state. The solution is compared to the immiscible case, which presents, as expected, a much earlier breakthrough time than the miscible process, and, consequently, a lower recovery factor at all times. The injection pressure declines faster up to the breakthrough time as the displacing fluid has higher mobility than the displaced oil. The overall pressure behavior of the miscible displacement is similar to the pressure behavior shown by an immiscible displacement. The solution presented here can be used for screening a miscible method for a given petroleum field.

Acknowledgements The authors wish to express their gratitude for the financial support provided by the Brazilian Government Agencies CAPES and CNPq, by Petrobras SIGER Research Network, and by the Universidade Estadual do Norte Fluminense (UENF).

References

[Be93] Bedrikovetsky, P. G.: *Mathematical Theory of Oil and Gas Recovery*, Kluwer Academic Publishers, London (1993).

[Bu42] Buckley, S. E., and Leverett, M. C.: Mechanisms of fluid displacement in sands. *Amer. Inst. Min. Metall. Pet. Eng.*, **146**, 107–116 (1942).

[CoEtAl56] Corey, A. T., Rathjens, C. H., Henderson, J. H., and Wyllie, M. R. J.: Three-phase relative permeability. *J. Can. Pet. Technol.*, **8**, 63–65 (1956).

[Ko14] Koottungal, L.: Survey: Miscible CO_2 continues to eclipse steam in US EOR production. *Oil & Gas Journal*, **112.4**, 78–91 (2014).

[La89] Lake, W. L.: *Enhanced Oil Recovery*, Prentice-Hall, Englewood Cliffs, NJ (1989).

[Ma00] Malik, M. M., and Islam, M. R.: CO_2 Injection in the Weyburn Field of Canada: Optimization of Enhanced Oil Recovery and Greenhouse Gas storage with horizontal wells. *In SPE/DOE Improved Oil Recovery Symposium*, Tulsa, OK, SPE 59327 (2000).

[Mc95] McGuire, P. L., and Stalkup F.I.: Performance analysis and optimization of the Prudhoe Bay miscible-gas project. *SPE Reservoir Engineering*, **10**, 88–93, SPE 22398 (1995).

[Mi92] Mizenko, G. J.: North Cross (Devonian) Unit CO_2 Flood: Status Report. *In SPE/DOE Improved Oil Recovery Symposium*, Tulsa, OK, SPE 24210 (1992).

[Or84] Orr Jr., F. M., and Taber, J. J.: Use of carbon dioxide in Enhanced Oil Recovery. *Science*, **24**, 563–569 (1984).

[Or07] Orr Jr., F. M.: *Theory of Gas Injection Processes*, Tie-Line Publications, Copenhagen, Denmark (2007).

[Pe03] Peres, A. M. M., and Reynolds, A. C.: Theory and analysis of injectivity tests on horizontal wells. *SPE J.*, **8(2)**, 147–159, SPE 84957 (2003).

[Pe06] Pedersen, K. S., and Christensen, P. L.: *Phase Behavior of Petroleum Reservoir Fluids*, Taylor & Francis Group, Boca Raton, FL (2006).

[Pi05] Pires, A. P., and Bedrikovetsky, P. G.: Analytical modeling of 1D n-component miscible displacement of ideal fluids. *In SPE Latin American and Caribbean Petroleum Engineering*, Rio de Janeiro, Brazil, SPE 94855 (2005).

[PR76] Peng, D. Y., and Robinson, D. B.: A new two-constant equation of state. *Industrial & Engineering Chemistry Fundamentals*, **15**, 59–64 (1976).

[PrEtAl86] Prausnitz, J. M., Lichtenthaler, R. N., and Azevedo, E. G.: *Molecular Thermodynamics of Fluid-Phase Equilibria*, Prentice-Hall, Englewood Cliffs, NJ (1986).

[Sc17] Scardini, R. B.: Utilizacão de um algoritmo genético para agrupamento de componentes de petróleo condicionada a experimentos PVT. *M.Sc. Thesis*, Universidade Estadual do Norte Fluminense, Macaé (2017).

[Sh02] Shaw, J., and Bachu, S.: Screening, evaluation, and ranking of oil reservoirs suitable for CO_2-flood EOR and carbon dioxide sequestration. *Journal of Canadian Petroleum Technology*, **41**, 51–61 (2002).

[Ta92] Tanner, C. S., Baxley, P. T., Crump, J. G., and Miller, W. C.: Production performance of the Wasson Denver unit CO_2 flood. *In SPE/DOE Improved Oil Recovery Symposium*, Tulsa, OK, SPE 24156 (1992).

[Va86] Varotsis, N., Stewart G., Todd, A. C., and Clancy, M.: Phase behavior of systems comprising North Sea reservoir fluids and injection gases. *Journal of Petroleum Technology*, **41**, 1221–1233, SPE 12647 (1986).

Chapter 8
Bending of Elastic Plates: Generalized Fourier Series Method for the Robin Problem

Christian Constanda and Dale Doty

8.1 Introduction

Generalized Fourier series methods are based on the availability of a suitable complete set of functions in the space of the solution. Here, we design such a set for the so-called interior Robin problem, by means of the details generated by the application of the boundary integral equation method to the mathematical model. A desirable complete set is thus constructed, which permits us to compute highly accurate approximations. The method is problem specific and has already been used in the case of the interior Dirichlet [CoDo17a] and Neumann [CoDo17b] boundary value problems.

Let S be a finite domain in $\mathbb{R}^2$ bounded by a simple, closed, C^2-curve ∂S, let x and y be generic points in S or on ∂S, and let $h_0 = \text{const} > 0$, $h_0 \ll \text{diam } S$. The three-dimensional region $(S \cup \partial S) \times [-h_0/2, h_0/2]$ is assumed to be filled with a homogeneous and isotropic material of Lamé constants λ and μ.

In the absence of body forces, the equilibrium state in the process of bending of an elastic plate with transverse shear deformation is described by the system [Co14]

$$A(\partial_x)u(x) = 0, \tag{8.1}$$

where $A(\partial_x) = A(\partial_1, \partial_2)$ is the matrix

$$\begin{pmatrix} h^2\mu\Delta + h^2(\lambda + \mu)\partial_1^2 - \mu & h^2(\lambda + \mu)\partial_1\partial_2 & -\mu\partial_1 \\ h^2(\lambda + \mu)\partial_1\partial_2 & h^2\mu\Delta + h^2(\lambda + \mu)\partial_2^2 - \mu & -\mu\partial_2 \\ \mu\partial_1 & \mu\partial_2 & \mu\Delta \end{pmatrix},$$

C. Constanda (✉) · D. Doty
The University of Tulsa, Tulsa, OK, USA
e-mail: christian-constanda@utulsa.edu; dale-doty@utulsa.edu

© Springer Nature Switzerland AG 2019
C. Constanda, P. Harris (eds.), *Integral Methods in Science and Engineering*,
https://doi.org/10.1007/978-3-030-16077-7_8

$u = (u_1, u_2, u_3)^\mathsf{T}$ is a vector characterizing the displacements, $h = h_0/\sqrt{12}$, and Δ is the two-dimensional Laplace operator.

It is not difficult to see that the columns $f^{(i)}$ of the matrix

$$f = \begin{pmatrix} 1 & 0 & 0 \\ 0 & 1 & 0 \\ -x_1 & -x_2 & 1 \end{pmatrix}$$

form a basis for the space F of rigid displacements. $D(x, y) = A^*(\partial_x)t(x, y)$ is a matrix of fundamental solutions for the operator $-A$, A^* is the adjoint of A, $t(x, y)$ is a solution of the equation

$$\Delta\Delta\left(\Delta - \frac{1}{h^2}\right)t(x, y) = -\frac{1}{h^4\mu^2(\lambda + 2\mu)}\,\delta(|x - y|),$$

and δ is the Dirac delta distribution. In the analytic handling of the problem, an important role is also played by the matrix of singular solutions

$$P(x, y) = \big(T(\partial_y)D(y, x)\big)^\mathsf{T}, \tag{8.2}$$

where $T(\partial_x) = T(\partial_1, \partial_2)$ is the boundary moment–force operator defined by the matrix

$$\begin{pmatrix} h^2(\lambda + 2\mu)v_1\partial_1 + h^2\mu v_2\partial_2 & h^2\mu v_2\partial_1 + h^2\lambda v_1\partial_2 & 0 \\ h^2\lambda v_2\partial_1 + h^2\mu v_1\partial_2 & h^2\mu v_1\partial_1 + h^2(\lambda + 2\mu)v_2\partial_2 & 0 \\ \mu v_1 & \mu v_2 & \mu(v_1\partial_1 + n_2\partial_2) \end{pmatrix},$$

$\partial_\alpha = \partial/\partial x_\alpha$, $\alpha = 1, 2$, $v = (v_1, v_2)^\mathsf{T}$ is the unit vector of the outward normal to ∂S, and a superscript T denotes matrix transposition. We can easily convince ourselves that the columns $D^{(i)}(x, y)$ of $D(x, y)$ and $P^{(i)}(x, y)$ of $P(x, y)$ satisfy system (8.1) for $x \neq y$.

8.2 The Boundary Value Problem

In the Robin problem, we solve the system (8.1) in S with the boundary condition

$$(T + \sigma(x))u(x) = \mathcal{R}(x), \quad x \in \partial S, \tag{8.3}$$

where σ is a symmetric, positive definite 3×3 matrix and $\mathcal{R}$ is a 3×1 vector function prescribed on ∂S. In what follows, we denote S by S^+ and $\mathbb{R}^2 \setminus \bar{S}^+ = S^-$.

Theorem 1 ([Co14]) *Problem* (8.1), (8.3) *has a unique solution* $u \in C^2(S^+) \cap C^1(\bar{S}^+)$ *for any* $\mathcal{R} \in C^{0,\alpha}(\partial S)$, $\alpha \in (0, 1)$, *which admits the integral representation formula (Somigliana formula)*

$$u(x) = \int_{\partial S} [D(x, y)(Tu)(y) - P(x, y)u(y)] \, ds(y), \quad x \in S^+,$$

$$\tag{8.4}$$

$$0 = \int_{\partial S} [D(x, y)(Tu)(y) - P(x, y)u(y)] \, ds(y), \quad x \in S^-.$$

Let ∂S_* be a simple, closed, C^2–curve surrounding $S \cup \partial S$, let S_*^+ and S_*^- be the interior and exterior domains to ∂S_*, and select a collection of points $\{x^{(k)}, \ k = 1, 2, \ldots\}$ densely distributed on ∂S_*. We consider the set of 3×1 vector functions on ∂S

$$G = \{\sigma f^{(i)}, \ \theta^{(jk)}, \ i, j = 1, 2, 3, \ k = 1, 2, \ldots\},$$

where

$$\theta^{(jk)}(x) = (T + \sigma(x))D^{(j)}(x, x^{(k)}), \quad j = 1, 2, 3, \ k = 1, 2, \ldots \tag{8.5}$$

and $D^{(j)}$ are the columns of the matrix D.

Theorem 2 *The set G is linearly independent on ∂S and complete in $L^2(\partial S)$.*

Proof Assuming the opposite, let N be a positive integer and c_i and c_{jk}, $i, j = 1, 2, 3$, $k = 1, 2, \ldots, N$, real numbers, not all zero, such that

$$\sum_{i=1}^{3} c_i \sigma(x) f^{(i)}(x) + \sum_{j=1}^{3} \sum_{k=1}^{N} c_{jk} \theta^{(jk)}(x) = 0, \quad x \in \partial S. \tag{8.6}$$

By (8.5) and (8.6),

$$\omega(x) = \sum_{i=1}^{3} c_i f^{(i)}(x) + \sum_{j=1}^{3} \sum_{k=1}^{N} c_{jk} D^{(j)}(x, x^{(k)})$$

is a solution of the (interior) Robin problem

$$A\omega = 0 \quad \text{in } S^+,$$

$$(T + \sigma)\omega = 0 \quad \text{on } \partial S.$$

Since the solution of this problem is unique [Co14], we have $\omega = 0$ in S^+ and then, by analyticity,

$$\omega = 0 \quad \text{in } S_*^+. \tag{8.7}$$

Written in terms of components, for $x \in S_*^+$ we can write

$$\omega_l(x) = \sum_{i=1}^{3} c_i f_{li}(x) + \sum_{j=1}^{3}\sum_{k=1}^{N} c_{jk} D_{lj}(x, x^{(k)}), \quad l = 1, 2, 3. \tag{8.8}$$

Let $x^{(q)}$ be any of the points $x^{(1)}, \ldots, x^{(N)}$. As $x \to x^{(q)}$ from within S_*^+, all the terms on the right-hand side in (8.8) remain bounded except $c_{lq} D_{ll}(x, x^{(q)})$, which (see [Co14]) is of order $O(\ln |x - x^{(q)}|)$. This contradicts (8.7) unless all $c_{lq} = 0$ for $l = 1, 2, 3$ and $q = 1, \ldots, N$. Then

$$\omega = \sum_{i=1}^{3} c_i f^{(i)} = 0 \quad \text{in } S^+,$$

which, since the $f^{(i)}$ are linearly independent, implies that $c_i = 0$, $i = 1, 2, 3$. Consequently, the set G is linearly independent on ∂S.

Suppose now that for all $i, j = 1, 2, 3$ and $k = 1, 2, \ldots$, a vector function $\varphi \in L^2(\partial S)$ satisfies

$$\int_{\partial S} (\sigma f^{(i)})^{\mathsf{T}} \varphi \, ds = \int_{\partial S} (\theta^{(jk)})^{\mathsf{T}} \varphi \, ds = 0. \tag{8.9}$$

Since σ is symmetric, we have $(\sigma f^{(i)})^{\mathsf{T}} = (f^{(i)})^{\mathsf{T}}\sigma^{\mathsf{T}} = (f^{(i)})^{\mathsf{T}}\sigma$, so (8.9) can be rewritten as

$$p(\sigma\varphi) = 0, \tag{8.10}$$

$$\int_{\partial S} [(T + \sigma(x))D^{(j)}(x, x^{(k)})]^{\mathsf{T}} \varphi(x) \, ds = 0, \quad k = 1, 2, \ldots, \tag{8.11}$$

where p is the vector-valued functional defined on vector functions $g \in L^2(\partial S)$ by [Co14]

$$pg = \int_{\partial S} f^{\mathsf{T}} g \, ds.$$

To simplify the notation, unless otherwise stipulated, in what follows we adopt the convention of summation from 1 to 3 over repeated subscripts.

By (8.2), $P_{ji}(x, y) = [T(y)D(y, x)]_{ij} = T_{il}(y)D_{lj}(y, x) = [T(y)D^{(j)}(y, x)]_i$, so

$$\int\limits_{\partial S} [TD^{(j)}(x, x^{(k)})]^T \varphi \, ds = \int\limits_{ds} [TD^{(j)}(x, x^{(k)})]_i \varphi_i \, ds$$

$$= \int\limits_{\partial S} P_{ji}(x^{(k)}, x)\varphi_i \, ds = \left[\int\limits_{\partial S} P(x^{(k)}, x)\varphi \, ds\right]_j.$$

Since $D(x, y) = [D(y, x)]^T$ [Co14] and σ is symmetric, we see that

$$[\sigma(x)D^{(j)}(x, x^{(k)})]^T \varphi(x) = [\sigma_{il}(x)D_{lj}(x, x^{(k)})]\varphi_i(x)$$

$$= [D_{jl}(x^{(k)}, x)\sigma_{li}(x)]\varphi_i(x) = [D(x^{(k)}, x)\sigma(x)\varphi(x)]_j. \qquad (8.12)$$

For $x \in S^+ \cup S^-$, we consider the single-layer and double-layer potentials

$$V(\sigma\varphi)(x) = \int\limits_{\partial S} D(x, y)\sigma(y)\varphi(y) \, ds(y), \quad W\varphi(x) = \int\limits_{\partial S} P(x, y)\varphi(y) \, ds(y),$$

and set $U = V(\sigma\varphi) + W\varphi$.

In the sequel, for any function $F(x)$ defined in $S^+ \cup S^-$, we denote by $F^+(x)$ and $F^-(x)$ the limiting values of $F(x)$ as x approaches ∂S from within S^+ and S^-, respectively (if appropriate, along the support line of the normal vector to ∂S), and by $F_0(x)$ the direct value of F on ∂S, if it exists.

By (8.2) and (8.12), for $j = 1, 2, 3$ and $k = 1, 2, \ldots$,

$$0 = \int\limits_{\partial S} (\theta^{(jk)})^T \varphi \, ds = \int\limits_{\partial S} [(T + \sigma(x))D^{(j)}(x, x^{(k)})]^T \varphi(x) \, ds$$

$$= (W\varphi)_j(x^{(k)}) + [V(\sigma\varphi)]_j(x^{(k)}) = U_j(x^{(k)}).$$

From the continuity of V and W on ∂S_*, (8.9), and the fact that the points $x^{(k)}$ are densely distributed on ∂S_*, it follows that $U = 0$ on ∂S_*. Also, by (8.10), $U \in \mathscr{A}$, where $\mathscr{A}$ is the class of functions with a specific far-field pattern indicated in [Co14]. Since, in addition [Co14], $AV = AW = 0$ in $S^+ \cup S^-$, we see that U is the solution of the homogeneous exterior Dirichlet problem

$$AU = 0 \quad \text{in } S_*^-,$$

$$U = 0 \quad \text{on } \partial S_*,$$

$$U \in \mathscr{A},$$

so $U = 0$ in $\bar{S}^-_*$ and, by analyticity, $U = 0$ in S^-. This means that

$$U^- = 0, \quad (TU)^- = 0, \tag{8.13}$$

which, in turn, by Theorem 4.20 in [Co14], implies that

$$\tfrac{1}{2}\varphi(x) + \int_{\partial S} D(x, y)\sigma(y)\varphi(y)\,ds(y) + \int_{\partial S} P(x, y)\varphi(y)\,ds(y) = 0 \quad \text{for a.a. } x \in \partial S,$$

where the second integral is understood as principal value. Since $D(x, y)\sigma(y) + P(x, y)$ is an α-regular singular kernel [Co14], repeating the proof of Theorem 6.12 in [Co14], we conclude that $\varphi \in C^{0,\alpha}(\partial S)$, $\alpha \in (0, 1)$. Then all the properties of the potentials with Hölder continuous densities proved in Chapter 4 in [Co14] apply to U, and we have

$$(TU)^+ = (TV(\sigma\varphi))^+ + (TW\varphi)^+ = \tfrac{1}{2}\sigma\varphi + (TV(\sigma\varphi))_0 + (TW\varphi)^+$$

$$= \tfrac{1}{2}\sigma\varphi + \left[\tfrac{1}{2}\sigma\varphi + (TV(\sigma\varphi))^-\right] = (TU)^- + \sigma\varphi,$$

$$(\sigma U)^+ = \sigma[(V(\sigma\varphi))^+ + (W\varphi)^+] = \sigma\left[(V(\sigma\varphi))^+ - \tfrac{1}{2}\varphi + (W\varphi)_0\right]$$

$$= \sigma\left[(V(\sigma\varphi))^- - \tfrac{1}{2}\varphi + \left(-\tfrac{1}{2}\varphi + (W\varphi)^-\right)\right] = \sigma U^- - \sigma\varphi,$$

which yields $[(T + \sigma)U]^+ = [(T + \sigma)U]^-$ on ∂S. Therefore, by (8.13), U is a solution of the homogeneous interior Robin problem

$$AU = 0 \quad \text{in } S^+,$$

$$[(T + \sigma)U]^+ = 0 \quad \text{on } \partial S,$$

so $U = 0$ in S^+, which implies that $(TU)^+ = 0$. Since, by (8.13), we also have $(TU)^- = 0$, we deduce that

$$(TV(\sigma\varphi))^+ + (TW\varphi)^+ = (TV(\sigma\varphi))^- + (TW\varphi)^-.$$

Given that $(TW\varphi)^+ = (TW\varphi)^-$, we now have $(TV(\sigma\varphi))^+ = (TV(\sigma\varphi))^-$, or [Co14]

$$\tfrac{1}{2}\sigma\varphi + (TV(\sigma\varphi))_0 = -\tfrac{1}{2}\sigma\varphi + (TV(\sigma\varphi))_0.$$

Hence, $\sigma\varphi = 0$, which, in view of the positive definiteness of σ, yields $\varphi = 0$. The completeness of G in $L^2(\partial S)$ now follows from the fact that $L^2(\partial S)$ is a Hilbert space and the orthogonal complement of G in it consists of the zero vector alone.

8.3   The Computational Algorithm

In what follows, $\langle \cdot, \cdot \rangle$ and $\| \cdot \|$ are the inner product and norm in $L^2(\partial S)$.
Denoting by ψ the trace of u on ∂S and setting $Tu = \mathcal{R} - \sigma\psi$ and

$$L(x) = \int_{\partial S} D(x, y)\mathcal{R}(y)\, ds(y), \quad x \in S^+ \cup S^-, \tag{8.14}$$

we rewrite Eqs. (8.4) as

$$u(x) = -\int_{\partial S} [D(x, y)\sigma(y)\psi(y) + P(x, y)\psi(y)]\, ds(y) + L(x), \quad x \in S^+,$$

$$\tag{8.15}$$

$$L(x) = \int_{\partial S} [D(x, y)\sigma(y)\psi(y) + P(x, y)\psi(y)]\, ds(y), \quad x \in S^-. \tag{8.16}$$

As all the $x^{(k)}$ lie in S^-, for $j = 1, 2, 3$ and $k = 1, 2, \ldots$, from (8.16) we have

$$L_j(x^{(k)}) = \int_{\partial S} [D_{jh}(x^{(k)}, x)(\sigma\psi)_h(x) + P_{jl}(x^{(k)}, x)\psi_l(x)]\, ds.$$

From the definition of P it follows that $P_{jl}(x^{(k)}, x) = T_{lh}D_{hj}(x, x^{(k)})$, so

$$L_j(x^{(k)}) = \int_{\partial S} [D_{jh}(x^{(k)}, x)\sigma_{hl}(x)\psi_l(x) + T_{lh}D_{hj}(x, x^{(k)})\psi_l(x)]\, ds$$

$$= \int_{\partial S} [\sigma_{lh}(x)D_{hj}(x, x^{(k)}) + T_{lh}D_{hj}(x, x^{(k)})]\psi_l(x)\, ds$$

$$= \int_{\partial S} [T + \sigma(x)]_{lh}(D^j)_h(x, x^{(k)})\psi_l(x)\, ds$$

$$= \int_{\partial S} [(T + \sigma(x))D^j(x, x^{(k)})]_l\, \psi_l(x)\, ds = \int_{\partial S} \theta_l^{(jk)}\psi_l\, ds$$

$$= \int_{\partial S} (\theta^{(jk)})^\mathsf{T}\psi\, ds.$$

Combining this with $L(x^{(k)})$ given by (8.14), we deduce that for $j = 1, 2, 3$ and $k = 1, 2, \ldots$,

$$\int_{\partial S} (\theta^{(jk)})^\mathsf{T}\psi\, ds = L_j(x^{(k)}) = \int_{\partial S} [D(x^{(k)}, x)\mathcal{R}(x)]_j\, ds. \tag{8.17}$$

Also,

$$\int_{\partial S} (\sigma f^{(i)})^{\mathsf{T}} \psi \, ds = \int_{\partial S} (f^{(i)})^{\mathsf{T}} \sigma^{\mathsf{T}} \psi \, ds = \int_{\partial S} (f^{(i)})^{\mathsf{T}} \sigma \psi \, ds = (p(\sigma \psi))_i, \quad i = 1, 2, 3.$$

Since $Tu = \mathcal{R} - \sigma \psi$ is the Neumann boundary data for u, we have $p(Tu) = 0$ [Co14], so $p(\sigma \psi) = p\mathcal{R}$, which means that

$$\int_{\partial S} (\sigma f^{(i)})^{\mathsf{T}} \psi \, ds = (p\mathcal{R})_i = \int_{\partial S} (f^{(i)})^{\mathsf{T}} \mathcal{R} \, ds, \quad i = 1, 2, 3. \tag{8.18}$$

We reorder the elements of G as the sequence

$$\{\sigma f^{(1)}, \; \sigma f^{(2)}, \; \sigma f^{(3)}, \; \theta^{(11)}, \; \theta^{(21)}, \; \theta^{(31)}, \; \theta^{(12)}, \; \theta^{(22)}, \; \theta^{(32)}, \; \ldots\}$$

and re-index them:

$$G = \{\theta^{(1)}, \; \theta^{(2)}, \; \theta^{(3)}, \; \theta^{(4)}, \; \theta^{(5)}, \; \theta^{(6)}, \ldots\}.$$

We orthonormalize the set G in $L^2(\partial S)$ to generate a sequence $\{\eta^{(i)}\}_{i=1}^{\infty}$. This orthonormalization is implemented by three different procedures: the classical Gram–Schmidt (CGS), the modified Gram–Schmidt (MGS), and the Householder reflections (HR), all of which give rise to an equality of the form

$$\eta^{(n)} = \sum_{m=1}^{n} k_{nm} \theta^{(m)}, \quad n = 1, 2, \ldots. \tag{8.19}$$

Then the approximation of the unknown vector function ψ is

$$\psi^{(n)} = \sum_{r=1}^{n} \langle \psi, \eta^{(r)} \rangle \eta^{(r)}$$

$$= \sum_{r=1}^{n} \left\{ \sum_{m=1}^{r} k_{rm} \langle \psi, \theta^{(m)} \rangle \left(\sum_{q=1}^{r} k_{rq} \theta^{(q)} \right) \right\}, \quad n = 1, 2, \ldots. \tag{8.20}$$

It is easily verified that for each $m = 4, 5, \ldots$, there is a unique pair (j, k), $j \in \{1, 2, 3\}, k \in \{1, 2, \ldots\}$, such that $m = j + 3k$. Then $\theta^{(m)} = \theta^{(j+3k)} = \theta^{(jk)}$, so, by (8.17),

$$\langle \psi, \theta^{(m)} \rangle = \langle \psi, \theta^{(jk)} \rangle = L_j(x^{(k)}) = \int_{\partial S} [D(x^{(k)}, x) \mathcal{R}(x)]_j \, ds, \quad m = 4, 5, \ldots.$$

$$\tag{8.21}$$

For $m = 1, 2, 3$, by (8.18),

$$\langle \psi, \theta^{(m)} \rangle = \langle \psi, \sigma f^{(m)} \rangle = \int_{\partial S} (f^{(m)})^{\mathsf{T}} \mathscr{R} \, ds. \tag{8.22}$$

In conclusion, $\psi^{(n)}$ is computed from (8.20) with the coefficients k_{rm} and $\langle \psi, \theta^{(m)} \rangle$ supplied, respectively, by the orthonormalization process (8.19) and equalities (8.21) and (8.22). For $x \in S^+$ and $n = 1, 2, \ldots$, this and (8.15) now yield the approximate solution

$$u^{(n)}(x) = - \int_{\partial S} [D(x, y)\sigma(y) + P(x, y)] \psi^{(n)}(y) \, ds(y) + L(x). \tag{8.23}$$

Theorem 3 *The sequence $\{u^{(n)}\}_{n=1}^{\infty}$ converges uniformly to the exact solution u on any closed subdomain S' of S as $n \to \infty$.*

Proof Let $x \in S'$ be arbitrary, and let $P_{(i)}(x, y)$, $i = 1, 2, 3$, be the rows of $P(x, y)$; then

$$u_i(x) = - \int_{\partial S} [P(x, y)\psi(y)]_i \, ds(y) + L_i(x)$$

$$= - \int_{\partial S} P_{(i)}(x, y)\psi(y) \, ds(y) + L_i(x) = -\langle P_{(i)}(x, \cdot), \psi \rangle + L_i(x).$$

We subtract from this the ith component of $u^{(n)}$ given by (8.23) to arrive at

$$|u_i(x) - u_i^{(n)}(x)| \le |\langle P_{(i)}(x, \cdot), \psi - \psi^{(n)} \rangle| \le \| P_{(i)}(x, \cdot) \| \, \| \psi - \psi^{(n)} \|.$$

Denoting by $| \cdot |_3$ the Euclidean norm in $\mathbb{R}^3$, we have

$$|u(x) - u^{(n)}(x)|_3 \le \sum_{i=1}^{3} |u_i(x) - u_i^{(n)}(x)| \le \left(\sum_{i=1}^{3} \| P_{(i)}(x, \cdot) \| \right) \| \psi - \psi^{(n)} \|,$$

which, combined with the fact that the $\| P_{(i)}(x, \cdot) \|$ are uniformly bounded on S' (since $x \ne y$) and $\| \psi - \psi^{(n)} \| \to 0$ as $n \to \infty$, proves the assertion.

Remark 1 It is obvious that each of the vector functions $u^{(n)}$ is a solution of (8.1).

8.4 Numerical Example

Let S^+ be the disk of radius 1 centered at the origin, and let (after suitable rescaling) $h = 0.5$ and $\lambda = \mu = 1$. We choose the matrix σ and the boundary data (written in terms of the polar angle θ with the pole at the origin)

$$
\sigma(x) = \begin{pmatrix} x_1^2 + x_2^2 & 0 & x_1^2 - x_2^2 \\ 0 & 2 & 0 \\ x_1^2 - x_2^2 & 0 & x_1^2 + x_2^2 \end{pmatrix},
$$

$$
\mathcal{R}(x) = \begin{pmatrix} \frac{1}{2}\left(8 + 18\cos\theta - 6\cos(2\theta) + 5\cos(3\theta) - 2\cos(4\theta) + \cos(5\theta)\right) \\ -10\sin\theta + 13\sin(2\theta) \\ -1 + 13\cos\theta + 3\cos(3\theta) - \cos(4\theta) \end{pmatrix},
$$

for which our boundary value problem has the exact solution

$$
u(x) = \begin{pmatrix} 4x_1 + 4x_2^2 \\ -4x_2 + 8x_1 x_2 \\ 6x_1 - 2x_1^2 + 2x_2^2 - 4x_1 x_2^2 \end{pmatrix}, \quad x \in S^+.
$$

We take the auxiliary curve ∂S_* to be the circle of radius 2 centered at the origin. This choice is recommended by the fact that if ∂S_* is too far from ∂S, the set G becomes "less linearly independent," and if ∂S_* is too close to ∂S, then the elements of G become too sensitive to the singularities of D and P.

The approximation accuracy depends on the selection of the dense set $\{x^{(k)} \in \partial S_*, \ k = 1, 2, \ldots\}$. For convenience, we choose uniformly spread points; that is, for any $n = 1, 2, \ldots$,

$$
\{x^{(k)} : k = 1, 2, \ldots, n\}_{\text{Cartesian}} = \{(2, 2\pi k/n) : k = 1, 2, \ldots, n\}_{\text{Polar}}.
$$

Remark 2 In this numerical example, floating-point computation is performed with machine precision of approximately 100 digits. As expected, the most sensitive element in the procedure is the evaluation of integrals in inner products, for which we set a target of 100 significant digits. However, as computation proceeds, this number deteriorates.

The three-dimensional physical displacement vector in this model [Co14] is $(x_3 u_1, x_3 u_2, u_3)^\mathsf{T}$. The graph of the functions $u_i^{(63)}$, $i = 1, 2, 3$, computed from $\psi^{(63)}$ with MGS for $0 \le r < 1.0$, $0 \le \theta < 2\pi$, are shown in Fig. 8.1.

Figure 8.2 displays the graphs of $u_i^{(63)}|_{\partial S}$, $i = 1, 2, 3$.

We are interested in evaluating computational errors. For $n = 20$ (our case), the three components of the absolute error $u^{(63)} - u$ are graphed in Fig. 8.3.

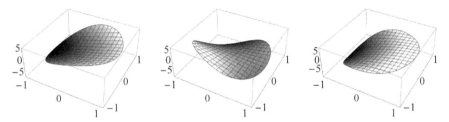

Fig. 8.1 The functions $u_i^{(63)}$, $i = 1, 2, 3$, in S^+

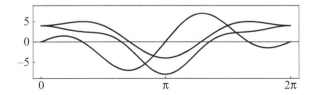

Fig. 8.2 The traces of $u_i^{(63)}$, $i = 1, 2, 3$, on ∂S

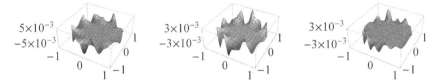

Fig. 8.3 The errors $u_i^{(63)} - u_i$, $i = 1, 2, 3$, in S^+

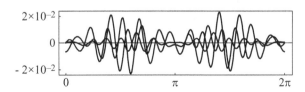

Fig. 8.4 The traces of $u_i^{(63)} - u_i$, $i = 1, 2, 3$, on ∂S

The approximation is 4–5 digits of accuracy near the boundary ∂S and improves significantly away from ∂S.

Figure 8.4 shows the graphs of the components of the absolute error $(u_i^{(63)} - u_i)|_{\partial S}$, $i = 1, 2, 3$.

The graph of the relative error

$$\frac{\|(u^{(3n+3)} - u)|_{\partial S}\|}{\|u|_{\partial S}\|}$$

in the approximation of the boundary trace $u|_{\partial S}$, as a function of n (best least square linear fit), is shown in Fig. 8.5.

The relative error decreases exponentially, approximately by half for each additional point added on ∂S_*. Our use of floating-point precision of 100 digits avoids contamination of the numerical result. The vertical axis above is displayed logarithmically to base 10. For our chosen example, the relative error is $179.5 \times 10^{-0.2957n}$.

Another type of error refers to the floating-point conditioning. We have added Row Reduction (RR) as a fourth computational method, which bypasses the need for orthonormalization. MGS, HR, and RR exhibit similar numerical accuracy; CGS has significantly reduced accuracy.

The number of significant digits remaining in the computation of $u^{(3n+3)}|_{\partial S}$ with MGS, as a function of n, when fixed floating-point of 100 significant digits is used, is shown in Fig. 8.6.

For $n = 200$, floating-point computation with 100 digits does not contaminate the corresponding computed relative error. Of course, this changes as n increases.

Remark 3 In CGS, each vector $\theta^{(k)}$ is projected on to the space orthogonal to span$\{\eta^{(1)}, \ldots, \eta^{(k-1)}\}$. The obvious drawback is that, as calculation proceeds, $\theta^{(k)}$ tends to become increasingly "parallel" to its projection on span$\{\eta^{(1)}, \ldots, \eta^{(k-1)}\}$, so the difference between them becomes increasingly smaller in magnitude, which leads to computational ill-conditioning. All the inner products $\langle \eta^{(i)}, \eta^{(j)} \rangle$ for $i \neq j$ should be zero. However, as n increases, the CGS process deteriorates until loss of orthogonality occurs.

In MGS, the mutual orthogonality of the $\eta^{(i)}$ is used to factor the projection operator of CGS in a different form (see [CoDo17b]), where the computation involves the full magnitude of the $\theta^{(k)}$, preventing the magnitude of the difference

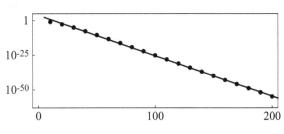

Fig. 8.5 The relative error as a function of n

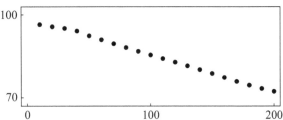

Fig. 8.6 The number of significant digits as a function of n

between $\eta^{(k)}$ and its projection on span$\{\eta^{(1)}, \ldots, \eta^{(k-1)}\}$ from becoming too small and, thus, avoiding computational ill-conditioning.

Unlike the CGS and MGS methods, in HR, the orthogonal projections used in the computational procedure are invertible and also norm-preserving. Details of this technique can be found in [Tr10]. CGS and MGS start with an incoming set $\{\theta^{(1)}, \ldots, \theta^{(n)}\}$ and predetermined rules for producing an orthonormal set $\{\eta^{(1)}, \ldots, \eta^{(n)}\}$. HR starts with both an incoming set $\{\theta^{(1)}, \ldots, \theta^{(n)}\}$ and a preselected orthonormal set $\{e^{(1)}, \ldots, e^{(n)}\}$, and computes the rules for the matrix that transforms the former into the latter. As a consequence, numerical computation with HR offers an improvement over the Gram–Schmidt procedures, subject to machine limitations.

RR is a pure algebraic procedure which, although not based on the use of orthonormalized systems, nevertheless yields a degree of accuracy comparable to that in MGS and HR.

Our specific choice of orthonormal vector functions $e^{(k)}$, which allows for periodicity on $0 \leq \theta < 2\pi$, is the same as that used in [CoDo17b].

References

[CoDo17a] Constanda, C. and Doty, D.: Bending of Elastic Plates: Generalized Fourier Series Method, in *Integral Methods in Science and Engineering. Vol.1: Theoretical Techniques*, Birkhäuser, New York (2017), pp. 71–81. https://doi.org/10.1007/978-3-319-59384-5_7.

[CoDo17b] Constanda, C. and Doty, D.: Bending of Elastic Plates with Transverse Shear Deformation: The Neumann Problem, *Mathematical Methods in the Applied Sciences*, 2017, https://doi.org/10.1002/mma.4704.

[Co14] Constanda, C.: *Mathematical Methods for Elastic Plates*, Springer, London (2014).

[Tr10] Trefethen, L.N.: Householder triangularization of a quasimatrix, *IMA Journal of Numerical Analysis* **30** (2010), 887–897.

Chapter 9
The Adjoint Spectral Green's Function Method Applied to Direct and Inverse Neutral Particle Source–Detector Problems

Jesús P. Curbelo, Odair P. da Silva, and Ricardo C. Barros

9.1 Introduction

The need to determine the state of a system from future observations or the identification of physical parameters from the observation of the system's evolution leads to define and solve and reverse the problem [EnEtAl96]. That is, while the solution of a direct problem is to find effects (reactions) based on a complete description of its causes, it is possible to state that solving an inverse problem consists in determining unknown causes (stimuli) from desired or observed effects [Al94, EnEtAl96, MoSi13]. In source–detector problems the causes are the boundary conditions, the sources of particles and the material properties of the domain. On the other hand, the effects are the detector response, as well as the profiles of particle flux (forward problem) or the importance function distribution (adjoint problem).

In direct transport problems the input parameters are geometry, distribution of external sources (including boundary conditions) and material properties. Nevertheless, in inverse transport problems, at least one of these parameters is unknown [HyAz11, Mc92]. In direct source–detector problems the use of the adjoint technique allows to obtain the detector response due to multiple sources by a single solution to the adjoint problem in each energy group. On the other hand, in inverse source–detector problems it is possible to calculate the intensity of the source in each energy group, given its location and the detector response.

This work is based on the application of the adjoint spectral Green's function method (SGF†) [MiEtAl12, CuEtAl17, CuEtAl18] for numerically solving slab-

J. P. Curbelo (✉) · O. P. da Silva · R. C. Barros
Instituto Politécnico, Universidade do Estado do Rio de Janeiro, Nova Friburgo, Brazil
e-mail: rcbarros@pq.cnpq.br

© Springer Nature Switzerland AG 2019
C. Constanda, P. Harris (eds.), *Integral Methods in Science and Engineering*,
https://doi.org/10.1007/978-3-030-16077-7_9

geometry direct and inverse source–detector transport problems in the energy multigroup discrete ordinates (S_N) formulation with arbitrary $L'th$-order of scattering anisotropy. The offered SGF† method, along with the one-region block inversion iterative scheme, generates numerical solutions that are completely free from spatial truncation errors; therefore a spatial reconstruction scheme is developed to analytically determine the detector response in direct problems and source intensities in inverse problems. In this chapter we describe how to estimate the intensity of a source of neutral particles Q_g ($g = 1 : G$) located in a specific region within the domain, knowing the detector response for each energy group g, the geometry and material properties of the domain. In addition, we also describe a technique to determine the intensity of incident particles by measuring the detector response due to sources placed outside the slab.

9.2   The Adjoint S_N Transport Problem

Let us consider a multilayer slab of thickness H where the regions Υ_j ($j = 1 : J$) have width h_j and constant material parameters. For steady-state source–detector problems in non-multiplying media, the energy multigroup, slab-geometry adjoint S_N equations, considering arbitrary $L'th$ order of scattering anisotropy, provided $L < N$, can be written as

$$
-\mu_m \frac{d}{dx} \psi_{mg}^\dagger(x) + \sigma_{T_{g,j}} \psi_{mg}^\dagger(x)
$$

$$
= \sum_{l=0}^{L} \frac{2l+1}{2} P_l(\mu_m) \sum_{g'=1}^{G} \sigma_{S_{g \to g',j}}^{(l)} \sum_{n=1}^{N} P_l(\mu_n) \omega_n \psi_{ng'}^\dagger(x) + Q_{g,j}^\dagger ,
$$

$$
x \in \Upsilon_j , \quad m = 1 : N , \quad g = 1 : G , \tag{9.1}
$$

with boundary conditions

$$
\psi_{mg}^\dagger(0) = \alpha \, \psi_{ng}^\dagger(0) , \quad \mu_m < 0 , \quad \mu_m = -\mu_n , \quad g = 1 : G
$$

and

$$
\psi_{mg}^\dagger(H) = \alpha \, \psi_{ng}^\dagger(H) , \quad \mu_m > 0 , \quad \mu_m = -\mu_n , \quad g = 1 : G ,
$$

where $\alpha = 1$ indicates that reflective boundary condition is considered and $\alpha = 0$ for zero adjoint flux in the exiting directions. We refer to the importance function $\psi_{mg}^\dagger(x)$ as the adjoint angular flux in the discrete ordinate direction μ_m in energy group g; the quantity $Q^\dagger$ is defined as the group macroscopic absorption cross section of the material the detector is made of [DuMa79, PrLa10]; otherwise, the notation is standard [LeMi93].

In order to determine the importance of neutral particles of all energy groups to the detector response only in the energy group g, the adjoint S_N problem must be solved considering the adjoint source numerically equal to the detector absorption macroscopic cross section for energy group g and equal to zero for all the other energy groups. Once the adjoint S_N equations are solved, we obtain the detector response for energy group g as [LeMi93]

$$
R_g^\dagger = \int_0^H \sum_{g'=1}^{G} Q_{g'}(x) \sum_{n=1}^{N} \omega_n \, \psi_{ng'}^{\dagger g}(x) \, dx
$$

$$
+ \sum_{g'=1}^{G} \sum_{\mu_n>0} \mu_n \, \omega_n \, \psi_{ng'}^{\dagger g}(0) \, \widetilde{\psi}_{g'}(0) + \sum_{g'=1}^{G} \sum_{\mu_n<0} |\mu_n| \, \omega_n \, \psi_{ng'}^{\dagger g}(H) \, \widetilde{\psi}_{g'}(H) \,,
$$

$$ (9.2) $$

where $g = 1 : G$, $\widetilde{\psi}_{g'}(0)$ and $\widetilde{\psi}_{g'}(H)$ are the forward fluxes, considering prescribed isotropic boundary conditions at $x = 0$ and $x = H$, respectively. In Eq. (9.2) the quantity $\psi_{ng'}^{\dagger g}(x)$ is read as the adjoint angular flux in the discrete ordinates direction μ_n, in energy group g', obtained by solving the adjoint S_N equations considering $Q^\dagger$ distinct from zero only for the energy group g.

In the next section we summarize the main ingredients of the generalized $\text{SGF}^\dagger$ method [CuEtAl18] that we use along with the one-region block inversion iterative scheme to obtain numerical solutions absolutely free from spatial truncation errors.

9.3   The Adjoint Spectral Green's Function ($\text{SGF}^\dagger$) Method

The $\text{SGF}^\dagger$ method was first presented by [MiEtAl12] for monoenergetic slab-geometry adjoint S_N problems with isotropic scattering. In the work by [CuEtAl18] is described the generalization of the $\text{SGF}^\dagger$ method to energy multigroup S_N problems considering arbitrary L'th order of scattering anisotropy, provided $L < N$, and non-zero prescribed boundary conditions for the forward S_N transport problem. The method uses the standard discretized spatial balance S_N equations for the adjoint problem

$$
- \frac{\mu_m}{\sigma_{T_{g,j}} h_j} \left(\psi_{mg,j}^\dagger - \psi_{mg,j-1}^\dagger \right) + \overline{\psi}_{mg,j}^\dagger
$$

$$
= \sum_{g'=1}^{G} \sum_{n=1}^{N} \omega_n \, \overline{\psi}_{ng',j}^\dagger \left\{ \sum_{l=0}^{L} \frac{2l+1}{2} c_{g \to g',j}^{(l)} P_l(\mu_m) \, P_l(\mu_n) \right\} + \frac{Q_{g,j}^\dagger}{\sigma_{T_{g,j}}} \,,
$$

$$
j = 1 : J \,, \quad m = 1 : N \,, \quad g = 1 : G \,, \qquad (9.3)
$$

which is exact as it is obtained by integrating Eq. (9.1) over a region Υ_j by using the operator $\dfrac{1}{h_j} \displaystyle\int_{x_{j-1}}^{x_j} (\cdot)\, dx$. We have defined $\overline{\psi}^{\dagger}_{mg,j}$ as the group region-average adjoint angular flux.

In order to solve the system of $N \times G$ algebraic linear equations in $3 \times N \times G$ unknowns represented in Eq. (9.3) by requiring uniqueness of the solution, we use the $N \times G$ equations for the continuity and/or boundary conditions and define $N \times G$ auxiliary equations

$$\overline{\psi}^{\dagger}_{mg,j} = \sum_{g'=1}^{G} \sum_{\mu_n<0} \Lambda^{j}_{ng'\to mg} \psi^{\dagger}_{ng',j-1} + \sum_{g'=1}^{G} \sum_{\mu_n>0} \Lambda^{j}_{ng'\to mg} \psi^{\dagger}_{ng',j} + B_{mg,j} ,$$

$$j = 1 : J , \quad m = 1 : N , \quad g = 1 : G . \qquad (9.4)$$

This auxiliary equation expresses each region-average adjoint angular flux as a combination of the adjoint source located within the region and all the region-edge adjoint angular fluxes in all exiting directions and energy groups. The quantities $\Lambda^{j}_{ng'\to mg}$ play the role of the Green's function and are constructed by using a spectral analysis based on local analytic solutions of Eq. (9.1) inside each region and $B_{mg,j}$ is a function of the interior adjoint source.

The general solution of the system of $N \times G$ ordinary differential equations represented in Eq. (9.1) can be written as

$$\psi^{\dagger}_{mg}(x) = \sum_{k=1}^{N \times G} \beta_k\, a^{\dagger}_{mg}(\xi_k)\, e^{\frac{-(x-\lambda_j)}{\xi_k}} + \psi^{\dagger\, P}_{g,j} ,$$

$$x \in \Upsilon_j , \quad m = 1 : N , \quad g = 1 : G , \qquad (9.5)$$

where $\psi^{\dagger\, P}_{g}(x)$ is a particular solution which is isotropic for each energy group, provided the adjoint source $Q^{\dagger}_{g,j}$ is independent of the angular directions, and first term of the right-hand side in Eq. (9.5) represents the homogeneous component of the local general solution of Eq. (9.1) in Υ_j. The quantities ξ_k and $a^{\dagger}_{mg}(\xi_k)$ are obtained by solving an eigenvalue problem of order $N \times G$ and β_k are arbitrary constants for each region.

The spatially discretized adjoint balance S_N Eqs. (9.3) together with the auxiliary Eqs. (9.4) and boundary and continuity conditions constitute the nodal equations in the SGF† method. In order to generate numerical solutions of SGF† equations, we use the partial one-region block inversion iterative scheme. More details about the SGF† method can be found in reference [CuEtAl18].

9.4 Spatial Reconstruction Scheme for the SGF† Solution

The SGF† method has the advantage of computational efficiency due to the possibility of setting up coarse-mesh in the discretization of the spatial variables; however, it has the disadvantage of not generating a detailed profile of the solution. An alternative is to perform an analytical reconstruction inside each homogenized region of the domain.

According to the general solution of the adjoint S_N Eqs. (9.5), we have the angular flux in the energy group g and in each direction μ_m in the position $x \in \Upsilon_j$. Therefore, using the adjoint angular flux in the energy group g and in the direction μ_m at the region edges, we obtain

$$\psi^\dagger_{mg}(x_{j-1}) = \sum_{n=1}^{N}\sum_{g'=1}^{G} \beta_{ng'}\, a^\dagger_{mg}(\xi_{ng'})\, e^{\frac{-(x_{j-1}-\lambda_j)}{\xi_{ng'}}} + \psi^{\dagger\,P}_{g,j-1}\ , \quad \mu_m < 0 \qquad (9.6a)$$

and

$$\psi^\dagger_{mg}(x_j) = \sum_{n=1}^{N}\sum_{g'=1}^{G} \beta_{ng'}\, a^\dagger_{mg}(\xi_{ng'})\, e^{\frac{-(x_j-\lambda_j)}{\xi_{ng'}}} + \psi^{\dagger\,P}_{g,j}\ , \quad \mu_m > 0\ , \qquad (9.6b)$$

where the values of $\psi^\dagger_{mg}(x_{j-1})$ e $\psi^\dagger_{mg}(x_j)$ are generated by the SGF† method. At this point, it is possible to form a system of $N \times G$ linear and algebraic equations in $N \times G$ unknowns to obtain the constants $\beta_{ng'}$ which appear in Eqs. (9.6).

Following this procedure, the set of coefficients $\beta_{ng'}$ for each region can be stored and used in calculations of local quantities of interest such as the adjoint scalar flux at any point of the slab. Then, it is possible to determine the detector response by substituting analytically the adjoint angular flux in Eq. (9.2). In the next section we present the procedure in order to obtain the intensity of a source of neutral particles located within the slab or on the boundaries, by solving an inverse problem.

9.5 Source–Detector Inverse Problems

Substituting the analytical general solution of the adjoint S_N Eqs. (9.5) into Eq. (9.2), we rewrite the detector response $R^\dagger$ for each energy group g located in region Υ_j due to a source of neutral particles Q within a limited region $x_0 < x < x_f$

$$R^\dagger_g = \sum_{g'=1}^{G}\sum_{n=1}^{N} \beta^{j,g}_{ng'}\, \xi^j_{ng'} \left[\sum_{g''=1}^{G}\sum_{m=1}^{N} Q^j_{g''}\, a^\dagger_{mg''}(\xi^j_{ng'})\, \omega_m \right] \times \left[e^{\frac{-(x_0-\lambda_j)}{\xi^j_{ng'}}} - e^{\frac{-(x_f-\lambda_j)}{\xi^j_{ng'}}} \right]$$

$$+ (x_f - x_0)\sum_{g''=1}^{G}\sum_{m=1}^{N} Q^j_{g''}\, \psi^{\dagger\,P,\,g}_{g''}\, \omega_m\ , \quad g = 1:G\ , \qquad (9.7)$$

where the quantities $\beta_{ng'}^{j,g}$ are calculated by using the spatial reconstruction scheme for the SGF† solution described in the previous section; also $\lambda_j = x_f$ and $\lambda_j = x_0$ for $\xi < 0$ and $\xi > 0$, respectively. Once we obtain the adjoint flux distribution and we know the detector response $R_g^\dagger$, it is possible to calculate the intensity of the source Q for each energy group g by solving the system of G linear equations represented in Eq. (9.7), which, in matrix form, appears as

$$\mathbf{M\,Q} = \mathbf{R}^\dagger \; ,$$

where $\mathbf{M}$ is the coefficient matrix of vector $\mathbf{Q}$ in Eq. (9.7) and $\mathbf{R}^\dagger$ is the vector whose entries are the multigroup detector responses due to interior source distribution.

An analogous procedure allows to determine the intensities of the sources located on one boundary of the domain. At this point we rewrite Eq. (9.2), only considering the terms corresponding to the prescribed boundary conditions, as indicated by the superscript BC

$$R_g^{\dagger\, BC} = \sum_{g'=1}^{G} \sum_{n=1}^{N/2} \mu_n \, \omega_n \, \psi_{ng'}^{\dagger\,g}(0) \, \tilde{\psi}_{g'}(0) + \sum_{g'=1}^{G} \sum_{n=\frac{N}{2}+1}^{N} |\mu_n| \, \omega_n \, \psi_{ng'}^{\dagger\,g}(H) \, \tilde{\psi}_{g'}(H) \; ,$$

$$g = 1 : G \; . \qquad (9.8)$$

For either left or right boundary, we consider the first or second term on the right-hand side of Eq. (9.8), which can be represented in matrix form as

$$\mathbf{M}^u \, \widetilde{\boldsymbol{\Psi}}(u) = \mathbf{R}^{\dagger\,u} \; .$$

Here $u = 0$ for the left boundary and $u = H$ for the right boundary, respectively; $\mathbf{M}^u$ is the coefficient matrix of vector $\widetilde{\boldsymbol{\Psi}}(u)$ and $\mathbf{R}^{\dagger\,u}$ is the vector whose entries are the multigroup detector responses due to incident flux of particles in one boundary of the slab.

In the next section we solve a classical source detector model problem by using the adjoint technique. First we solve the direct problem to determine the detector response due to sources of neutral particles located in the interior of the domain and in one boundary. Then, we perform an inverse problem to estimate the intensity of the sources to illustrate the application of the SGF† method in such type of problems.

9.6   Numerical Examples

Let us consider a model problem consisting of a multilayer slab (Fig. 9.1), 20 energy groups and anisotropic scattering of order $L = 10$. The total and scattering macroscopic cross sections can be found in references [GaSi83] and [CuEtAl18]. We set up two interior sources Q_g^1 and Q_g^2 located as illustrated in Fig. 9.1 and a unit isotropic source only in the first energy group as the prescribed left boundary condition. First we determine the response by a detector D_g located in the third region of the domain, as it is performed in the work by [CuEtAl18]. To solve the adjoint problem, the adjoint source is considered numerically equal to the detector absorption macroscopic cross section for each energy group. We used the S_{32} Gauss–Legendre angular quadrature set [LeMi93] and the stopping criterion for solving the direct problem required that the discrete maximum norm of the relative deviation between two consecutive estimates for the region-average adjoint scalar fluxes was less than 10^{-6}. All numerical results were obtained by running a computer code developed with the programming language $C++$ *Code::Blocks 13.12 IDE*. Table 9.1 shows the detector response for each energy group due to the two interior sources and the source located on the left boundary.

Using the results from Table 9.1 it is possible to perform the numerical experiment consisting in determining the intensities of the sources of particles by solving the inverse problem represented in Eqs. (9.7) and (9.8). In order to obtain the intensities of the interior sources Q_g^1 and Q_g^2, we solved two linear systems from Eq. (9.7) by forming the independent vector using the values of $R_g^{\dagger 1}$ and $R_g^{\dagger 2}$, respectively. Then, we solved the linear system from Eq. (9.8) considering only the first term on the right-hand side and defining the independent vector using the $R_g^{\dagger LBC}$ values.

Table 9.2 shows the intensities of the sources of particles by solving the inverse problem. It is possible to observe that for the first energy group the results agree with the expected from the conditions in the Model Problem. On the other hand, for the other energy groups, we observe numerical imprecision due to finite arithmetic.

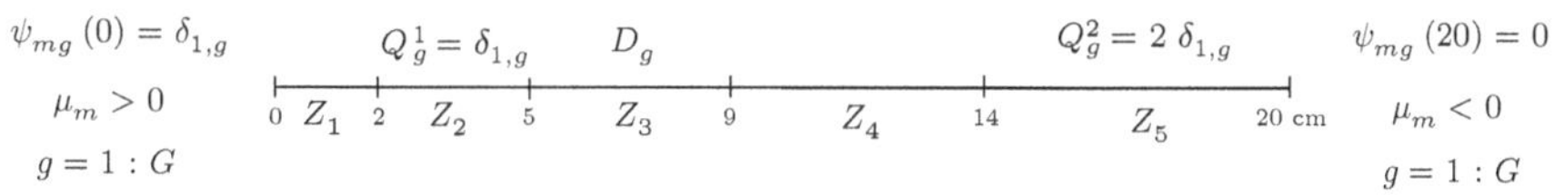

$\psi_{mg}(0) = \delta_{1,g}$	$Q_g^1 = \delta_{1,g}$	D_g		$Q_g^2 = 2\,\delta_{1,g}$	$\psi_{mg}(20) = 0$
$\mu_m > 0$					$\mu_m < 0$
$g = 1:G$					$g = 1:G$

$$0 \;\; Z_1 \;\; 2 \;\; Z_2 \;\; 5 \;\; Z_3 \;\; 9 \;\; Z_4 \;\; 14 \;\; Z_5 \;\; 20\text{ cm}$$

Fig. 9.1   Slab for the model problem

Table 9.1 Detector response for the model problem $(\text{cm}^{-2}\,\text{s}^{-1})$ (S_{32}) $(G = 20)$

Energy group	$R_g^{\dagger 1}$ a	$R_g^{\dagger 2}$	$R_g^{\dagger LBC}$
1	1.0201×10^0	3.9029×10^{-1}	9.4875×10^{-2}
2	4.3265×10^{-2}	2.1048×10^{-2}	5.0325×10^{-3}
3	3.1290×10^{-2}	1.3183×10^{-2}	3.4399×10^{-3}
4	2.4644×10^{-2}	9.7539×10^{-3}	2.6023×10^{-3}
5	1.8557×10^{-2}	7.5262×10^{-3}	2.0114×10^{-3}
6	1.7897×10^{-2}	6.8100×10^{-3}	1.8127×10^{-3}
7	1.5660×10^{-2}	5.9077×10^{-3}	1.5631×10^{-3}
8	1.3986×10^{-2}	5.2528×10^{-3}	1.3816×10^{-3}
9	1.2682×10^{-2}	4.7526×10^{-3}	1.2434×10^{-3}
10	1.1370×10^{-2}	4.2468×10^{-3}	1.1172×10^{-3}
11	1.0881×10^{-2}	4.0715×10^{-3}	1.0571×10^{-3}
12	1.0148×10^{-2}	3.7983×10^{-3}	9.8246×10^{-4}
13	9.5428×10^{-3}	3.5736×10^{-3}	9.2140×10^{-4}
14	9.0360×10^{-3}	3.3860×10^{-3}	8.7060×10^{-4}
15	8.6097×10^{-3}	3.2286×10^{-3}	8.2811×10^{-4}
16	8.2527×10^{-3}	3.0972×10^{-3}	7.9266×10^{-4}
17	7.9593×10^{-3}	2.9894×10^{-3}	7.6359×10^{-4}
18	7.7295×10^{-3}	2.9053×10^{-3}	7.4082×10^{-4}
19	7.5735×10^{-3}	2.8489×10^{-3}	7.2527×10^{-4}
20	7.5280×10^{-3}	2.8339×10^{-3}	7.2042×10^{-4}

$^a R_g^{\dagger 1}$, $R_g^{\dagger 2}$ and $R_g^{\dagger LBC}$ are the detector responses due to the sources Q_g^1, Q_g^2 and the prescribed left boundary condition, respectively

The higher imprecision values appear for the interior source Q_g^2, which is located in a region more distant from the detector. We note that the determinant of the coefficient matrix of each linear system in the inverse problem has a small value very close to zero. Since the numerical values displayed in Table 9.2 were generated using *double* precision, we decided to improve the accuracy by using *long double* precision and the results are displayed in Table 9.3. As we observe, the use of *long double* precision reduces the numerical imprecision in the results up to the order of 10^{-8}.

9.7 Conclusions and Perspectives

The solution of the adjoint transport equation gives the importance distribution of neutral particles within the domain. In this chapter we have used the SGF† method to

Table 9.2 Interior sources and prescribed boundary condition estimation (cm^{-3} s^{-1}) from solving the inverse problem by using *double* precision

Energy group	Q_g^1	Q_g^2	LBC
1	1.0000×10^0	2.0000×10^0	1.0000×10^0
2	1.7634×10^{-8}	5.7664×10^{-8}	8.2354×10^{-9}
3	1.7044×10^{-8}	-1.6843×10^{-8}	-9.5094×10^{-9}
4	8.8436×10^{-9}	3.1717×10^{-7}	1.2826×10^{-8}
5	4.3494×10^{-9}	-2.3686×10^{-8}	-1.8759×10^{-8}
6	9.1616×10^{-9}	1.2000×10^{-7}	1.0146×10^{-7}
7	1.5998×10^{-8}	4.6784×10^{-6}	-1.3000×10^{-7}
8	1.2149×10^{-8}	1.4126×10^{-5}	2.3886×10^{-7}
9	1.2684×10^{-8}	2.2510×10^{-5}	1.7519×10^{-7}
10	1.3279×10^{-9}	7.7176×10^{-6}	2.8223×10^{-7}
11	1.0746×10^{-8}	6.1387×10^{-4}	1.4375×10^{-6}
12	1.5123×10^{-8}	2.1579×10^{-3}	-7.9032×10^{-7}
13	7.8770×10^{-9}	-1.9296×10^{-3}	-3.8905×10^{-7}
14	8.6346×10^{-9}	-7.2322×10^{-3}	-2.2462×10^{-6}
15	9.6922×10^{-9}	3.3536×10^{-2}	1.2142×10^{-6}
16	1.0069×10^{-8}	-4.7418×10^{-2}	-6.2359×10^{-6}
17	9.6819×10^{-9}	3.7633×10^{-2}	4.5428×10^{-6}
18	6.1442×10^{-9}	7.1690×10^{-2}	-2.4602×10^{-7}
19	7.0298×10^{-9}	2.6944×10^0	8.9967×10^{-6}
20	7.9253×10^{-9}	1.3303×10^{-2}	1.1664×10^{-5}

numerically solve the adjoint multigroup slab-geometry S_N equations. The coarse-mesh $SGF^{\dagger}$ method generates numerical solutions that are completely free from spatial truncation errors. Then, we used the analytical solution to solve inverse source–detector problems in order to obtain the intensity of sources of particles, given their location and the detector responses.

According to the numerical results to the inverse problem presented in the previous section, we conclude that they are very sensitive to finite arithmetic calculations. To remedy this drawback of the present methodology we used *long double* precision.

We intend to apply the present adjoint technique to direct and inverse energy multigroup source–detector problems in multidimensional rectangular geometries.

Table 9.3 Interior sources and prescribed boundary condition estimation ($cm^{-3}\, s^{-1}$) from solving the inverse problem by using *long double* precision

Energy group	Q_g^1	Q_g^2	LBC
1	1.0000×10^0	2.0000×10^0	1.0000×10^0
2	-2.1118×10^{-17}	0.0000×10^0	0.0000×10^0
3	0.0000×10^0	-1.4745×10^{-16}	0.0000×10^0
4	6.5860×10^{-17}	-2.2151×10^{-15}	0.0000×10^0
5	-2.8303×10^{-17}	1.1687×10^{-15}	0.0000×10^0
6	2.5035×10^{-17}	6.7542×10^{-15}	0.0000×10^0
7	7.0931×10^{-18}	-7.8302×10^{-14}	0.0000×10^0
8	-2.7806×10^{-17}	3.2497×10^{-13}	0.0000×10^0
9	0.0000×10^0	-6.8479×10^{-13}	0.0000×10^0
10	4.2348×10^{-18}	6.2272×10^{-14}	0.0000×10^0
11	1.0509×10^{-17}	5.4287×10^{-12}	0.0000×10^0
12	3.9706×10^{-17}	-9.8525×10^{-12}	0.0000×10^0
13	-1.2768×10^{-16}	-9.8285×10^{-12}	0.0000×10^0
14	-3.2371×10^{-17}	-7.2851×10^{-11}	0.0000×10^0
15	2.0556×10^{-17}	-1.7133×10^{-10}	0.0000×10^0
16	1.4415×10^{-17}	1.2490×10^{-9}	0.0000×10^0
17	7.5341×10^{-18}	-1.0841×10^{-9}	0.0000×10^0
18	-7.8197×10^{-18}	-9.3299×10^{-9}	0.0000×10^0
19	-3.6191×10^{-17}	2.9434×10^{-8}	0.0000×10^0
20	2.0379×10^{-17}	-7.4047×10^{-8}	0.0000×10^0

Acknowledgements This study was financed in part by the Coordenação de Aperfeiçoamento de Pessoal de Nível Superior—Brasil (CAPES)—Finance Code 001. The authors also acknowledge the partial financial support of Fundação Carlos Chagas Filho de Amparo à Pesquisa do Estado do Rio de Janeiro—Brasil (FAPERJ) and Conselho Nacional de Desenvolvimento Científico e Tecnológico—Brasil (CNPq).

References

[EnEtAl96] Engl, H. W., Hanke, M., and Neubauer, A.: *Regularization of Inverse Problems*, Kluwer Academic Publishers, Dordrecht, The Netherlands (1996).

[Al94] Alifanov, O. M.: *Inverse Heat Transfer Problems*, Springer–Verlag, Berlin Heidelberg (1994).

[MoSi13] Moura Neto, F. D., and Silva Neto, A. J.: *An Introduction to Inverse Problems with Applications*, Springer–Verlag, Berlin Heidelberg (2013).

[HyAz11] Hykes, J. M., and Azmy, Y. Y.: Radiation source reconstruction with known geometry and materials using the adjoint. In *International Conference on Mathematics and Computational Methods Applied to Nuclear Science and Engineering (M&C) 2011*, Latin American Section (LAS) / American Nuclear Society (ANS), Rio de Janeiro, Brazil (2011).

[MiEtAl12] Militão, D. S., Alves, H. and Barros, R. C.: A numerical method for monoenergetic slab–geometry fixed-source adjoint transport problems in the discrete ordinates formulation with no spatial truncation error. *International Journal of Nuclear Energy Science and Technology*, **7**, 151–165 (2012).

[CuEtAl17] Curbelo, J. P., da Silva, O. P., García, C. R., and Barros, R. C.: Shifting Strategy in the Spectral Analysis for the Spectral Green's Function Nodal Method for Slab–Geometry Adjoint Transport Problems in the Discrete Ordinates Formulation. In *Integral Methods in Science and Engineering, Volume 2: Practical Applications*, C. Constanda et al. (eds.), Birkhäuser Basel (2017), Ch. 20, pp. 201–210.

[CuEtAl18] Curbelo, J. P., da Silva, O. P., and Barros, R. C.: An adjoint technique applied to slab–geometry source–detector problems using the generalized spectral Green's function nodal method. *Journal of Computational and Theoretical Transport*, (2018). (doi:10.1080/23324309.2018.1539403)

[Mc92] McCormick, N. J.: Inverse Radiative Transfer Problems: A Review. *Nuclear Science and Engineering*, **112**, 185–198 (1992).

[DuMa79] Duderstadt, J. J. and Martin, W. R.: Transport Theory. *Wiley–Interscience, New York, USA*, (1979).

[PrLa10] Prinja, A. K. and Larsen, E. W.: General Principles of Neutron Transport. *Cacuci, D. G. (Ed), Handbook of Nuclear Engineering, Ch. 5. Springer Science+Business Media, New York, USA* (2010).

[LeMi93] Lewis, E. E. and Miller, W. F.: Computational methods of neutron transport. *American Nuclear Society, Illinois, USA*, (1993).

[GaSi83] Garcia, R. D. M. and Siewert, C. E.: Multislab multigroup transport theory with L'th order anisotropic scattering. *Journal of Computational Physics*, **50**, 181–192 (1983).

Chapter 10
Relaxation of Periodic and Nonstandard Growth Integrals by Means of Two-Scale Convergence

Joel Fotso Tachago, Hubert Nnang, and Elvira Zappale

10.1 Introduction

In [F0Nn12], the authors extended the notion of two-scale convergence introduced by [Ng89] (see also [Al94, CiEtAl08, FoZa03, Vi06] among a wider literature for extensions and related notions) to the Orlicz-Sobolev setting and obtained, under strict convexity assumption on f and suitable boundary conditions, the existence of a unique minimizer for a suitable limit functional as the limit of the minimizers of the original functionals $\int_\Omega f(\frac{x}{\varepsilon}, Du)dx$ as $\varepsilon \to 0$.

In particular they proved (cf. [F0Nn12, Corollary 5.2]) that for every sequence $(u_\varepsilon)_\varepsilon \in W^1 L^B(\Omega; \mathbb{R})$ such that $(Du_\varepsilon)_\varepsilon$ weakly 2s-converges to $\mathbb{D}u_0 = Du + D_y u_1$, where $u_0 = (u, u_1) \in W^1 L^B(\Omega) \times L^1(\Omega; W^1_\sharp L^B_{\mathrm{per}}(Y))$. Then

$$\iint_{\Omega \times Y} f(y, \mathbb{D}u_0)dxdy \leq \liminf_{\varepsilon \to 0} \int_\Omega f\left(\frac{x}{\varepsilon}, Du_\varepsilon\right) dx, \tag{10.1}$$

where $Y := (0, 1)^d$ ($d \in \mathbb{N}$) and $\mathbb{D}u_0 := Du + D_y u_1$ (see Sect. 10.2 for the notations adopted in this introduction).

J. F. Tachago
Faculty of Sciences, Department of Mathematics and Computer Sciences, University of Bamenda, Bambili, Cameroon

Dipartimento di Ingegneria Industriale, Università degli Studi di Salerno, Fisciano, SA, Italy

H. Nnang
University of Yaoundé I and École Normale Supérieure de Yaoundé, Yaounde, Cameroon
e-mail: hnnang@u1.uninet.cm

E. Zappale (✉)
Dipartimento di Ingegneria Industriale, Università degli Studi di Salerno, Fisciano, SA, Italy
e-mail: ezappale@unisa.it

© Springer Nature Switzerland AG 2019   123
C. Constanda, P. Harris (eds.), *Integral Methods in Science and Engineering*,
https://doi.org/10.1007/978-3-030-16077-7_10

On the other hand, by the very nature of two-scale converge they obtained, under homogeneous boundary conditions on $\partial\Omega$, (the same proof can be performed for any boundary conditions), for u and u_1 regular, the existence of a suitable sequence $(\overline{u}_\varepsilon)_\varepsilon \subseteq W^1 L^B(\Omega)$ such that $\overline{u}_\varepsilon \rightharpoonup u$, and the opposite inequality holds:

$$\lim_{\varepsilon \to 0} \int_\Omega f\left(\frac{x}{\varepsilon}, D\overline{u}_\varepsilon\right) dx = \iint_{\Omega \times Y} f(y, \mathbb{D}u_0) dx dy.$$

Here, by means of two-scale convergence we aim to extend their result to any couple of functions $u_0 \equiv u + u_1$ (or (u, u_1)) $\in W^1 L^B(\Omega) \times L^1(\Omega; W^1_\sharp L^B_{\mathrm{per}}(Y))$, and also to obtain an integral representation result for

$$\inf\left\{\liminf_{\varepsilon \to 0} \int_\Omega f\left(\frac{x}{\varepsilon}, Du_\varepsilon\right) dx : u_\varepsilon \rightharpoonup u \text{ weakly in } W^1 L^B(\Omega)\right\}.$$

Indeed, after stating preliminary results in Sect. 10.2 on Orlicz-Sobolev spaces and homogenization theory, in Sect. 10.3 we will prove the following theorem:

Theorem 1 *Let Ω be a bounded open set with Lipschitz boundary and let $f : \Omega \times \mathbb{R}^d \to \mathbb{R}$ be a Carathéodory function such that*

$$f(x, \cdot) \text{ is convex for a.e.} x \in \Omega,$$

and there exist constants c, c' and $C \in \mathbb{R}^+$ such that for a.e. $x \in \Omega$ and every $\xi \in \mathbb{R}^d$,

$$cB'(|\xi|) - c' \leq f(x, \xi) \leq C(1 + B(|\xi|)) \tag{10.2}$$

with B, B' equivalent N-functions which satisfy the $\triangle_2$ condition. Then, it results that for every $u \in W^1 L^B(\Omega)$,

$$\inf\left\{\liminf_{\varepsilon \to 0} \int_\Omega f\left(\frac{x}{\varepsilon}, Du_\varepsilon\right) dx : u_\varepsilon \rightharpoonup u\right.$$

$$\left. \text{weakly in } W^1 L^B(\Omega)\right\} \tag{10.3}$$

$$= \inf\left\{\limsup_{\varepsilon \to 0} \int_\Omega f\left(\frac{x}{\varepsilon}, Du_\varepsilon\right) dx : u_\varepsilon \rightharpoonup u \text{ weakly in } W^1 L^B(\Omega)\right\}$$

$$= \int_\Omega f_{\mathrm{hom}}(Du) dx,$$

where $f_{\mathrm{hom}} : \mathbb{R}^d \to \mathbb{R}$ is the density defined by

$$f_{\mathrm{hom}}(\xi) := \inf\left\{\int_Y f(y, \xi + Du) dy : u \in W^1 L^B_{\mathrm{per}}(Y)\right\}. \tag{10.4}$$

We underline that the analysis presented in this paper holds also in the vectorial case, i.e. fields $u \in W^1 L^B(\Omega; \mathbb{R}^m)$, with the exact same techniques, provided that $f(x, \cdot)$ is convex.

Furthermore, in order to prove (10.3), we also obtain for every $u_0 \in W^1 L^B(\Omega) \times L^1(\Omega; W^1_\# L^B_{per}(Y))$, the following two-scale representation:

$$\inf\left\{\liminf_{\varepsilon \to 0} \int_\Omega f\left(\frac{x}{\varepsilon}, Du_\varepsilon\right) dx : u_\varepsilon \overset{2s}{\rightharpoonup} u_0\right\}$$

$$= \inf\left\{\limsup_{\varepsilon \to 0} \int_\Omega f\left(\frac{x}{\varepsilon}, Du_\varepsilon\right) dx : u_\varepsilon \overset{2s}{\rightharpoonup} u_0\right\} = \iint_{\Omega \times Y} f(y, \mathbb{D}u_0) dx dy.$$

10.2 Preliminaries

This section is devoted to fix notation adopted in the sequel and state preliminary results on Orlicz-Sobolev spaces and homogenization results that will be exploited in the next section. For more details concerning these latter results, for the sake of brevity, we will refer directly to [F0Nn12].

$\Omega \subset \mathbb{R}^d$ ($d \in \mathbb{N}$) denotes a bounded open set with Lipschitz boundary.

10.2.1 Orlicz-Sobolev Spaces

Let $B : [0, +\infty[\to [0, +\infty[$ be an N-function as in [Ad75], i.e., B is continuous, convex, with $B(t) > 0$ for $t > 0$, $\frac{B(t)}{t} \to 0$ as $t \to 0$, and $\frac{B(t)}{t} \to \infty$ as $t \to \infty$.

Equivalently, B is of the form $B(t) = \int_0^t b(\tau) d\tau$, where $b : [0, +\infty[\to [0, +\infty[$ is non decreasing, right continuous, with $b(0) = 0, b(t) > 0$ if $t > 0$ and $b(t) \to +\infty$ if $t \to +\infty$. We denote by $\tilde{B}$, the Fenchel's conjugate, also called the complementary N-function of B defined by

$$\tilde{B}(t) = \sup_{s \geq 0}\{st - B(s)\}, \quad t \geq 0.$$

It can be proven that (see [F0Nn12, Lemma 2.1]) if B is an N-function and $\tilde{B}$ is its conjugate, then for all $t > 0$, it results

$$\frac{tb(t)}{B(t)} \geq 1 (> 1 \text{ if } b \text{ is strictly increasing}),$$

$$\tilde{B}(b(t)) \leq tb(t) \leq B(2t).$$

An N-function B is of class $\triangle_2$ (denoted $B \in \triangle_2$) if there are $\alpha > 0$ and $t_0 \geq 0$ such that $B(2t) \leq \alpha B(t)$ for all $t \geq t_0$.

In all what follows B and $\tilde{B}$ are conjugates N-function*s* satisfying the (Δ_2) condition and c refers to a constant that may vary from line to line.

The Orlicz-space $L^B(\Omega) = \left\{ u : \Omega \to \mathbb{C} \text{ measurable, } \lim_{\delta \to 0^+} \int_\Omega B(\delta |u(x)|)\, dx = 0 \right\}$ is a Banach space for the Luxemburg norm:

$$\|u\|_{B,\Omega} = \inf \left\{ k > 0 : \int_\Omega B\left(\frac{|u(x)|}{k}\right) dx \leq 1 \right\} < +\infty.$$

It follows that: $C_c^\infty(\Omega)$ is dense in $L^B(\Omega)$, $L^B(\Omega)$ is separable and reflexive, the dual of $L^B(\Omega)$ is identified with $L^{\tilde{B}}(\Omega)$, and the norm induced on $L^{\tilde{B}}(\Omega)$ as a dual space is equivalent to $\|.\|_{\tilde{B},\Omega}$.

Analogously one can define the Orlicz-Sobolev space as follows:

$$W^1 L^B(\Omega) = \left\{ u \in L^B(\Omega) : \frac{\partial u}{\partial x_i} \in L^B(\Omega), 1 \leq i \leq d \right\},$$

where derivatives are taken in the distributional sense on Ω. Endowed with the norm $\|u\|_{W^1 L^B(\Omega)} = \|u\|_{B,\Omega} + \sum_{i=1}^d \left\| \frac{\partial u}{\partial x_i} \right\|_{B,\Omega}$, $u \in W^1 L^B(\Omega)$, $W^1 L^B(\Omega)$ is a reflexive Banach space. We denote by $W_0^1 L^B(\Omega)$, the closure of $C_c^\infty(\Omega)$ in $W^1 L^B(\Omega)$ and the semi-norm $u \to \|u\|_{W_0^1 L^B(\Omega)} = \|Du\|_{B,\Omega} = \sum_{i=1}^d \left\| \frac{\partial u}{\partial x_i} \right\|_{B,\Omega}$ is a norm on $W_0^1 L^B(\Omega)$ equivalent to $\|.\|_{W^1 L^B(\Omega)}$.

10.2.2 Homogenization

In order to deal with periodic integrands we will adopt the following notation.

Let $Y := (0,1)^d$. The letter ε throughout will denote a family of positive real numbers converging to 0. The set $\mathbb{R}_y^d$ will denote $\mathbb{R}^d$, but the subscript y emphasizes the fact that this is the set where the space variable y is. We also define

$$C_{\text{per}}(Y) = \{ v \in C(\mathbb{R}_y^d) : Y - \text{periodic} \},$$

and

$$L_{\text{per}}^B(Y) := \{ v \in L_{\text{loc}}^B(\mathbb{R}_y^N) : Y - \text{periodic} \}.$$

Moreover we observe that $L_{\text{per}}^B(Y)$ is a Banach space under the Luxemburg norm $\| \cdot \|_{B,Y}$, and $C_{\text{per}}(Y)$ is dense in $L_{\text{per}}^B(Y)$ (see [FONn12, Lemma 2.1]).

For $v \in L^B_{\mathrm{per}}(Y)$ let

$$v^\varepsilon(x) = v\left(\frac{x}{\varepsilon}\right), \ x \in \mathbb{R}^d.$$

Given $v \in L^B_{\mathrm{loc}}(\Omega \times \mathbb{R}^N_y)$ and $\varepsilon > 0$, we put

$$v^\varepsilon(x) = v\left(x, \frac{x}{\varepsilon}\right), \ x \in \mathbb{R}^d \text{ whenever it makes sense.}$$

We define the vector space

$$L^B(\Omega \times Y_{\mathrm{per}}) := \{u \in L^B_{\mathrm{loc}}(\Omega \times \mathbb{R}^N_y) : \text{ for a.e. } x \in \Omega, u(x, \cdot) \text{ is } Y - \text{ periodic}\}.$$

and observe that the embedding $L^B(\Omega, C_{\mathrm{per}}(Y)) \to L^B(\Omega \times Y_{\mathrm{per}})$ is continuous.
Moreover we will make use of the space

$$W^1 L^B_{\mathrm{per}}(Y) := \left\{u \in W^1 L^B_{\mathrm{loc}}(\mathbb{R}^N_y) : u, \frac{\partial u}{\partial x_i}, i = 1, \ldots, N, \ Y - \text{ periodic}\right\}$$

where the derivative $\frac{\partial u}{\partial x_i}$ is taken in the distributional sense on $\mathbb{R}^N_y$, and we endow
it with the norm $\|u\|_{W^1 L^B_{\mathrm{per}}} = \|u\|_{B,Y} + \sum_{i=1}^N \left\|\frac{\partial u}{\partial x_i}\right\|_{B,Y}$, which makes it a Banach
space.
We also consider the space

$$W^1_\sharp L^B_{\mathrm{per}}(Y) = \left\{u \in W^1 L^B_{\mathrm{per}}(Y) : \int_Y u(y)dy = 0\right\},$$

and we endow it with the gradient norm

$$\|u\|_{W^1_\sharp L^B_{\mathrm{per}}(Y)} = \sum_{i=1}^N \left\|\frac{\partial u}{\partial x_i}\right\|_{B,Y}.$$

Denoting by $C^\infty_{\mathrm{per}}(Y) = C_{\mathrm{per}}(Y) \cap C^\infty(\mathbb{R}^N)$, and recalling that the space
$C^\infty_{\sharp,\mathrm{per}}(Y; \mathbb{R}) = \left\{u \in C^\infty_{\mathrm{per}}(Y; \mathbb{R}) \int_Y : u(y)dy = 0\right\}$ is dense in $W^1_\sharp L^B_{\mathrm{per}}(Y)$, one
can deduce (cf. [F0Nn12]) the density of the embedding

$$C^\infty_c(\Omega; \mathbb{R}) \otimes C^\infty_{\sharp,\mathrm{per}}(Y; \mathbb{R}) \subseteq L^1(\Omega; W^1_\sharp L^B_{\mathrm{per}}(Y)). \tag{10.5}$$

In [F0Nn12] the notion of two-scale convergence introduced by [Ng89] and
developed by [Al94] (see also, among a wide literature, [CiEtAl08, FoZa03, Ne,
Vi06] for further developments and related notions like periodic unfolding method),
has been extended to the Orlicz-Sobolev setting.

Definition 1 A sequence of functions $(u_\varepsilon)_\varepsilon$ in $L^B(\Omega)$ is said to be:

- weakly two-scale convergent in $L^B(\Omega)$ to a function $u_0 \in L^B(\Omega \times Y_{\text{per}})$ if for every $\varepsilon \to 0$, we have

$$\int_\Omega u_\varepsilon f^\varepsilon dx \to \iint_{\Omega \times Y} u_0 f dx dy, \text{ for all } f \in L^{\widetilde{B}}(\Omega; C_{per}(Y)) \qquad (10.6)$$

- strongly two-scale convergent in $L^B(\Omega)$ to $u_0 \in L^B(\Omega \times Y_{\text{per}})$ if for $\eta > 0$ and $f \in L^B(\Omega; C_{per}(Y))$ verifying $\|u_0 - f\|_{L^B(\Omega \times Y)} \leq \frac{\eta}{2}$ there exist $\rho > 0$ such that $\|u_\varepsilon - f^\varepsilon\|_{L^B(\Omega)} \leq \eta$ for all $0 < \varepsilon \leq \rho$.

When (10.6) happens for all $f \in L^{\widetilde{B}}(\Omega; C_{per}(Y))$ we denote it by "$u_\varepsilon \overset{2s}{\rightharpoonup} u_0$ in $L^B(\Omega)$ weakly" or simply" $u_\varepsilon \overset{2s}{\to} u_0$ in $L^B(\Omega)$ two-scale weakly" and we will say that u_0 is the weak two-scale limit in $L^B(\Omega)$ of the sequence $(u_\varepsilon)_\varepsilon$. In order to denote strong two-scale convergence of $u_\varepsilon \to u_0$ we adopt the symbol $\|u_\varepsilon - u_0\|_{2s-L^B(\Omega \times Y)} \to 0$.

The following result, whose proof can be found in [F0Nn12], allows to extend the notion of weak two-scale convergence at Orlicz-Sobolev functions, guaranteeing, at the same time, a compactness result.

Proposition 1 *Let Ω be a bounded open set in $\mathbb{R}^d$ and let $(u_\varepsilon)_\varepsilon$ be bounded in $W^1 L^B(\Omega)$. There exist a subsequence, still denoted in the same way, and $u \in W^1 L^B(\Omega)$, $u_1 \in L^1(\Omega; W^1_\# L^B_{\text{per}}(Y))$ such that:*

(i) $u_\varepsilon \overset{2s}{\rightharpoonup} u$ in $L^B(\Omega)$,

(ii) $D_{x_i} u_\varepsilon \overset{2s}{\rightharpoonup} D_{x_i} u + D_{y_i} u_1$ in $L^B(\Omega)$, $1 \leq i \leq d$.

In the sequel we denote by $u_0(x, y)$ the function $u(x) + u_1(x, y)$, and by $\mathbb{D}u_0$ the vector $Du + D_y u_1$.

For the sake of brevity, we cannot explicitly quote all the results used throughout the paper but we will refer to [F0Nn12] for further necessary properties of Orlicz-Sobolev spaces, two-scale convergence and homogenization in the Orlicz setting.

10.3   Proof of Theorem 1

This section is devoted to the proof of Theorem 1. To this aim recall the definition of f_{hom} given by (10.4).

Proof (of Theorem 1) We start observing that the coercivity assumptions on f, the compactness result, given by Proposition 1, and (10.1) guarantee that for every $u_\varepsilon \rightharpoonup u \in W^1 L^B(\Omega)$

$$\liminf_{\varepsilon \to 0} \int_\Omega f\left(\frac{x}{\varepsilon}, Du_\varepsilon\right) dx \geq \iint_{\Omega \times Y} f(y, \mathbb{D}u_0) dx dy, \qquad (10.7)$$

where $u_0(x, y) = u(x) + u_1(x, y)$ is the weak two-scale limit of $(u_\varepsilon)_\varepsilon$. Clearly passing to the infimum on both sides of Eq. (10.7), and recalling that $\mathbb{D}u_0 = Du + D_y u_1$, we obtain

$$\inf\left\{\liminf_{\varepsilon\to 0}\int_\Omega f\left(\frac{x}{\varepsilon}, Du_\varepsilon\right)dx : u_\varepsilon \rightharpoonup u\right.$$
$$\left.\text{in } W^1 L^B(\Omega)\right\}$$
$$\geq \inf\left\{\iint_{\Omega\times Y} f(y, Du + D_y u_1)dxdy : u_1 \in L^1(\Omega; W^1_\sharp L^B_{\text{per}}(Y))\right\}$$
$$= \int_\Omega f_{\text{hom}}(Du)dx,$$

where one can replicate the same proof as [CiEtAl06, Lemma 2.2] replacing t by 1 and f_1 by f_{hom} in (10.4) therein and exploit the convexity of f to replace functions with null boundary datum on ∂Y, with periodic ones (see also end of [Ne, Chapter 3]).

The upper bound exploits an argument very similar to the one presented in [Ne], relying, in the present context, on the density result in (10.5). Indeed we can first observe that, as in [FONn12, Corollary 5.1] for any given $u \in C^\infty(\overline{\Omega})$ and $\phi_1 \in C^\infty_c(\Omega) \otimes C^\infty_{\text{per}}(Y)$ it results that, given $\phi_\varepsilon(x) := u(x) + \varepsilon\phi_1\left(x, \frac{x}{\varepsilon}\right)$

$$\lim_{\varepsilon\to 0}\int_\Omega f\left(\frac{x}{\varepsilon}, D\phi_\varepsilon(x)\right)dx = \iint_{\Omega\times Y} f(y, Du + D_y\phi_1(x, y))dxdy. \qquad (10.8)$$

On the other hand, given $u \in W^1 L^B(\Omega)$ and $u_1 \in L^1(\Omega; W^1_\sharp L^B_{\text{per}}(Y))$, (10.5) guarantees that for each $\delta > 0$ we can find maps $u_\delta \in C^\infty(\overline{\Omega})$ and $v_\delta \in C^\infty_c(\Omega; C^\infty_{\text{per}}(Y))$ (this latter with zero average) such that

$$\|u - u_\delta\|_{W^1 L^B(\Omega)} + \|u_1 - v_\delta\|_{L^1(\Omega; W^1 L^B_{\text{per}}(Y)} \leq \delta \qquad (10.9)$$

Next defining, for every δ, and for every $x \in \Omega$,

$$u_{\delta,\varepsilon}(x) := u_\delta(x) + \varepsilon v_\delta^\varepsilon(x),$$

one has

$$Du_{\delta,\varepsilon}(x) = Du_\delta(x) + \varepsilon D_x v_\delta\left(x, \frac{x}{\varepsilon}\right) + D_y v_\delta\left(x, \frac{x}{\varepsilon}\right).$$

Clearly, as $\varepsilon \to 0$, it results

$$u_{\delta,\varepsilon} \to u_\delta \text{ in } L^B(\Omega),$$

$$Du_{\delta,\varepsilon}(x) \xrightarrow{2s} Du_\delta(x) + D_y v_\delta(x, y) \text{ strongly in } L^B(\Omega \times Y_{\text{per}}).$$

Now we define

$$c_{\delta,\varepsilon} := \|u_{\delta,\varepsilon} - u\|_{W^1 L^B(\Omega)} + \Big| \big\|Du_{\delta,\varepsilon}\big\|_{L^B(\Omega)} - \big\|Du - D_y u_1\big\|_{L^B(\Omega \times Y)} \Big|, \tag{10.10}$$

having the aim of constructing, via a diagonalizing argument, a sequence strongly two-scale convergent to $u_0 = u + u_1$.

Thus, it is easily seen that

$$\lim_{\delta \to 0} \lim_{\varepsilon \to 0} c_{\delta,\varepsilon} = 0,$$

which allows us to apply H. Attouch Diagonalization Lemma, thus detecting a sequence $\delta(\varepsilon) \to 0$ as $\varepsilon \to 0$, such that $c_{\delta(\varepsilon),\varepsilon} \to 0$ and $u_{\delta(\varepsilon),\varepsilon} \to u$ in $L^B(\Omega)$, with

$$Du_{\delta(\varepsilon),\varepsilon}(x) \xrightarrow{2s} Du(x) + D_y u_1(x, y) \text{ strongly in } L^B(\Omega \times Y_{\text{per}}).$$

This latter convergence and [FONn12, Remark 4.1] ensure that $Du_{\delta(\varepsilon),\varepsilon} \rightharpoonup Du$ weakly in $L^B(\Omega)$, thus, (10.8), the continuity of f in the second variable, (10.2), guarantee that for every $u_1 \in L^1(\Omega; W^1_\sharp L^B_{\text{per}}(Y))$,

$$\lim_{\varepsilon \to 0} \int_\Omega f\left(\frac{x}{\varepsilon}, Du_{\delta(\varepsilon),\varepsilon}\right) dx = \iint_{\Omega \times Y} f(y, Du + D_y u_1(x, y)) dx dy.$$

as desired.

Thus we can conclude that

$$\inf \left\{ \limsup_{\varepsilon \to 0} \int_\Omega f\left(\frac{x}{\varepsilon}, Du_\varepsilon\right) dx : u_\varepsilon \rightharpoonup u \text{ in } W^1 L^B(\Omega) \right\}$$

$$\leq \lim_{\varepsilon \to 0} \int_\Omega f\left(\frac{x}{\varepsilon}, Du_{\delta(\varepsilon),\varepsilon}\right) dx \leq \iint_{\Omega \times Y} f(y, Du(x) + D_y u_1(x, y)) dx dy.$$

Hence

$$\inf \left\{ \limsup_{\varepsilon \to 0} \int_\Omega f\left(\frac{x}{\varepsilon}, Du_\varepsilon\right) dx : u_\varepsilon \rightharpoonup u \text{ in } W^{1,B}(\Omega; \mathbb{R}) \right\}$$

$$\leq \inf \left\{ \iint_{\Omega \times Y} f(y, Du + D_y u_1) dx dy : u_1 \in L^1(\Omega; W^1_\sharp L^B_{\text{per}}(Y)) \right\}.$$

This together with the last equality in (10.7) concludes the proof.

Acknowledgements The first and the third author acknowledge the support of the Programme ICTP-INdAM research in pairs 2018. Joel Fotso Tachago thanks Dipartimento di Ingegneria Industriale at University of Salerno for its hospitality. Elvira Zappale is a member of GNAMPA-

INdAM, whose support is gratefully acknowledged. This paper was written during a research stay of Joel Fotso Tachago at University of Salerno sponsored by ICTP-INdAM.

References

[Ad75] Adams R. A., Sobolev spaces, Pure and Applied Mathematics, Vol. 65, Academic Press [A subsidiary of Harcourt Brace Jovanovich, Publishers], New York-London, 1975, xviii+268.

[Al94] Allaire, G., Two-scale convergence: a new method in periodic homogenization, Nonlinear partial differential equations and their applications. Collège de France Seminar, Vol. XII (Paris, 1991–1993), Pitman Res. Notes Math. Ser., **302**, 1–14, Longman Sci. Tech., Harlow, 1994.

[CiEtAl06] Cioranescu, D., Damlamian, A., and De Arcangelis, R., Homogenization of quasiconvex integrals via the periodic unfolding method., SIAM J. Math. Anal., **37**, (2006),**5**,1435–1453.

[CiEtAl08] Cioranescu, D. Damlamian, A., and Griso G., The periodic unfolding method in homogenization, SIAM J. Math. Anal., **40**, (2008), n. 4, 1585–1620.

[F0Nn12] Fotso Tachago, J., and Nnang, H. Two-scale convergence of integral functionals with convex, periodic and nonstandard growth integrands, Acta Appl. Math., **121**, (2012), 175–196.

[FoZa03] Fonseca, I, and Zappale E., Multiscale relaxation of convex functionals, J. Convex Anal., **10**,(2003), n. 2, 325–350.

[Ne] Neukamm, S. Ph.d Dissertation

[Ng89] Nguentseng, A general convergence result for a functional related to the theory of homogenization. SIAM J. Math. Anal. **20**, (1989), 608–623.

[Vi06] Visintin, A. Towards a two-scale calculus, ESAIM Control Optim. Calc. Var., **12**, (2006), n. 3, 371–397.

Chapter 11
A Stiff Problem: Stationary Waves and Approximations

**Delfina Gómez, Santiago Navazo-Esteban,
and María-Eugenia Pérez-Martínez**

11.1   Introduction and Statement of the Problem

In this paper we revisit a Stiff problem, cf. (11.7), widely studied in the literature of Applied Mathematics using different techniques. It deals with the asymptotic behavior of the eigenvalues and eigenfunctions of a problem for the Laplace operator posed in a domain Ω of $\mathbb{R}^N$: this domain is composed of two parts in which the stiffness constants are of different order of magnitude, namely, $O(\varepsilon)$ and $O(1)$, respectively, where ε is a parameter $\varepsilon \ll 1$. Here, for each fixed ε, we give explicit formulas for the eigenvalues and the eigenfunctions which extend those in [Pe95, LoPe97] for the dimension 1 of the space, while, as $\varepsilon \to 0$, we provide approaches to solutions of the evolution problem (11.1) via "standing waves". These approaches are valid for long times which can also be estimated in terms of the parameter ε (cf. (11.34), (11.35), (11.38) and (11.39)). Due to the estimates (11.8), the so-called *low frequencies* are the eigenvalues λ^ε of (11.7) of order $O(\varepsilon)$, while the eigenvalues of order $O(\varepsilon^\alpha)$ for $\alpha < 1$ are referred to as the *high frequencies*. We mention [Gi82, Li85, Pa80, SaSa89] for the asymptotic analysis of the low frequencies, as $\varepsilon \to 0$, and [LoSa80, LoPe97, GoBa00, LoEtAl03] for that of the high frequencies. See [LoEtAl05] and [BaGo10] in connection with the elasticity system and a four order differential operator with double contrasts, respectively.

Let us consider Ω a bounded domain of $\mathbb{R}^N$, $N \geq 2$, with a Lipschitz boundary $\partial\Omega$, divided into two parts Ω_0 and Ω_1 by the interface Σ. Namely, $\overline{\Omega} = \overline{\Omega_0} \cup \overline{\Omega_1} \cup \Sigma$, where $\Omega_0, \Omega_1 \subset \mathbb{R}^N$ are two bounded domains which have a Lipschitz boundary and such that $\Omega_0 \cap \Omega_1 = \emptyset$, $\partial\Omega_0 \cap \partial\Omega_1 = \Sigma \neq \emptyset$, and $\partial\Omega \cap \Sigma \neq \emptyset$ (cf. Fig. 11.1).

D. Gómez · S. Navazo-Esteban · M.-E. Pérez-Martínez (✉)
Universidad de Cantabria, Santander, Spain
e-mail: gomezdel@unican.es; santiago.navazo@alumnos.unican.es; meperez@unican.es

© Springer Nature Switzerland AG 2019   133
C. Constanda, P. Harris (eds.), *Integral Methods in Science and Engineering*,
https://doi.org/10.1007/978-3-030-16077-7_11

Fig. 11.1 A possible division
of the domain Ω

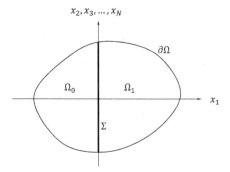

Let us consider the vibration problem

$$
\begin{cases}
\mathbf{u}_{tt}^{\varepsilon} - \Delta_{\mathbf{x}} \mathbf{u}^{\varepsilon} = 0 & \text{in } \Omega_0,\ t > 0, \\
\mathbf{u}_{tt}^{\varepsilon} - \varepsilon \Delta_{\mathbf{x}} \mathbf{u}^{\varepsilon} = 0 & \text{in } \Omega_1,\ t > 0, \\
\mathbf{u}^{\varepsilon} = 0 & \text{on } \partial\Omega,\ t > 0, \\
\mathbf{u}^{\varepsilon}|_{\partial\Omega_0 \cap \Sigma} = \mathbf{u}^{\varepsilon}|_{\partial\Omega_1 \cap \Sigma}, & t > 0, \\
\dfrac{\partial \mathbf{u}^{\varepsilon}}{\partial n}\Big|_{\partial\Omega_0 \cap \Sigma} = \varepsilon \dfrac{\partial \mathbf{u}^{\varepsilon}}{\partial n}\Big|_{\partial\Omega_1 \cap \Sigma} & t > 0, \\
\mathbf{u}^{\varepsilon}(\mathbf{x}, 0) = \phi_0(\mathbf{x}), \quad \mathbf{u}_t^{\varepsilon}(\mathbf{x}, 0) = \phi_1(\mathbf{x}) \text{ for } \mathbf{x} \in \Omega,
\end{cases}
\tag{11.1}
$$

where ε is a small positive parameter that we shall make to go to zero, and n is the outward normal vector to Ω_0 on Σ. (11.1) models the vibration of a body occupying the domain Ω, one part of which Ω_0 is very stiff with respect to the other.

For suitable initial data, $\phi_0 \in H_0^1(\Omega)$ and $\phi_1 \in L^2(\Omega)$, (11.1) admits a weak formulation: find $\mathbf{u}^{\varepsilon}(t)$ with values in $\mathbf{V}$ such that

$$
\left(\frac{d^2 \mathbf{u}^{\varepsilon}}{dt^2}, v \right)_{L^2(\Omega)} + a^{\varepsilon}(\mathbf{u}^{\varepsilon}, v) = 0 \qquad \forall v \in \mathbf{V},
$$
$$
\mathbf{u}^{\varepsilon}(\mathbf{x}, 0) = \phi_0, \qquad \frac{d\mathbf{u}^{\varepsilon}}{dt}(\mathbf{x}, 0) = \phi_1,
\tag{11.2}
$$

where $\mathbf{V} = H_0^1(\Omega)$, $\mathbf{H} = L^2(\Omega)$, and

$$
a^{\varepsilon}(u, v) = \int_{\Omega_0} \nabla_{\mathbf{x}} u \cdot \nabla_{\mathbf{x}} v \, dx + \varepsilon \int_{\Omega_1} \nabla_{\mathbf{x}} u \cdot \nabla_{\mathbf{x}} v \, dx \qquad \forall u, v \in \mathbf{V}.
\tag{11.3}
$$

Fixed $\varepsilon > 0$, as a^{ε} is a bilinear, symmetric, continuous and coercive form on $\mathbf{V}$, (11.2) is a standard vibration problem in the space $\mathbf{V}$, $\mathbf{H}$, $\mathbf{V} \subset \mathbf{H}$ with compact imbedding. As is well known, problem (11.2) has a unique solution $\mathbf{u}^{\varepsilon}(t)$,

$$
\mathbf{u}^{\varepsilon} \in L^{\infty}(0, \infty; \mathbf{V}), \qquad \frac{d\mathbf{u}^{\varepsilon}}{dt} \in L^{\infty}(0, \infty; \mathbf{H}),
$$

and it satisfies the conservation of the energy

$$a^\varepsilon(\mathbf{u}^\varepsilon(t), \mathbf{u}^\varepsilon(t)) + \left\| \frac{d\mathbf{u}^\varepsilon}{dt}(t) \right\|_{L^2(\Omega)}^2 = a^\varepsilon(\phi_0, \phi_0) + \|\phi_1\|_{L^2(\Omega)}^2, \qquad \forall t \in \mathbb{R};$$

$$(11.4)$$

see Section III.8 in [LiMa68], Sections I.6 and III.11 in [SaSa89] and Sections IV.5 and XII.3 in [Sa80] for the general theory.

The spectral problem associated with (11.2) is: find λ^ε, $u^\varepsilon \in \mathbf{V}$, $u^\varepsilon \neq 0$ such that

$$\int_{\Omega_0} \nabla_{\mathbf{x}} u^\varepsilon \cdot \nabla_{\mathbf{x}} v \, d\mathbf{x} + \varepsilon \int_{\Omega_1} \nabla_{\mathbf{x}} u^\varepsilon \cdot \nabla_{\mathbf{x}} v \, d\mathbf{x} = \lambda^\varepsilon \int_\Omega u^\varepsilon v \, d\mathbf{x} \qquad \forall v \in \mathbf{V}. \quad (11.5)$$

Fixed $\varepsilon > 0$, let us consider

$$0 < \lambda_1^\varepsilon \leq \cdots \leq \lambda_n^\varepsilon \leq \cdots \xrightarrow{n \to \infty} \infty \qquad (11.6)$$

the sequence of the eigenvalues, with the usual convention of the repeated eigenvalues. Let $\{u_n^\varepsilon\}_{n \geq 1}$ be the corresponding eigenfunctions, which are assumed to be an orthonormal basis in $L^2(\Omega)$, i.e. $\int_\Omega u_m^\varepsilon u_n^\varepsilon d\mathbf{x} = \delta_{m,n}$. Problem (11.5) is a variational formulation of the problem

$$\begin{cases} -\Delta_{\mathbf{x}} u^\varepsilon = \lambda^\varepsilon u^\varepsilon & \text{in } \Omega_0, \\ -\varepsilon \Delta_{\mathbf{x}} u^\varepsilon = \lambda^\varepsilon u^\varepsilon & \text{in } \Omega_1, \\ u^\varepsilon = 0 & \text{on } \partial\Omega, \\ u^\varepsilon|_{\partial\Omega_0 \cap \Sigma} = u^\varepsilon|_{\partial\Omega_1 \cap \Sigma}, \\ \dfrac{\partial u^\varepsilon}{\partial n}\Big|_{\partial\Omega_0 \cap \Sigma} = \varepsilon \dfrac{\partial u^\varepsilon}{\partial n}\Big|_{\partial\Omega_1 \cap \Sigma}. \end{cases} \qquad (11.7)$$

The asymptotic behavior, as $\varepsilon \to 0$, of the eigenvalues and the eigenfunctions of (11.5) has been widely studied with different techniques; see, for example, [SaSa89, LoPe97] and the references therein. We state here some bounds for the eigenvalues as well as the main convergence result for the *low frequencies*, which will be useful throughout the work.

Lemma 1 *Let $\{\lambda_n^\varepsilon\}_{n \geq 1}$ be the eigenvalues of* (11.5). *For each fixed $n \in \mathbb{N}$, we have*

$$C\varepsilon \leq \lambda_n^\varepsilon \leq C_n\varepsilon, \qquad (11.8)$$

where C, C_n are constants independent of ε, C independent of n, and $C_n \to \infty$ as $n \to \infty$. Moreover, for each $n \in \mathbb{N}$, the values $\lambda_n^\varepsilon/\varepsilon$ converge with conservation of the multiplicity, when $\varepsilon \to 0$, towards the eigenvalues λ_n^* of the Dirichlet problem

$$\begin{cases} -\Delta_{\mathbf{x}} u = \lambda u & \text{in } \Omega_1, \\ u = 0 & \text{on } \partial\Omega_1. \end{cases} \qquad (11.9)$$

The corresponding eigenfunctions u_n^ε, $\|u_n^\varepsilon\|_{L^2(\Omega)} = 1$, converge towards u_n^* in $L^2(\Omega)$, where u_n^* is an eigenfunction associated with λ_n^* of problem (11.9), u_n^* extended by 0 to Ω_0.

Let us notice that on the one hand, it has been shown in [Pe95, LoPe97] that in the simplest case of dimension $N = 1$, it may be not easy to obtain explicit formulas for the eigenvalues and eigenfunctions of (11.7). Also, dealing with the dimension $N = 1$ and explicit computations, we refer to [GoEtAl98, CaEtAl05] for vibrating systems with concentrated masses. On the other hand, it is well known that, for $\phi_0(\mathbf{x}) = \alpha u_k^\varepsilon(\mathbf{x})$, and $\phi_1(\mathbf{x}) = \beta u_k^\varepsilon(\mathbf{x})$, where $\alpha, \beta \in \mathbb{R}$, k fixed and $\{\lambda_n^\varepsilon, u_n^\varepsilon\}_{n \geq 1}$ are the eigenelements of (11.7), the solution of (11.1) is the standing wave

$$\mathbf{u}^\varepsilon(\mathbf{x}, t) = u_k^\varepsilon(\mathbf{x}) \left(\alpha \cos(\sqrt{\lambda_k^\varepsilon} t) + \frac{\beta}{\sqrt{\lambda_k^\varepsilon}} \sin(\sqrt{\lambda_k^\varepsilon} t) \right). \tag{11.10}$$

However, to consider the eigenfunctions (11.7), $u_k^\varepsilon(\mathbf{x})$, as initial data in (11.1) may not be useful to localize certain kinds of vibrations such as those affecting only the less stiff part of the body, namely, localized in Ω_1.

In this way, the aim of this paper is double. First, for fixed ε, we obtain explicit formulas for the eigenelements of (11.7) when $N \geq 2$ and a certain separation of variables is allowed. These formulas strongly depend on the range of variation of the spectral parameter λ^ε (cf. Remark 1). Second, using the eigenfunctions of the Dirichlet problem (11.9) as initial data in (11.1), namely, for $\phi_0(\mathbf{x}) = \alpha u_k^*(\mathbf{x})$, and $\phi_1(\mathbf{x}) = \beta u_k^*(\mathbf{x})$ we show that its solution can be approached by

$$u_k^*(\mathbf{x}) \left(\alpha \cos(\sqrt{\lambda_k^*\varepsilon} t) + \frac{\beta}{\sqrt{\lambda_k^*\varepsilon}} \sin(\sqrt{\lambda_k^*\varepsilon} t) \right) \tag{11.11}$$

for $t \leq t_{\varepsilon,k}$, where (λ_k^*, u_k^*) is defined in Lemma 1, and $t_{\varepsilon,k}$ depends on ε and the convergence rate of the eigenvalues (i.e., $|\lambda_k^\varepsilon - \lambda_k^*\varepsilon|$), and fixed k, $t_{\varepsilon,k} \xrightarrow[\varepsilon \to 0]{} \infty$. Note that, in this case, the solution (11.10) can also be approached by (11.11) for $t \leq t_{\varepsilon,k}$ (cf. Theorems 1 and 2). In any case, the associated vibrations leave the stiffer part Ω_0 almost at rest.

In order to approach the solutions of the evolution problem (11.1), we need to obtain improved convergence results for the eigenfunctions of (11.7) which somewhat implies obtaining convergence rates for the eigenfunctions in Ω_0 (cf. Proposition 1). Then, we use the technique developed in [Pe08] within a general abstract framework for second order evolution problems, which among other things imply using the energy conservation law (11.4). This technique can be applied in our problem to localize other kinds of vibrations associated with the high frequencies, for instance, those affecting the stiffer part of the body Ω_0 (cf. Remark 2). In this connection, we obtain approaches to solutions via "standing waves" in which the eigenfunction (cf. (11.10)) must be replaced by "groups of eigenfunctions" associated with eigenvalues λ^ε in "small intervals". In this connection, let us refer to

[Pe08] and [LoPe10] for long time approximations for solutions of wave equations via standing waves for vibrating systems with concentrated masses or for a Steklov problem, respectively.

11.2   Some Explicit Computations for Standing Waves

When the domains Ω_0 and Ω_1 appearing in (11.1) and (11.7) are such that a separation of variables can be performed, and ε is fixed, some explicit computations can be performed. This is the case when Ω, Ω_0, and Ω_1 are prisms in $\mathbb{R}^N$. See [Pe95, LoPe97] to compare with the explicit computations of the eigenelements of (11.7) when $N = 1$.

In this section, we write the main formulas for the dimension $N = 2$, and extend the results for the dimension $N > 2$. The computations become much more complicated than when $N = 1$ and, for the sake of simplicity, when $N = 2$, we take $\Omega_0 = (-1, 0) \times (0, 1)$, $\Omega_1 = (0, 1) \times (0, 1)$ while, for $N > 2$, we consider $\Omega_0 = (-1, 0) \times \tilde{\Omega}$, $\Omega_1 = (0, 1) \times \tilde{\Omega}$, with $\tilde{\Omega}$ a bounded domain of $\mathbb{R}^{N-1}$.

11.2.1   Results for the Dimension $N = 2$

Let us consider problems (11.1) and (11.7) where $\Omega_0 = (-1, 0) \times (0, 1)$, $\Omega_1 = (0, 1) \times (0, 1)$, $\Omega = (-1, 1) \times (0, 1)$, $\mathbf{x} = (x, y) \in \mathbb{R}^2$. Using separation of variables, we look for the eigenelements $(\lambda^\varepsilon, u^\varepsilon)$ of (11.7) in the form

$$u^\varepsilon(x, y) = X^\varepsilon(x) Y^\varepsilon(y). \tag{11.12}$$

Replacing (11.12) in (11.7) we get

$$\begin{cases} \dfrac{X^{\varepsilon''}(x)}{X^\varepsilon(x)} + \lambda^\varepsilon = \dfrac{-Y^{\varepsilon''}(y)}{Y^\varepsilon(y)} = \mu^\varepsilon \ \text{ for } (x, y) \in \Omega_0, \\[2mm] \dfrac{X^{\varepsilon''}(x)}{X^\varepsilon(x)} + \dfrac{\lambda^\varepsilon}{\varepsilon} = \dfrac{-Y^{\varepsilon''}(y)}{Y^\varepsilon(y)} = \hat{\mu}^\varepsilon \ \text{ for } (x, y) \in \Omega_1, \\[2mm] X^\varepsilon(-1) = X^\varepsilon(1) = 0, \qquad\qquad Y^\varepsilon(0) = Y^\varepsilon(1) = 0, \\[2mm] X^\varepsilon(0^-)Y^\varepsilon(y) = X^\varepsilon(0^+)Y^\varepsilon(y) \quad \text{for } y \in (0, 1), \\[2mm] X^{\varepsilon'}(0^-)Y^\varepsilon(y) = \varepsilon X^{\varepsilon'}(0^+)Y^\varepsilon(y) \ \text{for } y \in (0, 1), \end{cases} \tag{11.13}$$

where μ^ε and $\hat{\mu}^\varepsilon$ are constants to be determined. It is easy to check that the only values μ^ε and $\hat{\mu}^\varepsilon$ satisfying (11.13) with $u^\varepsilon(x, y) = X^\varepsilon(x)Y^\varepsilon(y) \neq 0$ are

$$\mu_j^\varepsilon = \hat{\mu}_j^\varepsilon = j^2 \pi^2, \text{ with } j \in \mathbb{N}, \tag{11.14}$$

and, consequently,

$$Y_j^\varepsilon(y) = \sin(j\pi y) \quad \text{for } y \in (0, 1). \tag{11.15}$$

Thus, for $j \in \mathbb{N}$ fixed, $(\lambda^\varepsilon, X^\varepsilon)$ verify

$$\begin{cases} X^{\varepsilon''}(x) = (j^2\pi^2 - \lambda^\varepsilon)X^\varepsilon(x) & \text{for } x \in (-1, 0), \\ X^{\varepsilon''}(x) = \left(j^2\pi^2 - \dfrac{\lambda^\varepsilon}{\varepsilon}\right)X^\varepsilon(x) & \text{for } x \in (0, 1), \\ X^\varepsilon(0^-) = X^\varepsilon(0^+), & X^{\varepsilon'}(0^-) = \varepsilon X^{\varepsilon'}(0^+), \\ X^\varepsilon(-1) = X^\varepsilon(1) = 0. \end{cases} \tag{11.16}$$

For $0 < \varepsilon < 1$ and $j \in \mathbb{N}$ fixed, we distinguish five cases depending on the value of λ^ε; $0 < \lambda^\varepsilon < \varepsilon j^2\pi^2$, $\lambda^\varepsilon = \varepsilon j^2\pi^2$, $\varepsilon j^2\pi^2 < \lambda^\varepsilon < j^2\pi^2$, $\lambda^\varepsilon = j^2\pi^2$ and $\lambda^\varepsilon > j^2\pi^2$. We analyze the different cases.

Cases I and II: It is easy to check that in the two first cases, i.e., $0 < \lambda^\varepsilon < \varepsilon j^2\pi^2$ or $\lambda^\varepsilon = \varepsilon j^2\pi^2$, the only function satisfying (11.16) is $X^\varepsilon \equiv 0$ and there are no eigenfunctions of (11.7) in the form (11.12) for $\lambda^\varepsilon \leq \varepsilon j^2\pi^2$ and $j \in \mathbb{N}$.

Case III: If $\varepsilon j^2\pi^2 < \lambda^\varepsilon < j^2\pi^2$ with $j \in \mathbb{N}$ fixed, it can be proved that only the values λ roots of the equation

$$\tanh\left(\sqrt{j^2\pi^2 - \lambda}\right)\cos\left(\sqrt{\frac{\lambda}{\varepsilon} - j^2\pi^2}\right) + \sqrt{\frac{j^2\pi^2 - \lambda}{\varepsilon\lambda - \varepsilon^2 j^2\pi^2}}\sin\left(\sqrt{\frac{\lambda}{\varepsilon} - j^2\pi^2}\right) = 0, \tag{11.17}$$

verify (11.16) with $X^\varepsilon \not\equiv 0$. Moreover, in this case

$$X_j^\varepsilon(x) = \begin{cases} \tanh\left(\sqrt{j^2\pi^2 - \lambda}\right)\cosh\left(\sqrt{j^2\pi^2 - \lambda}\,x\right) \\ + \sinh\left(\sqrt{j^2\pi^2 - \lambda}\,x\right) & \text{for } x \in (-1, 0), \\ \tanh\left(\sqrt{j^2\pi^2 - \lambda}\right)\cos\left(\sqrt{\dfrac{\lambda}{\varepsilon} - j^2\pi^2}\,x\right) \\ + \sqrt{\dfrac{j^2\pi^2 - \lambda}{\varepsilon\lambda - \varepsilon^2 j^2\pi^2}}\sin\left(\sqrt{\dfrac{\lambda}{\varepsilon} - j^2\pi^2}\,x\right) & \text{for } x \in (0, 1). \end{cases} \tag{11.18}$$

Thus, any λ root of (11.17), $\lambda \in (\varepsilon j^2\pi^2, j^2\pi^2)$, is an eigenvalue of (11.7) and the corresponding eigenfunction is

$$u_j^\varepsilon(x, y) = \alpha^\varepsilon X_j^\varepsilon(x)\sin(j\pi y) \tag{11.19}$$

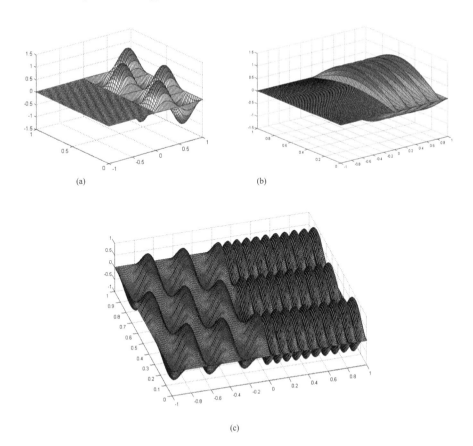

Fig. 11.2 Some examples of eigenfunctions of (11.7) in the ranges III, IV and V, respectively, for different values of ε, $\Omega_0 = (-1, 0) \times (0, 1)$, $\Omega_1 = (0, 1) \times (0, 1)$. (a) $\varepsilon = 0.01$; $\lambda^\varepsilon \approx 1.973289$. (b) $\varepsilon \approx 0.010094$; $\lambda^\varepsilon = \pi^2$. (c) $\varepsilon \approx 0.091439$; $\lambda^\varepsilon \approx 387.382$

where α^ε is a constant such that $\|u_j^\varepsilon\|_{L^2(\Omega)} = 1$, and X_j^ε is given by (11.18); see, for example, Fig. 11.2a.

Case V: If $\lambda^\varepsilon > j^2\pi^2$ with $j \in \mathbb{N}$ fixed, and $\cos(\sqrt{\lambda^\varepsilon - j^2\pi^2})\cos\left(\sqrt{\dfrac{\lambda^\varepsilon}{\varepsilon} - j^2\pi^2}\right)$

$\neq 0$, simple calculations show that the values λ^ε satisfying (11.16) with $X^\varepsilon \not\equiv 0$ are the roots of the equation

$$\sqrt{\varepsilon\lambda - \varepsilon^2 j^2\pi^2}\tan(\sqrt{\lambda - j^2\pi^2}) + \sqrt{\lambda - j^2\pi^2}\tan\sqrt{\frac{\lambda}{\varepsilon} - j^2\pi^2} = 0, \quad (11.20)$$

and the corresponding functions are

$$
\tilde{X}_j^\varepsilon(x) = \begin{cases}
\tan\left(\sqrt{\lambda - j^2\pi^2}\right)\cos\left(\sqrt{\lambda - j^2\pi^2}x\right) \\
\quad + \sin\left(\sqrt{\lambda - j^2\pi^2}x\right) & \text{for } x \in (-1, 0), \\[2mm]
\tan\left(\sqrt{\lambda - j^2\pi^2}\right)\cos\left(\sqrt{\dfrac{\lambda}{\varepsilon} - j^2\pi^2}x\right) \\[2mm]
\quad + \sqrt{\dfrac{\lambda - j^2\pi^2}{\varepsilon\lambda - \varepsilon^2 j^2\pi^2}}\sin\left(\sqrt{\dfrac{\lambda}{\varepsilon} - j^2\pi^2}x\right) & \text{for } x \in (0, 1).
\end{cases}
$$

$$(11.21)$$

Thus, any λ root of (11.20) with $\lambda > j^2\pi^2$ (recall $j \in \mathbb{N}$ fixed) is an eigenvalue of (11.7) and the corresponding eigenfunction is

$$u_j^\varepsilon(x, y) = \alpha^\varepsilon \tilde{X}_j^\varepsilon(x)\sin(j\pi y)$$

where α^ε is a constant such that $\|u_j^\varepsilon\|_{L^2(\Omega)} = 1$, and $\tilde{X}_j^\varepsilon$ is given by (11.21). We observe that the possible eigenvalues $\lambda = \lambda^\varepsilon$ such that $\cos(\sqrt{\lambda^\varepsilon - j^2\pi^2}) = 0$ and $\cos\left(\sqrt{\dfrac{\lambda^\varepsilon}{\varepsilon} - j^2\pi^2}\right) = 0$, with $j \in \mathbb{N}$ fixed, are not included in (11.20). Each one of these values, $\lambda = j^2\pi^2 + \frac{(2k+1)^2\pi^2}{4}$ and $\lambda = \varepsilon j^2\pi^2 + \frac{\varepsilon(2l+1)^2\pi^2}{4}$ with $k, l \in \mathbb{N}_0$, is an eigenvalue of (11.16) (and, consequently, of (11.7)) only for certain values of ε; those of the sequence

$$\varepsilon_{k,l} = \frac{4j^2 + (2k + 1)^2}{4j^2 + (2l + 1)^2}, \quad k, l \in \mathbb{N}_0, k < l.$$

In this case, the corresponding eigenfunctions are

$$
\tilde{X}_{j,k}^\varepsilon(x) = \begin{cases}
\cos\left(\dfrac{(2k + 1)\pi}{2}x\right) & \text{for } x \in (-1, 0), \\[3mm]
\cos\left(\sqrt{\dfrac{j^2 + (\frac{2k+1}{2})^2}{\varepsilon}} - j^2\pi x\right) & \text{for } x \in (0, 1),
\end{cases}
$$

$$(11.22)$$

and

$$u_{j,k}^\varepsilon(x, y) = \alpha^\varepsilon \tilde{X}_{j,k}^\varepsilon(x)\sin(j\pi y),$$

respectively, where α^ε is a constant such that $\|u_{j,k}^\varepsilon\|_{L^2(\Omega)} = 1$, and $\tilde{X}_{j,k}^\varepsilon$ is given by (11.22); see, for example, Fig. 11.2c where $j = 3, k = 5, l = 20$.

Case IV: Finally, we note that the values $\lambda^\varepsilon = j^2\pi^2$ with $j \in \mathbb{N}$ fixed, are eigenvalues of (11.16) (and, hence of (11.7)) only for certain values of ε which are the roots of the equation

$$\sin\left(\sqrt{\frac{j^2\pi^2}{\varepsilon} - j^2\pi^2}\right) + \cos\left(\sqrt{\frac{j^2\pi^2}{\varepsilon} - j^2\pi^2}\right) j\pi\sqrt{\varepsilon - \varepsilon^2} = 0.$$

In this case, the corresponding eigenfunctions are

$$\tilde{X}^\varepsilon_{j,0}(x) = \begin{cases} j\pi\sqrt{\varepsilon - \varepsilon^2}(1+x) & \text{for } x \in (-1,0), \\[2ex] j\pi\sqrt{\varepsilon - \varepsilon^2}\cos\left(\sqrt{\frac{j^2\pi^2}{\varepsilon} - j^2\pi^2 x}\right) \\[2ex] + \sin\left(\sqrt{\frac{j^2\pi^2}{\varepsilon} - j^2\pi^2 x}\right) & \text{for } x \in (0,1). \end{cases}$$

$$(11.23)$$

and

$$u^\varepsilon_{j,0}(x,y) = \alpha^\varepsilon \tilde{X}^\varepsilon_{j,0}(x)\sin(j\pi y), \qquad (11.24)$$

respectively, where α^ε is a constant such that $\|u^\varepsilon_{j,0}\|_{L^2(\Omega)} = 1$, and $\tilde{X}^\varepsilon_{j,0}$ is given by (11.23); see, for example, Fig. 11.2b.

Once the eigenelements of (11.7) are determined, we get the standing waves

$$\mathbf{u}^\varepsilon(x,y,t) = X^\varepsilon(x)\sin(j\pi y)\left(\alpha\cos(\sqrt{\lambda^\varepsilon}t) + \frac{\beta}{\sqrt{\lambda^\varepsilon}}\sin(\sqrt{\lambda^\varepsilon}t)\right),$$

where $\alpha, \beta \in \mathbb{R}$, X^ε is defined by (11.18), (11.21), (11.22) or (11.23) and λ^ε is the corresponding eigenvalue (root of (11.17), (11.20), $\lambda = j^2\pi^2 + \frac{(2k+1)^2\pi^2}{4}$ or $\lambda = j^2\pi^2$, respectively).

Remark 1 Note that with the explicit formulas we do not determine the eigenvalues ordered as in the sequence (11.6). However, the low frequencies belong to the range III and Fig. 11.2a illustrates that the eigenfunctions almost vanish in the stiffer part.

Remark 2 Note that Fig. 11.2c shows the graphic of an eigenfunction associated with a *high frequency* and, further specifying, with an eigenvalue λ^ε which coincides with an eigenvalue of the following problem in Ω_0.

$$\begin{cases} -\Delta_{\mathbf{x}} u = \lambda u & \text{in } \Omega_0, \\ u = 0 & \text{on } \partial\Omega_0 \setminus \Sigma, \\ \dfrac{\partial u}{\partial n} = 0 & \text{on } \Sigma. \end{cases}$$

This is in good agreement with the results obtained in [LoPe97] for the high frequencies. In this case, for certain sequences of ε, the associated eigenfunctions are far from vanishing in Ω_0. This does not happen with the eigenfunctions in Fig. 11.2a, b, which vanish asymptotically in Ω_0, as it can be verified in formulas (11.18)–(11.19) and (11.23)–(11.24), respectively.

11.2.2 Results for the Dimension $N > 2$

The explicit computations performed in Sect. 11.2.1 for dimension $N = 2$ can be extended for $N > 2$ when the domains Ω_0, Ω_1 are defined by $\Omega_0 = (-1, 0) \times \tilde{\Omega}$, $\Omega_1 = (0, 1) \times \tilde{\Omega}$, respectively, where $\tilde{\Omega}$ is a bounded domain of $\mathbb{R}^{N-1}$; we briefly outline the main differences here.

Using separation of variables, we look for the eigenelements $(\lambda^\varepsilon, u^\varepsilon)$ of (11.7) in the form

$$u^\varepsilon(\mathbf{x}) = X^\varepsilon(x) Y^\varepsilon(\mathbf{y}). \tag{11.25}$$

where $\mathbf{x} = (x, \mathbf{y})$ with $x \in \mathbb{R}$ and $\mathbf{y} = (x_2, \cdots, x_N) \in \mathbb{R}^{N-1}$. Replacing (11.25) in (11.7) we get

$$\begin{cases} \dfrac{X^{\varepsilon\prime\prime}(x)}{X^\varepsilon(x)} + \lambda^\varepsilon = \dfrac{-\Delta_{\mathbf{y}} Y^\varepsilon(\mathbf{y})}{Y^\varepsilon(\mathbf{y})} = \mu^\varepsilon & \text{for } (x, \mathbf{y}) \in \Omega_0, \\ \dfrac{X^{\varepsilon\prime\prime}(x)}{X^\varepsilon(x)} + \dfrac{\lambda^\varepsilon}{\varepsilon} = \dfrac{-\Delta_{\mathbf{y}} Y^\varepsilon(\mathbf{y})}{Y^\varepsilon(\mathbf{y})} = \hat{\mu}^\varepsilon & \text{for } (x, \mathbf{y}) \in \Omega_1, \\ X^\varepsilon(-1) = X^\varepsilon(1) = 0, \qquad\qquad Y^\varepsilon = 0 \quad \text{on } \partial\tilde{\Omega}, \\ X^\varepsilon(0^-) Y^\varepsilon(\mathbf{y}) = X^\varepsilon(0^+) Y^\varepsilon(\mathbf{y}) & \text{for } \mathbf{y} \in \tilde{\Omega}, \\ X^{\varepsilon\prime}(0^-) Y^\varepsilon(\mathbf{y}) = \varepsilon X^{\varepsilon\prime}(0^+) Y^\varepsilon(\mathbf{y}) & \text{for } \mathbf{y} \in \tilde{\Omega}, \end{cases} \tag{11.26}$$

where μ^ε and $\hat{\mu}^\varepsilon$ are constants to be determined. It is easy to check that the only values μ^ε and $\hat{\mu}^\varepsilon$ satisfying (11.26) with $u^\varepsilon(\mathbf{x}) = X^\varepsilon(x) Y^\varepsilon(\mathbf{y}) \neq 0$ are $\mu^\varepsilon = \hat{\mu}^\varepsilon = \mu$ the eigenvalues of the Dirichlet problem

$$\begin{cases} -\Delta_{\mathbf{y}} Y = \mu Y & \text{in } \tilde{\Omega}, \\ Y = 0 & \text{on } \partial\tilde{\Omega}. \end{cases} \tag{11.27}$$

Let us denote by $0 < \mu_1 \leq \mu_2 \leq \cdots \leq \mu_j \leq \cdots \xrightarrow[j \to \infty]{} \infty$ the sequence of
the eigenvalues of (11.27) and let $\{Y_j(\mathbf{y})\}_{j \geq 1}$ be the corresponding eigenfunctions,
which are assumed to be an orthonormal basis in $L^2(\tilde{\Omega})$. Thus, for $j \in \mathbb{N}$
fixed, $(\lambda^\varepsilon, X^\varepsilon)$ verify (11.16) replacing $j^2 \pi^2$ by the eigenvalues μ_j of (11.27).
Consequently, the different expressions obtained in Sect. 11.2.1 for X^ε, u^ε and $\mathbf{u}^\varepsilon$
in the two-dimensional case are also valid for $N > 2$ replacing the values $j^2 \pi^2$
and the functions $\sin(j \pi y)$ (cf. (11.14) and (11.15)) by the eigenvalues μ_j and the
corresponding eigenfunctions $Y_j(\mathbf{y})$ of (11.27), respectively.

11.3   On Approaches to Solutions of the Evolution Problem

In this section we use the convergence result for the low frequencies of problem (11.7) (cf. Lemma 1 and Proposition 1) to get some approaches by standing
waves for certain solutions of the evolution problem (11.1). We also provide
estimates for the time t in which these standing waves approach their solutions as
$\varepsilon \to 0$.

First, we introduce the following result which improves the convergence given in
Lemma 1.

Proposition 1 *Let* $\{\lambda_n^\varepsilon, u_n^\varepsilon\}_{n \geq 1}$ *be the eigenelements of* (11.5) *such that*
$\|u_n^\varepsilon\|_{L^2(\Omega)} = 1$. *Then, for each fixed* $n \in \mathbb{N}$, u_n^ε *converges towards* u_n^* *in* $H_0^1(\Omega)$,
as $\varepsilon \to 0$, *where* u_n^* *is an eigenfunction associated with the n-th eigenvalue* λ_n^* *of*
problem (11.9), u_n^* *extended by 0 to* Ω_0. *Moreover,*

$$\frac{1}{\varepsilon} \|\nabla u_n^\varepsilon\|_{L^2(\Omega_0)}^2 \to 0 \quad \text{as } \varepsilon \to 0. \tag{11.28}$$

Proof Taking into account (11.5) for $\lambda^\varepsilon = \lambda_n^\varepsilon$, $u^\varepsilon = u_n^\varepsilon$, and $v = u_n^\varepsilon$, the
normalization of u_n^ε and estimate (11.8), we get

$$\|\nabla u_n^\varepsilon\|_{L^2(\Omega_0)}^2 + \varepsilon \|\nabla u_n^\varepsilon\|_{L^2(\Omega_1)}^2 = \lambda_n^\varepsilon \|u_n^\varepsilon\|_{L^2(\Omega)}^2 \leq C_n \varepsilon, \tag{11.29}$$

where C_n is a constant independent of ε. Then, fixed n, there exists a subsequence,
still denoted by ε, such that u_n^ε converges towards u^* weakly in $H_0^1(\Omega)$, for some
$u^* \in H_0^1(\Omega)$. By Lemma 1, $u^* = u_n^*$, where u_n^* is an eigenfunction associated with
the n-th eigenvalue λ_n^* of problem (11.9), u_n^* extended by 0 to Ω_0. By (11.29), it
is clear that $\|\nabla u_n^\varepsilon\|_{L^2(\Omega_0)}^2 \to 0$ as $\varepsilon \to 0$ and, consequently, u_n^ε converges towards
zero strongly in $H^1(\Omega_0)$. Thus, if we prove that

$$\|\nabla u_n^\varepsilon\|_{L^2(\Omega_1)}^2 \to \|\nabla u_n^*\|_{L^2(\Omega_1)}^2 \quad \text{as } \varepsilon \to 0, \tag{11.30}$$

the strong convergence of u_n^ε towards u_n^* in $H_0^1(\Omega)$ follows.

Let us prove (11.30). Since u_n^ε converges towards u_n^* weakly in $H_0^1(\Omega)$, as $\varepsilon \to 0$,

$$\|\nabla u_n^*\|_{L^2(\Omega)}^2 \le \liminf_{\varepsilon \to 0} \|\nabla u_n^\varepsilon\|_{L^2(\Omega)}^2. \tag{11.31}$$

Besides, from (11.5) for $\lambda^\varepsilon = \lambda_n^\varepsilon$, $u^\varepsilon = u_n^\varepsilon$ and $v = u_n^\varepsilon/\varepsilon$, and Lemma 1, we get

$$\|\nabla u_n^\varepsilon\|_{L^2(\Omega)}^2 \le \frac{1}{\varepsilon}\|\nabla u_n^\varepsilon\|_{L^2(\Omega_0)}^2 + \|\nabla u_n^\varepsilon\|_{L^2(\Omega_1)}^2 = \frac{\lambda_n^\varepsilon}{\varepsilon}\|u_n^\varepsilon\|_{L^2(\Omega)}^2 \xrightarrow[\varepsilon \to 0]{} \lambda_n^*\|u_n^*\|_{L^2(\Omega)}^2 \tag{11.32}$$

and, using the fact that u_n^* is an eigenfunction associated with λ_n^* of problem (11.9), u_n^* extended by 0 to Ω_0, yields

$$\lambda_n^*\|u_n^*\|_{L^2(\Omega)}^2 = \lambda_n^*\|u_n^*\|_{L^2(\Omega_1)}^2 = \|\nabla u_n^*\|_{L^2(\Omega_1)}^2 = \|\nabla u_n^*\|_{L^2(\Omega)}^2. \tag{11.33}$$

Consequently, gathering (11.31), (11.32), and (11.33), we conclude (11.30).

Finally, we observe that the proof of (11.28) is included in the proof of (11.30). Indeed, by (11.32), (11.30), and (11.33), we have

$$\frac{1}{\varepsilon}\|\nabla u_n^\varepsilon\|_{L^2(\Omega_0)}^2 = \frac{\lambda_n^\varepsilon}{\varepsilon}\|u_n^\varepsilon\|_{L^2(\Omega)}^2 - \|\nabla u_n^\varepsilon\|_{L^2(\Omega_1)}^2 \xrightarrow[\varepsilon \to 0]{} 0,$$

which concludes the proof.

Theorem 1 *Let (λ_k^*, u_k^*) be an eigenelement of (11.9), u_k^* extended by zero to Ω_0. Let $\mathbf{U}^\varepsilon$ be the solution of (11.1) for $(\phi_0, \phi_1) = (u_k^*, 0)$. Then, for $t > 0$ and ε small enough*

$$\left\|\mathbf{U}^\varepsilon(t) - u_k^* \cos(\sqrt{\lambda_k^*\varepsilon}\, t)\right\|_\varepsilon \le C\varepsilon^{1/2}\left(\delta_\varepsilon + \left|\sqrt{\lambda_k^\varepsilon} - \sqrt{\lambda_k^*\varepsilon}\right| t\right), \tag{11.34}$$

$$\left\|\frac{d\mathbf{U}^\varepsilon}{dt}(t) + \sqrt{\lambda_k^*\varepsilon}\, u_k^* \sin(\sqrt{\lambda_k^*\varepsilon}\, t)\right\|_{L^2(\Omega)} \le C\varepsilon^{1/2}\delta_\varepsilon + C\left|\sqrt{\lambda_k^\varepsilon} - \sqrt{\lambda_k^*\varepsilon}\right|(1 + \varepsilon^{1/2}t), \tag{11.35}$$

where C is a constant that may depend on (λ_k^*, u_k^*) but is independent of ε and t, δ_ε is a constant that converges towards zero as $\varepsilon \to 0$, and $\|\cdot\|_\varepsilon$ is the norm associated with the scalar product $a_\varepsilon(\cdot, \cdot)$ defined by (11.3). In particular, if $t \le O(1/\sqrt{\varepsilon})$, we have

$$\|\nabla \mathbf{U}^\varepsilon(t)\|_{L^2(\Omega_0)} + \left\|\nabla\left(\mathbf{U}^\varepsilon(t) - u_k^* \cos(\sqrt{\lambda_k^*\varepsilon}\, t)\right)\right\|_{L^2(\Omega_1)} \to 0, \quad as\ \varepsilon \to 0,$$

$$(11.36)$$

$$\left\|\frac{d\mathbf{U}^\varepsilon}{dt}(t)\right\|_{L^2(\Omega_0)} + \left\|\frac{d\mathbf{U}^\varepsilon}{dt}(t) + \sqrt{\lambda_k^*\varepsilon}\, u_k^* \sin(\sqrt{\lambda_k^*\varepsilon}\, t)\right\|_{L^2(\Omega_1)} \to 0, \quad as\ \varepsilon \to 0.$$

$$(11.37)$$

Proof Let $\mathbf{u}^\varepsilon$ be the solution of (11.1) for $(\phi_0, \phi_1) = (u_k^\varepsilon, 0)$. It is clear that $\mathbf{u}^\varepsilon = u_k^\varepsilon \cos(\sqrt{\lambda_k^\varepsilon}\, t)$. Thus,

$$\left\|\mathbf{U}^\varepsilon(t) - u_k^* \cos(\sqrt{\lambda_k^*\varepsilon}\, t)\right\|_\varepsilon \le \|\mathbf{U}^\varepsilon - \mathbf{u}^\varepsilon\|_\varepsilon + \|u_k^\varepsilon - u_k^*\|_\varepsilon \left|\cos(\sqrt{\lambda_k^\varepsilon}\, t)\right|$$

$$+ \|u_k^*\|_\varepsilon \left|\cos(\sqrt{\lambda_k^\varepsilon}\, t) - \cos(\sqrt{\lambda_k^*\varepsilon}\, t)\right|.$$

Now, using Eq. (11.4) for $(\phi_0, \phi_1) = (u_k^* - u_k^\varepsilon, 0)$, the definition of $\|\cdot\|_\varepsilon$, and the fact that $u_k^* = 0$ in Ω_0, we get

$$\left\|\mathbf{U}^\varepsilon(t) - u_k^* \cos(\sqrt{\lambda_k^*\varepsilon}\, t)\right\|_\varepsilon \le 2\|u_k^\varepsilon - u_k^*\|_\varepsilon + \|u_k^*\|_\varepsilon \left|\cos(\sqrt{\lambda_k^\varepsilon}\, t) - \cos(\sqrt{\lambda_k^*\varepsilon}\, t)\right|$$

$$\le 2\left(\|\nabla u_k^\varepsilon\|_{L^2(\Omega_0)}^2 + \varepsilon\|\nabla(u_k^\varepsilon - u_k^*)\|_{L^2(\Omega_1)}^2\right)^{1/2}$$

$$+ \varepsilon^{1/2}\|\nabla u_k^*\|_{L^2(\Omega_1)} \left|\cos(\sqrt{\lambda_k^\varepsilon}\, t) - \cos(\sqrt{\lambda_k^*\varepsilon}\, t)\right|.$$

Thus, by Proposition 1, (11.34) holds.

Similarly, we can write

$$\left\|\frac{d\mathbf{U}^\varepsilon}{dt}(t) + \sqrt{\lambda_k^*\varepsilon}\, u_k^* \sin(\sqrt{\lambda_k^*\varepsilon}\, t)\right\|_{L^2(\Omega)} \le \left\|\frac{d(\mathbf{U}^\varepsilon - \mathbf{u}^\varepsilon)}{dt}\right\|_{L^2(\Omega)}$$

$$+ \|u_k^\varepsilon - u_k^*\|_{L^2(\Omega)}\sqrt{\lambda_k^\varepsilon}\left|\sin(\sqrt{\lambda_k^\varepsilon}\, t)\right| + \|u_k^*\|_{L^2(\Omega)}\left|\sqrt{\lambda_k^\varepsilon} - \sqrt{\lambda_k^*\varepsilon}\right|\left|\sin(\sqrt{\lambda_k^\varepsilon}\, t)\right|$$

$$+ \|u_k^*\|_{L^2(\Omega)}\sqrt{\lambda_k^*\varepsilon}\left|\sin(\sqrt{\lambda_k^\varepsilon}\, t) - \sin(\sqrt{\lambda_k^*\varepsilon}\, t)\right|,$$

and using (11.4) for $(\phi_0, \phi_1) = (u_k^* - u_k^\varepsilon, 0)$, (11.8), the definition of $\|\cdot\|_\varepsilon$, and Proposition 1, (11.35) holds.

Finally, (11.36) and (11.37) is a direct consequence of (11.34) and (11.35) and the convergence of $\lambda_k^\varepsilon/\varepsilon$ to λ_k^* as $\varepsilon \to 0$ (cf. Lemma 1).

Theorem 2 *Let (λ_k^*, u_k^*) be an eigenelement of (11.9), u_k^* extended by zero to Ω_0. Let $\mathbf{U}^\varepsilon$ be the solution of (11.1) for $(\phi_0, \phi_1) = (0, u_k^*)$. Then, for $t > 0$ and ε small enough*

$$\left\| \mathbf{U}^\varepsilon(t) - \frac{1}{\sqrt{\lambda_k^*\varepsilon}} u_k^* \sin(\sqrt{\lambda_k^*\varepsilon}\, t) \right\|_\varepsilon \le \delta_\varepsilon + C \left| \sqrt{\lambda_k^\varepsilon} - \sqrt{\lambda_k^*\varepsilon} \right| (\varepsilon^{-1/2} + t), \qquad (11.38)$$

$$\left\| \frac{d\mathbf{U}^\varepsilon}{dt}(t) - u_k^* \cos(\sqrt{\lambda_k^*\varepsilon}\, t) \right\|_{L^2(\Omega)} \le \delta_\varepsilon + C \left| \sqrt{\lambda_k^\varepsilon} - \sqrt{\lambda_k^*\varepsilon} \right| t, \qquad (11.39)$$

where C is a constant that may depend on (λ_k^*, u_k^*) but is independent of ε and t, δ_ε is a constant that converges towards zero as $\varepsilon \to 0$, and $\|\cdot\|_\varepsilon$ is the norm associated with the scalar product $a_\varepsilon(\cdot,\cdot)$ defined by (11.3). In particular, if $t \le O(1/\sqrt{\varepsilon})$, we have

$$\left\| \mathbf{U}^\varepsilon(t) - \frac{1}{\sqrt{\lambda_k^*\varepsilon}} u_k^* \sin(\sqrt{\lambda_k^*\varepsilon}\, t) \right\|_\varepsilon \to 0, \quad \text{as } \varepsilon \to 0, \qquad (11.40)$$

$$\left\| \frac{d\mathbf{U}^\varepsilon}{dt}(t) - u_k^* \cos(\sqrt{\lambda_k^*\varepsilon}\, t) \right\|_{L^2(\Omega)} \to 0, \quad \text{as } \varepsilon \to 0. \qquad (11.41)$$

Proof Let $\mathbf{u}^\varepsilon$ be the solution of (11.1) for $(\phi_0, \phi_1) = (0, u_k^\varepsilon)$. It is clear that $\mathbf{u}^\varepsilon = \frac{1}{\sqrt{\lambda_k^\varepsilon}} u_k^\varepsilon \sin(\sqrt{\lambda_k^\varepsilon}\, t)$. Thus,

$$\left\| \mathbf{U}^\varepsilon(t) - \frac{1}{\sqrt{\lambda_k^*\varepsilon}} u_k^* \sin(\sqrt{\lambda_k^*\varepsilon}\, t) \right\|_\varepsilon \le \| \mathbf{U}^\varepsilon - \mathbf{u}^\varepsilon \|_\varepsilon + \frac{1}{\sqrt{\lambda_k^\varepsilon}} \| u_k^\varepsilon - u_k^* \|_\varepsilon \left| \sin(\sqrt{\lambda_k^\varepsilon}\, t) \right|$$

$$+ \| u_k^* \|_\varepsilon \left| \frac{1}{\sqrt{\lambda_k^\varepsilon}} - \frac{1}{\sqrt{\lambda_k^*\varepsilon}} \right| \left| \sin(\sqrt{\lambda_k^\varepsilon}\, t) \right| + \frac{\| u_k^* \|_\varepsilon}{\sqrt{\lambda_k^*\varepsilon}} \left| \sin(\sqrt{\lambda_k^\varepsilon}\, t) - \sin(\sqrt{\lambda_k^*\varepsilon}\, t) \right|.$$

Now, from (11.4) for $(\phi_0, \phi_1) = (0, u_k^* - u_k^\varepsilon)$ and (11.8), we get

$$\left\| \mathbf{U}^\varepsilon(t) - \frac{1}{\sqrt{\lambda_k^*\varepsilon}} u_k^* \sin(\sqrt{\lambda_k^*\varepsilon}\, t) \right\|_\varepsilon \le \| u_k^\varepsilon - u_k^* \|_{L^2(\Omega)} + C\varepsilon^{-1/2} \| u_k^\varepsilon - u_k^* \|_\varepsilon$$

$$+ C\varepsilon^{-1} \| u_k^* \|_\varepsilon \left| \sqrt{\lambda_k^\varepsilon} - \sqrt{\lambda_k^*\varepsilon} \right| + C\varepsilon^{-1/2} \| u_k^* \|_\varepsilon \left| \sin(\sqrt{\lambda_k^\varepsilon}\, t) - \sin(\sqrt{\lambda_k^*\varepsilon}\, t) \right|,$$

where C is a constant independent of ε and t (but may depend on k). Thus, using the definition of $\|\cdot\|_\varepsilon$, the fact that $u_k^* = 0$ in Ω_0, and Proposition 1, (11.38) holds.

Similarly, we can write

$$
\left\| \frac{d\mathbf{U}^\varepsilon}{dt}(t) - u_k^* \cos(\sqrt{\lambda_k^*\varepsilon}\, t) \right\|_{L^2(\Omega)} \leq \left\| \frac{d(\mathbf{U}^\varepsilon - \mathbf{u}^\varepsilon)}{dt} \right\|_{L^2(\Omega)}
$$
$$
+ \|u_k^\varepsilon - u_k^*\|_{L^2(\Omega)} \left| \cos(\sqrt{\lambda_k^\varepsilon}\, t) \right| + \|u_k^*\|_{L^2(\Omega)} \left| \cos(\sqrt{\lambda_k^\varepsilon}\, t) - \cos(\sqrt{\lambda_k^*\varepsilon}\, t) \right|,
$$

and using (11.4) for $(\phi_0, \phi_1) = (0, u_k^* - u_k^\varepsilon)$, and Proposition 1, (11.39) holds.

Finally, (11.40) and (11.41) is a direct consequence of (11.38) and (11.39) and the convergence of $\lambda_k^\varepsilon / \varepsilon$ to λ_k^* as $\varepsilon \to 0$ (cf. Lemma 1).

References

[BaGo10] Babych, N., and Golovaty, Yu.: Low and high frequency approximations to eigenvibrations of string with double contrasts. *J. Comput. Appl. Math.*, **234**, 1860–1867 (2010).

[CaEtAl05] Caínzos, J., Vilasánchez, M., and Pérez, E.: Comportamiento asintótico de soluciones de un problema espectral de Neumann (1-D) con masa concentrada. In: *XIX CEDYA / IX Congress of Applied Mathematics*, Universidad Carlos III de Madrid, Madrid (2005), pp. 5.

[Gi82] Gibert, P. : Les basses et les moyennes fréquences dans des structures fortement hétérogenes. *C.R. Acad. Sci. Paris Ser. II*, **295**, 951–954 (1982).

[GoBa00] Golovaty, Yu., and Babych, N.: On WKB asymptotic expansions of high frequency vibrations in stiff problems. In: *International Conference on Differential Equations*, Vol. 1, 2, World Sci. Publ., River Edge, NJ (2000), pp. 103–105.

[GoEtAl98] Gómez, D., Lobo, M., and Pérez, E.: Sobre vibraciones de un sistema con una masa concentrada. In: *XV CEDYA / V Congress of Applied Mathematics*, Universidad de Vigo, Vigo (1998), pp. 453–458.

[Li85] Lions, J.L.: Remarques sur les problemes d'homogénéisation dans les milieux a structure périodique et sur les problemes raides. In: *Les Méthodes de l'Homogénéisation*, D. Bergman and J.L. Lions (eds.), Eyrolles, Paris (1985), pp. 129–228.

[LiMa68] Lions, J.L. and Magenes, E.: *Problèmes aux Limites non Homogenènes*, V. I., Dunod, Paris (1968).

[LoEtAl03] Lobo, M., Nazarov, S., and Pérez, E.: Asymptotically sharp uniform estimates in a scalar stiff problem. *Comptes Rendues de Mecanique*, **331**, 325–330 (2003).

[LoEtAl05] Lobo, M., Nazarov, S., and Pérez, E.: Eigenoscillations of contrasting non-homogeneous elastic bodies. Asymptotic and uniform estimates for eigenvalues. *IMA Journal of Applied Mathematics*, **70**, 419–458 (2005).

[LoPe97] Lobo, M., and Pérez, E.: High frequency vibrations in a stiff problem. *Math. Models Methods Appl. Sci.*, **7**, 291–311 (1997).

[LoPe10] Lobo, M., and Pérez, E.: Long time approximations for solutions of wave equations associated with the Steklov spectral homogenization problems. *Math. Methods Appl. Sci.*, **11**, 1356–1371 (2010).

[LoSa80] Lobo, M., and Sánchez-Palencia, E.: Low and high frequency vibration in stiff problems. In: *Partial Differential Equations and the Calculus of Variations*, Vol. II, F. Colombini, A. Marino, L. Modica, S. Spagnolo (eds.), Birkhäuser Boston, Boston, MA (1989), pp. 729–742.

[Pa80] Panasenko, G.P.: Asymptotic behavior of solutions and eigenvalues of elliptic equations with strongly varying coefficients. *Dokl. Akad. Nauk SSSR*, **252**, 1320–1325 (1980).

[Pe95] Pérez, E.: Altas frecuencias en un problema "stiff" relativo a las vibraciones de una cuerda. In: *XIV CEDYA / IV Congress of Applied Mathematics*, Universidad de Barcelona, Barcelona (1995), (electronic) pp. 7.

[Pe08] Pérez, E.: Long time approximations for solutions of wave equations via standing waves from quasimodes. *J. Math. Pures Appl.*, **90**, 387–411 (2008).

[SaSa89] Sanchez-Hubert, J., and Sanchez-Palencia, E.: *Vibration and Coupling of Continuous Systems. Asymptotic Methods*, Springer, Heidelberg (1989).

[Sa80] Sanchez-Palencia, E: *Non-homogeneous Media and Vibration Theory*, Lecture Notes in Physics, vol. 127. Springer, New York (1980).

Chapter 12
Modelling Creep in Concrete Under a Variable External Load

Layal Hakim

12.1 Introduction

Fracture mechanics is a mature subject. It is an essential topic for understanding and predicting the failure of buildings and machines. For many centuries, engineers have been studying the durability of materials; the main approach for this is to identify when failure will occur. This mainly concerns nucleation and propagation of cracks in a body. Cracks appear or grow due to stresses caused by external forces. The external load applied on the material could cause a crack to initiate, or could cause growth of an already existing crack and the presence of a crack could lead to rupture when the crack propagates directly through the body. To predict the strength of the material and where fracture will occur, the local approach or the non-local approach can be used. The local approach uses the stress (or stress history) at the point where the crack appears, while the non-local approach uses not only the stress (history) at the point but also in the vicinity of that point.

Creep is a time dependent permanent deformation of a solid material under the influence of an external load. The long-term exposure of a material to high stress, and high temperature are the main reasons for creep to occur. Real life examples such as steam turbine power plants, nuclear power plants, and heat exchangers are a few examples where creep deformation is critical. Analysing the material behaviour in such examples reduces the probability of manufacturing defects. Importantly many real materials do not behave in a perfectly linear or idealised fashion and issues such as creep, damage and time dependence contain open problems of high relevance, where the assumptions used in linear elastic fracture mechanics do not apply. In such cases, visco-elastic-plastic fracture mechanics has to be considered instead. Cohesive Zone (CZ) models have been used as a tool to study the behaviour of cracks in elasto-plastic materials.

L. Hakim (✉)
The University of Exeter, Exeter, UK
e-mail: l.hakim@exeter.ac.uk

© Springer Nature Switzerland AG 2019
C. Constanda, P. Harris (eds.), *Integral Methods in Science and Engineering*,
https://doi.org/10.1007/978-3-030-16077-7_12

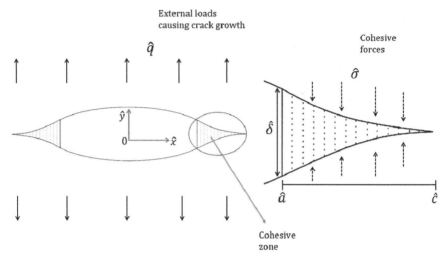

Fig. 12.1 Cohesive zones

Assume we have a crack, of length $\hat{a}$, present in a material. A CZ in a material is the region ahead of the crack tip, illustrated by the shaded area in Fig. 12.1. The CZ tip coordinate is denoted by $\hat{c}$. The material is subject to an external load $\hat{q}$, however cohesive stresses $\hat{\sigma}$ present at each CZ are opposing the effect of the external load, in the attempt to prevent crack growth. The crack tip opening, denoted by $\hat{\delta}$, increases with time as the CZ propagates. The Dugdale-Leonov-Panasyuk [Du60] and [LePa59] model was the first model to consider perfectly plastic CZs in an elastic solid. In this model the stress acting on the CZs has the form $\hat{\sigma} = \sigma_y$, where σ_y is a constant. Since this model was introduced in 1959, many other modifications and variations of this model have been introduced and used in many applications in fracture mechanics. A contemporary example where CZ modelling is being used is the modelling of wind turbine blades [NeEtAl12]. The rupture depends highly on the configuration of the crack tip as high local stress will lead to faster crack propagation. Generally, the three main components needed to study CZ models of the nonlinear Dugdale-Leonov-Panasyuk type are (1) what are the constitutive equations for the bulk of the material? (2) what is the criteria on the stress history on the CZ for the CZ to appear and propagate? (3) what is the criteria for the CZ to break and the crack to propagate?

12.1.1 Objectives

The work carried out considers visco-elastic-plastic fracture mechanics while incorporating a time and history dependent model. The essence of this objective is to manifest the behaviour of real life materials. In this paper, we will give a modification of the classical Dugdale-Leonov-Panasyuk model and use it to study the behaviour in the CZs and the rate of crack growth for the case of a time and history dependent material. The results will be obtained for the case when the bulk of the material is elastic, and when it is viscoelastic. For the viscoelastic case, we will consider using a creep function that models the behaviour of concrete materials as well as PMMA. The external load, denoted by $\hat{q}$ in Fig. 12.1 generally depends on time. In this paper, we will extend cases implemented in the current literature by taking other polynomial time-dependent external loads. Namely, in [HaMi18], a linear function $\hat{q}(t) = \frac{t}{\alpha}$ was considered, here we will take $\hat{q}(t) = \frac{t^p}{\alpha}$ where $p = \left[0, \frac{1}{2}, 1, \frac{3}{2}, 2\right]$.

12.2 Problem Formulation

A detailed formulation is given in [HaMi15] and [HaMi18] where the formulas for the normalisations can also be found. Here, we will give a brief summary of the principal equations used in our model. We will replace the Dugdale-Leonov-Panasyuk CZ condition $\hat{\sigma} = \sigma_y$ with the condition

$$\underline{\Lambda}(\hat{\sigma}; \hat{t}) = 1,$$

where $\underline{\Lambda}(\hat{\sigma}; \hat{t})$ is the history-dependent normalised equivalent stress given by

$$\underline{\Lambda}(\hat{\sigma}; \hat{t}) = \left(\frac{\beta}{b\sigma_0^\beta} \int_0^{\hat{t}} |\hat{\sigma}(\hat{\tau})|^\beta (\hat{t} - \hat{\tau})^{\frac{\beta}{b}-1} d\hat{\tau}\right)^{\frac{1}{\beta}} \qquad (12.1)$$

where $|\hat{\sigma}|$ is the maximal principal stress, $\hat{t}$ is time, and σ_0, β and b are dimensionless material parameters. More details on the physical interpretation of these parameters can be found in [MiNa11]. The parameters σ_0 and b are material constants in the assumed power-type relation

$$\hat{t}_\infty(\hat{\sigma}) = \left(\frac{\hat{\sigma}}{\sigma_0}\right)^{-b} \qquad (12.2)$$

between the physical rupture time $\hat{t}_\infty$ and the constant uniaxial tensile stress $\hat{\sigma}$ applied to a sample without cracks. The normalised equivalent stress given in Eq. (12.1) was implemented in [HaMi11] and [MiNa03] where a crack growth problem was studied which does not assume the presence of a CZ.

For the *constant loading* case, $\hat{q}(\hat{t}) = \hat{q}_0$ is independent of time, and using Eq. (12.2) we obtain

$$\hat{t}_\infty = \left(\frac{\hat{q}_0}{\sigma_0}\right)^{-b}.$$

Then we can introduce the normalised time, coordinate variable, crack tip coordinate, CZ tip coordinate, and stress as follows:

$$t = \frac{\hat{t}}{\hat{t}_\infty}, \quad x = \frac{\hat{x}}{\hat{a}_0}, \quad a(t) = \frac{\hat{a}(t\,\hat{t}_\infty)}{\hat{a}_0}, \quad c(t) = \frac{\hat{c}(t\,\hat{t}_\infty)}{\hat{a}_0}, \quad \sigma(x,t) = \frac{\hat{\sigma}(x\,\hat{a}_0, t\,\hat{t}_\infty)}{\hat{q}_0},$$

(12.3)

and the normalised external load becomes $q(t) = \hat{q}(\hat{t})/\hat{q}_0 = 1$.

Now, we will introduce a time step denoted by $t_c(x)$. This is the time when the point x becomes part of the CZ. Thus, preceding that step, the stress at those points, ahead of the CZ, are known.

Equating Eq. (12.1) to 1, and splitting the integral into two parts—namely from $\tau = [0, t_c(x)]$ and $\tau = [t_c(x), t]$, we arrive at the following CZ condition

$$\int_{t_c(x)}^{t} \sigma^\beta(x, \tau)(t - \tau)^{\frac{\beta}{b} - 1} d\tau = \frac{b}{\beta} - \int_0^{t_c(x)} \sigma^\beta(x, \tau)(t - \tau)^{\frac{\beta}{b} - 1} d\tau$$

(12.4)

where we have used the normalisation formulas given in Eq. (12.3). Equation (12.4) is an inhomogeneous nonlinear Volterra integral equation of the Abel type with unknown function $\sigma(x, \tau)$ for $t_c(x) \le \tau \le t$.

The formula for the stress ahead of the CZ was obtained using the results by Muskhelishvili [Mu54]. For an infinite elastic or viscoelastic plane under the traction $q(t)$ at infinity, the normalised stress ahead of the CZ is given by

$$\sigma(x, t) = \frac{x}{\sqrt{x^2 - c^2(t)}} \left(q(t) - \frac{2}{\pi} \int_{a(t)}^{c(t)} \frac{\sqrt{c^2(t) - \xi^2}}{x^2 - \xi^2} \sigma(\xi, t) d\xi \right)$$

(12.5)

for $|x| > c(t)$.

A sufficient condition for the normalised equivalent stress, Λ, to be bounded at the CZ tip is that the stress intensity factory is zero there. Denoted by K, the normalised stress intensity factor is given by

$$K(c(t), t) = \sqrt{\frac{c(t)}{2}} \left(1 - \frac{2}{\pi q(t)} \int_{a(t)}^{c(t)} \frac{\sigma(\xi, t)}{\sqrt{c^2(t) - \xi^2}} d\xi \right).$$

(12.6)

While keeping the crack length $a(t)$ as constant in time, we use Eqs. (12.4)–(12.6) simultaneously to obtain the rate of the CZ. This was done by introducing a time mesh and solving Eq. (12.6) at each time step. The detailed steps of the algorithm used, as well as the solutions, are presented in [HaMi15] and [HaMi18] where various sets of parameters were considered as well as an analysis of the numerical solution indicating the robustness of our numerical method.

Also using the representations by Muskhelishvili [Mu54], we deduce that the normalised crack opening in the elastic case is given by

$$[u_e(x;t)] = \frac{q(t)(1+\varkappa)}{2\mu_0}\sqrt{c(t)^2 - x^2} + \frac{1+\varkappa}{2\pi\mu_0}\left(\int_{a(t)}^{c(t)} \sigma(\xi,t)\Gamma(x,\xi;c(t))d\xi\right)$$

where

$$\Gamma(x,\xi;c) = \ln\left[\frac{2c^2 - \xi^2 - x^2 - 2\sqrt{(c^2 - x^2)(c^2 - \xi^2)}}{2c^2 - \xi^2 - x^2 + 2\sqrt{(c^2 - x^2)(c^2 - \xi^2)}}\right].$$

Here, $\varkappa = (3-\nu)/(1+\nu)$ under the plane stress condition, $\mu_0 = E_0/[2(1+\nu)]$ is the shear modulus; E_0 is Young's modulus of elasticity and ν is Poisson's ratio. The details of the derivation and normalisation can be found in [HaMi15].

Thus, the crack tip opening, for $x = a(t)$ in $[u_e(x;t)]$ is

$$\delta_e(t) = \frac{1+\varkappa}{2\mu_0}\left(q(t)\sqrt{c^2(t) - a^2(t)} + \frac{1}{\pi}\int_{a(t)}^{c(t)} \sigma(\xi,t)\Gamma(a(t),\xi,c(t))d\xi\right).$$

In the case of having a viscoelastic material, we will replace $\frac{1}{\mu_0}$ with a second kind Volterra integral operator given by

$$\delta_v(t) = \left(\mu^{-1}\mu_0\delta_e\right)(t)$$

$$= \left(\delta_e(t) + \int_0^t \dot{J}(t-\tau)\delta_e(\tau)d\tau\right). \tag{12.7}$$

In [HaMi15] and [HaMi18], we used the creep function of a polymer, to model creep in PMMA (also known as plexiglas)

$$\dot{J}(t-\tau) = \frac{\mu_0}{\eta}e^{-\frac{1}{\theta}(t-\tau)},$$

where $J(t-\tau)$ is the creep function of the material. Here, θ is the viscoelastic relaxation time with η denoting the viscosity. After normalisation, we have

$$\delta_e(t) = \sqrt{c^2(t) - a^2(t)} + \frac{1}{\pi}\int_{a(t)}^{c(t)} \sigma(\xi,t)\Gamma(a(t),\xi;c(t))d\xi,$$

for the elastic case, and for the viscoelastic case we have

$$\delta_v(t) = \delta_e(t) + \frac{\mu_0 t_\infty}{\eta}\int_0^t e^{-\frac{t-\tau}{\theta}}\delta_e(\tau)d\tau \tag{12.8}$$

The results of the crack tip opening for both the elastic and viscoelastic cases were obtained and presented in [HaMi15] and [HaMi18]. However, at a particular time, the crack will also start to propagate. This time is known as the delay time, denoted by t_d, and is reached when the critical crack tip opening reaches a critical value. For PMMA, we have used $\delta_c = 1.13$ which corresponds to the physical value of $\hat{\delta}_c = 0.0016\,\text{mm}$. Therefore, during the crack propagation stage, we set $\delta = \delta_c$ for $t_i \geq t_d$, at each time step and iteratively solve the system of equations to obtain the rate of the CZ growth and the rate of the crack growth. In [HaMi15], solutions before crack growth started, and during crack growth, were obtained for cases when the bulk of the material is elastic, and when the bulk of the material is viscoelastic. For the viscoelastic case, we used the creep function for PMMA to obtain the results.

12.3 Viscoelastic Model Applied to Creeping Concrete

Many materials, such as concrete, exhibit creep visco-elastic behaviour, and time-dependency of the stresses is a vital factor which should be taken into consideration. The first serious application of the theory of linear viscoelasticity was the study of creep in concrete. Proposed by Arutyunyan in 1952 [Ar52], the creep function in concrete can be modelled using the following equation

$$\dot{J}(t, \tau) = \frac{\partial}{\partial \tau} \left[\frac{1}{E(\tau)} + C(t, \tau) \right]$$

where

$$C(t, \tau) = \varphi(\tau)g(t - \tau),$$

where, on the basis of test results, [Ra69], $C(t, \tau)$ was given by

$$g(t - \tau) = \left(1 - e^{-\frac{t-\tau}{\theta}} \right)$$

which models the viscoelastic behaviour of the non-ageing material, and

$$\varphi(\tau) = c_0 + \frac{A_1}{\tau}$$

modelling the ageing effect of the concrete, where θ, c_0, A_1 are material constants. The creep function modelling concrete is of particular interest as it leads to a non-convolution kernel for the viscoelastic operator, unlike the standard creep functions used for polymers. The Arutyunyan model is a special case of the time-variant Boltzmann's model. A special case of the Arutyunyan model is known as the Whitney-Dischinger Model. This model is referred to as the time-variant Maxwell model since it can approximately interpret the creep of concrete, and in this model, $c_0 = 0$, [Ba66]. Substituting the creep function, in the Whitney-Dischinger model, into Eq. (12.7), the expression for the viscoelastic crack tip opening, as given in Eq. (12.8) for PMMA, will now be given by

$$\delta_v(t) = \delta_e(t) - \frac{A_1\mu}{\theta} \int_0^t e^{-\frac{\tau}{\theta}} \delta_e(\tau) d\tau$$

where $\frac{A_1\mu}{\theta}$ is dimensionless. We will take the critical crack tip opening for concrete to be $\delta_c = 0.03$ mm, see [WaEtAl11]. Furthermore, for this particular creep model, we have $A_1 = \frac{1}{E_0}$, where E_0 is the effective basic creep modulus for concrete, and occasionally referred to as Young's modulus of concrete, see [Ba66].

As well as on the age of concrete, Young's modulus of elasticity, and hence the viscosity depend highly on the composition of concrete, specifically on the aggregate to water ratio. However, using approximate values of $E_0 = 30$ GPa and $\mu = 12.7$ GPa, and using experimental results such as in [CaEtAl97], with an external load of $\hat{q} = 50$ MPa and an initial crack length of $\hat{a}_0 = 0.1$ mm, we obtain $\hat{t}_\infty \approx 236{,}637$ h, $b \approx 5$, and $\theta \approx 3.15*10^{-18}$.

The numerical algorithms used here to solve the problems for PMMA and concrete are given in Section 5.2 in [HaMi18], and the details of the algorithm are given therein (Fig. 12.2).

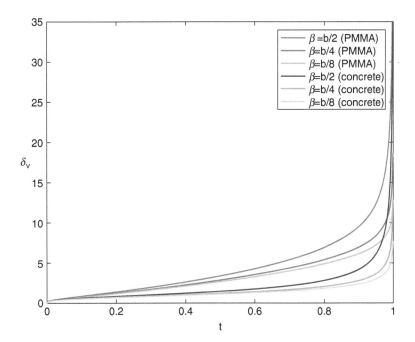

Fig. 12.2 Crack tip opening displacement vs. time for $b = 4$

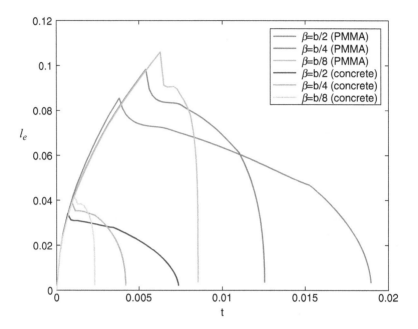

Fig. 12.3 CZ length vs. time for $b = 4$ (elastic)

12.3.1 Solutions

Here, we will present the solutions of the crack tip opening against time for the case $b = 4$ while comparing the results for PMMA and concrete. Shown in Figs. 12.3 and 12.4 is the evolution of the CZ with time. The maximum point of the curves indicates the points when the delay time is reached followed by the crack propagation stage.

12.4 A Polynomial Function for the External Load

We will consider the case when the external load acts as a polynomial function in time. Namely, when $\hat{q}(\hat{t}) = \dot{q}\,\hat{t}^p$, with $p \geq 0$ where $\dot{q}$ is a constant. Let $t_{\bullet\infty p}$ denote the time when the CZ spreads over the infinite plane without crack, under the variable load $\hat{q}(\hat{t})$. We will find $\hat{t}_{\bullet\infty p}$ by solving the following equation for $\hat{t} = \hat{t}_c$

$$\dot{q}^\beta \int_0^{\hat{t}} \hat{\tau}^{p\beta}(\hat{t} - \hat{\tau})^{\frac{\beta}{b}-1}d\hat{\tau} = \frac{b\sigma_0^\beta}{\beta}. \tag{12.9}$$

which was obtained form Eq. (12.4) for $\hat{\sigma}(\hat{x}, \hat{t}) = \hat{q}(\hat{t}) = \dot{q}\,\hat{\tau}^p$. Solving Eq. (12.9) for $\hat{t} = \hat{t}_{\bullet\infty}$ gives

$$\hat{t}_{\bullet\infty p} = \left(\frac{\sigma_0}{\dot{q}\alpha_p}\right)^{\frac{b}{1+pb}}, \tag{12.10}$$

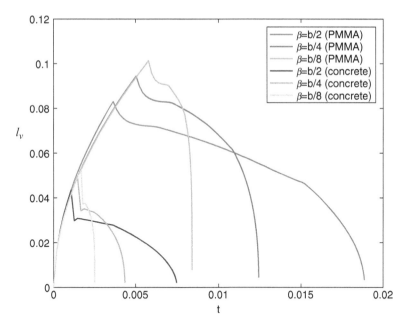

Fig. 12.4 CZ length vs. time for $b = 4$ (viscoelastic)

where

$$\alpha_p := \left(\frac{b}{\beta\, B\left[\frac{\beta}{b}, 1 + \beta \right]} \right)^{-1/\beta},$$

and B is the Beta-function. The load maximum, reached before rupture in the infinite plane without a crack, is evidently $\hat{q}(\hat{t}_{\bullet\infty p}) = \dot{q}\hat{t}_{\bullet\infty p}$. Also, let $\hat{q}_0 = \sigma_0 \hat{t}_{\bullet\infty p}^{-1/b}$ denote the reference constant load under which the plane ruptures at time $\hat{t}_{\bullet\infty p}$. Expressing $\dot{q}$ and $\hat{t}_{\bullet\infty p}$ in terms of $\hat{q}(\hat{t}_{\bullet\infty p})$ and $\hat{q}_0$ from these two equations and substituting them in Eq. (12.10), we obtain $\hat{q}_0 = \alpha\hat{q}(\hat{t}_{\bullet\infty p})$. We will use the following normalised time, coordinate variable, crack tip coordinate, CZ tip coordinate, and stress

$$t = \frac{\hat{t}}{\hat{t}_{\bullet\infty p}}, \quad t_c = \frac{\hat{t}_c}{\hat{t}_{\bullet\infty p}}, \quad x = \frac{\hat{x}}{\hat{a}_0},$$

$$a(t) = \frac{\hat{a}(t\,\hat{t}_{\bullet\infty p})}{\hat{a}_0}, \quad c(t) = \frac{\hat{c}(t\,\hat{t}_{\bullet\infty p})}{\hat{a}_0}, \quad \sigma(x, t) = \frac{\hat{\sigma}(x\,\hat{a}_0, t\,\hat{t}_{\bullet\infty p})}{\hat{q}_0},$$

and the normalised external load becomes

$$q(t) = \frac{\hat{q}(\hat{t})}{\hat{q}_0} = \frac{\hat{t}^p \dot{q}}{\hat{q}_0} = \frac{t\,\hat{t}_{\bullet\infty p}\dot{q}}{\hat{q}_0} = \frac{t^p}{\alpha_p}.$$

12.4.1 Solutions

For the case when $p = 0$ and $p = 1$, the results were obtained and presented in [HaMi18]. Here, we will consider other cases for p, and collectively compare the solutions for $p = \left[0, \frac{1}{2}, 1, \frac{3}{2}, 2\right]$. We present the solutions of the CZ tip coordinate with time, before the crack propagation stage (Fig. 12.5), the crack tip opening displacement against time for PMMA and concrete (Figs. 12.6, 12.7, and 12.8), and the CZ length against time (Figs. 12.9, 12.10, 12.11, and 12.12). To demonstrate the results, we have taken $b = 4$ and $\beta = 2$.

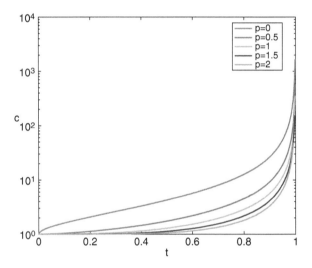

Fig. 12.5 CZ tip coordinate vs. time

Fig. 12.6 Crack tip opening displacement (elastic) vs. time

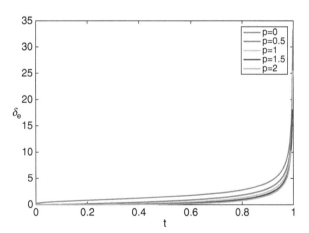

Fig. 12.7 Viscoelastic crack tip opening (PMMA)

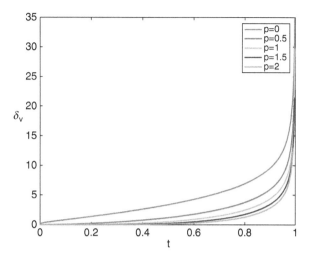

Fig. 12.8 Viscoelastic crack tip opening (concrete)

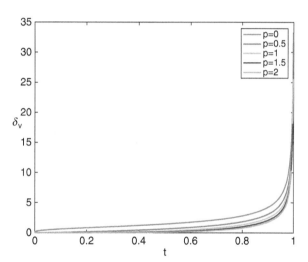

12.5 Conclusions and Ideas for Future Work

We can draw many conclusions form the findings. Similarly to PMMA, in a concrete material, after the crack starts, the crack growth rate increases, while the CZ length decreases with time. The time when the CZ length becomes 0 coincides with the time when the crack length becomes infinite and can be associated with the complete fracture of the body. The delay time for a concrete material is less than that of a polymer, which leads to faster rupture of the material. This is expected as concrete is very brittle and in most concrete structures, once the crack begins to grow, total failure occurs almost instantly. For both cases, PMMA and concrete, we can see

Fig. 12.9 Elastic, PMMA

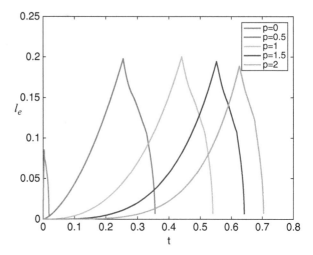

Fig. 12.10 Viscoelastic, PMMA

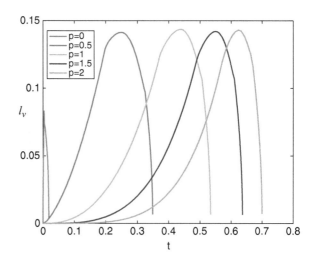

that as the order, of the external load, increases, the delay time also increases, and the CZ decreases monotonically during the crack growth stage for all cases of the external load considered.

One of the main future goals is to compare our theoretical results with experiential observations. We may need to make several assumptions in order to carry out suitable experiments, since our CZ model contains many time and temperature affected parameters which we assumed to be constant. However, comparing the trend of the CZ length from experimental results may be adequate in confirming that the CZ length does indeed decrease with time during the crack growth stage.

Due to the composition of concrete, studying crack propagation leads to difficulties as when the crack grows, it tends to pass through the interface between the

Fig. 12.11 Elastic, concrete

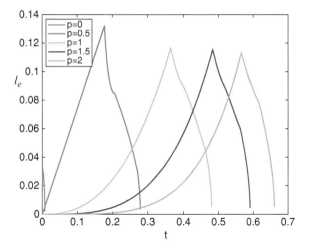

Fig. 12.12 Viscoelastic, concrete

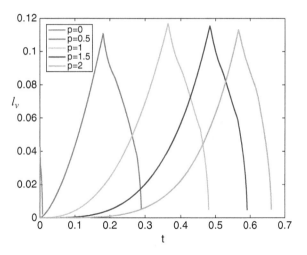

aggregate and the water-sand paste that is keeping the rock pieces together. Due to the complexity of concrete, several other creep functions have been proposed, such as those proposed in [Ba88], where in some cases, experimental results were used to justify the theoretical behaviour of the crack. An interesting future goal is to implement our CZ model to study the behaviour of steel since this material has a large plastic dominion so the deformation in the plastic region is more visible, both analytically and experimentally.

As well as considering higher values of p for the order of the external load, it is also interesting to analyse a time and history dependent CZ model while considering other time-dependent external loads. Such dependence of the variable load on time could involve sinusoidal functions to demonstrate a periodic effect of the load on the material.

References

[Ar52] Arutyunyan, N.Kh.: *Some Problems in the Theory of Creep*, Engl. transl.: Pergamon, Oxford, Moscow (1952).

[Ba66] Bazant, Z.P.: Phenomenological theories for creep of concrete based on rheological models, *Acta Technica CSAV*, **11**, 82–109 (1966).

[Ba88] Bazant, Z.P.: *Mathematical modelling of creep and shrinkage of concrete*, John Wiley and Sons Ltd (1988).

[CaEtAl97] Carpinteri, A., Valente, S., Zhou, F.P. Ferrara, G., Melchiorri, G.: Tensile and flexural creep rupture tests on partially-damaged concrete specimens, *J. of Materials and Structures*, **30**, 269–276 (1997).

[Du60] Dugdale, D.S.: Yielding of steel sheets containing slits, *J. Mech. Phys. Solids*, **8**, 100–101 (1960).

[HaMi11] Hakim, L., and Mikhailov, S.E.: Nonlinear Abel type integral equation in modelling creep crack propagation, *Integral Methods in Science and Engineering: Computational and Analytic Aspects*, C. Constanda C. and P. Harris, Springer, 191–201 (2011).

[HaMi15] Hakim, L., and Mikhailov, S.E.: Integral equations of a cohesive zone model for history-dependent materials and their numerical solution, *The Quarterly Journal of Mechanics and Applied Mathematics*, **68**, 387–419 (2015).

[HaMi18] Hakim, L., and Mikhailov, S.E.: A history-dependent cohesive zone model in elastic and visco-elastic materials under constant and variable loading, *International Journal of Mechanical Sciences*, **144**, 518–525 (2018).

[LePa59] Leonov, M.Ya., Panasyuk, V.V.: Development of the smallest cracks in the solid, *Applied Mechanics (Prikladnaya Mekhanika)*, **5**, No. 4, 391–401 (1959).

[MiNa03] Mikhailov, S.E., and Namestnikova, I.V.: Local and non-local approaches to creep crack initiation and propagation, *Proceedings of the 9th International Conference on the Mechanical Behaviour of Materials*, Geneva, Switzerland, (2003).

[MiNa11] Mikhailov, S.E., and Namestnikova, I.V.: History-sensitive accumulation rules for lifetime prediction under variable loading, *Archive of Applied Mechanics*, **81**, 1679–1696 (2011).

[Mu54] Muskhelishvili, N.I.: *Some Basic Problems of the Mathematical Theory of Elasticity*, Noordhoff International Publishing, The Netherlands (1954).

[NeEtAl12] Nelson, J., Cairns, D., Riddle, T., Workman, J., Composite wind turbine blade effects of defects: part B- progressive damage modeling of berglass/epoxy laminates with manufacturing induced flaws, *Structural Dynamics and Materials Conference*, Hawaii (2012)

[Ra69] Rabotnov, Iu N.: *Creep problems in structural members*, North-Holland Publishing Co, Amsterdam (1969).

[WaEtAl11] Wang, B., Zhang, X.F., Dai J.G.: Critical crack tip opening displacement of different strength concrete, *J. Central South University of Technology*, **18** (2011).

Chapter 13
A Combined Boundary Element and Finite Element Model of Cell Motion due to Chemotaxis

Paul J. Harris

13.1 Introduction

It is well known that biological cells can secrete chemical signals, in the form of proteins, into the fluid medium in which they are immersed in order to signal and attract other nearby cells. A summary of the ways in which bacteria can secrete proteins is given in [GrMe16]. When the cell is stationary the secreted chemical will spread out uniformly in all directions and this situation can mathematically modeled using the standard diffusion equation. However if the cell is moving the secreted chemical will move with the cell and the surrounding fluid. The work in this paper proposes a mathematical model based on coupling the convection–diffusion equation for simulating how the secreted chemical spreads out from a cell with a Stokes-flow model of fluid motion.

The convection–diffusion equation (and the closely related convection–diffusion-reaction equation) has been widely used to mathematically model and simulate the behavior of biological systems (see [DeEtA116, IsZa16, RiEtA116] and the references therein for example). In particular, the diffusion–reaction equation has been used to model how the concentrations of a type of cell change in response to changes in the concentration of a chemical signal (Keller-Segel models, see [ChEtA112, GaZa98, KeSe71, LaSc74] and the references therein for example).

P. J. Harris (✉)
The University of Brighton, Brighton, UK
e-mail: p.j.harris@brighton.ac.uk

© Springer Nature Switzerland AG 2019
C. Constanda, P. Harris (eds.), *Integral Methods in Science and Engineering*,
https://doi.org/10.1007/978-3-030-16077-7_13

13.2 Mathematical Model

The main mathematical model presented in this paper consists of solving two simultaneous and coupled problems: (1) the concentrations of the chemical secreted from the cells and (2) the motion of the cells and fluid. Mathematically, the two problems are coupled as a term depending on the velocity of the fluid appears in the convection–diffusion equation for modelling the spread of the chemical, and the cells experience as force that is proportional to gradient of the concentrations of the chemical which will influence their motion and the resulting motion of the fluid.

For both the fluid flow and chemical concentration parts of the model let Ω denote the region containing both the fluid and the cells. Let Ω_F denote the fluid-filled region exterior the cells. Let Ω_i and Γ_i denote the interior and boundary of the ith cell respectively, and let $\Omega_C = \bigcup \Omega_i$ be the union of the parts of the domain occupied by the cells. Finally, it is noted that $\Omega_C \bigcup \Omega_F = \Omega$.

13.2.1 Finite Element Method for the Chemical Concentrations

The concentration c of a chemical which is spreading through a domain containing a fluid which is moving with velocity $\mathbf{u}(\mathbf{x}, t)$ can be modeled using the convection–diffusion equation [Ma99]

$$\frac{\partial c(\mathbf{x}, t)}{\partial t} = \nabla \cdot (D(\mathbf{x}, t)\nabla c(\mathbf{x}, t)) - \nabla \cdot (c(\mathbf{x}, t)\mathbf{u}(\mathbf{x}, t)) + f(\mathbf{x}, t) \tag{13.1}$$

where $D(\mathbf{x}, t)$ is the diffusion constant which controls how quickly the chemical spreads out over time and $f(\mathbf{x}, t)$ is a source term. The differential equation (13.1) is solved on the whole of the domain Ω with different values of the parameters $D(\mathbf{x}, t)$ and $\mathbf{u}(\mathbf{x}, t)$ depending on whether $\mathbf{x}$ is in a cell or in the fluid. In the model presented here, the value of D will be constant within each sub-domain Ω_C and Ω_F and hence

$$D(\mathbf{x}, t) = \begin{cases} D_F & \mathbf{x} \in \Omega_F \\ D_C & \mathbf{x} \in \Omega_C \end{cases}$$

For points inside one of the cells, the velocity term $\mathbf{u}(\mathbf{x}, t)$ in (13.1) will be simply taken as $\mathbf{v}$, the velocity of the cell, whilst in the exterior fluid the velocity will be calculated from that of the cell as described in Sect. 13.2.2 below. Under these conditions the concentrations c will be continuous at the cell boundary Γ, but if the value of D in the cell is different from that in the fluid, the normal derivative of the concentrations will have a jump discontinuity at the boundary.

The finite element method has been used extensively to obtain the approximate solutions of differential equations, and there are many texts on the subject, such as [ZiTa89]. The use of the method to solve the convection–diffusion equation (and the related convection–diffusion-reaction equation) has been widely reported in the literature, see [DeEtAl16, IsZa16, RiEtAl16] for example.

For the problem under consideration here the domain Ω of the governing differential equations is very large when compared to the size of a typical cell and, in general, it is not computationally feasible to apply the finite element method to the whole of Ω. In order to apply the finite element method to this problem, the large domain Ω is replaced by a much smaller approximate domain Ω_a which has an outer boundary Γ_a. An approximate boundary condition is needed on the outer boundary Γ_a. Here it is simply assumed that the outer boundary of Ω_a is far enough away from the cell so that the diffusing chemical does not reach this boundary at the final time at which the concentrations are required, and hence there is no flow of the chemical across the boundary Γ_a. This leads to the boundary condition

$$(D(\mathbf{x}, t)\nabla c(\mathbf{x}, t) - c(\mathbf{x}, t)\mathbf{u}(\mathbf{x}, t)) \cdot \mathbf{n} = 0 \quad \mathbf{x} \in \Gamma_a.$$

Whilst this is not an ideal situation, it makes solving the problem computationally feasible and it is possible to check that the concentrations near to the outer boundary are not significantly different from zero.

Let $\mathbf{c}(t)$ denote the vector of the time-dependent nodal values of the concentration. and let an over-dot denote differentiation with respect to time. The finite element method for solving (13.1) can be written in matrix notation as [ZiTa89]

$$M\dot{\mathbf{c}} = K(t)\mathbf{c} + \mathbf{f}(t) \tag{13.2}$$

where

$$M_{ij} = \int_{\Omega_a} \psi_i(\mathbf{x})\psi_j(\mathbf{x}) \, d\mathbf{x}$$

$$K_{ij}(t) = -\int_{\Omega_a} \left[D(\mathbf{x}, t)\nabla\psi_j(\mathbf{x}) - \psi_j(\mathbf{x})\mathbf{u}(\mathbf{x}, t) \right] \cdot \nabla\psi_i(\mathbf{x}) \, d\mathbf{x}$$

$$\mathbf{f}_i = \int_{\Omega_a} f(\mathbf{x}, t)\psi_i(\mathbf{x}) \, d\mathbf{x}$$

and $\{\psi_i(\mathbf{x})\}$ denotes the set of finite element basis functions. Here the simple linear triangular elements have been used. See [ZiTa89], for example, for more details on the finite element method and for other choices of basis functions.

13.2.2   Boundary Integral Method for the Fluid Flow

If at least one of the cells is moving, this will set up a flow in the exterior fluid. Assuming that the fluid is incompressible, the equations for the velocity $\mathbf{u}$ of the fluid can be expressed as the continuity equation [Li86]

$$\nabla \cdot \mathbf{u} = 0$$

for the conservation of mass, and the Navier-Stokes equation

$$\rho \left(\frac{\partial \mathbf{u}}{\partial t} + \mathbf{u} \cdot \nabla \mathbf{u} \right) = -\nabla p + \mu \nabla^2 \mathbf{u} + \rho \mathbf{b} \qquad (13.3)$$

which can be considered as an expression of Newton's second law for a small particle of the fluid. Here ρ and μ are used to denote the density and dynamic viscosity of the fluid, p is the pressure and $\mathbf{b}$ is a known body force (which is assumed to be zero in this work, but can be used to include effects such as gravity).

For the flows under consideration here the Reynolds number is very small, and the terms on the left-hand side of (13.3) can be neglected leading to the Stokes-flow equations [Li86]

$$\begin{aligned} -\nabla p + \mu \nabla^2 \mathbf{u} &= \mathbf{0} \\ \nabla \cdot \mathbf{u} &= 0 \end{aligned} \qquad (13.4)$$

The boundary conditions for the fluid flow problem are

$$\begin{aligned} \mathbf{u}(\mathbf{x}) &= \mathbf{v}_i + \omega_i \begin{bmatrix} -(y - y_i) \\ (x - x_i) \end{bmatrix} \quad \mathbf{x} \in \Gamma_i \\ \mathbf{u}(\mathbf{x}) &= \mathbf{0} \qquad\qquad\qquad\qquad \mathbf{x} \in \Gamma_O \end{aligned}$$

where Γ_O is an outer boundary that is required to avoid the problems associated with Stokes paradox. Here $\mathbf{v}_i$ and ω_i denote the velocity and angular velocity of the ith cell, where the rotation is assumed to be around the cells centre of mass located at (x_i, y_i). In this work the outer boundary is taken to be a square with sides of length 5000 cell radii which corresponds to the cells being in a petri dish with sides length 5 cm.

It can be shown that in two space dimensions, the fluid velocity which satisfies (13.4) in a closed domain Ω_F with a piecewise smooth boundary curve Γ also satisfies the boundary integral equation [Po92]

$$\oint_\Gamma T(\mathbf{x}, \mathbf{x}_0) \mathbf{u}(\mathbf{x}) \, d\Gamma_{\mathbf{x}} - \oint_\Gamma G(\mathbf{x}, \mathbf{x}_0) \mathbf{F}(\mathbf{x}) \, d\Gamma_{\mathbf{x}} = \begin{cases} \mathbf{u}(\mathbf{x}_0) & \mathbf{x}_0 \in \Omega_F \\ \frac{1}{2}\mathbf{u}(\mathbf{x}_0) & \mathbf{x}_0 \in \Gamma \\ \mathbf{0} & \mathbf{x} \in \Omega_C \end{cases} \qquad (13.5)$$

where **F** are the surface forces,

$$T_{ij}(\mathbf{x}, \mathbf{x}_0) = -\frac{\mathbf{r} \cdot \mathbf{n}}{\pi r^4} r_i r_j \quad G_{ij}(\mathbf{x}, \mathbf{x}_0) = \frac{1}{4\pi\mu}\left(-\delta_{ij}\ln(r) + \frac{r_i r_j}{r^2}\right)$$

$\mathbf{r} = \mathbf{x} - \mathbf{x}_0$, $r = |\mathbf{r}|$, $\mathbf{n}$ is the unit normal to Γ directed onto the fluid domain Ω_F and δ_{ij} is the Kronecker delta function. Here Γ is the union of the boundaries of the cells and the outer fluid boundary. It should be noted that (13.5) for $\mathbf{x}_0 \in \Gamma$ is strictly only valid for points $\mathbf{x}_0$ which are on a smooth part of Γ and not at vertex of boundary curve. For the piecewise constant boundary element method used in this work this is not a problem as $\mathbf{x}_0$ will always be chosen to be on a smooth part of the curve. However, if a higher-order approximation to the solution, such a piecewise linear or piecewise quadratic, were used then this would need to be considered as the collocation points for such methods can potentially be located at a vertex of the boundary curve.

From the fluid velocity boundary condition, the value of $\mathbf{u}$ is known on the whole of Γ, and hence (13.5) for $\mathbf{x}_0 \in \Gamma$ is a first kind Fredholm integral equation for the unknown force $\mathbf{F}$ on the boundary. Once the forces have been found, (13.5) for $\mathbf{x}_0 \in \Omega_F$ can be used to calculate the fluid velocity at any point in the fluid.

In this work the integral equation (13.5) is solved using the boundary element method with a piecewise linear approximations to the boundary Γ and a piecewise constant approximation to the unknown forces $\mathbf{F}$. A complete description of the boundary element method used to solve the integral equation will not be given here as there are numerous texts available which give all the details of how to implement the boundary element method.

It is noted here that the boundary integral equation (13.5) can be used to compute the fluid velocity at any given point in time. However, if the fluid velocities are subsequently required at a different point in time, then the full boundary integral/element calculations will have to be carried out for the new time.

The boundary element method described above has been used to model the motion of cell clusters moving through a viscous fluid where the chemical concentrations were simulated using a simple formula, and the results reported in [Ha18].

13.2.3   Time Integration Method

The translational and angular accelerations of each cell are given by

$$\frac{d\mathbf{v}_i}{dt} = \mathbf{a}_i(t) = \frac{1}{m_i}\left[\int_{\Gamma_i} \mathbf{F}(\mathbf{x}, t)\, d\Gamma_{\mathbf{x}} + \int_{\Omega_i} k\nabla c(\mathbf{x}, t)\, d\Omega_{\mathbf{x}}\right]$$

and

$$\frac{d\omega_i}{dt} = \alpha_i(t) = \frac{1}{I_i} \left[\int_{\Gamma_i} \left((x - x_i) f_y(\mathbf{x}) - (y - y_i) f_x(\mathbf{x}) \right) d\Gamma_{\mathbf{x}} \right.$$
$$\left. + \int_{\Omega_C} \left(k(x - x_c) \frac{\partial C}{\partial y} - k(y - y_c) \frac{\partial C}{\partial x} \right) d\Omega_{\mathbf{x}} \right]$$

respectively, where m_i and I_i denote the mass and moment of inertia of the cell, and k is a constant which controls how strong the cells react to the gradient of the concentrations of the chemical. The value of k can be different for each cell, but here it is assumed to be the same for all cells.

For each cell, the velocity $\mathbf{v}_i$, angular velocity ω_i, the location of its centre of mass $\mathbf{p}_i$ and rotation θ_i can be integrated through time using

$$\mathbf{v}_i(t + h) = \mathbf{v}_i(t) + h\mathbf{a}_i$$
$$\omega_i(t + h) = \omega_i(t) + h\alpha_i$$
$$\mathbf{p}_i(t + h) = \mathbf{p}_i(t) + \frac{h}{2} (\mathbf{v}_i(t) + \mathbf{v}_i(t + h)) \qquad (13.6)$$
$$\theta_i(t + h) = \theta_i(t) + \frac{h}{2} (\omega_i(t) + \omega_i(t + h))$$

Using the values calculated in (13.6) it is possible to calculate the positions of the boundaries of the cells at the new time-step, and hence use the boundary integral method to calculate the fluid velocities at the new time-step. Once these have been found the finite element matrix $K(t + h)$ can be calculated so that the Crank-Nicholson [CrNi47] method can be applied to (13.2) to give

$$M \left(\frac{\mathbf{c}(t + h) - \mathbf{c}(t)}{h} \right) = \frac{K(t + h)\mathbf{c}(t + h) + K(t)\mathbf{c}(t)}{2} + \frac{\mathbf{F}(t) + \mathbf{F}(t + h)}{2}.$$
$$(13.7)$$

Rearranging (13.7) gives

$$(2M - hK(t + h)) \mathbf{c}(t+h) = (2M + hK(t)) \mathbf{c}(t) + h (\mathbf{F}(t) + \mathbf{F}(t + h)) \qquad (13.8)$$

which is a system of linear algebraic equations that can be solved for $\mathbf{c}(t + h)$.

The cell locations and velocities at the new time-step can be refined by using

$$\mathbf{v}_i(t + h) = \mathbf{v}_i(t) + \frac{h}{2} (\mathbf{a}_i(t) + \mathbf{a}_i(t + h))$$
$$\omega_i(t + h) = \omega_i(t) + \frac{h}{2} (\alpha_i(t) + \alpha_i(t + h))$$
$$\mathbf{p}_i(t + h) = \mathbf{p}_i(t) + \frac{h}{2} (\mathbf{v}_i(t) + \mathbf{v}_i(t + h)) \qquad (13.9)$$
$$\theta_i(t + h) = \theta_i(t) + \frac{h}{2} (\omega_i(t) + \omega_i(t + h))$$

where the accelerations at time $t + h$ are calculated using the concentrations calculated at time $t + h$. The refined cell locations and velocities can then be used in the boundary integral method and (13.8) to refine the concentrations at the new time-step. This iterative process can be continued until the locations and concentrations at the new time-step converge, although numerical experiments indicate that one refinement is usually sufficient.

13.3   Numerical Results

This paper considers cells moving in a square petri-dish with sides of length 5 cm, which corresponds to domain approximately 5000 times greater than the radius of a typical cell.

The results presented here are for two elongated cells where Cell 1 is secreting a chemical which attracts Cell 2. The production of the left-hand cell is modeled by using the source term

$$f(\mathbf{x}, t) = \begin{cases} 1 \ \mathbf{x} \in \Omega_1 \text{ and } t \leq 1 \\ 0 \ \text{Otherwise} \end{cases}$$

where Ω_1 denotes the interior of Cell 1. Figure 13.1a–d shows the spread of the chemical signal on the left and the resulting fluid motion on the right, where the arrows indicate the direction of the fluid flow. In the diagrams blue denotes low values of the chemical concentrations or fluid velocity, and red indicates high values.

In order to check the accuracy of the results, the displacements and rotations of the two cells were calculated using two different meshes. In the coarse meshes there were 15,488 elements with 7921 finite element nodes and 28 boundary elements on each cell. For the fine meshes there were 61,952 finite elements with 31,329 nodes and 56 boundary elements on each cell. The results for Cell 1 are shown in Fig. 13.2a and those for Cell 2 are shown in Fig. 13.2b. The graphs of the various quantities computed using both cells as almost identical, indicating that the model is yielding accurate results.

13.4   Conclusions and Future Work

This paper describes a combined finite element and boundary element which models how cells move through a viscous fluid in response to chemical signals secreted by other nearby cells. The numerical results show that the methods described in the paper are yielding accurate results since when the finite element and boundary element meshes are refined there is little difference in the calculated positions and rotations of the cells.

Fig. 13.1 The spread of the
chemical signal and resulting
fluid motion for two
elongated cells. (**a**) $t = 0.0$.
(**b**) $t = 1.5$. (**c**) $t = 3.0$. (**d**)
$t = 4.5$

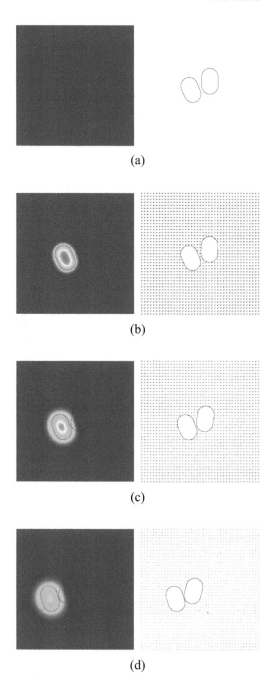

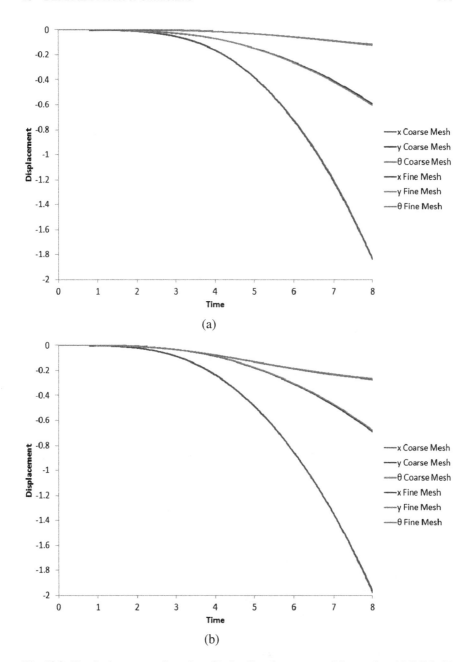

Fig. 13.2 The displacement and rotation of both cells using coarse and fine meshes. (**a**) Cell 1. (**b**) Cell 2

The present model can only continue a calculation up to the point where two (or more) cells collide. There is some experimental evidence that the time-scale of the collision is shorter than the time-scale of the overall cell motion which means that the collision can be considered to be instantaneous. This means that the collision can be simulated in a separate calculation and then the original cells are replaced by the new connected cells in the model. Suitable conservation laws can be used to calculate quantities such as the velocity of the new combined cells from the velocities of the individual cells before they collided.

Acknowledgements The author would like to thank Matteo Santin from Brighton Centre for Regenerative Medicine and Devices for his help and advice with some of the biological aspects of this paper.

References

[ChEtAl12] Chertock, A., Kurganov, A., Wang, X. and Wu, Y.: On a chemotaxis model with saturated chemotactic flux. *Kin. and Rel. Mod*, **5**, 51–95, (2012).

[CrNi47] Crank J. and Nicolson P.: A practical method for numerical evaluation of solutions of partial differential equations of the heat-conduction type. *Mathematical Proceedings of the Cambridge Philosophical Society*, **43**, 50–67, (1947).

[DeEtAl16] Deleuze Y., Chiang, C., Thiriet, M. and Sheu, T. W. H.: Numerical study of plume patterns in a chemotaxis-diffusion-convection coupling system. *Computers & Fluids*, **126**, 58–70, (2016).

[GaZa98] Gajewski, H. and Zacharias, K.: Global behaviour of a reaction - diffusion system modelling chemotaxis. *Math. Nachr*, **195**, 77–114, (1998).

[GrMe16] Green, E. R. and Mecsas, J.: Bacterial Secretion Systems: An Overview. *Microbiology Spectrum*, **4**, (2016).

[Ha18] Harris P. J.: Modelling the motion of clusters of cells in a viscous fluid using the boundary integral method. *Mathematical Biosciences*, **306**, 145–152, (2018).

[KeSe71] Keller, E. F. and Segel, L. A.: Model for chemotaxis. *J. Theor. Biol.*, **30**, 225–234, (1971).

[IsZa16] Islam, S. and Zaman, R.: A computational modeling and simulation of spatial dynamics in biological systems. *App. Math. Mod.*, **40**, 4524–4542, (2016).

[LaSc74] Lapidus, I. R. and Schiller, R.: A mathematical model for bacterial chemotaxis. *Biophys. J.*, **14**, 825–834, (1974).

[Li86] Lighthill, M. J.: *An informal introduction to theoretical fluid mechanics.* Clarendon Press, Oxford. (1986).

[Ma99] Mazumdar, J. The mathematics of diffusion., *An Introduction to Mathematical Physiology and Biology.*, Cambridge University Press, (1999).

[Po92] Pozrikidis, C.: *Boundary Integral and Singularity Methods for Linearized Viscous Flow.* Cambridge Texts in Applied Mathematics, Cambridge University Press, (1992).

[RiEtAl16] Ritter, J., Klar, A. and Schneider, F.: Partial-moment minimum-entropy models for kinetic chemotaxis equations in one and two dimensions. *J. Comp. and App. Maths*, **306**, (2016).

[ZiTa89] Zienkiewicz, O. C. and Taylor R.L.: *The Finite Element Method, fourth Edition* McGraw-Hill Book Company Europe, London, (1989).

Chapter 14
Numerical Calculation by Quadruple Precision Higher Order Taylor Series Method of the Pythagorean Problem of Three Bodies

Hiroshi Hirayama

14.1 Introduction

Usually, explicit calculation methods [HaWa91] such as Euler method and Runge-Kutta method are used for numerical solution of ordinary differential equations. Through these methods, we consider the following differential equation of initial value problem.

$$\mathbf{y}' = \mathbf{f}(x, \mathbf{y}) \qquad \mathbf{y}(x_0) = \mathbf{y}_0$$

The s-stage explicit Runge-Kutta method can be written as follows:

$$\begin{cases} \mathbf{k}_1 &= \mathbf{f}(x_n, \mathbf{y}_n) \\ \mathbf{k}_2 &= \mathbf{f}(x_n + c_2 h, \mathbf{y}_n + a_{21} h \mathbf{k}_1) \\ &\vdots \\ \mathbf{k}_s &= \mathbf{f}(x_n + c_s h, \mathbf{y}_n + a_{s1} h \mathbf{k}_1 + a_{s2} h \mathbf{k}_2 + \cdots + a_{s,s-1} h \mathbf{k}_{s-1}) \\ \mathbf{y}_{n+1} &= \mathbf{y}_n + \sum_{i=1}^{s} b_i \mathbf{k}_i \end{cases}$$

In these methods, the higher order formulas contain a large number of constants proportional to the square of the number of stages, so there is a problem that it is difficult to use. Moreover, there is a problem that it is very difficult to obtain these constants.

Many formulas below the tenth degree are introduced in the literature [EnUh96]. It is known that it is very necessary to calculate [Oh06] for the creation of higher-order formulas.

H. Hirayama (✉)
Kanagawa Institute of Technology, Atsugi, Kanagawa, Japan
e-mail: hirayama@sd.kanagawa-it.ac.jp

© Springer Nature Switzerland AG 2019
C. Constanda, P. Harris (eds.), *Integral Methods in Science and Engineering*,
https://doi.org/10.1007/978-3-030-16077-7_14

Fig. 14.1 The initial state of
the Pythagorean three-body
problem

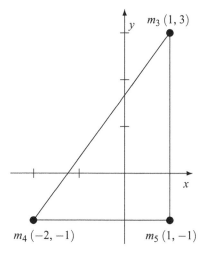

It is required to calculate a large scale system described by an ordinary differential equation with high accuracy over a long period of time. In these methods, the order of calculation is limited, and calculation with high precision for a long time becomes very difficult. To solve these problems, Taylor expansion method which is an arbitrary degree and variable step calculation method seems to be most suitable.

In this paper, as an example, we tried solving astronomical three-body problem using Taylor series method, examined the calculation accuracy, and investigated the performance of Taylor series method. We chose Pythagoras' three-body problem [NaHi92, SzBu67] which is considered to be difficult to solve as an ordinary differential equation to calculate. The Pythagoras three-body problem is the problem of mass points 3, 4, 5 which are proportional to the length of each opposite side, at the vertexes of the right triangle with side lengths 3, 4, and 5 like Fig. 14.1, and using that state as an initial condition, these mass points are pursuing how to move after this by mutual attractive force. Defining the distance between the mass points m_3 and m_4 as r_{34}

$$r_{34} = \sqrt{(x_3 - x_4)^2 + (y_3 - y_4)^2} \qquad r_{35} = \sqrt{(x_3 - x_5)^2 + (y_3 - y_5)^2}$$

$$r_{45} = \sqrt{(x_4 - x_5)^2 + (y_4 - y_5)^2}$$

Writing the equation of motion using these symbols is as follows:

$$\frac{d^2 x_3}{dt^2} = \frac{4(x_4 - x_3)}{r_{34}^3} + \frac{5(x_5 - x_3)}{r_{35}^3} \qquad \frac{d^2 y_3}{dt^2} = \frac{4(y_4 - y_3)}{r_{34}^3} + \frac{5(x_5 - x_3)}{r_{35}^3}$$

$$\frac{d^2 x_4}{dt^2} = \frac{3(x_3 - x_4)}{r_{34}^3} + \frac{5(x_5 - x_4)}{r_{45}^3} \qquad \frac{d^2 y_4}{dt^2} = \frac{3(y_3 - y_4)}{r_{34}^3} + \frac{5(x_5 - x_4)}{r_{45}^3}$$

$$\frac{d^2x_5}{dt^2} = \frac{3(x_3 - x_5)}{r_{35}^3} + \frac{4(x_4 - x_5)}{r_{45}^3} \qquad \frac{d^2y_5}{dt^2} = \frac{3(y_3 - y_5)}{r_{35}^3} + \frac{4(x_4 - x_5)}{r_{45}^3}$$

$$(14.1)$$

The initial condition is

$$x_3 = 1, \frac{dx_3}{dt} = 0, y_3 = 3, \frac{dy_3}{dt} = 0 \quad x_4 = -2, \frac{dx_4}{dt} = 0, y_4 = -1, \frac{dy_4}{dt} = 0$$

$$x_5 = 1, \frac{dx_5}{dt} = 0, y_5 = -1, \frac{dy_5}{dt} = 0 \qquad (14.2)$$

For the calculation below, Visual Studio 2015 C ++ was used as a compiler, and the personal computer (Intel Core i7 7700 K 4.2 GHz) was used as a calculator.

14.2 Taylor Series of Ordinary Differential Equations

Here, we explain Taylor series method of ordinary differential equations easily. Since higher order ordinary differential equations can be written in the first order differential equations without loss of generality, we assume that they have the following form:

$$\frac{d\mathbf{y}}{dx} = \mathbf{f}(x, \mathbf{y}(x)) \qquad \text{Initial condition:} \quad \mathbf{y}(x_0) = \mathbf{y}_0$$

Here, $\mathbf{f}$, $\mathbf{y}$ is generally a vector function, it is sufficiently smooth and can be differentiated as many times as necessary. The initial condition $\mathbf{y}_0$ is a constant vector. Such a differential equation can be solved by the following Picard's successive approximation method [Sa93].

$$\mathbf{y}_0(x) = \mathbf{y}_0 \qquad \mathbf{y}_{k+1}(x) = \mathbf{y}_0 + \int_{x_0}^{x} \mathbf{f}(t, \mathbf{y}_k(t))dt$$

Substitute the Taylor expansion equation into the integrand of the above equation and expand the integrand to the Taylor series. Compute the Taylor series of the order of k by iterating k times [HiEtAl02].

Integrate the Taylor expansion of the k order and add the constant term $\mathbf{y}_0$ to calculate the solution of $k + 1$ order. A high solution of at least 1 degree order is obtained by one calculation.

14.2.1   Solution of Simple Differential Equations by Taylor Series

Solve the following simple ordinary differential equation as an example.

$$\frac{dy}{dx} = 1 + \sqrt{y} \qquad y(0) = 1$$

Since $y_0(x) = 1$ from the initial condition, this is substituted into Picard's successive approximation [Sa93]. We get

$$y_1(x) = 1 + \int_0^x (1 + \sqrt{1})dt = 1 + 2x$$

The integrand is a zero-order constant, and the final calculation result is a linear expression. Further, calculate this result by substituting it into Picard's successive approximation formula. Expand the integrand and take the first order equation. It is $1 + 2x$. By integrating this and calculate

$$y_2(x) = 1 + \int_0^x (1 + \sqrt{1 + 2t})dt = 1 + 2x + 0.5x^2 + O(x^3)$$

By repeating such calculation twice, solutions up to the fourth order are obtained as follows:

$$y_3(x) = 1 + \int_0^x (1 + \sqrt{1 + 2t + 0.5t^2})dt$$

$$= 1 + 2x + 0.5x^2 - 0.08333333x^3 + O(x^4)$$

$$y_4(x) = 1 + \int_0^x (1 + \sqrt{1 + 2t + 0.5t^2 - 0.08333333t^3})dt$$

$$= 1 + 2x + 0.5x^2 - 0.08333333x^3 + 0.0520833x^4 + O(x^5)$$

The above calculation can be calculated by the following program of C++ language [El90].

```
// template program of Taylor series

#include "taylor_template.h"

// coefficient is double precision

typedef taylor_template<double> taylor ;

int main()
{
```

```
double y0=1, h=0.1 ;              // initial value, step size
taylor y ;                       // define Taylor series
y = y0 ;                         // set initial value to y

// compute Taylor series upto 5 order

for( int i=1 ; i<=5 ; i++ )
{
  set_degree(i) ;                         // computing order i
  y = y0 + integrate( 1.0+sqrt(y) ) ; // compute solution
  cout << y << endl ;                    // out put solution
}
}
```

If this calculation is performed a necessary number of times, a Taylor series solution of an arbitrary order is obtained.

Using this Taylor series solution, calculate the function value in the next step. Let h be the width of the next step to calculate $y(h)$. Using this value as the initial value of the next step, the function value of the next step is obtained in the same way as the previous method. Solve the differential equations by repeating this.

Here, although it was calculated using Picard's successive approximation method, as a method of calculating the coefficients of the Taylor series, Picard's successive approximation method is the same as the series expansion method.

14.2.2   Calculation of Square Root of Taylor Series

In the previous example we are calculating the Taylor series of the square root of the Taylor series. This calculation can be calculated by solving the following differential equation by a series expansion method. In general, we consider to multiply α. $\alpha = \frac{1}{2}$, it becomes square root.

Let $f(x)$ and $g(x)$ be the following Taylor series.

$$f(x) = f_0+f_1x+\cdots+f_nx^n+\cdots \qquad g(x) = g_0+g_1x+\cdots+g_nx^n+\cdots \qquad (14.3)$$

The following differential equation is obtained by differentiating $g(x) = f(x)^\alpha$.

$$g'(x) = \alpha f(x)^{\alpha-1} f'(x) \quad f(x)g'(x) = \alpha g(x)f'(x) \qquad (14.4)$$

Substituting the Taylor series (14.3) and (14.4), as follows:

$$(f_0 + f_1x + f_2x^2 + \cdots)(g_1 + 2g_2x + 3g_3x^2 + \cdots)$$
$$= \alpha(g_0 + g_1x + g_2x^2 + \cdots)(f_1 + 2f_2x + 3f_3x^2 + \cdots)$$

When comparing the terms of x^{n-1}, the following equation is obtained.

$$g_0 = f_0^\alpha, \qquad g_n = \frac{1}{n f_0} \sum_{i=1}^{n} \{(\alpha + 1)i - n\} f_i g_{n-i} \quad (n >= 1)$$

The coefficient can be obtained from the above recurrence formula. In this way, it is possible to calculate power of Taylor series, exponential logarithm, trigonometric function, etc. by using differential equations.

14.3 Quad Precision Calculation

C ++ language used for this calculation cannot handle quadruple precision numbers. In this case, Bailey's double–double algorithm [KoEtAl08] performs quadruple precision calculations. In the double–double algorithm, a quadruple-precision floating-point number (real16) is represented by two double precision floating-point numbers and the upper digit is denoted by m0, and the lower digit is denoted by m1 and represented by the following structure.

```
class real16 {  double  m0, m1 ; }
```

The quadruple precision variable a is expressed as follows using two double precision variables a.m0 (upper digit) and a.m1 (lower digit).

$$a = a.m0 + a.m0 \quad (\frac{1}{2} ulp(a.m0) \geq |a.m1|)$$

Here, $ulp(x)$ means the minimum bit (unit in the last place) of x. At this time, a.m0 and a.m1 are double precision floating point numbers. For this reason, the precision of the mantissa part is 53 bits, and by using two double precision floating point numbers, it can be expressed with accuracy of 106 bits. Therefore, the double–double algorithm is inferior in accuracy by 8 bits when compared with the quadruple precision of IEEE 754-2008. However, since the quadruple precision of IEEE 754-2008 is created by software, the calculation speed can be calculated quickly by double–double quadruple precision because there are many parts to calculate hardware, so a practical method it is said to be [HiEtAl00, YaEtal12].

Quadruple precision addition and multiplication can be programmed using the double–double algorithm as follows:

```
Addition of double--double numbers      Multiplication of double--double numbers
real16 add( const real16 &a,            real16 mul( const real16 &a,
            const real16 &b )                       const real16 &b )
{                                       {
  real16 c ;                              quad c ;
  double sh, eh, v ;                      double sl ;
  sh = a.m0 + b.m0 ;                      c.m0 = a.m0 * b.m0 ;
  v = sh - a.m0 ;                         c.m1 = fma(a.m0, b.m0, -c.m0) ;
```

```
eh = (a.m0 - (sh-v)) + (b.m0 - v) ;        c.m1 = c.m1 + (high * b.m1 +low * b.m0) ;
eh = eh + a.m1 + b.m1 ;                     s1 = c.m0 +c.m1 ;
c.m0 = sh + eh ;                            c.m1 =c.m1 - (s1 - c.m0) ;
c.m1 = eh-(c.m0-sh) ;                       c.m0 =s1 ;
return c ;                                  return c ;
}                                          }
```

By using this addition and multiplication, division, square root, exponential log function, and trigonometric function, etc. can be calculated [HiEtAl00].

As a calculation example, we introduce a quadruple precision program to solve the following quadratic equation. The following program is C++ program for solving quadratic equations ($ax^2 + bx + c = 0$).

```
//Solving quadruple-precision quadratic equations
#include "real16.h"
int main()
{
    real16 a, b, c, d, x1, x2 ;
    a=2 ; b = 7.5 ; c=real16("-12.2") ;
    d=b*b-4*a*c ; d = sqrt(d) ;
    x1=(-b+d)/(2*a) ; x2=(-b-d)/(2*a) ;
    set_format("%35.32g") ; // set formatting
    cout << "x1=" << x1 << endl ;
    cout << "x2=" << x2 << endl ;
}
```

The calculation result is a solution of $2x^2 + 7.5x - 12.2 = 0$, which is as follows:

```
x1 =   1.2259071253425182195488491564024
x2 = -4.9759071253425182195488491564024
```

14.4 Calculation of Three-Body Problem of Pythagoras

We solved Eq. (14.1) using the initial condition (14.2) by using the Taylor series of the order of n. The step width h is determined so that the absolute value of the coefficient of the highest degree term is smaller than the required accuracy ϵ.

If the coefficient of the highest degree term is zero, the coefficients of the low order terms of the first degree are used. If the coefficient of that term is zero, repeat similarly again.

If a_n is a coefficient of n order, then $|a_n|h^n <= \epsilon$. That is, for each Taylor series, we calculate $h = \sqrt[n]{\frac{\epsilon}{|a_n|}}$. Let the minimum value be h.

In this calculation, we calculated the required accuracy $\epsilon = 10^{-28}$ using the 24th order Taylor expansion formula ($n = 24$). The resulting 11 decimal places are shown in Table 14.1. The result of this calculation is a result with higher accuracy than the result calculated using Levi-Civita conversion of Hiyama [NaHi92], but it agrees perfectly within the accuracy range.

Table 14.1 Three celestial coordinates

t	x_3	y_3	x_4	y_4	x_5	y_5
0.0	1.00000000000	3.00000000000	−2.00000000000	−1.00000000000	1.00000000000	−1.00000000000
10.0	0.77848041014	0.14139230029	−2.02509247798	0.09721938415	1.15298573630	−0.16261088749
20.0	3.00429263670	0.51192523502	−1.38862653751	−0.47047605025	−0.69167435201	0.06922569919
30.0	0.85634049890	2.28709366370	−0.87798387903	−0.86596382772	0.18858280389	−0.67948513605
40.0	−0.62200369180	1.85831815790	0.17354455682	−2.36841044328	0.23436656963	0.77973745989
50.0	−2.70146140927	−3.79722268272	1.50593819242	0.96081342494	0.41612629162	1.50968286968
60.0	0.74380750012	1.93994795109	0.26401033469	−0.73162439487	−0.65749276782	−0.57866925476
70.0	6.93346359901	20.26180430595	−2.00300600260	−6.87246343582	−2.55767335732	−6.65911183491
80.0	12.44744289203	36.64230872512	−3.55586671500	−12.35478120556	−4.62377236322	−12.10156027062

In order to confirm this calculation result, the path was calculated with the initial condition in which the velocity was reversed ($\mathbf{v} = -\mathbf{v}$) at the final time t. We calculated how closely it returned to the initial value of the original problem. It was necessary to calculate 10,633 times Taylor expansion with maximum step size $h_{max} = 0.1362$, minimum step width $h_{min} = 1.7092 \times 10^{-7}$.

A numerical value obtained by changing only the sign of the speed at the last time point was calculated up to $t = 0$ as an initial value. It was necessary to develop Taylor 10,635 times in this calculation. The number of calculations of Taylor expansion is almost the same, but the inverse calculation was twice more. The difference between the initial value calculated here and the original initial value was up to 2.23×10^{-18} and it matched about 17 digits. From this result, the calculation seems to be correct by about 17 figures.

It was also checked if energy on the way was preserved. Energy $E = -\frac{769}{60}$ has a maximum relative error of 1.2×10^{-26}. Energy was constant with accuracy of almost 25 orders of magnitude. It can be seen that the high precision energy conservation law is established.

Calculate the time and distance at which the object is closest. During the calculation of the differential equation, it is found that the distance $r45$ between object 4 and object 5 becomes the minimum in the vicinity of $t = 15.829920$. At this point it expands $r45$ to the Taylor series of time. The Taylor series is a series of 24th order as follows:

$$r45(t) = 4.1403728 \times 10^{-4} - 4.7226778t + 2.6220212 \times 10^7 t^2$$
$$+ 3.9888548 \times 10^{11} t^3 + \cdots - 1.6455306 \times 10^{128} t^{24}$$

The constant term is small as 4.1404×10^{-4}, but the 24th order coefficient is very large as -1.6455×10^{128}. From this fact, it is found that the step size h needs to be very small.

In order to find the closest point, solving $\frac{d}{dt}r45(t) = 0$ gives the time of the closest point. Differentiating $r45(t)$ yields following 23 orders Taylor series.

$$\frac{d}{dt}r45(t) = -4.7226778 + 5.2440424 \times 10^7 t + 1.1966564^{12} t^2$$
$$- 4.4087319 \times 10^{18} t^3 + \cdots - 3.9492734 \times 10^{129} t^{23}$$

To find the zero of this Taylor series, calculate the Taylor series of the inverse function ($i45(s)$) of $\frac{d}{dt}r45(t)$.

The Taylor series of the inverse function $y = f^{-1}(x)$ of the function $y = f(x)$ can be easily obtained by solving differential equation $\frac{d}{dx}y = \frac{1}{f'(y)}$ by Taylor series method. Since $y = f(0)$ when $x = 0$, the initial condition is reversed and solved as $y = 0$ when $x = f(0)$. If you write the function $\frac{d}{dt}r45(t)$ as $DR45(t)$, the inverse function of $DR45(t)$ becomes

$$DR45^{-1}(s) = 1.9069259 \times 10^{-8}(s-a) - 8.2979517 \times 10^{-12}(s-a)^2$$
$$+ 5.9019532 \times 10^{-13}(s-a)^3 + \cdots - 2.3526647 \times 10^{-56}(s-a)^{24}$$

where $a = \frac{d}{dt}r45(t)|_{t=0} = -4.7226778$.

Substituting $s = 0$ into the inverse function and finding the relative time dt of the zero point yields $dt = 8.99347711832073120 \times 10^{-8}$. Therefore, the position of the zero point is $t_{zero} = 15.8299202715809$. Calculating the distance $r45$ between objects yields $r45 = 4.13824836258701 \times 10^{-4}$. As you can see, the results of Szebehely [SzBu67] ($t_{zero} = 15.8230, r45 = 4 \times 10^{-4}$) were able to be highly accurate.

14.5 Conclusion

By using higher order formulas and quadruple precision numbers without coordinate transformation such as Levi-Civita transformation, it was possible to calculate the three-body problem of Pythagoras with sufficient accuracy.

In addition to the calculation on the three-body problem of Pythagoras, this method seems to be able to solve the ordinary differential equations of bad condition with sufficient accuracy with high accuracy.

The Taylor series method can easily calculate higher order, it seems to be optimal for such problems with features not found in Runge-Kutta.

Currently mainstream CPUs are fast, but unfortunately quadruple precision calculations are not implemented as hardware. It seems to be an indispensable function to facilitate calculation of many ranges.

It was calculated the required accuracy as 10^{-28}, expecting about 25 digits of result, but the result was about 17 digits. About eight digits of precision is lost. The future task is to investigate this reason.

References

[El90] Ellis M.A. and Stroustrup B., *The Annotated C++ Reference Manual*, Addison-Wesley, 1990

[EnUh96] G. Engeln-Mullges G. and Uhlig, F. *Numerical Algorithms with Fortran*, Springer (1996)

[HaWa91] Hairer E. and Wanner G., *Solving Ordinary Differential Equations II*, Springer-Verlag, 1991

[HiEtAl02] Hirayama H., Komiya S. and Satou S., Solving Ordinary Differential Equations by Taylor Series, *Trans. of JSIAM*, Vol. 12, No. 1, 1–8, (2002) (Japanese)

[HiEtAl00] Hida Y., Xiaoye S. Li and Bailey, D., H., *Quad-Double Arithmetic:Algorithms, Implementation, and Application*, Lawrence Berkeley National Laboratory Technical Report LBNL-46996 (2000)

[NaHi92] Nagasawa K. and Hiyama S., *Motion of celestial bodies seen on a personal computer*, Chijinkan (1992) (Japanese)

[KoEtAl08] Kotakemori H., Fujii A., Hasegawa H. and Nishida A, Implementation of Fast Quad Precision Operation and Acceleration with SSE2 for Iterative Solver Library, *IPSJ Computing System*, Vol. 1 No. 1 73–84 (2008) (Japanese)

[YaEtal12] Yamada S., Sasa N., Imamura and T., Machida M., *Introduction of quadruple precision of Basic Linear Algebra Subprograms QPBLAS and its application*, IPSJ SIG Technical Report, Vol. 2012-HPC-137, No. 23, pp. 1–6, (2012) (Japanese)

[Oh06] Ohno H., On the 25 stage 12th order explicit Runge-Kutta method, *Trans. of JSIAM*, Vol. 16, 177–186 (2006)

[Sa93] Sano O., *Keypoint Differential Equation*, Iwanami-Shoten, Tokyo, (1993) (Japanese)

[SzBu67] Szebehely, V., Burrau's Problem of Three Bodies, *Proceedings of the National Academy of Sciences of the United States of America*, vol. 58, Issue 1, 60–65 (1967)

Chapter 15
Shape Optimization for Interior Neumann and Transmission Eigenvalues

Andreas Kleefeld

15.1 Introduction

The task is to optimize the shape of a domain $\Omega \subset \mathbb{R}^2$ with respect to the k-th eigenvalue under the constraint that the area $|\Omega|$ of the domain is constant, say A. Here, the domain is an open and bounded set with smooth boundary $\partial\Omega$ which is also allowed to be disconnected. In the sequel, we consider two different problems.

First, we deal with the maximization of interior Neumann eigenvalues (INEs). Precisely, one has to find numbers $\lambda > 0$ such that

$$\Delta u + \lambda u = 0 \text{ in } \Omega\,, \qquad \partial_\nu u = 0 \text{ on } \partial\Omega$$

is satisfied for non-trivial u, where ν denotes the normal pointing in the exterior. It is well-known that this problem is elliptic and the eigenvalues are discrete. The case $\lambda = 0$ which corresponds to a constant function is not considered here. It has been shown in 1954 and 1956 that the first INE is maximized by a circle (see [Sz54, We56]) and recently that the second INE is maximized by two disjoint circles of the same size (see [GiNaPo09]). However, the existence and uniqueness of a shape maximizer for higher INEs is from the theoretically point of view still unknown. But numerical results suggest that such a maximizer might exist. We refer the reader to [AnFr12, AnOu17] for recent results and a good overview over who has already worked in this direction. In Fig. 15.1 we show numerically the shape maximizer for the first six INEs.

The optimal values $\lambda_k \cdot A$ for $k = 1, \ldots, 6$ are 10.66, 21.28, 32.79, 43.43, 54.08, 67.04 (see [AnFr12]) which have been improved recently to 10.66, 21.28, 32.90,

A. Kleefeld (✉)
Forschungszentrum Jülich GmbH, Jülich Supercomputing Centre, Jülich, Germany
e-mail: a.kleefeld@fz-juelich.de

© Springer Nature Switzerland AG 2019   185
C. Constanda, P. Harris (eds.), *Integral Methods in Science and Engineering*,
https://doi.org/10.1007/978-3-030-16077-7_15

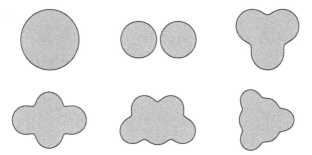

Fig. 15.1 Shape maximizer for the first six INEs obtained numerically. The recent optimal values $\lambda_k \cdot A$ for $k = 1, \ldots, 6$ are 10.66, 21.28, 32.90, 43.86, 55.17, 67.33 (see [AnOu17])

43.86, 55.17, 67.33 (see [AnOu17]). This paper reports improved values for the third and fourth INE and at the same time the boundary of the shape maximizer is described explicitly in terms of two parameters.

The second problem under consideration is the interior transmission problem. Interior transmission eigenvalues (ITEs) are numbers $\lambda \in \mathbb{C}\setminus\{0\}$ such that

$$\Delta w + \lambda n w = 0 \quad \text{in } \Omega ,$$

$$\Delta v + \lambda \quad v = 0 \quad \text{in } \Omega ,$$

$$v = w \quad \text{on } \partial\Omega ,$$

$$\partial_\nu v = \partial_\nu w \text{ on } \partial\Omega ,$$

has a non-trivial solution $(v, w) \neq (0, 0)$, where n is the given index of refraction. However, this is a non-elliptic and non-self-adjoint problem appearing first in 1986 (see [Ki86]). Existence and discreteness for real-valued λ has been shown in [CaGiHa10]. But, the existence is still open for complex-valued λ except for special geometries (see [SlSt16, CoLe17]). The computation of ITEs for a given shape is therefore a very challenging task (see [KlPi18] for an excellent overview of existing methods). It is also noteworthy that neither theoretical nor numerical results are available for a shape optimizer of the first two ITEs. Within this paper we give numerical evidence for a shape minimizer of the first two ITEs and state a conjecture which researcher in this field might want to prove in the future.

15.1.1 Contribution of the Paper

The contribution of this paper is twofold. First, improved numerical results for the maximization of some interior Neumann eigenvalues are presented using a simplified parametrization of the boundary. Second, the previous concept is transferred in order to obtain numerical results for the minimization of interior transmission eigenvalues for the first time for which no single theoretical result is yet available.

15.1.2 Outline of the Paper

The paper is organized as follows: In Sect. 15.2, it is explained in detail how to compute interior Neumann eigenvalues using a boundary integral equation followed by its discretization. Then, it is described how the resulting non-linear eigenvalue problem is solved numerically. Further, the new parametrization is introduced and used to obtain improved numerical results for the maximization of some interior Neumann eigenvalues. In Sect. 15.3, the concept of the previous section is applied for the minimization of interior transmission eigenvalues for which neither numerical results nor theoretical results are yet available. Finally, a short summary and an outlook is given in Sect. 15.4.

15.2 Shape Optimization for Interior Neumann Eigenvalues

Recall that interior Neumann eigenvalues (INEs) are numbers $\lambda = \kappa^2$ such that

$$\Delta u + \kappa^2 u = 0 \text{ in } \Omega, \qquad \partial_\nu u = 0 \text{ on } \partial\Omega$$

is satisfied. Note that this problem is elliptic and it is well-known that the eigenvalues are discrete and positive real-valued numbers. In the sequel, we ignore $\kappa = 0$ which corresponds to the constant function. To find such INEs for a given domain Ω, we use a boundary integral equation approach. A single layer ansatz with unknown density ψ given by

$$u(X) = \int_{\partial\Omega} \Phi_\kappa(X, y)\psi(y)\,\mathrm{d}s(y), \qquad X \in \Omega$$

is used, where $\Phi_\kappa(X, y) = \mathrm{i}\,H_0^{(1)}(\kappa\|X - y\|)/4$ is the fundamental solution of the Helmholtz equation. Taking the normal derivative, $\Omega \ni X \to x \in \partial\Omega$, and using the jump condition yields the following boundary integral equation of the second kind

$$\frac{1}{2}\psi(x) + \underbrace{\int_{\partial\Omega} \partial_{\nu(x)}\Phi_\kappa(x, y)\psi(y)\,\mathrm{d}s(y)}_{K(\kappa)} = 0. \tag{15.1}$$

Note that the operator $K(\kappa) : H^{-1/2}(\partial\Omega) \to H^{-1/2}(\partial\Omega)$ is compact assuming a regular boundary (see [Mc00]). Hence, $Z(\kappa) = I/2 + K(\kappa)$ is Fredholm of index zero for $\kappa \in \mathbb{C}\backslash\mathbb{R}_{\leq 0}$ and thus the theory of eigenvalue problems for holomorphic Fredholm operator-valued functions applies to $Z(\kappa)$.

The integral equation (15.1) is discretized via the boundary element collocation method. Precisely, we subdivide the boundary into $n/2$ pieces, approximate it by quadratic interpolation (the approximated boundary is denoted by $\widetilde{\partial\Omega}$), and define on each piece a quadratic interpolation for ψ. This leads to

$$\underbrace{\left(\frac{1}{2}\mathbf{I} + \mathbf{M}(\kappa)\right)}_{\mathbf{Z}(\kappa)\in\mathbb{C}^{n\times n}} \vec{\psi} = \vec{0},$$

where the matrix entries of $\mathbf{M}$ are numerically calculated with the Gauss-Kronrad quadrature (see [KlLi12] for details in the three-dimensional case). The resulting non-linear eigenvalue problem of the form

$$\mathbf{Z}(\kappa)\vec{\psi} = \vec{0}$$

is solved with the method of Beyn [Be12]. This method can find all eigenvalues κ including their multiplicities within any contour $\mathscr{C} \subset \mathbb{C}$ which is based on Keldysh's theorem. Precisely, one integrates the resolvent over the given contour whereas the integral is approximated with the trapezoidal rule (see [Be12] for more details). Hence, we are now able to compute highly accurate INEs for a given shape Ω. Next, it is explained how to choose a parametrization for the boundary of Ω. The idea is to use an implicit curve rather than an explicit representation of the curve. Equipotentials are implicit curves of the form

$$\sum_{i=1}^{m} \frac{1}{\|x - P_i\|} = c, \tag{15.2}$$

where the parameter c and the centers P_i are given. Here, $\|\cdot\|$ denotes the Euclidean norm. Precisely, all points $x \in \mathbb{R}^2$ satisfying (15.2) for given points $P_i, i = 1, \ldots, m$ and parameter c describe the implicit curve.

Example 1 We choose three points $(-\sqrt{3}/2, 1/2), (\sqrt{3}/2, 1/2), (0, -1)$ for $m = 3$ and $(-3/2, 0), (3/2, 0), (0, -\sqrt{3}/2), (0, \sqrt{3}/2)$ for $m = 4$. The edge length of the following geometric shapes as shown in Fig. 15.2 is $\sqrt{3}$.

Next, we show the influence of the parameter c. As one can see in Fig. 15.3 the larger the parameter c gets, the more constricting the boundary gets. Additionally, one can see that we are almost able to obtain a possible shape of the maximizer

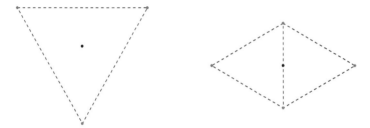

Fig. 15.2 The choice of the points for $m = 3$ are $(-\sqrt{3}/2, 1/2)$, $(\sqrt{3}/2, 1/2)$, $(0, -1)$ and for $m = 4$ are $(-3/2, 0)$, $(3/2, 0)$, $(0, -\sqrt{3}/2)$, $(0, \sqrt{3}/2)$ shown as a red dot. The origin is shown as a black dot

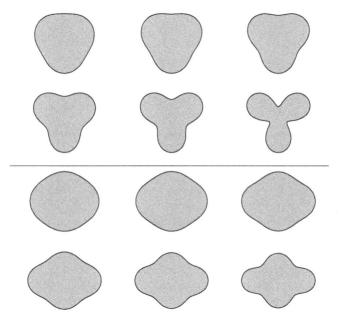

Fig. 15.3 The influence of the parameter $c = 1.75, 2.00, 2.25, 2.50, 2.75$, and 3.00 for $m = 3$ (first and second row) and for $m = 4$ (third and fourth row)

for the third and fourth INE. To add more flexibility, we introduce the additional parameter α. The modified equipotentials are given in the form

$$\sum_{i=1}^{m} \frac{1}{\|x - P_i\|^{2\alpha}} = c \tag{15.3}$$

We introduce the two in front of the parameter α in order to avoid the computation of the square root in the norm definition. In Fig. 15.4 we show the influence of the

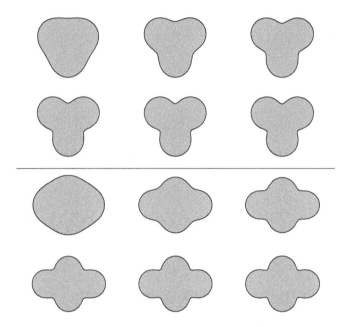

Fig. 15.4 The influence of the parameter $\alpha = 0.5$, 1.0, 1.5, 2.0, 2.5, and 3.0 with fixed $c = 2$ for $m = 3$ (first and second row) and for $m = 4$ (third and fourth row)

parameter α fixing $c = 2$. As one can see, we have enough flexibility to obtain very good approximations for a possible shape maximizer for the third and fourth INE. Thus, we have seen the influence of the parameters α and c. We shortly explain how to generate n points on the boundary for the given parameters α and c. This is done as follows. First, Eq. (15.3) is rewritten in polar coordinates. Then, $n + 1$ equidistant angles ϕ_i in the interval $[0, 2\pi]$ are generated. Next, for each angle ϕ_i the implicit equation is solved for the unique r_i via a root finding algorithm. Finally, the points given in polar coordinates (r_i, ϕ_i), $i = 1, \ldots, n + 1$ are transformed back to rectangular coordinates $(x_i, y_i) = (r_i \cos(\phi_i), r_i \sin(\phi_i))$, $i = 1, \ldots, n + 1$. Hence, we obtain n different points on the boundary of the scatterer (the $(n + 1)$-th point is the same as the first point by construction). Those n points can now be used in the boundary element collocation method.

In order to calculate the value $\lambda_k \cdot A$, we need to numerically approximate the area enclosed by the given implicit curve (see (15.3)). That is, we have n points distributed on the boundary $\partial \Omega$. With these points and the approximation via quadratic interpolation, the domain $\widetilde{\Omega}$ with the boundary $\widetilde{\partial \Omega}$ is defined. To approximate the area of this region, we compute the area of the non-self-intersecting polygon spanned by choosing $p \gg n$ points including an additional point (the first point is the additional $(p + 1)$-th point). The approximate area is given by

$$A \approx A_{\widetilde{\Omega}} = \frac{1}{2} \left| \sum_{i=1}^{p} (x_i - x_{i+1})(y_i + y_{i+1}) \right|$$

which is an easy consequence of the formula [Zw12, 4.6.1, p. 206]

$$\frac{1}{2} \left| \left| \begin{matrix} x_1 & x_2 \\ y_1 & y_2 \end{matrix} \right| + \left| \begin{matrix} x_2 & x_3 \\ y_2 & y_3 \end{matrix} \right| + \ldots + \left| \begin{matrix} x_p & x_1 \\ y_p & y_1 \end{matrix} \right| \right| .$$

The exterior normals on the boundary given implicitly by (15.3) are given by $v = \tilde{v}/\|\tilde{v}\|$ with

$$\tilde{v} = -2\alpha \sum_{i=1}^{m} \frac{(x - P_i)}{\|x - P_i\|^{2(\alpha+1)}} .$$

Now, we have everything together in order to optimize with respect to the two parameters c and α. First, we consider the third INE. The reference value given by Antunes and Oudet is given by 32.90 using 37 unknown coefficients. The third eigenvalue has multiplicity three. If we fix $\alpha = 3/2$, then the optimization with respect to c yields the result $c = 1.8416$ with 32.8929, 32.8929, 32.8929 for the third, fourth, and fifth, respectively. As we observe, the reported numbers are more accurate. If we fix $\alpha = 2$, then we obtain $c = 1.6921$ with 32.9018, 32.9018, 32.9018 which improves the result slightly compared to the value 32.90. But remember that we have only one unknown describing the boundary. If we choose $\alpha = 5/2$, then we have $c = 1.6112$ with 32.8970, 32.8970, 32.8970. Optimizing with respect to both parameters yields $\alpha = 2.0171$ and $c = 1.6883$ with 32.9018, 32.9018, 32.9018. The situation slightly changes for the optimization of the fourth eigenvalue. The reference value of Antunes and Oudet is given by 43.86 with multiplicity three using 33 unknown coefficients. If we use $\alpha = 2$, we obtain $c = 2.0571$ with 43.6968, 43.6968, 44.2247. Using $\alpha = 5/2$ gives $c = 2.0794$ with 43.8586, 43.8586, 43.8935 which is close to the value of Antunes and Oudet, but we have room for more considering the last eigenvalue. Fixing $\alpha = 3$ yields $c = 2.0875$ with 43.7822, 43.7822, 44.0634. Optimizing with respect to the two parameters α and c gives $\alpha = 2.5426$ and $c = 2.0845$ with 43.8694, 43.8694, 43.8694. This is a much better result. In Fig. 15.5 we show the three eigenfunctions of the possible shape optimizers for the third and fourth INE.

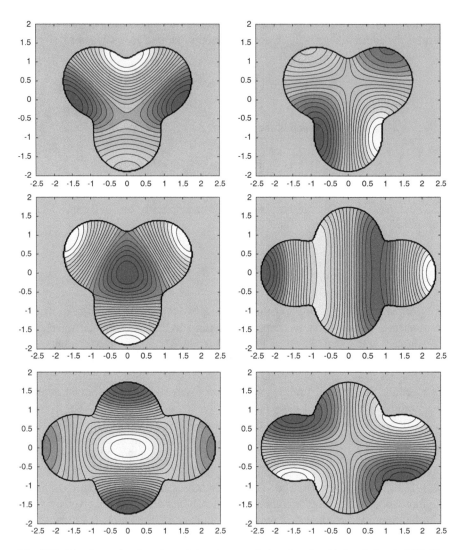

Fig. 15.5 The three eigenfunction of the shape optimizer for the third and fourth INE. The parameters are $\alpha = 2.0171$ and $c = 1.6883$ with 32.9018 having multiplicity three for the third INE and $\alpha = 2.5426$ and $c = 2.0845$ with 43.8694 having multiplicity three

Note that we used $n = 512$ for all numerical calculation to ensure that we have at least six digits accuracy for the values $\lambda_k \cdot A$. This is guaranteed since we almost have a convergence of order four due to the fact that we have approximated the boundary and the unknown density function by quadratic interpolation (refer to [KlLi12] for a superconvergence proof for three-dimensional scattering objects).

15.3   Shape Optimization for Interior Transmission Eigenvalues

Recall that interior transmission eigenvalues (ITEs) are numbers $\lambda = \kappa^2 \in \mathbb{C}\backslash\{0\}$ such that

$$\Delta w + \kappa^2 n w = 0 \quad \text{in } \Omega \,,$$

$$\Delta v + \kappa^2 \quad v = 0 \quad \text{in } \Omega \,,$$

$$v = w \quad \text{on } \partial\Omega \,,$$

$$\partial_\nu v = \partial_\nu w \text{ on } \partial\Omega \,,$$

has a non-trivial solution $(v, w) \neq (0, 0)$. Here, n is the given index of refraction. This is a non-elliptic and non-self-adjoint problem. Existence and discreteness for real-valued κ has already been established. However, the existence is still open for complex-valued κ except for special geometries. To compute such ITEs for a given shape is therefore very challenging. We use the same technique as presented before for the numerical calculation of interior Neumann eigenvalues; that is, reduce the problem to a system of boundary integral equations, discretize it via a boundary element collocation method, and solve the resulting non-linear eigenvalue problem via the method of Beyn (see [Be12]). For more details, we refer the reader to [Kl13, Kl15] where ITEs for three-dimensional domains are computed and to [KlPi18] for a good introduction for other methods to compute such ITEs. Straightforwardly looking at real-valued ITEs using the index of refraction $n = 4$ for different domains taken from [KlPi18] reveals that neither the circle is maximizing nor minimizing $\lambda_1 = A \cdot \kappa_1^2$. The values λ_1 for eight different domains are given in Fig. 15.6.

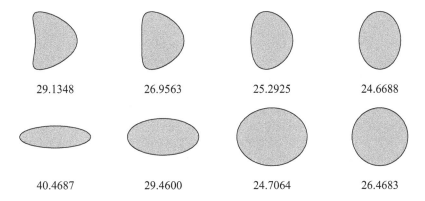

| 29.1348 | 26.9563 | 25.2925 | 24.6688 |

| 40.4687 | 29.4600 | 24.7064 | 26.4683 |

Fig. 15.6 The values λ_1 for eight different domains using $n = 4$

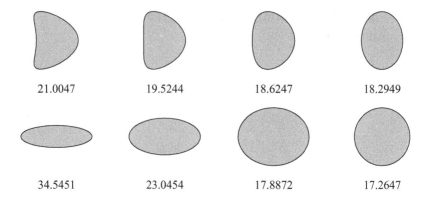

| 21.0047 | 19.5244 | 18.6247 | 18.2949 |
| 34.5451 | 23.0454 | 17.8872 | 17.2647 |

Fig. 15.7 The values $|\lambda_1|$ for eight different domains using $n = 4$

But recall that there might be complex-valued ITEs as well which are not taken into account. If we consider $|\lambda_1|$ instead of λ_1 using the same eight domains, we obtain the results as presented in Fig. 15.7.

As one can observe, it seems that the circle is minimizing $|\lambda_1|$. Hence, if we consider $|\lambda_1| \leq |\lambda_2| \leq |\lambda_3| \leq \cdots$, then we make the **conjecture** that the first absolute ITE is minimal for a circle for the index of refraction $n > 1$. If this is true, then it is also true for $0 < n < 1$ using the relation $\kappa(1/n) = \sqrt{n}\kappa(n)$. Further, since λ_1 is complex-valued, it comes in complex conjugate pairs. Hence, the second eigenvalue will be minimized by a circle as well.

Further investigation of shapes that minimize higher interior transmission eigenvalues is a very interesting and challenging topic.

15.4 Summary and Outlook

In this paper, it is shown how to efficiently compute interior Neumann eigenvalues for a given domain in two dimensions. Additionally, the value of the shape maximizer for the third and fourth interior Neumann eigenvalue has been improved from 32.90 and 43.86 to 32.9018 and 43.8694 with multiplicity three, respectively. At the same time, the number of parameters describing the boundary of a possible maximizer has been reduced to two parameters using modified equipotentials. The conjecture is that the third and fourth interior Neumann eigenvalue might be given by such modified equipotentials. This work presents very recent numerical results and a further investigation has to be carried out in order to validate whether the shape maximizer for higher interior Neumann eigenvalues can be found with modified equipotentials. This idea can easily be used for extending this approach to the three-dimensional case.

Moreover, for the first time numerical results are presented for the minimization of interior transmission eigenvalues in two dimensions although already the numerical calculation of those for a given domain is a very challenging task since the problem is neither elliptic nor self-adjoint and hence complex-valued interior transmission eigenvalues might exist. From the theoretical point of view, this fact is still open. Additionally, it is open whether there exists a unique minimizer for the first and second interior transmission eigenvalue. Here, we show numerically and hence conjecture that the first and second interior transmission eigenvalue is minimized by a circle. It remains to prove this observation, but it cannot be carried out by standard spectral arguments like for the Dirichlet, Neumann, Robin, or Steklov eigenvalue problem. Moreover, one can now try to investigate the three-dimensional case.

Above all, one could also investigate the electromagnetic and/or the elastic scattering case in two and three dimensions.

Acknowledgements I would like to thank the IMSE'18 steering committee for giving me the opportunity to present my recent results for the maximization of interior Neumann and minimization of interior transmission eigenvalues on July 19th, 2018. Further, I would like to thank Paul Harris for the organization of this nice event at the University of Brighton, UK.

References

[AnFr12] Antunes, P.R.S. and Freitas, P.: Numerical optimization of low eigenvalues of the Dirichlet and Neumann Laplacians. *J. Optim. Theory Appl.*, **154**, 235–257 (2012).

[AnOu17] Antunes, P.R.S. and Oudet, E.: Numerical results for extremal problem for eigenvalues of the Laplacian. In *Shape optimization and spectral theory*, A. Henrot (ed.), De Gruyter, Warzow/Berlin, (2017), pp. 398–412.

[Be12] Beyn, W.-J.: An integral method for solving nonlinear eigenvalue problems. *Linear Algebra Appl.*, **436**, 3839–3863 (2012).

[CaGiHa10] Cakoni, F., Gintides, D., and Haddar, H.: The existence of an infinite discrete set of transmission eigenvalues. *SIAM J. Math. Anal.*, **42**, 237–255 (2010).

[CoLe17] Colton, D. and Leung, Y.-J.: The existence of complex transmission eigenvalues for spherically stratified media. *Appl. Anal.*, **96**, 39–47 (2017).

[GiNaPo09] Girouard, A., Nadirashvili, N., and Polterovich, I.: Maximization of the second positive Neumann eigenvalue for planar domains. *J. Differ. Geom.*, **83**, 637–662 (2009).

[Ki86] Kirsch, A.: The denseness of the far field patterns for the transmission problem. *IMA J. Appl. Math.*, **37** 213–225 (1986).

[Kl13] Kleefeld, A.: A numerical method to compute interior transmission eigenvalues. *Inverse Problems*, **29**, 104012 (2013).

[Kl15] Kleefeld, A.: *Numerical methods for acoustic and electromagnetic scattering: Transmission boundary-value problems, interior transmission eigenvalues, and the factorization method*. Habilitation Thesis, Brandenburg University of Technology Cottbus-Senftenberg (2015).

[KlLi12] Kleefeld, A. and Lin, T.-C.:. Boundary element collocation method for solving the exterior Neumann problem for Helmholtz's equation in three dimensions. *Electron. Trans. Numer. Anal.*, **39**, 113–143 (2012).

[KlPi18] Kleefeld, A. and Pieronek, L.: The method of fundamental solutions for computing acoustic interior transmission eigenvalues. *Inverse Problems*, **34**, 035007 (2018).

[Mc00] McLean, W.: *Strongly elliptic systems and boundary integral operators*. Cambridge University Press, Cambridge (2000).

[SlSt16] Sleeman, B.D. and Stocks, D.C.: Interior transmission eigenvalues of a rectangle. *Inverse Problems*, **32**, 025010 (2016).

[Sz54] Szegö, G.: Inequalities for certain eigenvalues of a membrane of given area. *Arch. Ration. Mech. Anal.*, **3**, 343–356 (1954).

[We56] Weinberger, H.F.: An isoperimetric inequality for the N-dimensional free membrane problem. *Arch. Ration. Mech. Anal.*, **5**, 633–636 (1956).

[Zw12] Zwillinger, D.: *Standard mathematical tables and formulae*. CRC Press, Boca Raton (2012).

Chapter 16
On the Integro-Differential Radiative Conductive Transfer Equation: A Modified Decomposition Method

Cibele A. Ladeia, Bardo E. J. Bodmann, and Marco T. Vilhena

16.1 Introduction

The problem of integro-differential radiative conductive transfer in spherical geometry has been the subject of numerous research, including the radiation transfer in furnaces, nuclear reactors, combustion systems and the planetary atmosphere [Mo13, HoEtAl16]. In this context, we derive a solution for the radiative conductive transfer equation in spherical geometry. The solution allows us to simulate the radiation intensity and temperature field together with conductive and radiative energy transport. In general, the equation of radiative conductive transfer in spherical geometry is solved introducing some approximations, such as linearisation or discretising angular terms, that turn the construction of an acceptably precise solution to an approximate problem feasible. Solutions found in the literature are typically determined by numerical means, see, for instance, [ThOz85, TsEtAl89, Th90, MiEtAl10]. Recently, the authors solved the aforementioned problem in a semi-analytical fashion and proved consistency and convergence of the solution obtained by a decomposition method for the radiative conductive transfer equation in cylinder geometry [LaEtAl18, ViEtAl11]. The Laplace method is related to procedures for linear problems, while the decomposition method allows to treat the non-linear contribution as source terms in a linear recursive scheme and thus opens a pathway to determine a solution, in principle to any prescribed precision [ViEtAl11]. In the present discussion, we report on arithmetic stability issues for the recursive scheme. In order to stabilise convergence by virtue of limited arithmetic precision, we propose a modification of the Adomian Decomposition Method by splitting responsible terms for instability, i.e. the source terms incorporating the non-

C. A. Ladeia · B. E. J. Bodmann (⊠) · M. T. Vilhena
Department of Mechanical Engineering, Federal University of Rio Grande do Sul, Porto Alegre, Brazil
e-mail: bardo.bodmann@ufrgs.br; vilhena@mat.ufrgs.br

© Springer Nature Switzerland AG 2019
C. Constanda, P. Harris (eds.), *Integral Methods in Science and Engineering*,
https://doi.org/10.1007/978-3-030-16077-7_16

linearity and distributing them in a finite number of source terms. This procedure results then in a finite number of modified recursion steps controlled by a splitting parameter α, whereas all subsequent recursion steps follow the usual recursion scheme. Finally, we report on some case studies with numerical results for the solutions and convergence behaviour.

16.2 The Integro-Differential Radiative Conductive Transfer Equation

We consider the one-dimensional steady state problem in spherical geometry for a solid sphere. The problem of energy transfer is described in [Oz73] by the radiative conductive transfer equation coupled to the energy equation with spherical symmetry, with $\mathscr{T} = (1 - \omega(r)) \, \Theta^4(r)$,

$$\left[\mu \frac{\partial}{\partial r} + \frac{1-\mu^2}{r} \frac{\partial}{\partial \mu} + 1 \right] I(r, \mu) = \frac{\omega(r)}{2} \int_{-1}^{1} \mathscr{P}(\mu, \mu') I(r, \mu') d\mu' + \mathscr{T}, \quad (16.1)$$

for $r \in [0, R]$ and $\mu \in [-1, 1]$. Here, $\mu = \cos(\theta)$, θ is the polar angle, I is the radiation intensity, ω is the single scattering albedo and $\mathscr{P}(\mu, \mu')$ signifies the differential scattering coefficient, also called the phase function [Ch50]. The integral on the right-hand side of (16.1) is

$$\int_{-1}^{1} P(\mu, \mu') I(r, \mu') d\mu' = \sum_{l=0}^{\mathfrak{L}} \beta_l \int_{-1}^{1} P_l(\mu) P_l(\mu') I(r, \mu') d\mu' \, ,$$

where β_l are the expansion coefficients of the Legendre polynomials $P_l(\mu)$ and l refers to the degree of anisotropy, for details see [ViEtAl11]. The energy equation for the temperature that connects the radiative flux to a temperature gradient is

$$r^2 \frac{d^2}{dr^2} \Theta(r) + 2r \frac{d}{dr} \Theta(r) = \frac{1}{4\pi N_c} \frac{d}{dr} [r^2 q_r^*] - r^2 H \, . \quad (16.2)$$

Here, $N_c = \frac{k\beta_{ext}}{4\sigma n^2 T_r^3}$ is the radiation conduction parameter with k the thermal conductivity, β_{ext} the extinction coefficient, σ the Stefan-Boltzmann constant, n the refractive index, T_r is a reference temperature, $H = [k\beta_{ext}^2 T_r]^{-1} h$ is the normalised constant, and h is used to denote a prescribed heat generation in the medium that is independent of the radiation intensity. The dimensionless radiative heat flux is expressed in terms of the radiative intensity by

$$q_r^*(r) = 2\pi \int_{-1}^{1} I(r, \mu) \mu d\mu \, .$$

The boundary conditions of Eq. (16.1) are

$$I(0, \mu) = I(0, -\mu) \,,$$

$$I(R, \mu) = \epsilon(R)\Theta^4(R) + 2\rho^d(R) \int_0^1 I(R, \mu')\mu'd\mu' \,,$$

for $\mu \in [-1, 1]$, ρ^d is the diffuse reflectivity, ϵ is the emissivity, the thermal photon emission is according to the Stefan-Boltzmann law (see ref. [El09]). In Eq. (16.2), $r = 0$ is not a physical boundary, and for the solution $\Theta(r)$ to be physically meaningful we need to impose at $r = 0$ the condition that the solution shall be bounded at the origin. The boundary condition of Eq. (16.2) is

$$\Theta(r)|_{r=R} = \Theta_B \,. \tag{16.3}$$

In addition, if the radiative heat flux is known, we can solve (16.2) and use Eq. (16.3) to find

$$\Theta(r) = \Theta_B + \frac{H}{6}\left(R^2 - r^2\right) - \frac{1}{4\pi N_c} \int_r^R q_r^*(r')dr' \,. \tag{16.4}$$

16.3   Solution by the Modified Decomposition Method

In order to derive the solution, we use the S_N approximation in the angular variable $I_n \equiv I_n(r, \mu_n)$. The S_N approximation [Ch50] is based on the angular variable discretisation Ω in an enumerable set of angles or equivalently their direction cosines, in our work μ_n. Then Eqs. (16.1) and (16.2) can be simplified using an enumerable set of angles following the collocation method,

$$\mu_n \frac{\partial I_n}{\partial r} + \left(\frac{1 - \mu_n^2}{r}\right) \left.\frac{\partial I_n}{\partial \mu}\right|_{\mu=\mu_n} + I_n = \frac{\omega(r)}{2} \sum_{l=0}^{\mathcal{L}} \beta_l \mathcal{P}_l(\mu_n) \sum_{p=1}^N \varpi_p \mathcal{P}_l(\mu_p) I_p$$

$$+ (1 - \omega(r)) \, \Theta^4(r) \,, \tag{16.5}$$

$$\frac{d}{dr}\Theta(r) - \left.\frac{d}{dr}\Theta(r)\right|_{r=0} = \frac{1}{2N_c} \sum_{p=1}^N \varpi_p[I_p(r) - I_p(0)] \,.$$

Here n indicates a discrete direction of μ_n. The factor ϖ_p is the weight from the quadrature that approximates the integral explained in detail further down. The integration is carried out with $1 \le n \le N$. In this approximation Eq. (16.5) may be written as

$$\mu_n \frac{\partial I_n}{\partial r} + \left(\frac{1-\mu_n^2}{r}\right) \left.\frac{\partial I_n}{\partial \mu}\right|_{\mu=\mu_n} + I_n = \Upsilon(r, \mu_n) , \qquad (16.6)$$

$$\Upsilon(r, \mu_n) = \frac{\omega(r)}{2} \sum_{l=0}^{\mathcal{L}} \beta_l \mathcal{P}_l(\mu_n) \sum_{p=1}^{N} \varpi_p \mathcal{P}_l(\mu_p) I_p + (1 - \omega(r)) \Theta^4(r) .$$

Instead of $\mu_n \frac{\partial I_n}{\partial r} + \left(\frac{1-\mu_n^2}{r}\right) \left.\frac{\partial I_n}{\partial \mu}\right|_{\mu=\mu_n}$, we take as independent variable $\psi = r\xi_n$ for $1 \leq n \leq N$. Then Eq. (16.6) becomes $\frac{\partial I_n}{\partial \psi} + I_n = \Upsilon(\psi)$ or in the S_N representation

$$\frac{1}{\xi_n} \frac{\partial I_n}{\partial r} + I_n = \frac{\omega(r)}{2} \sum_{l=0}^{\mathcal{L}} \beta_l \mathcal{P}_l(\mu_n) \sum_{p=1}^{N} \varpi_p \mathcal{P}_l(\mu_p) I_p + (1 - \omega(r)) \Theta^4(r) ,$$

$$\frac{d}{dr}\Theta(r) - \left.\frac{d}{dr}\Theta(r)\right|_{r=0} = \frac{1}{2N_c} \sum_{p=1}^{N} \varpi_p [I_p(r) - I_p(0)] . \qquad (16.7)$$

Here μ_n are evaluation points, with $1 \leq n \leq N$ and subject to the following boundary conditions:

$$I_n(0) = -I_{N-n+1}(0) ,$$

$$I_{N-n+1}(R) = \epsilon(R)\Theta^4(R) + 2\rho^d(R) \sum_{p=1}^{N/2} \varpi_p I_p(R)\mu_p .$$

As already indicated above, the integral over the angular variable is replaced by Gauss-Legendre quadrature with weight ϖ_p, with Gauss-Legendre weight normalisation $\sum_{p=1}^{N} \varpi_p = 1$. The choice for the specific quadrature is due to the phase function representation in terms of Legendre polynomials. The equation system (16.7) may be cast in a first order differential equation system in matrix form

$$\mathbf{A}\frac{d}{dr}\mathbf{I} - \mathbf{BI} = \mathbf{\Psi}, \qquad (16.8)$$

where $\mathbf{A}$ is a diagonal matrix of order $N^2 \times N^2$, with $\mathbf{A}_{nn} = 1/\mu_n$, $\mathbf{B}$ is a square matrix of the same order as $\mathbf{A}$ with elements

$$B_{i,j} = \delta_{ij} + \omega_j(r) \left[\sum_{l=0}^{\mathcal{L}} \beta_l P_l(\mu_i) P_l(\mu_j)\right] ,$$

δ_{ij} is the usual Kronecker symbol and the vector of the intensity of order N^2 is defined by $\mathbf{I} = \left(I_1, \cdots, I_{\frac{N}{2}+1}, \cdots, I_N\right)^T$. The non-linear terms are N sequences of N identical angular term for the N directions.

$$\boldsymbol{\Psi} = \left((1 - \omega(r))\, \Theta^4(r), \ldots, (1 - \omega(r))\, \Theta^4(r)\right)^T.$$

According to [ViEtAl11] the radiation intensity is expanded in an infinite series $\mathbf{I} = \sum_{\ell=0}^{\infty} \mathbf{Y}_\ell$, thus introducing an infinite number of artificial degrees of freedom which may be used to set up a non-unique recursive scheme of linear differential equations, where the non-linear terms appear as source term containing known solutions $\mathbf{Y}_\ell$ from the previous recursion steps [ViEtAl11] and the solution of the linear differential equations is known. Equation (16.8) is then

$$\sum_{\ell=0}^{\infty} \left(\mathbf{A}\frac{d}{dr}\mathbf{Y}_\ell - \mathbf{B}\mathbf{Y}_\ell\right) = ((1 - \omega(r)), \ldots, (1 - \omega(r)))^T \underbrace{\sum_{\ell=0}^{\infty} G_{\ell-1}\left(\{\mathbf{Y}_\ell\}_{\ell=0}^{\ell-1}\right)}_{\Theta^4(r)}.$$

(16.9)

In order to solve the equation system (16.9) in a recursive fashion, the initialisation is chosen to be $\mathbf{A}\frac{d}{dr}\mathbf{Y}_0 - \mathbf{B}\mathbf{Y}_0 = 0$ further subject to the original boundary conditions and then making use of a recursive process of the equations for the remaining components $\mathbf{Y}_\ell$, with homogeneous boundary conditions.

$$\mathbf{A}\frac{d}{dr}\mathbf{Y}_\ell - \mathbf{B}\mathbf{Y}_\ell = ((1 - \omega(r)), \ldots, (1 - \omega(r)))^T G_{\ell-1}\left(\{\mathbf{Y}_\ell\}_{\ell=0}^{\ell-1}\right), \quad (16.10)$$

with $\ell \in \mathbb{N}^+$. Accordingly, the recursion initialisation corresponds to the homogeneous solution where all subsequent recursion steps result in particular solutions. The remaining $\mathscr{L}^{-1}$ denotes the inverse Laplace transform operator, s is the r dual complex variable from Laplace transform of Eq. (16.10), $\mathbf{U} = \mathbf{A}^{-1}\mathbf{B}$ and the decomposed matrix $\mathbf{U} = \mathbf{X}\mathbf{D}\mathbf{X}^{-1}$ with $\mathbf{D}$ the diagonal matrix with distinct eigenvalues and $\mathbf{X}$ the eigenvector matrix. The general solution for each decomposition term $\mathbf{Y}_\ell(r)$ is explicitly given by

$$\begin{aligned}\mathbf{Y}_\ell(r) &= \mathscr{L}^{-1}((s\mathbf{I} - \mathbf{U})\mathbf{Y}_\ell(0)) + \mathscr{L}^{-1}((s\mathbf{I} - \mathbf{U})\mathbf{A}^{-1}\overline{\boldsymbol{\Psi}}(s)) \\ &= \mathbf{X}e^{\mathbf{D}r}\mathbf{V}^\ell + \mathbf{X}e^{\mathbf{D}r}\mathbf{X}^{-1} * \mathbf{A}^{-1}\left((1 - \omega(r)), \ldots, (1 - \omega(r))\right)^T G_{\ell-1}.\end{aligned}$$

The non-linearity from the temperature term $\Theta^4(r) = \sum_{\ell=0}^{\mathscr{I}} G_\ell\left(\{\mathbf{Y}_\ell\}_{\ell=0}^{\ell-1}\right)$ is represented by Adomian polynomials $G_\ell\left(\{\mathbf{Y}_\ell\}_{\ell=0}^{\ell-1}\right)$. In the following the notation $f_0^{(\ell)}$ stands for the ℓ-th functional derivative of the non-linearity at $\mathbf{Y} = \mathbf{Y}_0$, so that

one may identify the first term $f_0^{(0)}$ and all the subsequent terms of the series that
define the Adomian polynomials G_ℓ are shown below (for details see [ViEtAl11]).

$$G_0(r) = f_0^{(0)} = f(\mathbf{Y}_0) = \left(\Theta_B + \frac{H}{6}(R^2 - r^2) - \frac{1}{4\pi N_c}\int_r^R q_{0,r}^*(r')dr'\right)^4 ,$$

$$G_1(r) = f_0^{(1)}\mathbf{Y}_1 = \mathbf{Y}_1 \frac{d}{d\mathbf{Y}_0} f(\mathbf{Y}_0) = \left(-\frac{1}{4\pi N_c}\int_r^R \left(q_{0,r}^*(r') + q_{1,r}^*(r')\right)dr'\right)^4$$

$$G_2(r) = f_0^{(1)}\mathbf{Y}_2 + f_0^{(2)}\mathbf{Y}_1^2 = \mathbf{Y}_2 \frac{d}{d\mathbf{Y}_0} f(\mathbf{Y}_0) + \frac{\mathbf{Y}_1^2}{2!}\frac{d}{d\mathbf{Y}_0} f(\mathbf{Y}_0)$$

$$= \left(-\frac{1}{4\pi N_c}\int_r^R \left(q_{0,r}^*(r') + q_{1,r}^*(r') + q_{2,r}^*(r')\right)dr'\right)^4$$

$$\vdots$$

$$G_\ell(r) = f_0^{(1)}\mathbf{Y}_\ell + \sum_{j=2}^{l} \frac{1}{j!} f_0^{(j)} \sum_{\substack{b_1,\ldots,b_{l-1} \\ \Sigma b_i = j}} \left(\binom{j}{\{b_i\}_1^{\ell-1}}\prod_{v=1}^{\ell-1} \mathbf{Y}_v^{b_v}\right). \qquad (16.11)$$

Here $\left(\binom{j}{\{b_i\}_1^{\ell-1}}\right)$ are the usual multinomial coefficients. In principle one has to solve an
infinite number of equations, so that in a computational implementation one has to
truncate the scheme according to a prescribed precision. In the present state of the
work the pertinent question of convergence as addressed in Sect. 16.4.2 is based on
heuristic arguments. From the last term in Eq. (16.11) one observes that for a non-
linearity with polynomial structure there is only a limited number of combinations in
the last term. Consequently, for j sufficiently large the factorial term $j!$ controls the
magnitude of the correction terms, that for increasing j tends to zero. One may now
determine the temperature profile. To this end and in order to stabilise convergence a
correction parameter α for the recursive scheme was introduced in Eq. (16.4), where
$\frac{1}{N_c} = Z\alpha$, with $Z \in \mathbb{Z}^+$,

$$\Theta_0(r) = \Theta_R(r) + \frac{H}{6}(R^2 - r^2) ,$$

$$\Theta_1(r) = -\frac{\alpha}{4\pi}\int_r^R q_{0,r}^*(r')dr' ,$$

$$\Theta_2(r) = -\frac{\alpha}{4\pi}\int_r^R (q_{0,r}^*(r') + q_{1,r}^*(r'))dr' ,$$

$$\vdots$$

$$\Theta_Z(r) = -\frac{\alpha}{4\pi} \sum_{\ell=1}^{Z} \int_r^R q_{\ell-1,r}^*(r')\, dr' \,,$$

$$\vdots$$

$$\Theta_{Z+1}(r) = -\frac{\alpha}{4\pi} \int_r^R q_{Z,r}^*(r')dr' - \frac{\alpha}{4\pi} \int_r^R q_{Z-1,r}^*(r')dr' - \cdots$$

$$-\frac{\alpha}{4\pi} \int_r^R q_{1,r}^*(r')dr' \,,$$

$$\Theta_{Z+j}(r) = -\frac{\alpha}{4\pi} \sum_{\ell=1+j}^{Z+j} \int_r^R q_{\ell-1,r}^*(r')dr', \quad , j = 1, 2, \ldots$$

Here $\delta_{0,\ell}$ is the Kronecker symbol, $\mathcal{M} = \min(Z, \ell)$ and

$$\Theta_\ell(r) = \delta_{0,\ell}\left[\Theta_B + \frac{H}{6}(R^2 - r^2)\right] - \frac{\alpha}{4\pi} \sum_{i=1}^{\mathcal{M}} \int_r^R q_{\ell-i,r}^*(r')dr'. \quad (16.12)$$

16.4   Numerical Results and Discussion

To check if the proposed method is appropriate for solving the radiative conductive transfer problem in a solid sphere, we evaluate the normalised temperature, the conductive, the radiative and the total heat flux, respectively.

$$Q_r(r) = \frac{1}{4\pi N_c} q_r^* \,, \qquad Q_c(r) = \frac{r}{3}H - \frac{1}{4\pi N_c} q_r^* \,, \qquad Q(r) = \frac{r}{3}H \quad (16.13)$$

For all the numerical results we consider isotropic scattering ($\mathfrak{L} = 0$), $N = 8$ directions and r in multiples of r/R that varies between 0 and 1.

Problem 1 The results in Fig. 16.1 are based on the parameter set with $\epsilon = 0.8$, $\rho^d = 0.2$, $\Theta_B = 1$, $\omega = 0.9$, $R = 5$, $N_c = 0.0005$, $H = 0$ and $\alpha = 1$, and show the temperature profile (Θ), the conductive heat flux ($Q_c(r)$), the radiative heat flux ($Q_r(r)$) and the total heat flux ($Q(r)$). We also show the temperature evolution dependence on the number of recursion steps. We used a stopping criterion such that the last twenty recursions of temperature changed less than 10^{-6}. Accordingly, the series of Adomian polynomials was truncated at $\mathcal{J} = 139$.

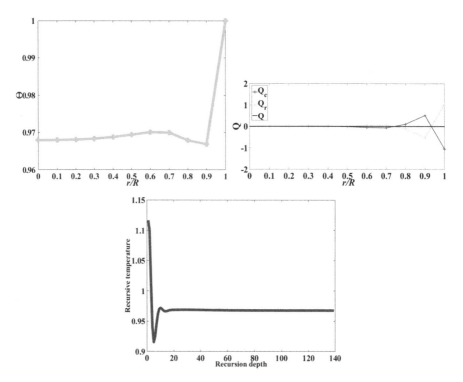

Fig. 16.1 The temperature profile Θ (top left) and the conductive Q_c, radiative Q_r, and total heat flux Q (top right) against the relative radial optical depth, and the temperature evolution at r/R depending on the recursion depth (bottom centre), (Problem 1)

Problem 2 The results in Fig. 16.2 are based on the parameter set with $\epsilon = 0.8$, $\rho^d = 0.2$, $\Theta_B = 1$, $\omega = 0.9$, $R = 1$, $N_c = 0.0005$, $H = 0$ and $\alpha = 10$, and show the temperature profile (Θ), the conductive heat flux $(Q_c(r))$, the radiative heat flux $(Q_r(r))$ and the total heat flux $(Q(r))$, moreover, we show the temperature for a sequence of recursion steps. We used the same stopping criterion as in the previous problem. Accordingly, the series of Adomian polynomials was truncated at $\mathcal{J} = 195$.

Problem 3 The results in Fig. 16.3 are based on the parameter set with $\epsilon = 0.8$, $\rho^d = 0.2$, $\Theta_B = 1$, $\omega = 0.9$, $R = 0.5$, $N_c = 0.0005$, $H = 0$ and $\alpha = 10$, and show the temperature profile (Θ), the conductive heat flux $(Q_c(r))$, the radiative heat flux $(Q_r(r))$ and the total heat flux $(Q(r))$, furthermore, we show the temperature evolution after a number of recursion steps. The same stopping criterion as in Problem 1 was applied. Accordingly, the series of Adomian polynomials was truncated at $\mathcal{J} = 1157$.

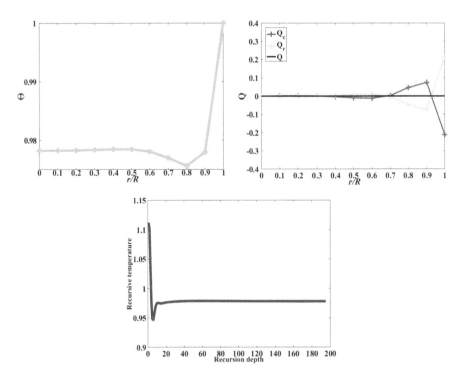

Fig. 16.2 The temperature profile Θ (top left) and the conductive Q_c, radiative Q_r, and total heat flux Q (top right) against the relative radial optical depth, and the temperature evolution at r/R depending on the recursion depth (bottom centre), (Problem 2)

Problem 4 The results in Fig. 16.4 are based on the parameter set with $\epsilon = 0.8$, $\rho^d = 0.2$, $\Theta_B = 1$, $\omega = 0.9$, $R = 0.5$, $N_c = 0.0005$, $H = 0$ and $\alpha = 1$, and show the temperature profile (Θ), the conductive heat flux ($Q_c(r)$), the radiative heat flux ($Q_r(r)$) and the total heat flux ($Q(r)$), moreover, we show the temperature after a number of recursion steps. The stopping criterion was the same as in Problem 1. Accordingly, the series of Adomian polynomials was truncated at $\mathscr{I} = 362$.

   Figures 16.1, 16.2, 16.3, and 16.4 show the influence of the boundary radius R. Note that in Figs. (16.1, 16.2, 16.3, and 16.4)—(top right) the conductive heat flux passes through the minimum whereas the radiative heat flux has a corresponding maximum and Eq. (16.13) is satisfied. The values of α are related to the number of source terms whereto contributions which cause arithmetic instability in the recursion steps are distributed. In order to stabilise convergence where necessary, α was changed in a decreasing succession from 10 to 1 until stable results were obtained. Large values ($\alpha = 10$) correspond to weak and small values ($\alpha = 1$) correspond to strong corrections. For example, in Problems 3 and 4 convergence is established using $\alpha = 10$ and $\alpha = 1$, respectively (see also Figs. 16.3—(bottom centre) and 16.4—(bottom centre)).

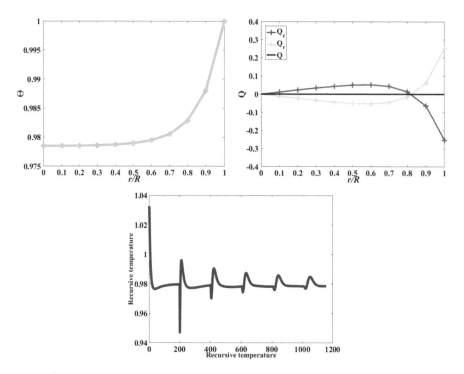

Fig. 16.3 The temperature profile Θ (top left) and the conductive Q_c, radiative Q_r, and total heat flux Q (top right) against the relative radial optical depth, and the temperature evolution at r/R depending on the recursion depth (bottom centre), (Problem 3)

16.4.1 Consistency

In order to evaluate the truncation error of the recursive scheme we calculate the residual term

$$\mathscr{R} = \left\| (1 - \omega(r)) \left(\sum_{l=0}^{\mathscr{J}} \Theta_l(r) \right)^4 \right\|_{\infty} ,$$

where $\| \cdot \|_{\infty}$ is the maximum norm and $\sum_{l=0}^{\mathscr{J}} \Theta_l(r)$ represents the recursive scheme that has been calculated from Eq. (16.12). Thus, in Fig. 16.5 are shown the residual terms for Problems 1, 2, 3 and 4.

One observes that the residual terms of Problems 1, 2, 3 and 4 describe a zero sequence, which implies that the truncation error also tends to zero.

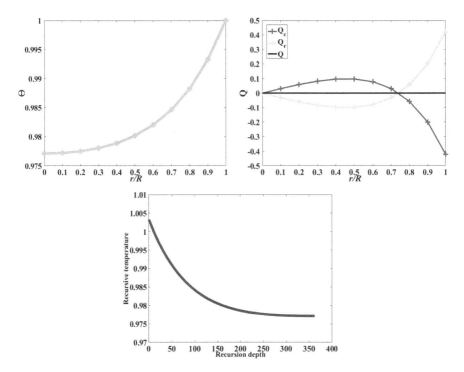

Fig. 16.4 The temperature profile Θ (top left) and the conductive Q_c, radiative Q_r, and total heat flux Q (top right) against the relative radial optical depth, and the temperature evolution at r/R depending on the recursion depth (bottom centre), (Problem 4)

16.4.2   A Convergence Criterion by Stability Analysis

In general convergence is not guaranteed by the decomposition method, so that the solution quality shall be tested by a convenient criterion. Since standard convergence criteria do not apply to non-linear problems, we modify the Lyapunov-Boichenko stability criterion [BoEtAl05] for dynamical systems into a stability criterion for convergence of recursive schemes. To this end let

$$\left| \delta \mathscr{L}_{\mathscr{J}} \right| = \left\| \sum_{l=\mathscr{J}+1}^{\infty} \Theta_l \right\|_{\infty} ,$$

represent the major estimated difference between the solution obtained with trun-cation $\mathscr{J}$, in this case, $\Upsilon_{\mathscr{J}} = \| \sum_{l=0}^{\mathscr{J}} \Theta_l \|_{\infty}$, and the true solution, with $\| \cdot \|_{\infty}$ the maximum norm. Then $\left| \delta \mathscr{L}_{\mathscr{J}} \right| = e^{\Upsilon_{\mathscr{J}} \lambda} \left| \delta \mathscr{L}_0 \right|$ shows how the difference of

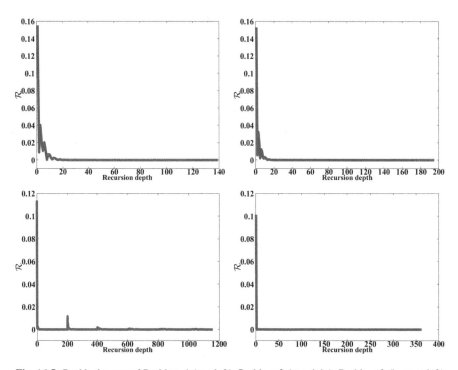

Fig. 16.5 Residual terms of Problem 1 (top left), Problem 2 (top right), Problem 3 (bottom left) and Problem 4 (bottom right)

the recursion initialisation to the true solution evolves until recursion depth $\mathscr{J}$. If the exponent remains negative $\lambda < 0$ for increasing $\mathscr{J}$ the series is exponentially convergent

$$\lambda = \frac{1}{\left\| \sum_{l=0}^{\mathscr{J}} \Theta_l \right\|_\infty} \ln \left(\frac{|\delta \mathscr{Z}_{\mathscr{J}}|}{|\delta \mathscr{Z}_0|} \right) .$$

Here $\left\| \sum_{l=0}^{\mathscr{J}} \Theta_l \right\|_\infty$ is known and $\ln \left(\frac{|\delta \mathscr{Z}_{\mathscr{J}}|}{|\delta \mathscr{Z}_0|} \right)$ is estimated. The general term Θ_l is expressed by Eq. (16.12). The last term in this Eq. (16.12) contains all the recursions of Θ_i for $i \geq 1$.

Figure 16.6 shows the stability of the proposed methodology for the four presented problems, where the Lyapunov criterion has been met for truncation orders $\mathscr{J} = 139$, $\mathscr{J} = 195$, $\mathscr{J} = 1157$ and $\mathscr{J} = 362$, respectively. Although the recursive scheme shows an oscillatory character, the solution is already stable before the applied truncation since λ is negative and successive corrections are bound by the exponential horn. The oscillatory character is probably due to the use of the maximum norm which yields a worst case estimate, nevertheless the average solution may improve from one recursion step to the next one.

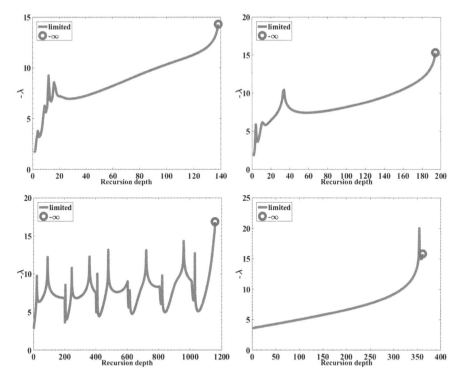

Fig. 16.6 Negative Lyapunov exponents with recursion depth; Problem 1 (top left), Problem 2 (top right), Problem 3 (bottom left) and Problem 4 (bottom right)

16.5   Conclusions

In the present work we presented a semi-analytical solution for the radiative conductive transfer equation in spherical geometry and in the S_N approximation. To this end the original non-linear problem was decomposed into a recursive scheme of equation systems by the use of a modified decomposition method, following the reasoning of reference [ViEtAl11]. The initialisation of the recursion is a linear equation system with known solution and subject to the original boundary conditions. All the subsequent equation systems to be solved are of linear type, where the non-linearity appears as source term that contains only terms from solutions of all previous recursion steps. It is noteworthy that the recursive scheme is not unique and convergence depends strongly on the recursion initialisation. Consistency was shown upon inserting the solution for each parameter set into the original integro-differential equation where the residual term was determined by the use of the maximum norm. It is noteworthy that standard convergence criteria do not apply for non-linear problems, so that we proposed a convergence criterion in analogy to Lyapunov's stability idea for dynamical systems [LaEtAl18]. In some of the considered problems, the specific parameter choice needed a corrective procedure for the recursion initialisation and the first number of recursion steps as indicated by the parameter α. Moreover, the recursion depth was fixed such that the Lyapunov exponent was strongly negative indicating exponential convergence [LaEtAl18] and the quality of the solution was analysed evaluating consistency and convergence.

References

[Mo13] Modest, M. F.: *Radiative Heat Transfer*, Academic Press, New York (2013).

[HoEtAl16] Howell, J. R., Menguc, M. P., and Siegel, R.: *Thermal Radiation Heat Transfer*, CRC Press, New York (2016).

[ThOz85] Thynell, S. T., and Ozisik, M. N.: Radiation Transfer in an Isotropically Scattering Solid Sphere With Space Dependent Albedo, $\omega(r)$. *J. Heat Transfer*, **104**, 732–734 (1985).

[TsEtAl89] Tsai, J. R., Ozisik, M. N., and Santarelli, F.: Radiation in spherical symmetry with anisotropic scattering and variable properties. *J. Quant. Spectrosc. Radiative Transfer*, **42**, 187–199 (1989).

[Th90] Thynell, S. T.: Interaction of conduction and radiation in anisotropically scattering, spherical media. *J. Thermophys Heat Transfer*, **4**, 299–304 (1990).

[MiEtAl10] Mishra, S. C., Krishna, Ch. H., and Kim, M. Y.: Lattice Boltzmann Method and Modified Discrete Ordinate Method Applied to Radiative Transport in a Spherical Medium with and without Conduction. *Numer. Heat Transfer, Part A*, **58**, 852–881 (2010).

[LaEtAl18] Ladeia, C. A., Bodmann, B. E. J., and Vilhena, M. T.: The radiative conductive transfer equation in cylinder geometry: Semi-analytical solution and a point analysis of convergence. *J. Quant. Spectrosc. Radiative Transfer*, **217**, 338–352 (2018).

[ViEtAl11] Vilhena, M. T. M. B., Bodmann, B. E. J., and Segatto, C. F.: Non-linear radiative-conductive heat transfer in a heterogeneous gray plane-parallel participating medium. *In: Ahasan, A. (ed.) Convection and Conduction Heat Transfer. InTech*, New York (2011).

[Oz73] Ozisik, M.N.: *Radiative Transfer and Interaction with Conductions and Convection*, John Wiley & Sons Inc., New York (1973).

[Ch50] Chandrasekhar, S.: *Radiative Transfer*, Oxford University Press, New York (1950).

[El09] Elghazaly, A.: Conductive-Radiative Heat Transfer in a Scattering Medium with Angle-Dependent Reflective Boundaries. *J. Nucl. Radiation Phys.* **4**, 31–41 (2009).

[BoEtAl05] Boichenko, V. A., Leonov, G. A., and Reitmann, V.: *Dimension theory for ordinary differential*, Teubner, Stuttgart (2005).

Chapter 17
Periodic Transmission Problems for the Heat Equation

Paolo Luzzini and Paolo Musolino

17.1 Introduction

This paper is devoted to the application of layer potential methods to the solution of some initial-boundary value problems for the heat equation in parabolic cylinders defined as the product of a bounded time interval and unbounded periodic domains.

Layer heat potentials have been systematically exploited in the analysis of boundary value problems for the heat equation. For example, we mention the well-known monographs Ladyženskaja, Solonnikov and Ural'ceva [LaEtAl68] and Friedman [Fr64], where a large variety of parabolic problems are solved. Moreover, by layer potential methods, Fabes and Rivière [FaRi97] have solved the Dirichlet and Neumann problem for the heat equation in C^1 cylinders with data in Lebesgue spaces. Later on, Brown [Br89, Br90] has considered the case of Lipschitz cylinders. Finally, we mention that Costabel [Co90] has obtained the solvability of some boundary value problem for the heat equation in Lipschitz cylinders with data in anisotropic Sobolev spaces.

In this paper, we are interested in developing potential theoretic techniques in order to solve transmission problems in spatially periodic domains. A first step has been done in [Lu18], where space-periodic layer heat potentials have been introduced. Moreover, as a consequence of the results of [LaLu18] for the classical layer heat potentials, regularizing properties for some boundary integral operators related to the space-periodic layer heat potentials have been shown in [Lu18]. Then, the space-periodic versions of the Dirichlet and the Neumann problems for the heat equation have been solved by means of space-periodic layer heat potentials. Here,

P. Luzzini (✉) · P. Musolino
Dipartimento di Matematica "Tullio Levi-Civita", University of Padova, Padova, Italy
e-mail: pluzzini@math.unipd.it; musolino@math.unipd.it

© Springer Nature Switzerland AG 2019   211
C. Constanda, P. Harris (eds.), *Integral Methods in Science and Engineering*,
https://doi.org/10.1007/978-3-030-16077-7_17

instead, we are interested in exploiting the results of [Lu18] in order to solve space-periodic transmission problems for the heat equation.

Regarding spatially periodic evolution problems, we mention that Rodríguez-Bernal [Ro17] has developed an L^q theory for the space-periodic heat equation.

We now introduce the geometry of our setting. We fix once for all a natural number $n \in \mathbb{N} \setminus \{0, 1\}$ and an n-tuple of positive real numbers $(q_{11}, \ldots, q_{nn}) \in]0, +\infty[^n$. Then we define the periodicity cell $Q \equiv \prod_{j=1}^{n}]0, q_{jj}[$ and the diagonal matrix $q \equiv \text{diag}(q_{11}, \ldots, q_{nn})$. Clearly, $q\mathbb{Z}^n \equiv \{qz : z \in \mathbb{Z}^n\}$ is the set of vertices of a periodic subdivision of $\mathbb{R}^n$ corresponding to the cell Q. Then we fix once and for all

$$\alpha \in]0, 1[, \qquad m \in \mathbb{N} \setminus \{0\}, \qquad T \in]0, +\infty[,$$

and a bounded open subset Ω of $\mathbb{R}^n$ of class $C^{m,\alpha}$ such that $\text{cl}\,\Omega \subseteq Q$. We denote by $\mathbf{n}_\Omega$ and by $\mathbf{n}_Q$ the outward unit normal fields to $\partial\Omega$ and to ∂Q, respectively. Then, we introduce the following q-periodic sets:

$$\mathbb{S}_q[\Omega] \equiv \bigcup_{z \in \mathbb{Z}^n} (qz + \Omega) = q\mathbb{Z}^n + \Omega, \qquad \mathbb{S}_q[\Omega]^- \equiv \mathbb{R}^n \setminus \text{cl}\,\mathbb{S}_q[\Omega].$$

In the pair of domains $\mathbb{S}_q[\Omega]$ and $\mathbb{S}_q[\Omega]^-$ we consider two transmission problems for the heat equation: problem (17.4), known as non-ideal transmission problem, and problem (17.8), which is called ideal transmission problem. The aim of this paper is to solve the two problems in parabolic Schauder spaces of space-periodic functions: more precisely, after some preliminaries (Sect. 17.2), in Sect. 17.3 we solve the non-ideal problem (17.4), while in Sect. 17.4 we consider the ideal problem (17.8).

17.2 Preliminaries and Notation

If $\mathbb{D} \subseteq \mathbb{R}^n$, then we set $\mathbb{D}_T \equiv] - \infty, T] \times \mathbb{D}$, and $\partial_T \mathbb{D} \equiv (\partial \mathbb{D})_T =] - \infty, T] \times \partial \mathbb{D}$. We have $(\text{cl}\,\mathbb{D})_T = \text{cl}\,\mathbb{D}_T$. For the definition of the parabolic Schauder spaces $C^{\frac{m+\alpha}{2};m+\alpha}$ we refer to Ladyženskaja, Solonnikov and Ural'ceva [LaEtAl68, Chapter 1] (see also [Lu18]). We now introduce two subspaces useful for initial-boundary value problems with zero initial condition. If $\tilde{\Omega}$ is a subset of $\mathbb{R}^n$ of class $C^{m,\alpha}$, we set

$$C_0^{\frac{m+\alpha}{2};m+\alpha}(\text{cl}\,\tilde{\Omega}_T)$$

$$\equiv \left\{ u \in C^{\frac{m+\alpha}{2};m+\alpha}(\text{cl}\,\tilde{\Omega}_T) : u(t, x) = 0 \ \forall t \in] - \infty, 0], \ x \in \text{cl}\,\tilde{\Omega} \right\},$$

which we regard as a Banach subspace of $C^{\frac{m+\alpha}{2};m+\alpha}(\mathrm{cl}\,\tilde{\Omega}_T)$. Moreover, we set

$$C_0^{\frac{m+\alpha}{2};m+\alpha}(\partial_T\tilde{\Omega})$$
$$\equiv \left\{u \in C^{\frac{m+\alpha}{2};m+\alpha}(\partial_T\tilde{\Omega}) : u(t,x) = 0 \ \forall t \in]-\infty,0], \ x \in \partial\tilde{\Omega}\right\},$$

which we regard as a Banach subspace of $C^{\frac{m+\alpha}{2};m+\alpha}(\partial_T\tilde{\Omega})$. Now let $\mathbb{D}$ be a subset of $\mathbb{R}^n$ such that $x \pm qe_i \in \mathbb{D}$ for all $x \in \mathbb{D}$ and for all $i \in \{1,\ldots,n\}$, where $\{e_1,\ldots,e_n\}$ denotes the canonical basis of $\mathbb{R}^n$. We say that a function u from $\mathbb{D}_T$ to $\mathbb{C}$ is q-periodic in space, or simply q-periodic, if $u(t,x) = u(t,x+qe_i)$ for all $(t,x) \in \mathbb{D}_T$, and for all $i \in \{1,\ldots,n\}$. Since we will consider space-periodic problems, we introduce the following subspaces of parabolic Schauder spaces.

$$C_q^{\frac{m+\alpha}{2};m+\alpha}(\mathrm{cl}\,\mathbb{S}_q[\Omega]_T) \equiv \left\{u \in C^{\frac{m+\alpha}{2};m+\alpha}(\mathrm{cl}\,\mathbb{S}_q[\Omega]_T) : u \text{ is } q\text{-periodic in space}\right\}, \tag{17.1}$$

which we regard as a Banach subspace of $C^{\frac{m+\alpha}{2};m+\alpha}(\mathrm{cl}\,\mathbb{S}_q[\Omega]_T)$, and

$$C_q^{\frac{m+\alpha}{2};m+\alpha}(\mathrm{cl}\,\mathbb{S}_q[\Omega]_T^-) \equiv \left\{u \in C^{\frac{m+\alpha}{2};m+\alpha}(\mathrm{cl}\,\mathbb{S}_q[\Omega]_T^-) : u \text{ is } q\text{-periodic in space}\right\}, \tag{17.2}$$

which we regard as a Banach subspace of $C^{\frac{m+\alpha}{2};m+\alpha}(\mathrm{cl}\,\mathbb{S}_q[\Omega]_T^-)$. Then we can define $C_{0,q}^{\frac{m+\alpha}{2};m+\alpha}(\mathrm{cl}\,\mathbb{S}_q[\Omega]_T)$ and $C_{0,q}^{\frac{m+\alpha}{2};m+\alpha}(\mathrm{cl}\,\mathbb{S}_q[\Omega]_T^-)$ replacing $C^{\frac{m+\alpha}{2};m+\alpha}(\mathrm{cl}\,\mathbb{S}_q[\Omega]_T)$ and $C^{\frac{m+\alpha}{2};m+\alpha}(\mathrm{cl}\,\mathbb{S}_q[\Omega]_T^-)$ in the right-hand side of (17.1) and (17.2) by the spaces $C_0^{\frac{m+\alpha}{2};m+\alpha}(\mathrm{cl}\,\mathbb{S}_q[\Omega]_T)$ and $C_0^{\frac{m+\alpha}{2};m+\alpha}(\mathrm{cl}\,\mathbb{S}_q[\Omega]_T^-)$, respectively.

Next, in order to build space-periodic layer heat potentials, we plan to replace the fundamental solution of the heat equation by a periodic analog. Therefore, we introduce the function $\Phi_{q,n}$ from $(\mathbb{R}\times\mathbb{R}^n)\setminus(\{0\}\times q\mathbb{Z}^n)$ to $\mathbb{R}$ defined by

$$\Phi_{q,n}(t,x) \equiv \begin{cases} \sum_{z\in\mathbb{Z}^n} \frac{1}{(4\pi t)^{\frac{n}{2}}} e^{-\frac{|x+qz|^2}{4t}} & \text{if } (t,x) \in]0,+\infty[\times\mathbb{R}^n, \\ 0 & \text{if } (t,x) \in (]-\infty,0]\times\mathbb{R}^n)\setminus(\{0\}\times q\mathbb{Z}^n) \end{cases}$$

(see [Lu18]). As it is known, $\Phi_{q,n}$ is a q-periodic analog of the (classical) fundamental solution of the heat equation. We are now ready to introduce the q-periodic analog of the single layer heat potential. Let $\mu \in C_0^{\frac{m-1+\alpha}{2};m-1+\alpha}(\partial_T\Omega)$.

Then, we set

$$v_q[\partial_T\Omega, \mu](t, x) \equiv \int_0^t \int_{\partial\Omega} (\Phi_{q,n}(t - \tau, x - y)\mu(\tau, y) \, d\sigma_y d\tau,$$

for all $(t, x) \in (\mathbb{R}^n)_T$, where $d\sigma$ denotes the area element of a manifold embedded in $\mathbb{R}^n$. Moreover, we set

$$w_{q,*}[\partial_T\Omega, \mu](t, x) \equiv \int_0^t \int_{\partial\Omega} \frac{\partial}{\partial \mathbf{n}_\Omega(x)} \Phi_{q,n}(t - \tau, x - y)\mu(\tau, y) \, d\sigma_y d\tau,$$

for all $(t, x) \in \partial_T\Omega$. The function $v_q[\partial_T\Omega, \mu]$ is the q-periodic in space single layer heat potential with density μ. If $\mu \in C_0^{\frac{m-1+\alpha}{2};m-1+\alpha}(\partial_T\Omega)$, then $v_q[\partial_T\Omega, \mu]$ is continuous in $(\mathbb{R}^n)_T$, is q-periodic and $v_q[\partial_T\Omega, \mu] \in C^\infty((\mathbb{R}^n \setminus \partial\mathbb{S}_q[\Omega])_T)$. Moreover $v_q[\partial_T\Omega, \mu]$ solves the heat equation in $(\mathbb{R}^n \setminus \partial\mathbb{S}_q[\Omega])_T$. We denote by $v_q^+[\partial_T\Omega, \mu]$ and $v_q^-[\partial_T\Omega, \mu]$ the restriction of $v_q[\partial_T\Omega, \mu]$ to $\mathrm{cl}\,\mathbb{S}_q[\Omega]_T$ and $\mathrm{cl}\,\mathbb{S}_q[\Omega]_T^-$, respectively. We have

$$\frac{\partial}{\partial \mathbf{n}_\Omega(x)} v_q^{\pm}[\partial_T\Omega, \mu](t, x) = \pm\frac{1}{2}\mu(t, x) + w_{q,*}[\partial_T\Omega, \mu](t, x), \tag{17.3}$$

for all $(t, x) \in \partial_T\Omega$ (see [Lu18]). We now recall some basic properties of the space-periodic single layer heat potential. For a proof, we refer to [Lu18].

Theorem 1 *The following statements hold.*

(i) *The operator from* $C_0^{\frac{m-1+\alpha}{2};m-1+\alpha}(\partial_T\Omega)$ *to* $C_{0,q}^{\frac{m+\alpha}{2};m+\alpha}(\mathrm{cl}\,\mathbb{S}_q[\Omega]_T)$ *which takes* μ *to* $v_q^+[\partial_T\Omega, \mu]$ *is linear and continuous. The same statement holds with* $v_q^+[\partial_T\Omega, \mu]$ *and* $\mathrm{cl}\,\mathbb{S}_q[\Omega]_T$ *replaced by* $v_q^-[\partial_T\Omega, \mu]$ *and* $\mathrm{cl}\,\mathbb{S}_q[\Omega]_T^-$, *respectively.*

(ii) *The operator* $w_{q,*}[\partial_T\Omega, \cdot]$ *is compact from* $C_0^{\frac{m-1+\alpha}{2};m-1+\alpha}(\partial_T\Omega)$ *to itself.*

Now we prove the following result on the invertibility of $v_q[\partial_T\Omega, \cdot]_{|\partial_T\Omega}$.

Theorem 2 *The operator* $v_q[\partial_T\Omega, \cdot]_{|\partial_T\Omega}$ *is a linear homeomorphism from the space* $C_0^{\frac{m-1+\alpha}{2};m-1+\alpha}(\partial_T\Omega)$ *to* $C_0^{\frac{m+\alpha}{2};m+\alpha}(\partial_T\Omega)$.

Proof By Theorem 1 (i) and by the continuity of the trace operator, $v_q[\partial_T\Omega, \cdot]_{|\partial_T\Omega}$ is linear and continuous from $C_0^{\frac{m-1+\alpha}{2};m-1+\alpha}(\partial_T\Omega)$ to $C_0^{\frac{m+\alpha}{2};m+\alpha}(\partial_T\Omega)$. Accordingly, by the Open Mapping Theorem, it suffices to show that $v_q[\partial_T\Omega, \cdot]_{|\partial_T\Omega}$ is a bijection. We first prove the injectivity. Let $\mu \in C_0^{\frac{m-1+\alpha}{2};m-1+\alpha}(\partial_T\Omega)$ be such that $v_q[\partial_T\Omega, \mu]_{|\partial_T\Omega} = 0$. The continuity of the single layer potential implies that both $v_q^+[\partial_T\Omega, \mu]_{|\partial_T\Omega} = v_q^-[\partial_T\Omega, \mu]_{|\partial_T\Omega} = 0$. Thus $v_q^+[\partial_T\Omega, \mu]$ solves a Dirichlet problem for the heat equation in $[0, T] \times \Omega$ with zero initial condition and with zero Dirichlet boundary condition. Accordingly, the uniqueness of the solution for the classical Dirichlet problem implies that $v_q^+[\partial_T \Omega, \mu] = 0$ in $[0, T] \times \mathrm{cl}\,\Omega$. Moreover, since $v_q^-[\partial_T \Omega, \mu]_{|\partial_T \Omega} = 0$, the function $v_q^-[\partial_T \Omega, \mu] = 0$ solves the periodic Dirichlet problem

$$\begin{cases} \partial_t u - \Delta u = 0 & \text{in }]0, T] \times \mathbb{S}_q[\Omega]^-, \\ u(t, x + q e_i) = u(t, x) \ \forall\,(t, x) \in [0, T] \times \mathrm{cl}\,\mathbb{S}_q[\Omega]^-, \ \forall\,i \in \{1, \ldots, n\}, \\ u = 0 & \text{on } [0, T] \times \partial\Omega, \\ u(0, \cdot) = 0 & \text{in } \mathrm{cl}\,\mathbb{S}_q[\Omega]^-. \end{cases}$$

Hence, by the maximum principle for the periodic heat equation we have that $v_q^-[\partial_T \Omega, \mu] = 0$ in $[0, T] \times \mathrm{cl}\,\mathbb{S}_q[\Omega]^-$ (see [Lu18]). Finally, the jump formula (17.3) implies that

$$\mu = \frac{\partial}{\partial \mathbf{n}_\Omega} v_q^+[\partial_T \Omega, \mu] - \frac{\partial}{\partial \mathbf{n}_\Omega} v_q^-[\partial_T \Omega, \mu] = 0 \quad \text{on } [0, T] \times \partial\Omega.$$

Next we prove the surjectivity. Let $\phi \in C_0^{\frac{m+\alpha}{2};m+\alpha}(\partial_T \Omega)$. By the results of [Lu18, Section 5] there exists a unique function $u_\phi^- \in C_0^{\frac{m+\alpha}{2};m+\alpha}(\mathrm{cl}\,\mathbb{S}_q[\Omega]_T)$ which solves

$$\begin{cases} \partial_t u - \Delta u = 0 & \text{in }]0, T] \times \mathbb{S}_q[\Omega]^-, \\ u(t, x + q e_i) = u(t, x) \ \forall\,(t, x) \in [0, T] \times \mathrm{cl}\,\mathbb{S}_q[\Omega]^-, \ \forall\,i \in \{1, \ldots, n\}, \\ u = \phi & \text{on } [0, T] \times \partial\Omega, \\ u(0, \cdot) = 0 & \text{in } \mathrm{cl}\,\mathbb{S}_q[\Omega]^-. \end{cases}$$

Since $\frac{\partial}{\partial \mathbf{n}_\Omega} u_\phi^- \in C_0^{\frac{m-1+\alpha}{2};m-1+\alpha}(\partial_T \Omega)$, the results of [Lu18, Section 5] imply that there exists a unique $\mu \in C_0^{\frac{m-1+\alpha}{2};m-1+\alpha}(\partial_T \Omega)$ such that $v_q^-[\partial_T \Omega, \mu]$ solves the problem

$$\begin{cases} \partial_t u - \Delta u = 0 & \text{in }]0, T] \times \mathbb{S}_q[\Omega]^-, \\ u(t, x + q e_i) = u(t, x) \ \forall\,(t, x) \in [0, T] \times \mathrm{cl}\,\mathbb{S}_q[\Omega]^-, \ \forall\,i \in \{1, \ldots, n\}, \\ \frac{\partial}{\partial \mathbf{n}_\Omega} u = \frac{\partial}{\partial \mathbf{n}_\Omega} u_\phi^- & \text{on } [0, T] \times \partial\Omega, \\ u(0, \cdot) = 0 & \text{in } \mathrm{cl}\,\mathbb{S}_q[\Omega]^-. \end{cases}$$

By the uniqueness of the solution for the periodic Neumann problem for the heat equation (see [Lu18]), we have that $v_q^-[\partial_T \Omega, \mu] = u_\phi^-$. In particular,

$$v_q[\partial_T \Omega, \mu]_{|\partial_T \Omega} = v_q^-[\partial_T \Omega, \mu]_{|\partial_T \Omega} = \phi \quad \text{on } [0, T] \times \partial\Omega,$$

and accordingly the statement follows.   □

17.3   A Periodic Non-ideal Transmission Problem

In this section we consider a periodic transmission problem which models the heat diffusion in a two-phase composite material with thermal resistance at the interface. We fix once and for all $\lambda^+, \lambda^-, \gamma \in]0, +\infty[$. Then we take $f, g \in C_0^{\frac{m-1+\alpha}{2};m-1+\alpha}(\partial_T \Omega)$ and we consider the following non-ideal transmission problem.

$$
\begin{cases}
\partial_t u^+ - \Delta u^+ = 0 & \text{in }]0, T] \times \mathbb{S}_q[\Omega], \\
\partial_t u^- - \Delta u^- = 0 & \text{in }]0, T] \times \mathbb{S}_q[\Omega]^-, \\
u^+(t, x + qe_i) = u^+(t, x) & \forall\, (t, x) \in [0, T] \times \mathrm{cl}\,\mathbb{S}_q[\Omega], \forall\, i \in \{1, \dots, n\}, \\
u^-(t, x + qe_i) = u^-(t, x) & \forall\, (t, x) \in [0, T] \times \mathrm{cl}\,\mathbb{S}_q[\Omega]^-, \forall\, i \in \{1, \dots, n\}, \\
\lambda^+ \frac{\partial}{\partial \mathbf{n}_\Omega} u^+ + \gamma(u^+ - u^-) = f & \text{on } [0, T] \times \partial\Omega, \\
\lambda^- \frac{\partial}{\partial \mathbf{n}_\Omega} u^- - \lambda^+ \frac{\partial}{\partial \mathbf{n}_\Omega} u^+ = g & \text{on } [0, T] \times \partial\Omega, \\
u^+(0, \cdot) = 0 & \text{in } \mathrm{cl}\,\mathbb{S}_q[\Omega], \\
u^-(0, \cdot) = 0 & \text{in } \mathrm{cl}\,\mathbb{S}_q[\Omega]^-.
\end{cases}
$$

$$(17.4)$$

The set $\mathrm{cl}\,\mathbb{S}_q[\Omega]^-$ plays the role of a matrix with thermal conductivity λ^- where the periodic array of inclusions $\mathrm{cl}\,\mathbb{S}_q[\Omega]$ with thermal conductivity λ^+ is inserted. The fifth condition of system (17.4) is the non-ideal transmission (or imperfect contact) condition, which models the thermal resistance at the interface. In particular this condition says that the temperature field at the interface displays a jump proportional to the normal heat flux. Concerning parabolic transmission problems, we mention the works of Donato and Jose [DoJo15], for the study of the asymptotic behavior of the approximate control of a parabolic transmission problem. For the stationary case, we mention [DaMu13], where the authors consider a singularly perturbed stationary version of the above transmission problem in order to study the effective conductivity of a periodic composite. Incidentally, we observe that the discontinuity of the temperature field is a well-known phenomenon in physics which has been studied since the work of Kapitza in 1941, in which the author has studied for the first time the thermal interface behavior in liquid helium (see, e.g., Swartz and Pohl [SwPo89], Lipton [Li98] and the references therein). We begin our analysis with the following uniqueness result for problem (17.4).

Proposition 1 *Let* $u^+ \in C_{0,q}^{\frac{1}{2};1}(\mathrm{cl}\,\mathbb{S}_q[\Omega]_T)$, $u^- \in C_{0,q}^{\frac{1}{2};1}(\mathrm{cl}\,\mathbb{S}_q[\Omega]_T^-)$ *be one time continuously differentiable with respect to the time variable and two times continuously differentiable with respect to the space variables in* $]0, T] \times \mathbb{S}_q[\Omega]$ *and* $]0, T] \times \mathbb{S}_q[\Omega]^-$, *respectively. Moreover, let the pair* (u^+, u^-) *solve the system* (17.4) *with* $f = g = 0$. *Then* $u^+ = 0$ *in* $[0, T] \times \mathrm{cl}\,\mathbb{S}_q[\Omega]$ *and* $u^- = 0$ *in* $[0, T] \times \mathrm{cl}\,\mathbb{S}_q[\Omega]^-$.

Proof Let e^+, e^- be the functions from $[0, T]$ to $[0, +\infty[$ defined by

$$e^+(t) \equiv \int_\Omega (u^+(t, y))^2 \, dy, \qquad e^-(t) \equiv \int_{Q \backslash \mathrm{cl}\, \Omega} (u^-(t, y))^2 \, dy, \qquad \forall t \in [0, T].$$

By the Dominated Convergence Theorem, $e^+, e^- \in C^0([0, T])$. In addition, classical differentiation theorems for integrals depending on a parameter and the approximation argument of Verchota [Ve84, Theorem 1.12, p. 581] if $m = 1$ (see [Lu18]) imply that $e^+, e^- \in C^1(]0, T[)$. Also, by the Divergence Theorem, we have that

$$\frac{d}{dt} e^+(t) = 2 \int_\Omega u^+(t, y) \partial_t u^+(t, y) \, dy = 2 \int_\Omega u^+(t, y) \Delta u^+(t, y) \, dy$$

$$= -2 \int_\Omega |Du^+(t, y)|^2 \, dy + 2 \int_{\partial \Omega} u^+(t, y) \frac{\partial}{\partial \mathbf{n}_\Omega(y)} u^+(t, y) \, d\sigma_y,$$

for all $t \in]0, T[$. Moreover, in a similar way, exploiting the Divergence Theorem and the q-periodicity of u we have that

$$\frac{d}{dt} e^-(t) = 2 \int_{Q \backslash \mathrm{cl}\, \Omega} u^-(t, y) \partial_t u^-(t, y) \, dy = 2 \int_{Q \backslash \mathrm{cl}\, \Omega} u^-(t, y) \Delta u(t, y) \, dy$$

$$= -2 \int_{Q \backslash \mathrm{cl}\, \Omega} |Du^-(t, y)|^2 \, dy - 2 \int_{\partial \Omega} u^-(t, y) \frac{\partial}{\partial \mathbf{n}_\Omega(y)} u^-(t, y) \, d\sigma_y,$$

for all $t \in]0, T[$. Then, if we set $e \equiv \lambda^+ e^+ + \lambda^- e^-$, we have that

$$\frac{d}{dt} e(t) = -2 \left(\lambda^+ \int_\Omega |Du^+(t, y)|^2 \, dy + \lambda^- \int_{Q \backslash \mathrm{cl}\, \Omega} |Du^-(t, y)|^2 \, dy \right)$$

$$+ 2\lambda^+ \int_{\partial \Omega} u^+(t, y) \frac{\partial}{\partial \mathbf{n}_\Omega(y)} u^+(t, y) \, d\sigma_y - 2\lambda^- \int_{\partial \Omega} u^-(t, y) \frac{\partial}{\partial \mathbf{n}_\Omega(y)} u^-(t, y) \, d\sigma_y$$

$$= -2 \left(\lambda^+ \int_\Omega |Du^+(t, y)|^2 \, dy + \lambda^- \int_{Q \backslash \mathrm{cl}\, \Omega} |Du^-(t, y)|^2 \, dy \right)$$

$$+ 2\lambda^+ \int_{\partial \Omega} (u^+(t, y) - u^-(t, x)) \frac{\partial}{\partial \mathbf{n}_\Omega(y)} u^+(t, y) \, d\sigma_y$$

$$= -2 \left(\lambda^+ \int_\Omega |Du^+(t, y)|^2 \, dy + \lambda^- \int_{Q \backslash \mathrm{cl}\, \Omega} |Du^-(t, y)|^2 \, dy \right)$$

$$- \frac{2}{\gamma} \int_{\partial \Omega} \left(\lambda^+ \frac{\partial}{\partial \mathbf{n}_\Omega(y)} u^+(t, y) \right)^2 d\sigma_y \qquad \forall t \in]0, T[.$$

Hence $\frac{d}{dt}e \leq 0$ in $]0, T[$. Since $e(0) = 0$ and $e \geq 0$ in $[0, T]$, then $e = 0$ in $[0, T]$. Then $u^+ = 0$ in $[0, T] \times \mathrm{cl}\,\Omega$, $u^- = 0$ in $[0, T] \times \mathrm{cl}\,Q \setminus \Omega$ and therefore the q-periodicity of u^+, u^- implies the validity of the statement. $\square$

Since we plan to solve the problem (17.4) with two space-periodic single layer heat potentials, we need to solve the related boundary integral equations. In order to do so, we show the invertibility of the operators which appear in such integral equations.

Lemma 1 *Let $J \equiv (J_1, J_2)$ be the operator from the space $(C_0^{\frac{m-1+\alpha}{2};m-1+\alpha}(\partial_T\Omega))^2$ to $(C_0^{\frac{m-1+\alpha}{2};m-1+\alpha}(\partial_T\Omega))^2$ defined by*

$$J_1[\mu^+, \mu^-] \equiv \lambda^+ \left(\frac{1}{2}\mu^+ + w_{q,*}[\partial_T\Omega, \mu^+]\right) \tag{17.5}$$
$$+ \gamma(v_q^+[\partial_T\Omega, \mu^+]_{|\partial_T\Omega} - v_q^-[\partial_T\Omega, \mu^-]_{|\partial_T\Omega}),$$

$$J_2[\mu^+, \mu^-] \equiv \lambda^- \left(-\frac{1}{2}\mu^- + w_{q,*}[\partial_T\Omega, \mu^-]\right) - \lambda^+ \left(\frac{1}{2}\mu^+ + w_{q,*}[\partial_T\Omega, \mu^+]\right),$$

for all $(\mu^+, \mu^-) \in (C_0^{\frac{m-1+\alpha}{2};m-1+\alpha}(\partial_T\Omega))^2$. Then J is a linear homeomorphism.

Proof Let $J^\# = (J_1^\#, J_2^\#)$ be the linear operator from $(C_0^{\frac{m-1+\alpha}{2};m-1+\alpha}(\partial_T\Omega))^2$ to $(C_0^{\frac{m-1+\alpha}{2};m-1+\alpha}(\partial_T\Omega))^2$ defined by

$$J_1^\#[\mu^+, \mu^-] \equiv \frac{\lambda^+}{2}\mu^+, \qquad J_2^\#[\mu^+, \mu^-] \equiv -\frac{\lambda^-}{2}\mu^- - \frac{\lambda^+}{2}\mu^+,$$

for all $(\mu^+, \mu^-) \in (C_0^{\frac{m-1+\alpha}{2};m-1+\alpha}(\partial_T\Omega))^2$. Clearly $J^\#$ is a linear homeomorphism. Moreover, let $\bar{J} = (\bar{J}_1, \bar{J}_2)$ be the linear operator from $(C_0^{\frac{m-1+\alpha}{2};m-1+\alpha}(\partial_T\Omega))^2$ to $(C_0^{\frac{m-1+\alpha}{2};m-1+\alpha}(\partial_T\Omega))^2$ defined by

$$\bar{J}_1[\mu^+, \mu^-] \equiv \lambda^+ w_{q,*}[\partial_T\Omega, \mu^+] + \gamma(v_q^+[\partial_T\Omega, \mu^+]_{|\partial_T\Omega} - v_q^-[\partial_T\Omega, \mu^-]_{|\partial_T\Omega}),$$

$$\bar{J}_2[\mu^+, \mu^-] \equiv \lambda^- w_{q,*}[\partial_T\Omega, \mu^-] - \lambda^+ w_{q,*}[\partial_T\Omega, \mu^+],$$

for all $(\mu^+, \mu^-) \in (C_0^{\frac{m-1+\alpha}{2};m-1+\alpha}(\partial_T\Omega))^2$. By Theorem 1 (ii), the map $w_{q,*}[\partial_T\Omega, \cdot]$ is compact in $C_0^{\frac{m-1+\alpha}{2};m-1+\alpha}(\partial_T\Omega)$. By Theorem 1 (i), $v_q^+[\partial_T\Omega, \cdot]$ and $v_q^-[\partial_T\Omega, \cdot]$ are linear and continuous from $C_0^{\frac{m-1+\alpha}{2};m-1+\alpha}(\partial_T\Omega)$ to $C_0^{\frac{m+\alpha}{2};m+\alpha}(\mathrm{cl}\,\mathbb{S}_q[\Omega]_T)$ and $C_0^{\frac{m+\alpha}{2};m+\alpha}(\mathrm{cl}\,\mathbb{S}_q[\Omega]_T^-)$, respectively. Then, by the continuity of the trace operators from $C_{0,q}^{\frac{m+\alpha}{2};m+\alpha}(\mathrm{cl}\,\mathbb{S}_q[\Omega]_T)$ to $C_0^{\frac{m+\alpha}{2};m+\alpha}(\partial_T\Omega)$ and from $C_{0,q}^{\frac{m+\alpha}{2};m+\alpha}(\mathrm{cl}\,\mathbb{S}_q[\Omega]_T^-)$ to $C_0^{\frac{m+\alpha}{2};m+\alpha}(\partial_T\Omega)$, and by the compactness of the embedding of $C_0^{\frac{m+\alpha}{2};m+\alpha}(\partial_T\Omega)$ into $C_0^{\frac{m-1+\alpha}{2};m-1+\alpha}(\partial_T\Omega)$, which is a consequence of the Ascoli-Arzelà Theorem, $v_q^+[\partial_T\Omega,\cdot]_{|\partial_T\Omega}$ and $v_q^-[\partial_T\Omega,\cdot]_{|\partial_T\Omega}$ are compact in $C_0^{\frac{m-1+\alpha}{2};m-1+\alpha}(\partial_T\Omega)$. Then the operator $\bar{J}$ is compact in $(C_0^{\frac{m-1+\alpha}{2};m-1+\alpha}(\partial_T\Omega))^2$. Since compact perturbations of linear homeomorphisms are Fredholm operators of index 0, we have that $J = J^{\#} + \bar{J}$ is a Fredholm operator of index 0. Thus, to show that J is a linear homeomorphism, it suffices to show that J is injective. Let $(\mu^+,\mu^-) \in (C_0^{\frac{m-1+\alpha}{2};m-1+\alpha}(\partial_T\Omega))^2$ be such that $J[\mu^+,\mu^-] = (0,0)$. Then, $v_q^+[\partial_T\Omega,\mu^+]$ and $v_q^-[\partial_T\Omega,\mu^-]$ satisfy the assumptions of Proposition 1 and thus $v_q^+[\partial_T\Omega,\mu^+] = 0$ in $[0,T] \times \mathrm{cl}\,\mathbb{S}_q[\Omega]$ and $v_q^-[\partial_T\Omega,\mu^-] = 0$ in $[0,T] \times \mathrm{cl}\,\mathbb{S}_q[\Omega]^-$. In particular, by the continuity of the periodic single layer heat potential, $v_q[\partial_T\Omega,\mu^+]_{|\partial_T\Omega} = v_q^+[\partial_T\Omega,\mu^+]_{|\partial_T\Omega} = 0$ and $v_q[\partial_T\Omega,\mu^-]_{|\partial_T\Omega} = v_q^-[\partial_T\Omega,\mu^-]_{|\partial_T\Omega} = 0$. Accordingly, Theorem 2 implies that $\mu^+ = \mu^- = 0$ on $[0,T] \times \partial\Omega$, and the statement follows. □

Finally we are ready to prove the following solvability result for the non-ideal transmission problem (17.4).

Theorem 3 *Let $f,g \in C_0^{\frac{m-1+\alpha}{2};m-1+\alpha}(\partial_T\Omega)$. Then problem (17.4) has a unique solution $(u^+,u^-) \in C_{0,q}^{\frac{m+\alpha}{2};m+\alpha}(\mathrm{cl}\,\mathbb{S}_q[\Omega]_T) \times C_{0,q}^{\frac{m+\alpha}{2};m+\alpha}(\mathrm{cl}\,\mathbb{S}_q[\Omega]_T^-)$. Moreover,*

$$u^+ = v_q^+[\partial_T\Omega,\mu^+] \quad in\ \mathrm{cl}\,\mathbb{S}_q[\Omega]_T, \qquad u^- = v_q^-[\partial_T\Omega,\mu^-] \quad in\ \mathrm{cl}\,\mathbb{S}_q[\Omega]_T^-,$$
(17.6)

where (μ^+,μ^-) is the unique solution in $(C_0^{\frac{m-1+\alpha}{2};m-1+\alpha}(\partial_T\Omega))^2$ of

$$J[\mu^+,\mu^-] = (f,g) \quad on\ \partial_T\Omega.$$
(17.7)

Proof Proposition 1 implies that problem (17.4) has at most one solution. Then we only need to show that the pair (u^+,u^-) defined by (17.6) is a solution of problem (17.4). Lemma 1 implies that there exists a unique solution (μ^+,μ^-) in $(C_0^{\frac{m-1+\alpha}{2};m-1+\alpha}(\partial_T\Omega))^2$ of (17.7). Then by Theorem 1 (i) and by the definition (17.5) of J, the functions u^+,u^- defined by (17.6) are q-periodic functions which solve the heat equation and which satisfy all the transmission conditions in (17.4). Thus (u^+,u^-) is a solution of problem (17.4). □

17.4   A Periodic Ideal Transmission Problem

In this section we consider a periodic transmission problem which models the heat diffusion in a two-phase composite material with perfect contact at the interface. We fix once and for all $\lambda^+, \lambda^- \in]0, +\infty[$. Then we take $f \in C_0^{\frac{m+\alpha}{2};m+\alpha}(\partial_T \Omega)$ and $g \in C_0^{\frac{m-1+\alpha}{2};m-1+\alpha}(\partial_T \Omega)$ and we consider the following ideal transmission problem.

$$
\begin{cases}
\partial_t u^+ - \Delta u^+ = 0 & \text{in }]0, T] \times \mathbb{S}_q[\Omega], \\
\partial_t u^- - \Delta u^- = 0 & \text{in }]0, T] \times \mathbb{S}_q[\Omega]^-, \\
u^+(t, x + q e_i) = u^+(t, x) & \forall (t, x) \in [0, T] \times \text{cl}\,\mathbb{S}_q[\Omega], \ \forall i \in \{1, \ldots, n\}, \\
u^-(t, x + q e_i) = u^-(t, x) & \forall (t, x) \in [0, T] \times \text{cl}\,\mathbb{S}_q[\Omega]^-, \ \forall i \in \{1, \ldots, n\}, \\
u^+ - u^- = f & \text{on } [0, T] \times \partial\Omega, \\
\lambda^- \frac{\partial}{\partial \mathbf{n}_\Omega} u^- - \lambda^+ \frac{\partial}{\partial \mathbf{n}_\Omega} u^+ = g & \text{on } [0, T] \times \partial\Omega, \\
u^+(0, \cdot) = 0 & \text{in } \text{cl}\,\mathbb{S}_q[\Omega], \\
u^-(0, \cdot) = 0 & \text{in } \text{cl}\,\mathbb{S}_q[\Omega]^-.
\end{cases}
$$

$$(17.8)$$

The set $\text{cl}\,\mathbb{S}_q[\Omega]^-$ plays the role of a matrix with thermal conductivity λ^- where the periodic array of inclusions $\text{cl}\,\mathbb{S}_q[\Omega]$ with thermal conductivity λ^+ is inserted. The fifth and sixth conditions of system (17.8) are the ideal transmission (or perfect contact) conditions, which say that heat flux and the temperature field are continuous at the interface between the two materials. We mention Hofmann, Lewis, and Mitrea [HoEtAl03] for the study of the non-periodic version of this transmission problem, in case Ω is a Lipschitz domain and the boundary conditions are in suitable Lebesgue spaces. For the study of the stationary version of ideal transmission problems we mention Ammari, Kang, and Touibi [AmEtAl05] for the computation of the effective conductivity of a material with periodic inclusions and Pukhtaievych [Pu18A, Pu18B] for the asymptotic behavior when the diameter of the periodic inclusions tends to zero. We start our analysis of problem (17.8) with the following uniqueness result that can be proved as the one of Proposition 1.

Proposition 2 *Let* $u^+ \in C_{0,q}^{\frac{1}{2};1}(\text{cl}\,\mathbb{S}_q[\Omega]_T)$, $u^- \in C_{0,q}^{\frac{1}{2};1}(\text{cl}\,\mathbb{S}_q[\Omega]_T^-)$ *be one time continuously differentiable with respect to the time variable and two times continuously differentiable with respect to the space variables in* $]0, T] \times \mathbb{S}_q[\Omega]$ *and* $]0, T] \times \mathbb{S}_q[\Omega]^-$, *respectively. Moreover, let the pair* (u^+, u^-) *solve the system* (17.8) *with* $f = g = 0$. *Then* $u^+ = 0$ *in* $[0, T] \times \text{cl}\,\mathbb{S}_q[\Omega]$ *and* $u^- = 0$ *in* $[0, T] \times \text{cl}\,\mathbb{S}_q[\Omega]^-$.

Our aim is to provide a solution of problem (17.8) in terms of space-periodic single layer heat potentials. By exploiting the potential theoretic method, we will convert problem (17.8) into integral equations. Therefore, in order to prove the solvability of the integral equations, we need to perform a preliminary study of an auxiliary integral operator. We do so in the following lemma.

Lemma 2 *Let K be the map from $C_0^{\frac{m-1+\alpha}{2};m-1+\alpha}(\partial_T\Omega)$ to $C_0^{\frac{m-1+\alpha}{2};m-1+\alpha}(\partial_T\Omega)$ defined by*

$$K[\mu] = -\frac{1}{2}\mu + \frac{\lambda^- - \lambda^+}{\lambda^- + \lambda^+}w_{q,*}[\partial_T\Omega, \mu]$$

for all $\mu \in C_0^{\frac{m-1+\alpha}{2};m-1+\alpha}(\partial_T\Omega)$. Then K is a linear homeomorphism.

Proof By Theorem 1 (ii), $w_{q,*}[\partial_T\Omega, \cdot]$ is compact in $C_0^{\frac{m-1+\alpha}{2};m-1+\alpha}(\partial_T\Omega)$. Accordingly, the Fredholm Alternative implies that it suffices to show that K is injective. Let $\mu \in C_0^{\frac{m-1+\alpha}{2};m-1+\alpha}(\partial_T\Omega)$ be such that $K[\mu] = 0$. Then

$$\lambda^+ \frac{\partial}{\partial \mathbf{n}_\Omega} v_q^+[\partial_T\Omega, \mu]_{|\partial_T\Omega} - \lambda^- \frac{\partial}{\partial \mathbf{n}_\Omega} v_q^-[\partial_T\Omega, \mu]_{|\partial_T\Omega} = 0.$$

Accordingly, the pair $(v_q^+[\partial_T\Omega, \mu], v_q^-[\partial_T\Omega, \mu])$ satisfies all the assumptions of Proposition 2 and then $(v_q^+[\partial_T\Omega, \mu], v_q^-[\partial_T\Omega, \mu]) = (0, 0)$. In particular, by the continuity of the periodic single layer potential we have that $v_q[\partial_T\Omega, \mu]_{|\partial_T\Omega} = 0$ and accordingly Theorem 2 implies the validity of the statement. □

In the following lemma, we prove the next step, which consists in showing the invertibility of an operator which appears in the integral equations associated with the transmission problem (17.8).

Lemma 3 *Let $H \equiv (H_1, H_2)$ be the map from the space $(C_0^{\frac{m-1+\alpha}{2};m-1+\alpha}(\partial_T\Omega))^2$ to $C_0^{\frac{m+\alpha}{2};m+\alpha}(\partial_T\Omega) \times C_0^{\frac{m-1+\alpha}{2};m-1+\alpha}(\partial_T\Omega)$ defined by*

$$H_1[\mu^+, \mu^-] \equiv v_q^+[\partial_T\Omega, \mu^+]_{|\partial_T\Omega} - v_q^-[\partial_T\Omega, \mu^-]_{|\partial_T\Omega},$$

$$H_2[\mu^+, \mu^-] \equiv \lambda^- \left(-\frac{1}{2}\mu^- + w_{q,*}[\partial_T\Omega, \mu^-]\right) - \lambda^+ \left(\frac{1}{2}\mu^+ + w_{q,*}[\partial_T\Omega, \mu^+]\right),$$

for all $(\mu^+, \mu^-) \in (C_0^{\frac{m-1+\alpha}{2};m-1+\alpha}(\partial_T\Omega))^2$. Then H is a linear homeomorphism.

Proof Theorem 1 implies that H is linear and continuous. Accordingly, by the Open Mapping Theorem, it suffices to show that it is a bijection. Let (ϕ, ψ) be in $C_0^{\frac{m+\alpha}{2};m+\alpha}(\partial_T\Omega) \times C_0^{\frac{m-1+\alpha}{2};m-1+\alpha}(\partial_T\Omega)$. We show that there exists a unique pair $(\mu^+, \mu^-) \in (C_0^{\frac{m-1+\alpha}{2};m-1+\alpha}(\partial_T\Omega))^2$ such that

$$H[\mu^+, \mu^-] = (\phi, \psi). \tag{17.9}$$

We first show the uniqueness. Let $(\mu^+, \mu^-) \in (C_0^{\frac{m-1+\alpha}{2};m-1+\alpha}(\partial_T\Omega))^2$ be such that (17.9) holds. Theorem 2 implies that there exists a unique $\mu^\# \in C_0^{\frac{m-1+\alpha}{2};m-1+\alpha}(\partial_T\Omega)$ such that $v_q[\partial_T\Omega, \mu^\#]_{|\partial_T\Omega} = \phi$. Accordingly, $H_1[\mu^+, \mu^-] = \phi$ implies that

$$\mu^\# = \mu^+ - \mu^-. \tag{17.10}$$

By substituting the previous equality in the equality $H_2[\mu^+, \mu^-] = \psi$ we get

$$-\frac{1}{2}\mu^+ + \frac{\lambda^- - \lambda^+}{\lambda^- + \lambda^+}w_{q,*}[\partial_T\Omega, \mu^+] = \frac{\lambda^-}{\lambda^- + \lambda^+}\left(-\frac{1}{2}\mu^\# + w_{q,*}[\partial_T\Omega, \mu^\#]\right) \tag{17.11}$$

$$+ \frac{1}{\lambda^- + \lambda^+}\psi.$$

Since the right-hand side of the previous equality belongs to $C_0^{\frac{m-1+\alpha}{2};m-1+\alpha}(\partial_T\Omega)$, Lemma 2 implies that there exists a unique $\mu^+ \in C_0^{\frac{m-1+\alpha}{2};m-1+\alpha}(\partial_T\Omega)$ such that (17.11) holds, and accordingly μ^+ is uniquely determined. Then the equality (17.10) uniquely determines μ^- and thus uniqueness follows. On the other hand, by reading backward the argument above, one deduces the existence of a pair $(\mu^+, \mu^-) \in (C_0^{\frac{m-1+\alpha}{2};m-1+\alpha}(\partial_T\Omega))^2$ such that (17.9) holds.   □

Finally, by exploiting Proposition 2, Lemma 3, and the properties of the space-periodic single layer heat potential, we can deduce the following result concerning the solvability of the ideal transmission problem (17.8).

Theorem 4 *Let* $f \in C_0^{\frac{m+\alpha}{2};m+\alpha}(\partial_T\Omega), g \in C_0^{\frac{m-1+\alpha}{2};m-1+\alpha}(\partial_T\Omega)$. *Then problem (17.8) has a unique solution* $(u^+, u^-) \in C_{0,q}^{\frac{m+\alpha}{2};m+\alpha}(\mathrm{cl}\,\mathbb{S}_q[\Omega]_T) \times C_{0,q}^{\frac{m+\alpha}{2};m+\alpha}(\mathrm{cl}\,\mathbb{S}_q[\Omega]_T^-)$. *Moreover,*

$$u^+ = v_q^+[\partial_T\Omega, \mu^+] \quad \text{in } \mathrm{cl}\,\mathbb{S}_q[\Omega]_T, \quad u^- = v_q^-[\partial_T\Omega, \mu^-] \quad \text{in } \mathrm{cl}\,\mathbb{S}_q[\Omega]_T^-,$$

where (μ^+, μ^-) *is the unique solution in* $(C_0^{\frac{m-1+\alpha}{2};m-1+\alpha}(\partial_T\Omega))^2$ *of*

$$H[\mu^+, \mu^-] = (f, g) \quad \text{on } \partial_T\Omega.$$

Acknowledgements The authors acknowledge the support of the project BIRD168373/16 'Singular perturbation problems for the heat equation in a perforated domain' of the University of Padova. The authors are members of the 'Gruppo Nazionale per l'Analisi Matematica, la Probabilità e le loro Applicazioni' (GNAMPA) of the 'Istituto Nazionale di Alta Matematica' (INdAM).

References

[AmEtAl05] Ammari, H., Kang, H., and Touibi, K.: Boundary layer techniques for deriving the effective properties of composite materials. *Asymptot. Anal.* **41** (2005), no. 2, 119–140.

[Br89] Brown, R. M.: The method of layer potentials for the heat equation in Lipschitz cylinders. *Amer. J. Math.* **111** (1989), no. 2, 339–379.

[Br90] Brown, R. M.: The initial-Neumann problem for the heat equation in Lipschitz cylinders. *Trans. Amer. Math. Soc.* **320** (1990), no. 1, 1–52.

[Co90] Costabel, M.: Boundary integral operators for the heat equation. *Integral Equations Operator Theory*, **13** (1990), no. 4, 498–552.

[DaMu13] Dalla Riva, M., and Musolino, P.: A singularly perturbed nonideal transmission problem and application to the effective conductivity of a periodic composite. *SIAM J. Appl. Math.*, **73** (2013), no. 1, 24–46.

[DoJo15] Donato, P., and Jose, E. C.: Asymptotic behavior of the approximate controls for parabolic equations with interfacial contact resistance, *ESAIM: Control Optim. Calc. Var.*, **21** (2015), no. 1, 138–164.

[FaRi97] Fabes, E. B., and Rivière,N. M.: Dirichlet and Neumann problems for the heat equation in C^1-cylinders. *Harmonic analysis in Euclidean spaces* (Proc. Sympos. Pure Math., Williams Coll., Williamstown, Mass., 1978), Part 2, pp. 179–196, Proc. Sympos. Pure Math., XXXV, Part, Amer. Math. Soc., Providence, R.I., 1979.

[Fr64] Friedman, A.: *Partial differential equations of parabolic type.* Partial differential equations of parabolic type. Prentice-Hall, Inc., Englewood Cliffs, N.J. 1964.

[HoEtAl03] Hofmann, S., Lewis, J., and Mitrea, M.: Spectral properties of parabolic layer potentials and transmission boundary problems in nonsmooth domains. *Illinois J. Math.* **47** (2003), no. 4, 1345–1361.

[LaEtAl68] Ladyženskaja, O. A., Solonnikov, V. A., and Ural'ceva, N. N.: *Linear and quasilinear equations of parabolic type.* (Russian) Translated from the Russian by S. Smith. Translations of Mathematical Monographs, Vol. 23 American Mathematical Society, Providence, R.I. 1968.

[LaLu18] Lanza de Cristoforis, M., and Luzzini, P.: Tangential derivatives and higher order regularizing properties of the double layer heat potential. *Analysis (Berlin)*, **38** (2018), no. 4, 167–193.

[Li98] Lipton, R.: Heat conduction in fine scale mixtures with interfacial contact resistance, *SIAM J. Appl. Math.*, **58** (1998), no. 1, pp. 55–72.

[Lu18] Luzzini, P.: Regularizing properties of space-periodic layer heat potentials and applications to boundary value problems in periodic domains, *submitted* (2018).

[Pu18A] Pukhtaievych, R.: Asymptotic behavior of the solution of singularly perturbed transmission problems in a periodic domain. *Math. Methods Appl. Sci.* **41** (2018), no. 9, 3392–3413.

[Pu18B] Pukhtaievych, R.: Effective conductivity of a periodic dilute composite with perfect contact and its series expansion. *Z. Angew. Math. Phys.* **69** (2018), no. 3, Art. 83, 22 pp.

[Ro17] Rodríguez-Bernal, A.: The heat equation with general periodic boundary conditions. Potential Anal. **46** (2017), no. 2, 295–321.

[SwPo89] Swartz, E. T., and Pohl, R. O.: Thermal boundary resistance, *Rev. Mod. Phys.*, **61** (1989), pp. 605–668.

[Ve84] Verchota G., Layer potentials and regularity for the Dirichlet problem for Laplace's equation in Lipschitz domains. *J. Funct. Anal.* **59** (1984), no. 3, 572–611.

Chapter 18
On United Boundary-Domain Integro-Differential Equations for Variable Coefficient Dirichlet Problem with General Right-Hand Side

Sergey E. Mikhailov and Zenebe W. Woldemicheal

18.1 Introduction

In this paper, the Dirichlet boundary value problem (BVP) for the linear stationary diffusion partial differential equation with a variable coefficient is considered. The PDE right-hand side belongs to the Sobolev spaces $H^{-1}(\Omega)$, when neither classical nor canonical co-normal derivatives are well defined. Using an appropriate parametrix (Levi function) the problem is reduced to a direct boundary-domain integro-differential equation (BDIDE) or to a domain integral equation supplemented by the original boundary condition thus constituting a boundary-domain integro-differential problem (BDIDP). Solvability, solution uniqueness, and equivalence of the BDIDE/BDIDP to the original BVP are analysed in Sobolev (Bessel potential) spaces.

Let Ω be a bounded open three-dimensional region of $\mathbb{R}^3$. For simplicity, we assume that the boundary $\partial\Omega$ is a simply connected, closed, infinitely smooth surface. Let $a \in C^\infty(\overline{\Omega})$, $a(x) > 0$ for $x \in \overline{\Omega}$.

We consider the scalar elliptic differential equation, which for sufficiently smooth u has the following strong form

$$Au(x) := A(x, \partial_x)u(x) := \sum_{i=1}^{3} \frac{\partial}{\partial x_i} \left(a(x) \frac{\partial u(x)}{\partial x_i} \right) = f(x), \quad x \in \Omega \quad (18.1)$$

where u is an unknown function and f is a given function in Ω.

S. E. Mikhailov (✉)
Brunel University London, Uxbridge, UK
e-mail: sergey.mikhailov@brunel.ac.uk

Z. W. Woldemicheal
Addis Ababa University, Addis Ababa, Ethiopia
e-mail: zenebe.wogderesegn@aau.edu.et

© Springer Nature Switzerland AG 2019
C. Constanda, P. Harris (eds.), *Integral Methods in Science and Engineering*,
https://doi.org/10.1007/978-3-030-16077-7_18

In what follows $\mathscr{D}(\Omega) := C^{\infty}_{comp}(\Omega)$ denotes the space of Schwartz test functions, $H^s(\Omega) = H^s_2(\Omega)$, $H^s(\partial\Omega) = H^s_2(\partial\Omega)$ are the Bessel potential spaces, where $s \in \mathbb{R}$ (see, e.g., [LiMa72, Mc00]). We recall that H^s coincide with the Sobolev-Slobodetski spaces W^s_2 for any non-negative s. We denote by $\widetilde{H}^s(\Omega)$ the subspace of $H^s(\mathbb{R}^3)$,

$$\widetilde{H}^s(\Omega) := \{g : g \in H^s(\mathbb{R}^3), \text{ supp } g \subset \overline{\Omega}\}.$$

And the space $H^s(\Omega)$ denotes the space of restriction on Ω of distributions from $H^s(\mathbb{R}^3)$,

$$H^s(\Omega) = \{r_\Omega g : g \in H^s(\mathbb{R}^3)\}$$

where r_Ω denotes the restriction operator on Ω.

18.2 Co-normal Derivatives and the Boundary Value Problem

For $u \in H^1(\Omega)$, the partial differential operator A is understood in the sense of distributions,

$$\langle Au, v \rangle_\Omega := -\mathscr{E}(u, v) \quad \forall v \in \mathscr{D}(\Omega) \tag{18.2}$$

where

$$\mathscr{E}(u, v) := \int_\Omega a(x)\nabla u(x) \cdot \nabla v(x)dx$$

and the duality brackets $\langle g, \cdot \rangle_\Omega$ denote the value of a linear functional (distribution) g, extending the usual L_2 dual product.

Since the set $\mathscr{D}(\Omega)$ is dense in $\widetilde{H}^1(\Omega)$, formula (18.2) defines (cf. e.g. [Mi11, Section 3.1]) the continuous linear operator $A : H^1(\Omega) \to H^{-1}(\Omega) = [\widetilde{H}^1(\Omega)]^*$, where

$$\langle Au, v \rangle_\Omega := -\mathscr{E}(u, v) \quad \forall v \in \widetilde{H}^1(\Omega).$$

Let us also consider the different operator, $\check{A} : H^1(\Omega) \to \widetilde{H}^{-1}(\Omega) = [H^1(\Omega)]^*$

$$\langle \check{A}u, v \rangle_\Omega = -\mathscr{E}(u, v) = -\int_\Omega a(x)\nabla u(x) \cdot \nabla v(x)dx$$

$$= -\int_{\mathbb{R}^3} \mathring{E}[a\nabla u](x) \cdot \nabla V(x)dx$$

$$= \langle \nabla \cdot \mathring{E}[a\nabla u], V \rangle_{\mathbb{R}^3}$$
$$= \langle \nabla \cdot \mathring{E}[a\nabla u], v \rangle_{\Omega}, \quad \forall u \in H^1(\Omega), \ v \in H^1(\Omega),$$

(18.3)

which is evidently continuous and can be written as

$$\check{A}u := \nabla \cdot \mathring{E}[a\nabla u].$$

Here $V \in H^1(\mathbb{R}^3)$ is such that $r_\Omega V = v$ and $\mathring{E}$ denotes the operator of extension of functions, defined in Ω, by zero outside Ω in $\mathbb{R}^3$. For any $u \in H^1(\Omega)$, the functional $\check{A}u$ belongs to $\tilde{H}^{-1}(\Omega)$ and is an extension of the functional $Au \in H^{-1}(\Omega)$ which domain is thus extended from $\tilde{H}^1(\Omega)$ to the domain $H^1(\Omega)$ for $\check{A}u$.

From the trace theorem (see, e.g., [LiMa72, DaLi90, Mc00]) for $u \in H^1(\Omega)$, it follows that $\gamma^+ u \in H^{\frac{1}{2}}(\partial\Omega)$, where $\gamma^+ := \gamma^+_{\partial\Omega}$ is the trace operator on $\partial\Omega$ from Ω. Let also $\gamma^{-1} : H^{\frac{1}{2}}(\partial\Omega) \longrightarrow H^1(\Omega)$ denote a (non-unique) continuous right inverse to the trace operator γ^+, i.e., $\gamma^+ \gamma^{-1} w = w$ for any $w \in H^{\frac{1}{2}}(\partial\Omega)$, and $(\gamma^{-1})^* : \tilde{H}^{-1}(\Omega) \longrightarrow H^{-\frac{1}{2}}(\partial\Omega)$ is the continuous operator dual to $\gamma^{-1} : H^{\frac{1}{2}}(\partial\Omega) \longrightarrow H^1(\Omega)$, i.e., $\langle (\gamma^{-1})^* \tilde{f}, w \rangle_{\partial\Omega} := \langle \tilde{f}, \gamma^{-1} w \rangle_\Omega$ for any $\tilde{f} \in \tilde{H}^{-1}(\Omega)$ and $w \in H^{\frac{1}{2}}(\partial\Omega)$.

For $u \in H^2(\Omega)$, we can denote by T^{c+} the corresponding classical (strong) co-normal derivative operator on $\partial\Omega$ in the sense of traces,

$$T^{c+} u(x) := \sum_{i=1}^{3} a(x) n_i^+(x) \gamma^+ \left(\frac{\partial u(x)}{\partial x_i} \right) = a(x) \gamma^+ \left(\frac{\partial u(x)}{\partial n(x)} \right),$$

where $n^+(x)$ is the outward (to Ω) unit normal vectors at the point $x \in \partial\Omega$. However the classical co-normal derivative operator is generally not well defined if $u \in H^1(\Omega)$ (cf. an example in [Mi15, Appendix A]).

Definition 1 Let $u \in H^1(\Omega)$ and $\tilde{f} \in \tilde{H}^{-1}(\Omega)$. Then the *formal co-normal derivative* $T^+(\tilde{f}, u) \in H^{-\frac{1}{2}}(\partial\Omega)$ is defined as

$$\langle T^+(\tilde{f}, u), w \rangle_{\partial\Omega} := \langle \tilde{f}, \gamma^{-1} w \rangle_\Omega + \mathscr{E}(u, \gamma^{-1} w)$$
$$= \langle \tilde{f} - \check{A}u, \gamma^{-1} w \rangle_\Omega \quad \forall w \in H^{\frac{1}{2}}(\partial\Omega).$$

that is,

$$T^+(\tilde{f}, u) := (\gamma^{-1})^*(\tilde{f} - \check{A}u) = (\gamma^{-1})^* \tilde{f} + T^+(0, u). \qquad (18.4)$$

If, in addition, $Au = r_\Omega \tilde{f}$ in Ω, then $T^+(\tilde{f}, u)$ becomes the *generalised co-normal derivative*, cf. Definition 3.1 in [Mi11] and Definition 5.2 in [Mi13]. Note that the formal co-normal derivative generally depends on the chosen right inverse, γ^{-1}, of the trace operator; however, the generalised co-normal derivative does not. Some other properties of the generalised conormal derivative also hold true for the formal conormal derivative. In particular, similarly to [Mc00, Lemma 4.3], [Mi11, Theorem 5.3], we have the estimate

$$\left\| T^+(\tilde{f}, u) \right\|_{H^{-\frac{1}{2}}(\partial\Omega)} \leqslant C_1 \|u\|_{H^1(\Omega)} + C_2 \|\tilde{f}\|_{\tilde{H}^{-1}(\Omega)}.$$

The first Green identity holds in the following form for $u \in H^1(\Omega)$ such that $Au = r_\Omega \tilde{f}$ in Ω for some $\tilde{f} \in \tilde{H}^{-1}(\Omega)$,

$$\langle T^+(\tilde{f}, u), \gamma^+ v \rangle_{\partial\Omega} = \langle \tilde{f}, v \rangle_\Omega + \mathscr{E}(u, v) = \langle \tilde{f} - \check{A}u, v \rangle_\Omega \quad \forall v \in H^1(\Omega). \tag{18.5}$$

As follows from Definition 1, the formal and generalised co-normal derivatives are non-linear with respect to u for a fixed $\tilde{f}$, but still linear with respect to the couple $(\tilde{f}, u)$.

We will consider the following Dirichlet boundary value problem:
Find a function $u \in H^1(\Omega)$ satisfying the conditions

$$Au = f \quad \text{in } \Omega, \tag{18.6}$$

$$\gamma^+ u = \varphi_0 \quad \text{on } \partial\Omega, \tag{18.7}$$

where $f \in H^{-1}(\Omega)$ and $\varphi_0 \in H^{\frac{1}{2}}(\partial\Omega)$.

Equation (18.6) is understood in the distributional sense (18.2) and the Dirichlet boundary condition (18.7) in the trace sense.

The following assertion is well-known and can be proved, e.g., using variational settings and the Lax-Milgram lemma.

Theorem 1 *The Dirichlet problem (18.6)–(18.7) is uniquely solvable in $H^1(\Omega)$. The solution is $u = (A^D)^{-1}(f, \varphi_0)^T$ where the inverse operator $(A^D)^{-1}$: $H^{\frac{1}{2}}(\partial\Omega) \times H^{-1}(\Omega) \longrightarrow H^1(\Omega)$ to the left-hand side operator, $A^D : H^1(\Omega) \longrightarrow H^{\frac{1}{2}}(\partial\Omega) \times H^{-1}(\Omega)$, of the Dirichlet problem (18.6)–(18.7) is continuous.*

18.3 Parametrix and Potential Type Operators

We will say, a function $P(x, y)$ of two variables $x, y \in \Omega$ is a parametrix (the Levi function) for the operator $A(x, \partial_x)$ in $\mathbb{R}^3$ if (see, e.g., [Mi02, Mi70, Po98a, Po98b])

$$A(x, \partial_x)P(x, y) = \delta(x - y) + R(x, y), \tag{18.8}$$

where $\delta(.)$ is the Dirac distribution and $R(x, y)$ possesses a weak (integrable) singularity at $x = y$, i.e.,

$$R(x, y) = \mathcal{O}(|x - y|^{-\varkappa}) \text{ with } \varkappa < 3. \tag{18.9}$$

It is easy to see that for the operator $A(x, \partial_x)$ given by the left-hand side in (18.1), the function

$$P(x, y) = \frac{-1}{4\pi a(y)|x - y|}, \qquad x, y \in \mathbb{R}^3, \tag{18.10}$$

is a parametrix and the corresponding remainder function is

$$R(x, y) = \sum_{i=1}^{3} \frac{x_i - y_i}{4\pi a(y)|x - y|^3} \frac{\partial a(x)}{\partial x_i}, \qquad x, y \in \mathbb{R}^3, \tag{18.11}$$

and satisfies estimate (18.9) with $\varkappa = 2$, due to smoothness of the function $a(x)$. Evidently, the parametrix $P(x, y)$ given by (18.10) is related with the fundamental solution to the operator $A(y, \partial_x) := a(y)\Delta(\partial_x)$ with the "frozen" coefficient $a(x) = a(y)$ and $A(y, \partial_x)P(x, y) = \delta(x - y)$.

Let $a \in C^{\infty}(\mathbb{R}^3)$ and $a > 0$ a.e. in $\mathbb{R}^3$. For scalar functions g, for which the integrals have sense, the parametrix-based volume potential operator and the remainder potential operator, corresponding to parametrix (18.10) and remainder (18.11) are defined as

$$\mathbf{P}g(y) := \int_{\mathbb{R}^3} P(x, y)g(x)dx, \ y \in \mathbb{R}^3$$

$$\mathscr{P}g(y) := \int_{\Omega} P(x, y)g(x)dx, \ y \in \Omega$$

$$\mathscr{R}g(y) := \int_{\Omega} R(x, y)g(x)dx, \ y \in \Omega$$

The single and double layer surface potential operators are defined as

$$Vg(y) := -\int_{\partial\Omega} P(x, y)g(x)dS_x, \ y \notin \partial\Omega \tag{18.12}$$

$$Wg(y) := -\int_{\partial\Omega} [T(x, n(x), \partial_x)P(x, y)]g(x)dS_x, \ y \notin \partial\Omega \tag{18.13}$$

where the integrals are understood in the distributional sense if g is not integrable.

The corresponding boundary integral (pseudodifferential) operators of direct surface values of the single layer potential $\mathscr{V}$ and of the double layer potential $\mathscr{W}$, and the co-normal derivatives of the single layer potential $\mathscr{W}'$ and of the double layer potential $\mathscr{L}^+$, for $y \in \partial\Omega$ are

$$\mathscr{V}g(y) := -\int_{\partial\Omega} P(x, y)g(x)dS_x, \tag{18.14}$$

$$\mathscr{W}g(y) := -\int_{\partial\Omega} [T_x^+ P(x, y)]g(x)dS_x \tag{18.15}$$

$$\mathscr{W}'g(y) := -\int_{\partial\Omega} [T_y^+ P(x, y)]g(x)dS_x, \tag{18.16}$$

$$\mathscr{L}^+ g(y) := T^+ Wg(y). \tag{18.17}$$

When integrals in (18.12)–(18.16) are not well defined, they can be understood, e.g., as pseudo-differential operators or dual forms.

From definitions (18.10), (18.12), (18.13) one can obtain representations of the parametrix-based potential operators in terms of their counterparts for $a = 1$ (i.e. associated with the Laplace operator Δ), which we equip with the subscript Δ, cf. [CMN09],

$$\mathbf{P}g = \frac{1}{a}\mathbf{P}_\Delta g, \quad \mathscr{P}g = \frac{1}{a}\mathscr{P}_\Delta g, \quad \mathscr{R}g = -\frac{1}{a}\sum_{i=1}^{3} \partial_i \mathscr{P}_\Delta [g(\partial_i a)], \tag{18.18}$$

$$Vg = \frac{1}{a}V_\Delta g, \quad Wg = \frac{1}{a}W_\Delta(ag), \tag{18.19}$$

$$\mathscr{V}g = \frac{1}{a}\mathscr{V}_\Delta g, \quad \mathscr{W}g = \frac{1}{a}\mathscr{W}_\Delta(ag), \tag{18.20}$$

$$\mathscr{W}'g = \mathscr{W}'_\Delta g + \left[a\frac{\partial}{\partial n}\left(\frac{1}{a}\right)\right]\mathscr{V}_\Delta g, \tag{18.21}$$

$$\mathscr{L}^\pm g = \mathscr{L}_\Delta(ag) + \left[a\frac{\partial}{\partial n}\left(\frac{1}{a}\right)\right]\mathscr{W}_\Delta^\pm(ag). \tag{18.22}$$

Hence

$$\Delta(aVg) = 0, \quad \Delta(aWg) = 0 \text{ in } \Omega, \quad \forall g \in H^s(\partial\Omega) \; \forall s \in \mathbb{R},$$

$$\Delta(a\mathscr{P}g) = g \text{ in } \Omega, \quad \forall g \in \widetilde{H}^s(\Omega) \; \forall s \in \mathbb{R}$$

The jump relations as well as mapping properties of potentials and operators are well known for the case $a = \text{const}$. They were extended to the case of variable coefficient $a(x)$ in [CMN09].

18.4   The Third Green Identity and Integral Relations

For $u \in H^1(\Omega)$ and $v(x) = P(x, y)$, where the parametrix $P(x, y)$ is given by (18.10), the following *generalised third Green identity* can be obtained from (18.5), (18.3), (18.8), see [Mi15, Theorem 4.1], [Mi18, Theorem 4.1],

$$u + \mathcal{R}u + W\gamma^+ u = \mathcal{P}\check{A}u \quad \text{in} \quad \Omega,$$

where

$$\mathcal{P}\check{A}u(y) := \langle \check{A}u, P(., y) \rangle_\Omega = -\mathcal{E}(u, P(., y)) = -\int_\Omega a(x)\nabla u(x) \cdot \nabla_x P(x, y)dx.$$

If $r_\Omega Au = \tilde{f}$ in Ω, where $\tilde{f} \in \tilde{H}^{-1}(\Omega)$, then the generalised third Green identity takes the following form,

$$u + \mathcal{R}u - VT^+(\tilde{f}, u) + W\gamma^+ u = \mathcal{P}\tilde{f} \quad \text{in} \quad \Omega, \tag{18.23}$$

For some functions $\tilde{f}, \Psi$ and Φ, let us consider a more general "indirect" integral relation associated with Eq. (18.23),

$$u + \mathcal{R}u - V\Psi + W\Phi = \mathcal{P}\tilde{f} \quad \text{in } \Omega \tag{18.24}$$

The following statement proved in [Mi15, Lemma 4.2] (see also [Mi18, Lemma 4.2] for Lipschitz domains and more general spaces and coefficients) extends Lemma 4.1 from [CMN09], where the corresponding assertion was proved for $\tilde{f} \in L_2(\Omega)$.

Lemma 1 *Let* $u \in H^1(\Omega), \Psi \in H^{-\frac{1}{2}}(\partial\Omega), \Phi \in H^{\frac{1}{2}}(\partial\Omega),$ *and* $\tilde{f} \in \tilde{H}^{-1}(\Omega)$ *satisfy Eq. (18.24). Then*

$$Au = r_\Omega \tilde{f} \text{ in } \Omega, \tag{18.25}$$

$$r_\Omega V(\Psi - T^+(\tilde{f}, u)) - r_\Omega W(\Phi - \gamma^+ u) = 0 \quad \text{in } \Omega. \tag{18.26}$$

The following statement was proved in [CMN09, Lemma 4.2].

Lemma 2

(i) *If* $\Psi^* \in H^{-\frac{1}{2}}(\partial\Omega)$ *and* $r_\Omega V\Psi^* = 0$ *in* Ω, *then* $\Psi^* = 0$.
(ii) *If* $\Phi^* \in H^{\frac{1}{2}}(\partial\Omega)$ *and* $r_\Omega W\Phi^* = 0$ *in* Ω, *then* $\Phi^* = 0$.

Let us now generalise Theorem 5.1 from [Mi06] to the right-hand side $\tilde{f} \in \tilde{H}^{-1}(\Omega)$.

Theorem 2 *Let $\tilde{f} \in \tilde{H}^{-1}(\Omega)$. A function $u \in H^1(\Omega)$ is a solution of PDE $Au = r_\Omega \tilde{f}$ in Ω if and only if it is a solution of boundary-domain integro-differential equation* (18.23).

Proof If $u \in H^1(\Omega)$ solves PDE $Au = r_\Omega \tilde{f}$ in Ω, then it satisfies (18.23). On the other hand, if $u \in H^1(\Omega)$ solves boundary-domain integro-differential Eq. (18.23), then using Lemma 1 with $\Psi = T^+(\tilde{f}, u)$ and $\Phi = \gamma^+ u$, we obtain that u satisfies (18.25), which completes the proof. $\quad\square$

18.5 United Boundary-Domain Integro-Differential Equations

Let us consider reduction of the Dirichlet problem (18.6)–(18.7) with $f \in H^{-1}(\Omega)$, for $u \in H^1(\Omega)$, to a united boundary-domain integro-differential problem or to a united boundary-domain integro-differential equation. Formulations for the mixed problem for $u \in H^{1,0}(\Omega; \Delta)$ with $f \in L_2(\Omega)$ were introduced and analysed in [Mi06]. Let $\tilde{f} \in \tilde{H}^{-1}(\Omega)$ be an extension of $f \in H^{-1}(\Omega)$ (i.e., $f = r_\Omega \tilde{f}$), which always exists, see [Mi11, Lemma 2.15 and Theorem 2.16].

18.5.1 United Boundary-Domain Integro-Differential Problem

Supplementing BDIDE (18.23) in the domain Ω, where we take into account (18.4), with the original Dirichlet condition (18.7) on the boundary $\partial\Omega$, we arrive at the following united boundary-domain integro-differential problem, BDIDP, for u in Ω,

$$\mathcal{G}^D u = \mathcal{F}^D \tag{18.27}$$

where

$$\mathcal{G}^D u = \begin{bmatrix} u + \mathcal{R}u - VT^+(0, u) + W\gamma^+ u \\ \gamma^+ u \end{bmatrix}, \quad \mathcal{F}^D = \begin{bmatrix} \mathcal{P}\tilde{f} + V(\gamma^{-1})^* \tilde{f} \\ \varphi_0 \end{bmatrix} \tag{18.28}$$

and we invoked representation (18.4). Note also that by (18.12),

$$V(\gamma^{-1})^* \tilde{f}(y) = -\langle \gamma P(\cdot, y), (\gamma^{-1})^* \tilde{f}\rangle_{\partial\Omega} = -\langle \gamma^{-1}\gamma P(\cdot, y), \tilde{f}\rangle_\Omega$$
$$= -\langle P(\cdot, y), \gamma^*(\gamma^{-1})^* \tilde{f}\rangle_{\partial\Omega} = -\mathcal{P}\gamma^*(\gamma^{-1})^* \tilde{f}.$$

BDIDP (18.27) is equivalent to the Dirichlet boundary value problem (18.6)–(18.7) in Ω, in the following sense.

Theorem 3 *Let $f \in H^{-1}(\Omega)$, $\varphi_0 \in H^{\frac{1}{2}}(\partial\Omega)$, and $\tilde{f} \in \tilde{H}^{-1}(\Omega)$ be such that $r_\Omega \tilde{f} = f$. A function $u \in H^1(\Omega)$ solves the Dirichlet BVP (18.6)–(18.7) in Ω if and only if u solves BDIDP (18.27). Such solution does exist and is unique.*

Proof A solution of BVP (18.6)–(18.7) does exist and is unique due to Theorem 1 and provides a solution to BDIDP (18.27) due to Theorem 2. On the other hand, due to the same Theorem 2, any solution of BDIDP (18.27) satisfies also BVP (18.6)–(18.7), which is unique. □

Due to the mapping properties of operators V, W, $\mathscr{P}$ and $\mathscr{R}$, cf. [CMN09], we have $\mathscr{F}^D \in H^1(\Omega) \times H^{\frac{1}{2}}(\partial\Omega)$ and the operator $\mathscr{G}^D : H^1(\Omega) \to H^1(\Omega) \times H^{\frac{1}{2}}(\partial\Omega)$ is continuous. It is also injective due to Theorem 3.

18.5.2 United Boundary-Domain Integro-Differential Equation

Substituting the Dirichlet boundary condition (18.7) and relation (18.4) into (18.23), we arrive at the following boundary-domain integro-differential equation, BDIDE, for $u \in H^1(\Omega)$:

$$\mathscr{G}^2 u := u + \mathscr{R}u - VT^+(0, u) = \mathscr{F}^2 \quad in \quad \Omega \tag{18.29}$$

where

$$\mathscr{F}^2 = \mathscr{P}\tilde{f} + V(\gamma^{-1})^*\tilde{f} - W\varphi_0 \tag{18.30}$$

Let us prove the equivalence of BDIDE (18.29) to BVP (18.6)–(18.7).

Theorem 4 *Let $f \in H^{-1}(\Omega)$, $\varphi_0 \in H^{\frac{1}{2}}(\partial\Omega)$, and $\tilde{f} \in \tilde{H}^{-1}(\Omega)$ be such that $r_\Omega \tilde{f} = f$. A function $u \in H^1(\Omega)$ solves the Dirichlet BVP (18.6)–(18.7) in Ω if and only if u solves BDIDE (18.29) with right-hand side (18.30). Such solution does exist and is unique.*

Proof Any solution of BVP (18.6)–(18.7) solves BDIDE (18.29) due to the third Green formula (18.23). On the other hand, if u is a solution of BDIDE (18.29) then Lemma 1 implies that u satisfies Eq. (18.6) and $r_\Omega W(\varphi_0 - \gamma^+u) = 0$ in Ω. Lemma 2 (ii) then implies that $\varphi_0 - \gamma^+u = 0$, i.e., the Dirichlet boundary condition (18.7) is satisfied. Thus any solution of BDIDE (18.29) satisfies BVP (18.6)–(18.7). The unique solvability of BVP (18.6)–(18.7) and hence of BDIDE (18.29) is implied by Theorem 1. □

The mapping properties of operators V, W, $\mathscr{P}$ and $\mathscr{R}$ imply the membership $\mathscr{F}^2 \in H^1(\Omega)$ and continuity of the operator $\mathscr{G}^2$ in $H^1(\Omega)$, while Theorem 4 implies its injectivity.

Note that Theorems 3 and 4 imply that the non-uniqueness of extension of $f \in H^{-1}(\Omega)$ to $\tilde{f} \in \tilde{H}^{-1}(\Omega)$ and the non-uniqueness of the right inverse to the trace operator, γ^{-1}, involved in the definition of $T^{+}(\tilde{f}, u)$, do not compromise the uniqueness of solutions u of BDIDP (18.27) and BDIDE (18.29).

18.6   Conclusion

A Dirichlet BVP for a variable-coefficient second order PDE with general right-hand side function from $H^{-1}(\Omega)$ and with the Dirichlet data from the space $H^{\frac{1}{2}}(\partial \Omega)$ was considered in this paper. It was shown that the BVP can be equivalently reduced to a united boundary-domain integro-differential problem, or to a united boundary-domain integro-differential equation of the second kind.

Similarly one can also consider the united BDIEs for the Neumann and mixed problems in interior and exterior domains for the general right-hand side as well as the united versions of other BDIEs formulated and analysed in [AyMi11, ADM17, Mi02, Mi06].

Acknowledgements The first author acknowledges the support of the grant EP/M013545/1: 'Mathematical Analysis of Boundary-Domain Integral Equations for Nonlinear PDEs' from the EPSRC, UK. The second author's work on this paper was supported by ISP, Sweden. He would like also to thank his PhD advisor Dr. Tsegaye Gedif Ayele for discussing the results.

References

[AyMi11] Ayele, T.G. and Mikhailov, S.E.: Analysis of two-operator boundary-domain integral equations for variable-coefficient mixed BVP. *Eurasian Math. J.,* **2**(3): 20–41 (2011).

[ADM17] Ayele, T.G., Dufera, T.T. and Mikhailov, S.E.: Analysis of boundary-domain integral equations for variable-coefficient Neumann BVP in 2D. In: *Integral Methods in Science and Engineering, Vol.1 Theoretical Techniques. C. Constanda et al, eds*, Chapter 3, Springer (Birkhäuser), Boston (2017).

[CMN09] Chkadua, O., Mikhailov, S.E. and Natroshvili, D.: Analysis of direct boundary-domain integral equations for a mixed BVP with variable coefficient, I: Equivalence and Invertibility. *J. Integral Equat. and Appl.* **21**: 499–543 (2009).

[DaLi90] Dautray, R. and Lions, J.: *Mathematical Analysis and Numerical Methods for Science and Technology. Integral Equations and Numerical Methods*, Vol. 4. Springer: Berlin, Heidelberg-New York (1990).

[LiMa72] Lions, J.L., Magenes, E.: *Non-Homogeneous Boundary Value Problems and Applications*, Vol. 1. Springer, Berlin, Heidelberg, New York (1972).

[Mc00] McLean, W.: *Strongly Elliptic Systems and Boundary Integral Equations*. Cambridge University Press, UK (2000).

[Mi02] Mikhailov, S.E.: Localized boundary-domain integral formulation for problems with variable coefficients. *Internat. J. of Engineering Analysis with Boundary Elements,* **26**: 681–690 (2002).

[Mi06] Mikhailov, S.E.: Analysis of united boundary domain integral and integro differential equations for a mixed BVP with variable coefficients. *Math. Meth. Appl. Sci.* **29**: 715–739 (2006).

[Mi11] Mikhailov, S.E.: Traces, extensions and co-normal derivatives for elliptic systems on Lipschitz domains. *J. Math. Analysis and Appl.*, **378**: 324–342 (2011).

[Mi13] Mikhailov, S.E.: Solution regularity and co-normal derivatives for elliptic systems with non-smooth coefficients on Lipschitz domains. *J. Math. Anal. Appl.*, **400**: 48–67 (2013).

[Mi15] Mikhailov, S.E.: Analysis of segregated Boundary-Domain Integral Equations for Variable-Coefficient Dirichlet and Neumann Problems with General Data. *ArXiv*, 1509.03501:1–32 (2015).

[Mi18] Mikhailov, S.E.: Analysis of segregated boundary-domain integral equations for BVPs with non-smooth coefficients on Lipschitz domains. Boundary Value Problems, **2018**:87 (2018) 1–52.

[Mi70] Miranda, C.: *Partial differential equations of elliptic type,* Second revised edition. Springer-Verlag, Berlin (1970).

[Po98a] Pomp, A.: *The boundary-domain integral method for elliptic systems. With applications in shells.* Lecture notes in Mathematics. Vol. 1683, Springer, Berlin-Heidelberg- New York (1998).

[Po98b] Pomp, A.: Levi functions for linear elliptic systems with variable coefficients including shell equations. *Comput. Mech.,* **22**: 93–99 (1998).

Chapter 19
Rescaling and Trace Operators in Fractional Sobolev Spaces on Bounded Lipschitz Domains with Periodic Structure

Sergey E. Mikhailov, Paolo Musolino, and Julia Orlik

19.1 Introduction

We develop tools, which can be useful for elliptic boundary value problems on domains with a periodic structure with holes involving some linear or non-linear Robin-type conditions on the oscillating interface [GrEtAl12, KhEtAl17, GaEtAl16], or contact problems (see [GaMe18, GrOr18]).

This paper considers rescaling of functions from the Bessel potential, Riesz potential, and Sobolev–Slobodetskii spaces on the boundary or in the domain and also rescaling of the boundary trace operator.

Denote by Ω a bounded domain in $\mathbb{R}^n$ with Lipschitz boundary. Let $Y := (0, 1)^n$ be the reference cell. We denote by T a hole, that is, an open set, which closure is strictly included in Y and let $Y^* := Y \setminus \overline{T}$ (see Fig. 19.1). Let ∂T be the Lipschitz boundary of T and ν be the outward to Y^* unit normal vector on the boundary ∂T. Recall, e.g., from [CiEtAl12] that in the periodic setting, every point $z \in \mathbb{R}^n$ can be written as $z = [z] + \{z\}$, $[z] \in \mathbb{Z}^n$, $\{z\} \in Y$. Here the integer function $[\cdot]$ for a vector means the floor function $\lfloor \cdot \rfloor$ for each of its components. Denote

$$\Xi_\varepsilon = \left\{\xi \in \mathbb{Z}^n \mid \varepsilon\xi + \varepsilon Y \subset \Omega\right\}, \quad \widehat{\Omega}_\varepsilon = \text{interior}\left\{\bigcup_{\xi \in \Xi_\varepsilon} \left(\varepsilon\xi + \varepsilon\overline{Y}\right)\right\}, \quad \Lambda_\varepsilon = \Omega \setminus \overline{\widehat{\Omega}_\varepsilon},$$

i.e., the set Λ_ε contains the parts of the cells intersecting the boundary $\partial\Omega$.

S. E. Mikhailov (✉)
Brunel University London, Uxbridge, UK
e-mail: sergey.mikhailov@brunel.ac.uk

P. Musolino
University of Padua, Padua, Italy
e-mail: musolino@math.unipd.it

J. Orlik
Fraunhofer ITWM, Kaiserslautern, Germany
e-mail: orlik@itwm.fhg.de

© Springer Nature Switzerland AG 2019   237
C. Constanda, P. Harris (eds.), *Integral Methods in Science and Engineering*,
https://doi.org/10.1007/978-3-030-16077-7_19

Fig. 19.1 Bounded domain
with periodically distributed
holes

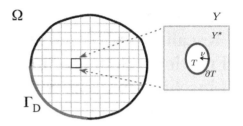

Let us introduce the notations for the unions of all holes in the interior cells,
$T_\varepsilon := \left\{ x \in \widehat{\Omega}_\varepsilon \;\middle|\; \left\{ \frac{x}{\varepsilon} \right\} \in T \right\}$, for the hole boundaries, $\partial T_\varepsilon := \left\{ x \in \widehat{\Omega}_\varepsilon \;\middle|\; \left\{ \frac{x}{\varepsilon} \right\} \in \partial T \right\}$, in $\widehat{\Omega}_\varepsilon$ and for the remaining part, $\widehat{\Omega}_\varepsilon^* = \widehat{\Omega}_\varepsilon \setminus \overline{T_\varepsilon}$. Let also $\Omega_\varepsilon^* = \Omega \setminus \overline{T_\varepsilon}$.

19.2 Function Spaces

For an arbitrary non-empty open subspace Ω' of $\mathbb{R}^m$, let $W_2^s(\Omega')$, $s \geq 0$ denote the Sobolev–Slobodetskii space, cf. e.g., [Mc00]. If s is an integer, the space coincides with the Sobolev space and

$$\|u\|_{W_2^s(\Omega')}^2 := \sum_{|\alpha| \leq s} \int_{\Omega'} |\partial^\alpha u(x)|^2 dx.$$

If s is not an integer,

$$\|u\|_{W_2^s(\Omega')}^2 := \|u\|_{W_2^{[s]}(\Omega')}^2 + \|u\|_{W_2^s(\Omega')}'^2,$$

and the Slobodetskii seminorm is defined as

$$\|u\|_{W_2^s(\Omega')}'^2 := \sum_{|\alpha|=[s]} \int_{\Omega'} \int_{\Omega'} \frac{|\partial^\alpha u(x) - \partial^\alpha u(y)|^2}{|x-y|^{m+2\mu}} dxdy, \quad \mu := s - [s].$$

$$(19.1)$$

Let $S(\mathbb{R}^m)$ be the Schwartz space of all complex-valued, rapidly decreasing, infinitely differentiable functions on $\mathbb{R}^m$. Let $S^*(\mathbb{R}^m)$ denote the space of sequentially continuous linear functionals on $S(\mathbb{R}^m)$ (temperate distributions). Let us denote by

$$\hat{u}(\eta) \equiv \mathscr{F}u(\eta) := \int_{\mathbb{R}^m} u(x) e^{-i2\pi \eta \cdot x} \, dx,$$

$$u(x) \equiv \mathscr{F}^{-1}\hat{u}(x) := \int_{\mathbb{R}^m} \hat{u}(\eta) e^{i2\pi \eta \cdot x} \, d\eta,$$

the direct and inverse Fourier transforms, respectively. These definitions, applicable to functions from $L_1(\mathbb{R}^m)$, are easily extended to $S^*(\mathbb{R}^m)$, see, e.g., [Mc00, p. 72].

Let us denote $\rho(\eta) := (1 + |\eta|^2)^{1/2}$. For $s \in \mathbb{R}$, we define the Bessel potential operator of order s, $J^s : S(\mathbb{R}^m) \to S(\mathbb{R}^m)$,

$$J^s u(x) := \mathscr{F}^{-1}(\rho^s \mathscr{F} u)(x) = \int_{\mathbb{R}^m} (1 + |\eta|^2)^{s/2} \hat{u}(\eta) e^{i2\pi \eta \cdot x} d\eta \quad \text{for } x \in \mathbb{R}^m,$$

which is extended, in the distribution sense, to the operator $J^s : S^*(\mathbb{R}^m) \to S^*(\mathbb{R}^m)$.

Let $s \in \mathbb{R}$, $H^s(\mathbb{R}^m) = \{u \in S^*(\mathbb{R}^m) : J^s u(x) \in L_2(\mathbb{R}^m)\}$ denote the Bessel potential space equipped with the norm $\|u\|_{H^s(\mathbb{R}^m)} = \|J^s u\|_{L_2(\mathbb{R}^m)} = \|\rho^s \hat{u}\|_{L_2(\mathbb{R}^m)}$, cf., e.g., [Mc00, p. 75–76].

Similarly, let us define the Riesz potential operator

$$\mathscr{I}^s u(x) := \mathscr{F}^{-1}(|\eta|^s \mathscr{F} u)(x) := \int_{\mathbb{R}^m} |\eta|^s \hat{u}(\eta) e^{i2\pi \eta \cdot x} d\eta \quad \text{for } x \in \mathbb{R}^m.$$

Let $h^s(\mathbb{R}^m)$ denote the Riesz potential space, i.e., the completion in the norm $\|u\|_{h^s(\mathbb{R}^m)} = \|\mathscr{I}^s u\|_{L_2(\mathbb{R}^m)} = \||\eta|^s \hat{u}\|_{L_2(\mathbb{R}^m)}$ of the space of infinitely smooth functions having compact supports in $\mathbb{R}^m$.

If $s > 0$, then $H^s(\mathbb{R}^m) \subset h^s(\mathbb{R}^m)$ and $\|u\|_{h^s(\mathbb{R}^m)}$ is equivalent to the Sobolev–Slobodetskii semi-norm $\|u\|'_{W_2^s(\mathbb{R}^m)}$, see, e.g., Theorem 4 in [Ma11, Section 10.1.2]. Particularly, if $0 < s < 1$, then

$$\|u\|'^2_{W_2^s(\mathbb{R}^m)} = a_{s,m} \|u\|^2_{h^s(\mathbb{R}^m)}, \tag{19.2}$$

where $a_{s,m}$ is a number depending only on s and m, which is finite and positive for any $s \in (0, 1)$, see, e.g., [Mc00, Lemma 3.15], hence

$$\|u\|^2_{W_2^s(\mathbb{R}^m)} = \|u\|^2_{L_2(\mathbb{R}^m)} + \|u\|'^2_{W_2^s(\mathbb{R}^m)} = \|u\|^2_{L_2(\mathbb{R}^m)} + a_{s,m} \|u\|^2_{h^s(\mathbb{R}^m)}. \tag{19.3}$$

On the other hand, from the inequality

$$2^{s-1}(1 + \psi^s) \le (1 + \psi)^s \le 1 + \psi^s, \quad \forall \psi \in (0, \infty), \quad s \in [0, 1]$$

we obtain the following norm equivalence inequalities for any $s \in [0, 1]$,

$$2^{s-1}[\|u\|^2_{L_2(\mathbb{R}^m)} + \|u\|^2_{h^s(\mathbb{R}^m)}] \le \|u\|^2_{H^s(\mathbb{R}^m)} \le \|u\|^2_{L_2(\mathbb{R}^m)} + \|u\|^2_{h^s(\mathbb{R}^m)}. \tag{19.4}$$

For any non-empty open set $\Omega \subset \mathbb{R}^m$, $H^s(\Omega) := \{u = U|_\Omega \text{ for some } U \in H^s(\mathbb{R}^m)\}$. This space is equipped with the norm $\|u\|_{H^s(\Omega)} = \inf_{U|_\Omega = u, U \in H^s(\mathbb{R}^m)} \|U\|_{H^s(\mathbb{R}^m)}$. Also for the space $h^s(\Omega)$ we define the norm in the similar way, $\|u\|_{h^s(\Omega)} = \inf_{U|_\Omega = u, U \in h^s(\mathbb{R}^m)} \|U\|_{h^s(\mathbb{R}^m)}$. Moreover, one can prove that $h^s(\Omega) = H^s(\Omega)$ if the domain Ω is bounded and $-m/2 < s < m/2$, cf. [Du77, Section 1.3].

We further follow the notations of [Mc00, p. 98] for the definition of Bessel potential spaces on Lipschitz manifolds. Let ∂T be a Lipschitz boundary, and for a partition of unity $\{\phi_j\}$, ∂T is locally a Lipschitz hypograph of some function ζ_j up to some rigid motion $\kappa_j \equiv \omega_j(\cdot - a_j)\colon \mathbb{R}^n \to \mathbb{R}^n$, where $a_j \in \mathbb{R}^n$ and ω_j is a rotation. Let $u_j := u\phi_j$. Then $u = \sum_j u_j$ and for $s \in [-1, 1]$ we have the following definition of the Bessel potential norm on the boundary,

$$\|u\|_{H^s(\partial T)}^2 \equiv \sum_j \|u_j(\kappa_j^{(-1)}(\cdot, \zeta_j(\cdot)))\sqrt{1 + |\nabla \zeta_j(\cdot)|^2}\,\|_{H^s(\mathbb{R}^{n-1})}^2, \tag{19.5}$$

We recall that H^s coincides with the Sobolev–Slobodetskii spaces W_2^s for any nonnegative s. Replacing H^s with h^s or W_2^s in (19.5), we arrive at definitions of the norms in $h^s(\partial T)$ and $W_2^s(\partial T)$, respectively, in terms of their counterparts on $\mathbb{R}^{n-1}$. The same argument is valid, of course, if we replace ∂T with ∂T_ε. This implies an extension of the equality (19.2) to T and T_ε,

$$\|u\|_{W_2^s(T)}^{\prime 2} = a_{s,n-1}\|u\|_{h^s(T)}^2, \qquad \|u\|_{W_2^s(T_\varepsilon)}^{\prime 2} = a_{s,n-1}\|u\|_{h^s(T_\varepsilon)}^2, \tag{19.6}$$

where $a_{s,n-1}$ is a number still depending only on s and $n-1$, which is finite and positive for any $s \in (0, 1)$.

Let $\widehat{v(\varepsilon^{-1}\cdot)}(\eta) := \int_{\mathbb{R}^m} v(\varepsilon^{-1}x)e^{-i2\pi\eta\cdot x}\,dx$, $\hat{v}(\varepsilon\eta) := \int_{\mathbb{R}^m} v(y)e^{-i2\pi\varepsilon\eta\cdot y}\,dy$.

Then, changing the variables, we evidently have

$$\widehat{v(\varepsilon^{-1}\cdot)}(\eta) = \varepsilon^m \hat{v}(\varepsilon\eta), \qquad \widehat{v(\varepsilon\cdot)}(\eta) = \varepsilon^{-m}\hat{v}(\varepsilon^{-1}\eta), \quad \forall \eta \in \mathbb{R}^m. \tag{19.7}$$

We will employ these relations for $m = n$ and $m = n - 1$.

Let $\varepsilon \in (0, \infty)$. If $\alpha \in \mathbb{R}$, and $v \in h^\alpha(\mathbb{R}^m)$, then the substitution $\bar{\eta} = \varepsilon\eta$ gives

$$\|v(\varepsilon^{-1}\cdot)\|_{h^\alpha(\mathbb{R}^m)}^2 = \int_{\mathbb{R}^m} |\widehat{v(\varepsilon^{-1}\cdot)}(\eta)|^2 |\eta|^{2\alpha}\,d\eta = \varepsilon^{2m}\int_{\mathbb{R}^m} |\hat{v}(\varepsilon\eta)|^2 |\eta|^{2\alpha}\,d\eta$$

$$= \varepsilon^{m-2\alpha}\int_{\mathbb{R}^m} |\hat{v}(\bar{\eta})|^2 |\bar{\eta}|^{2\alpha}\,d\bar{\eta} = \varepsilon^{m-2\alpha}\|v\|_{h^\alpha(\mathbb{R}^m)}^2. \tag{19.8}$$

Replacing ε with $1/\varepsilon$ we obtain $\|v(\varepsilon\cdot)\|_{h^\alpha(\mathbb{R}^m)}^2 = \varepsilon^{-m+2\alpha}\|v\|_{h^\alpha(\mathbb{R}^m)}^2$.

Definition 1 For $s \in \mathbb{R}$, let us introduce the following ε-dependent norm in the Bessel potential space $H^s(\mathbb{R}^m)$,

$$\|\phi\|_{H_\varepsilon^s(\mathbb{R}^m)}^2 := \int_{\mathbb{R}^m} [\rho(\varepsilon\eta)]^{2s} |\hat{\phi}(\eta)|^2\,d\eta,$$

where $\rho(\varepsilon\eta) = (1 + |\varepsilon\eta|^2)^{1/2}$, $\varepsilon \neq 0$.

For a domain $\Omega \subset \mathbb{R}^m$, this norm generates the corresponding ε-dependent norm in the space $H^s(\Omega)$, $\|\phi\|_{H_\varepsilon^s(\Omega)}^2 := \inf_{\Phi \in H^s(\mathbb{R}^m): r_\Omega \Phi = \phi} \|\Phi\|_{H_\varepsilon^s(\mathbb{R}^m)}^2$, $s \in \mathbb{R}$.

It is easy to verify that due to the first relation in (19.7),

$$\|\phi(\varepsilon^{-1}\cdot)\|^2_{H^s_\varepsilon(\mathbb{R}^m)} = \varepsilon^m \|\phi\|^2_{H^s(\mathbb{R}^m)}. \tag{19.9}$$

Recall that the volume of the unit cell Y is $|Y| = 1$. Let us provide the following two definitions and two propositions from [CiEtАl12].

Definition 2 Let $p \in [1, +\infty]$ and ϕ be Lebesgue-measurable on $\widehat{\Omega}^*_\varepsilon$ and extended by zero in $\Omega^*_\varepsilon \setminus \widehat{\Omega}^*_\varepsilon$. The unfolding operator $\mathcal{T}_\varepsilon$ from $L_p(\widehat{\Omega}^*_\varepsilon)$ into $L_p(\Omega \times Y^*)$ is defined by

$$\begin{cases} \mathcal{T}_\varepsilon(\phi)(x, y) = \phi(\varepsilon[x/\varepsilon] + \varepsilon y) & \text{for a.e. } (x, y) \in \widehat{\Omega}_\varepsilon \times Y^*, \\ \mathcal{T}_\varepsilon(\phi)(x, y) = 0 & \text{for a.e. } (x, y) \in \Lambda_\varepsilon \times Y^*. \end{cases}$$

Proposition 1 Let $p \in [1, +\infty]$.

(i) If $\phi \in L_p(\Omega^*_\varepsilon)$, then $\|\mathcal{T}_\varepsilon(\phi)\|_{L_p(\Omega \times Y^*)} = \|1_{\widehat{\Omega}^*_\varepsilon}\phi\|_{L_p(\Omega^*_\varepsilon)} \le \|\phi\|_{L_p(\Omega^*_\varepsilon)}$.
(ii) If $\phi \in W^1_p(\Omega^*_\varepsilon)$, then $\nabla_y \mathcal{T}_\varepsilon(\phi)(x, y) = \varepsilon \mathcal{T}_\varepsilon(\nabla\phi)(x, y)$ for a.e. $(x, y) \in \Omega \times Y^*$ and $\|\nabla_y \mathcal{T}_\varepsilon(\phi)\|_{L_p(\Omega \times Y^*)} = \varepsilon \|1_{\widehat{\Omega}^*_\varepsilon}\nabla\phi\|_{L_p(\Omega^*_\varepsilon)}$.

Here $1_{\widehat{\Omega}^*_\varepsilon}$ is the characteristic function of the set $\widehat{\Omega}^*_\varepsilon$.

Definition 3 Let $p \in [1, +\infty]$. The operator $\mathcal{T}^b_\varepsilon$ from $L_p(\partial T_\varepsilon)$ into $L_p(\Omega \times \partial T)$ is defined by

$$\begin{cases} \mathcal{T}^b_\varepsilon(\phi)(x, y) = \phi(\varepsilon[x/\varepsilon] + \varepsilon y) & \text{for a.e. } (x, y) \in \widehat{\Omega}_\varepsilon \times \partial T, \\ \mathcal{T}^b_\varepsilon(\phi)(x, y) = 0 & \text{for a.e. } (x, y) \in \Lambda_\varepsilon \times \partial T. \end{cases} \quad , \quad \forall \phi \in L_p(\partial T_\varepsilon).$$

Proposition 2 Let $p \in [1, +\infty]$ and $\phi \in L_p(\partial T_\varepsilon)$. Then

$$\int_{\Omega \times \partial T} \mathcal{T}^b_\varepsilon(\phi)(x, y)\, dx d\sigma_y = \varepsilon \int_{\partial T_\varepsilon} \phi(x)\, d\sigma_x,$$

$$\|\mathcal{T}^b_\varepsilon(\phi)\|_{L_p(\Omega \times \partial T)} = \varepsilon^{1/p}\|\phi\|_{L_p(\partial T_\varepsilon)}.$$

19.3 Rescaling Norms on Oscillating Lipschitz Manifold

Definition 4 Similar to (19.5), we will employ the following norms on $\partial T_\varepsilon = \bigcup_{\xi \in \Xi_\varepsilon}(\varepsilon\xi + \varepsilon\partial T)$,

$$\|u\|^2_{h^\alpha(\partial T_\varepsilon)} := \sum_{\xi \in \Xi_\varepsilon} \sum_j \|u_{\varepsilon,\xi,j}(\kappa^{(-1)}_{\varepsilon,\xi,j}(\cdot, \zeta_{\varepsilon,\xi,j}(\cdot)))\sqrt{1 + |\nabla\zeta_{\varepsilon,\xi,j}(\cdot)|^2} \,\|^2_{h^\alpha(\mathbb{R}^{n-1})},$$

$$\|u\|^2_{H^\alpha_\varepsilon(\partial T_\varepsilon)} := \sum_{\xi \in \Xi_\varepsilon} \sum_j \|u_{\varepsilon,\xi,j}(\kappa^{(-1)}_{\varepsilon,\xi,j}(\cdot, \zeta_{\varepsilon,\xi,j}(\cdot)))\sqrt{1 + |\nabla \zeta_{\varepsilon,\xi,j}(\cdot)|^2}\|^2_{H^\alpha_\varepsilon(\mathbb{R}^{n-1})}.$$

$$(19.10)$$

It is evident that the norms $\| \cdot \|_{H^s_\varepsilon(\partial T)}$ and $\| \cdot \|_{H^s(\partial T)}$ are equivalent if $\varepsilon \neq 0$, although with the equivalence inequality constants depending on ε.

In Definition 4, $u_{\varepsilon,j}(x) := u(x)\phi_{\varepsilon,j}(x)$, while $\phi_{\varepsilon,j}, \kappa_{\varepsilon,j}, \zeta_{\varepsilon,j}$ are some periodic families of partitions of unity, local rigid rotations and local Lipschitz hypographs. To this end, we can exploit the families $\phi_j, \kappa_j, \zeta_j$, associated with ∂T, and set

$$\phi_{\varepsilon,j}(x) := \phi_j(\{x/\varepsilon\}), \quad \zeta_{\varepsilon,j}(x) := \varepsilon \zeta_j(\{x/\varepsilon\}).$$

where, as before, $\{\cdot\}$ denotes the fractional part of the vector (components). Moreover, if $\kappa_j(x) = \omega_j(x - a_j)$, we also set $\kappa_{\varepsilon,j}(x) := \varepsilon \omega_j(\{x/\varepsilon\} - a_j)$. Note that

$$x = \kappa^{(-1)}_{\varepsilon,\xi,j}(\bar{x}) = \varepsilon\left[\omega^{(-1)}_j(\bar{x}/\varepsilon) + a_j\right] + \varepsilon\xi = \varepsilon\kappa^{(-1)}_j(\bar{x}/\varepsilon) + \varepsilon\xi.$$

As a consequence,

$$\kappa^{(-1)}_{\varepsilon,\xi,j}\left(\overline{x}', \zeta_{\varepsilon,j}(\overline{x}')\right) = \varepsilon\left[\kappa^{(-1)}_j\left(\overline{x}'/\varepsilon, \frac{1}{\varepsilon}\varepsilon\zeta_j(\overline{x}'/\varepsilon)\right) + \xi\right]$$

$$= \varepsilon\xi + \varepsilon\kappa^{(-1)}_j\left(\overline{x}'/\varepsilon, \zeta_j(\overline{x}'/\varepsilon)\right).$$

Moreover,

$$\phi_{\varepsilon,j}\left(\kappa^{(-1)}_{\varepsilon,\xi,j}\left(\overline{x}', \zeta_{\varepsilon,j}(\overline{x}')\right)\right) = \phi_{\varepsilon,j}\left(\varepsilon\xi + \varepsilon\kappa^{(-1)}_j\left(\overline{x}'/\varepsilon, \zeta_j(\overline{x}'/\varepsilon)\right)\right)$$

$$= \phi_j\left(\kappa^{(-1)}_j\left(\overline{x}'/\varepsilon, \zeta_j(\overline{x}'/\varepsilon)\right)\right).$$

Finally, $\sqrt{1 + |\nabla \zeta_{\varepsilon,j}(\overline{x}')|^2} = \sqrt{1 + \varepsilon^2 \frac{1}{\varepsilon^2}|\nabla_{\overline{y}'}\zeta_j(\overline{x}'/\varepsilon)|^2} = \sqrt{1 + |\nabla_{\overline{y}'}\zeta_j(\overline{x}'/\varepsilon)|^2}$.

Let us return to the geometric setting from Sect. 19.1 and prove the following proposition.

Theorem 1 *Let* $u \in H^\alpha(\partial T_\varepsilon)$, $-1 \leq \alpha \leq 1$. *Then*

$$\|u\|^2_{h^\alpha(\partial T_\varepsilon)} = \varepsilon^{-1-2\alpha}\|\mathscr{T}^b_\varepsilon(u)\|^2_{L_2(\Omega, h^\alpha(\partial T))} := \varepsilon^{-1-2\alpha}\int_\Omega \|\mathscr{T}^b_\varepsilon(u)(x, \cdot)\|^2_{h^\alpha(\partial T)}\,dx.$$

$$(19.11)$$

Proof By (19.8) and taking into account that $|\varepsilon Y| = \varepsilon^n$, we obtain

$$\sum_j \|u_{\varepsilon,j}(\kappa^{(-1)}_{\varepsilon,\xi,j}(\cdot, \zeta_{\varepsilon,j}(\cdot)))\sqrt{1 + |\nabla \zeta_{\varepsilon,j}(\cdot)|^2}\|^2_{h^\alpha(\mathbb{R}^{n-1})}$$

$$= \sum_j \| u \left(\varepsilon \xi + \varepsilon \kappa_j^{(-1)}(\cdot/\varepsilon, \zeta_j(\cdot/\varepsilon)) \right) \phi_j \left(\kappa_j^{(-1)}(\cdot/\varepsilon, \zeta_j(\cdot/\varepsilon)) \right)$$

$$\times \sqrt{1 + |\nabla \zeta_j(\cdot/\varepsilon)|^2} \, \|_{h^\alpha(\mathbb{R}^{n-1})}^2$$

$$= \sum_j \varepsilon^{n-1-2\alpha} \| u \left(\varepsilon \xi + \varepsilon \kappa_j^{(-1)}(\cdot, \zeta_j(\cdot)) \right) \phi_j \left(\kappa_j^{(-1)}(\cdot, \zeta_j(\cdot)) \right)$$

$$\times \sqrt{1 + |\nabla \zeta_j(\cdot)|^2} \, \|_{h^\alpha(\mathbb{R}^{n-1})}^2$$

$$= \varepsilon^{n-1-2\alpha} \| u(\varepsilon \xi + \varepsilon \cdot) \|_{h^\alpha(\partial T)}^2 = \frac{\varepsilon^{n-1-2\alpha}}{|\varepsilon Y|} \int_{\varepsilon(\xi+Y)} \| u(\varepsilon \lfloor x/\varepsilon \rfloor + \varepsilon \cdot) \|_{h^\alpha(\partial T)}^2 \, dx$$

$$= \varepsilon^{-1-2\alpha} \int_{\varepsilon(\xi+Y)} \| \mathcal{T}_\varepsilon^b(u)(x, \cdot) \|_{h^\alpha(\partial T)}^2 \, dx.$$

Finally, summing up in $\xi \in \Xi_\varepsilon$, and taking into account that $\mathcal{T}_\varepsilon^b(u)(x, y) = 0$ at $x \in \Lambda_\varepsilon = \Omega \setminus \widehat{\widetilde{\Omega}_\varepsilon}$, we obtain (19.11). $\qquad \square$

Theorem 2 *Let $u \in H^\alpha(\partial T_\varepsilon)$, $-1 \le \alpha \le 1$. Then*

$$\| u \|_{H_\varepsilon^\alpha(\partial T_\varepsilon)}^2 = \varepsilon^{-1} \| \mathcal{T}_\varepsilon^b(u) \|_{L_2(\Omega, H^\alpha(\partial T))}^2 . \tag{19.12}$$

Proof We will follow the same pattern as in the proof of Theorem 1. By (19.9), (19.10) and taking into account that $|\varepsilon Y| = \varepsilon^n$, we obtain

$$\sum_j \| u_{\varepsilon,j}(\kappa_{\varepsilon,\xi,j}^{(-1)}(\cdot, \zeta_{\varepsilon,j}(\cdot))) \sqrt{1 + |\nabla \zeta_{\varepsilon,j}(\cdot)|^2} \, \|_{H_\varepsilon^\alpha(\mathbb{R}^{n-1})}^2$$

$$= \sum_j \| u \left(\varepsilon \xi + \varepsilon \kappa_j^{(-1)}(\cdot/\varepsilon, \zeta_j(\cdot/\varepsilon)) \right) \phi_j \left(\kappa_j^{(-1)}(\cdot/\varepsilon, \zeta_j(\cdot/\varepsilon)) \right)$$

$$\times \sqrt{1 + |\nabla \zeta_j(\cdot/\varepsilon)|^2} \, \|_{H_\varepsilon^\alpha(\mathbb{R}^{n-1})}^2$$

$$= \sum_j \varepsilon^{n-1} \| u \left(\varepsilon \xi + \varepsilon \kappa_j^{(-1)}(\cdot, \zeta_j(\cdot)) \right) \phi_j \left(\kappa_j^{(-1)}(\cdot, \zeta_j(\cdot)) \right) \sqrt{1 + |\nabla \zeta_j(\cdot)|^2} \, \|_{H^\alpha(\mathbb{R}^{n-1})}^2$$

$$= \varepsilon^{n-1} \| u(\varepsilon \xi + \varepsilon \cdot) \|_{H^\alpha(\partial T)}^2 = \frac{\varepsilon^{n-1}}{|\varepsilon Y|} \int_{\varepsilon(\xi+Y)} \| u(\varepsilon \lfloor x/\varepsilon \rfloor + \varepsilon \cdot) \|_{H^\alpha(\partial T)}^2 \, dx$$

$$= \varepsilon^{-1} \int_{\varepsilon(\xi+Y)} \| \mathcal{T}_\varepsilon^b(u)(x, \cdot) \|_{H^\alpha(\partial T)}^2 \, dx.$$

Finally, summing up in $\xi \in \Xi_\varepsilon$, and again taking into account that $\mathcal{T}_\varepsilon^b(u)(x, y) = 0$ at $x \in \Lambda_\varepsilon = \Omega \setminus \widehat{\widetilde{\Omega}_\varepsilon}$, we obtain (19.12). $\qquad \square$

Remark 1 Rescaling (19.12) coincides with (19.9), i.e., passing from the hyperplane to the Lipschitz boundaries and rescaling the parametrization and its Jacobian does not influence the order of the norm rescaling.

Let us now obtain some inequalities for standard norms.

Theorem 3 *Let* $u \in H^{\alpha}(\partial T_{\varepsilon})$. *For* $\alpha \in [0, 1]$, *the following norm equivalence holds*

$$2^{\alpha-1}\varepsilon^{-1}\left[\|\mathcal{T}_{\varepsilon}^{b}(u)\|_{L_2(\Omega \times \partial T)}^2 + \varepsilon^{-2\alpha}\|\mathcal{T}_{\varepsilon}^{b}(u)(x,\cdot)\|_{L_2(\Omega, h^{\alpha}(\partial T))}^2\right] \leq$$

$$\|u\|_{H^{\alpha}(\partial T_{\varepsilon})}^2 \leq \varepsilon^{-1}\left[\|\mathcal{T}_{\varepsilon}^{b}(u)\|_{L_2(\Omega \times \partial T)}^2 + \varepsilon^{-2\alpha}\|\mathcal{T}_{\varepsilon}^{b}(u)(x,\cdot)\|_{L_2(\Omega, h^{\alpha}(\partial T))}^2\right].$$

Proof Owing to (19.4), for $\alpha \in [0, 1]$,

$$2^{\alpha-1}[\|u\|_{L_2(\partial T_{\varepsilon})}^2 + \|u\|_{h^{\alpha}(\partial T_{\varepsilon})}^2] \leq \|u\|_{H^{\alpha}(\partial T_{\varepsilon})}^2 \leq \|u\|_{L_2(\partial T_{\varepsilon})}^2 + \|u\|_{h^{\alpha}(\partial T_{\varepsilon})}^2.$$

This gives the equivalence of the norms. It suffices to note that by Proposition 2,

$$\|u\|_{L_2(\partial T_{\varepsilon})}^2 = \varepsilon^{-1}\int_{\Omega}\|\mathcal{T}_{\varepsilon}^{b}(u)(x,\cdot)\|_{L_2(\partial T)}^2 \, dx = \varepsilon^{-1}\|\mathcal{T}_{\varepsilon}^{b}(u)\|_{L_2(\Omega \times \partial T)}^2,$$

and by Theorem 1,

$$\|u\|_{h^{\alpha}(\partial T_{\varepsilon})}^2 = \varepsilon^{-1-2\alpha}\int_{\Omega}\|\mathcal{T}_{\varepsilon}^{b}(u)(x,\cdot)\|_{h^{\alpha}(\partial T)}^2 \, dx = \varepsilon^{-1-2\alpha}\|\mathcal{T}_{\varepsilon}^{b}(u)\|_{L_2(\Omega, h^{\alpha}(\partial T))}^2.$$

$\square$

Definition 5 On pair with Definition 1, we also define the following ε-dependent norms equivalent to the standard ones for the Sobolev–Slobodetskii spaces $W_2^s(\widehat{\Omega}_{\varepsilon}^*)$, $s \geq 0$. If s is integer, then let

$$\|v\|_{W_{2,\varepsilon}^s(\widehat{\Omega}_{\varepsilon}^*)}^2 := \sum_{|\alpha|\leq s}\varepsilon^{2|\alpha|}\|\partial^{\alpha}v\|_{L_2(\widehat{\Omega}_{\varepsilon}^*)}^2.$$

If s is not integer, then let

$$\|v\|_{W_{2,\varepsilon}^s(\widehat{\Omega}_{\varepsilon}^*)}^2 := \sum_{|\alpha|\leq[s]}\varepsilon^{2|\alpha|}\|\partial^{\alpha}v\|_{L_2(\widehat{\Omega}_{\varepsilon}^*)}^2 + \varepsilon^{2s}\|v\|_{W_2^s(\widehat{\Omega}_{\varepsilon}^*)}^{'2}$$

Similarly, we define the equivalent ε-dependent norm in the space $W_2^s(\partial T_{\varepsilon})$, $s \in (0, 1)$, cf. [GrOr18],

$$\|v\|_{W_{2,\varepsilon}^s(\partial T_{\varepsilon})}^2 := \|v\|_{L_2(\partial T_{\varepsilon})}^2 + \varepsilon^{2s}\|v\|_{W_2^s(\partial T_{\varepsilon})}^{'2}. \tag{19.13}$$

Note that the semi-norm in (19.13) can be expressed as $\|v\|^{'2}_{W^{\alpha}_2(\partial T_{\varepsilon})} :=$ $a_{\alpha}\|v\|^2_{h^{\alpha}(\partial T_{\varepsilon})}$ with the constant a_{α} depending on α but not on ε, cf. (19.3).

The following assertion is an immediate consequence of Theorem 1, Proposition 2 and relations (19.6).

Corollary 1 *Let* $u \in H^{\alpha}(\partial T_{\varepsilon})$, $\alpha \in (0, 1)$. *Then*

$$\|u\|^2_{W^{\alpha}_2(\partial T_{\varepsilon})} = \varepsilon^{-1}\left[\int_{\Omega} \|\mathscr{T}^b_{\varepsilon}(u)(x, \cdot)\|^2_{L_2(\partial T)}\, dx\right.$$

$$\left. + \varepsilon^{-2\alpha}\int_{\Omega} \|\mathscr{T}^b_{\varepsilon}(u)(x, \cdot)\|^{'2}_{W^{\alpha}_2(\partial T)}\, dx\right],$$

and in terms of the ε-dependent norm,

$$\|u\|^2_{W^{\alpha}_{2,\varepsilon}(\partial T_{\varepsilon})} = \varepsilon^{-1}\|\mathscr{T}^b_{\varepsilon}u\|^2_{L_2(\Omega, W^{\alpha}_2(\partial T))}. \tag{19.14}$$

Remark 2 By [GrOr18, Lem. 4.1], equality (19.14) is also valid for negative α in the sense of the dual to the Sobolev–Slobodetskii spaces.

19.4 Unfolding in Sobolev–Slobodetskii Spaces in Perforated Domains

Theorem 4

(i) *If* $\phi \in W^s_2(\widehat{\Omega}^*_{\varepsilon})$, $0 < s < 1$, *then*

$$\|\mathscr{T}_{\varepsilon}(\phi)\|^{'2}_{L_2(\Omega, W^s_2(Y^*))} \le \varepsilon^{2s}\|\phi\|^{'2}_{W^s_2(\widehat{\Omega}^*_{\varepsilon})}, \tag{19.15}$$

$$\|\mathscr{T}_{\varepsilon}(\phi)\|^2_{L_2(\Omega, W^s_2(Y^*))} \le \|\phi\|^2_{L_2(\widehat{\Omega}^*_{\varepsilon})} + \varepsilon^{2s}\|\phi\|^{'2}_{W^s_2(\widehat{\Omega}^*_{\varepsilon})} = \|\phi\|^2_{W^s_{2,\varepsilon}(\widehat{\Omega}^*_{\varepsilon})}, \tag{19.16}$$

where $\|\mathscr{T}_{\varepsilon}(\phi)\|^{'2}_{L_2(\Omega, W^s_2(Y^*))} := \int_{\Omega} \|\mathscr{T}_{\varepsilon}(\phi)(x, \cdot)\|^{'2}_{W^s_2(Y^*)}\, dx.$

(ii) *If* $\phi \in W^1_2(\widehat{\Omega}^*_{\varepsilon})$, *i.e.,* $s = 1$, *then*

$$\|\nabla \mathscr{T}_{\varepsilon}(\phi)\|^2_{L_2(\Omega, L_2(Y^*))} = \varepsilon^2\|\nabla\phi\|^2_{L^s_2(\widehat{\Omega}^*_{\varepsilon})}, \tag{19.17}$$

$$\|\mathscr{T}_{\varepsilon}(\phi)\|^2_{L_2(\Omega, W^1_2(Y^*))} = \|\phi\|^2_{L_2(\widehat{\Omega}^*_{\varepsilon})} + \varepsilon^2\|\nabla\phi\|^2_{L_2(\widehat{\Omega}^*_{\varepsilon})} = \|\phi\|^2_{W^1_{2,\varepsilon}(\widehat{\Omega}^*_{\varepsilon})}. \tag{19.18}$$

(iii) *If $\phi \in W_2^s(\widehat{\Omega}_\varepsilon^*)$, $1 < s < 2$, then* (19.15) *still holds and*

$$\|\mathscr{T}_\varepsilon(\phi)\|_{L_2(\Omega, W_2^s(Y^*))}^2 \leq \|\phi\|_{L_2(\widehat{\Omega}_\varepsilon^*)}^2 + \varepsilon^2 \|\nabla\phi\|_{L_2(\widehat{\Omega}_\varepsilon^*)}^2 + \varepsilon^{2s} \|\phi\|_{W_2^s(\widehat{\Omega}_\varepsilon^*)}^{\prime 2}$$

$$= \|\phi\|_{W_{2,\varepsilon}^s(\widehat{\Omega}_\varepsilon^*)}^2. \qquad (19.19)$$

(iv) *If $\phi \in W_2^s(\widehat{\Omega}_\varepsilon^*)$, $0 < s < 1/2$, then*

$$\varepsilon^{2s} \|\phi\|_{W_2^s(\widehat{\Omega}_\varepsilon^*)}^{\prime 2} \leq C_1 \|\mathscr{T}_\varepsilon(\phi)\|_{L_2(\Omega, W_2^s(Y^*))}^2, \qquad (19.20)$$

$$\|\phi\|_{L_2(\widehat{\Omega}_\varepsilon^*)}^2 + \varepsilon^{2s} \|\phi\|_{W_2^s(\widehat{\Omega}_\varepsilon^*)}^{\prime 2} = \|\phi\|_{W_{2,\varepsilon}^s(\widehat{\Omega}_\varepsilon^*)}^2 \leq C_2 \|\mathscr{T}_\varepsilon(\phi)\|_{L_2(\Omega, W_2^s(Y^*))}^2, \qquad (19.21)$$

where C_1 and C_2 are independent on ε and ϕ.

(v) *If $\phi \in W_2^s(\widehat{\Omega}_\varepsilon^*)$, $1 < s < 3/2$, then* (19.20) *still holds and*

$$\|\phi\|_{L_2(\widehat{\Omega}_\varepsilon^*)}^2 + \varepsilon^2 \|\nabla\phi\|_{L_2(\widehat{\Omega}_\varepsilon^*)}^2 + \varepsilon^{2s} \|\phi\|_{W_2^s(\widehat{\Omega}_\varepsilon^*)}^{\prime 2}$$

$$= \|\phi\|_{W_{2,\varepsilon}^s(\widehat{\Omega}_\varepsilon^*)}^2 \leq C_3 \|\mathscr{T}_\varepsilon(\phi)\|_{L_2(\Omega, W_2^s(Y^*))}^2, \qquad (19.22)$$

where C_3 is independent on ε and ϕ.

Proof

(i) Let $s \in (0, 1)$. Then

$$\|\phi\|_{W_2^s(\widehat{\Omega}_\varepsilon^*)}^{\prime 2} = \int_{\widehat{\Omega}_\varepsilon^*} \int_{\widehat{\Omega}_\varepsilon^*} \frac{|\phi(x) - \phi(y)|^2}{|x - y|^{n+2s}} \, dx \, dy$$

$$= \sum_{\xi_1 \in \Xi_\varepsilon} \sum_{\xi_2 \in \Xi_\varepsilon} \int_{\varepsilon(\xi_1 + Y^*)} \int_{\varepsilon(\xi_2 + Y^*)} \frac{|\phi(x) - \phi(y)|^2}{|x - y|^{n+2s}} \, dx \, dy$$

$$\geq \sum_{\xi \in \Xi_\varepsilon} \int_{\varepsilon(\xi + Y^*)} \int_{\varepsilon(\xi + Y^*)} \frac{|\phi(x) - \phi(y)|^2}{|x - y|^{n+2s}} \, dx \, dy = \sum_{\xi \in \Xi_\varepsilon} \|\phi\|_{W_2^s(\varepsilon\xi + \varepsilon Y^*)}^{\prime 2}$$

$$= \sum_{\xi \in \Xi_\varepsilon} \varepsilon^{n-2s} \|\phi(\varepsilon\xi + \varepsilon\cdot)\|_{W_2^s(Y^*)}^{\prime 2}$$

$$= \sum_{\xi \in \Xi_\varepsilon} \frac{1}{|Y|} \varepsilon^{-2s} \|\phi(\varepsilon\xi + \varepsilon\cdot)\|_{W_2^s(Y^*)}^{\prime 2} \int_{\varepsilon\xi + \varepsilon Y} dx$$

$$= \sum_{\xi \in \Xi_\varepsilon} \frac{1}{|Y|} \varepsilon^{-2s} \int_{\varepsilon\xi + \varepsilon Y} \|\phi(\varepsilon[x/\varepsilon] + \varepsilon \cdot)\|_{W_2^s(Y^*)}'^2 \, dx$$

$$= \frac{1}{|Y|} \varepsilon^{-2s} \int_\Omega \|\mathscr{T}_\varepsilon(\phi)(x, \cdot)\|_{W_2^s(Y^*)}'^2 \, dx \,.$$

Since $|Y| = 1$, we obtain (19.15).

Taking into account that $\|\phi\|_{L_2(\widehat{\Omega}_\varepsilon^*)}^2 = \|\mathscr{T}_\varepsilon(\phi)\|_{L_2(\Omega \times Y^*)}^2$ by item (i) of Proposition 1, the definition $\|u\|_{W_2^s(\Omega')}^2 := \|u\|_{L_2(\Omega')}^2 + \|u\|_{W_2^s(\Omega')}'^2$, employed with $\Omega' = \widehat{\Omega}_\varepsilon^*$ and $\Omega' = Y^*$, implies (19.16).

(ii) Equalities (19.17) and (19.18) for the case $s = 1$ immediately follow from Proposition 1.

(iii) Let now $s \in (1, 2)$ and $\mu = s - 1$. Then by (19.1), item (i) of Proposition 1 and inequality (19.15) with ϕ replaced by $\nabla\phi$ and s by μ, we obtain

$$\|\mathscr{T}_\varepsilon(\phi)\|_{L_2(\Omega, W_2^s(Y^*))}'^2 = \|\nabla\mathscr{T}_\varepsilon(\phi)\|_{L_2(\Omega, W_2^\mu(Y^*))}'^2$$

$$= \|\varepsilon\mathscr{T}_\varepsilon(\nabla\phi)\|_{L_2(\Omega, W_2^\mu(Y^*))}'^2 \leq \varepsilon^{2\mu}\varepsilon^2 \|\nabla\phi\|_{W_2^\mu(\widehat{\Omega}_\varepsilon^*)}'^2 = \varepsilon^{2s} \|\phi\|_{W_2^s(\widehat{\Omega}_\varepsilon^*)}'^2,$$

which implies inequality (19.15) also for $s \in (1, 2)$.

Definition of the Sobolev–Slobodetskii space $\|\mathscr{T}_\varepsilon(\phi)\|_{L_2(\Omega, W_2^s(Y^*))}^2$ for $1 < s < 2$ together with relations (19.18) and (19.15) implies (19.19).

(iv) Let $s \in (0, 1/2)$. Then

$$\|\phi\|_{W_2^s(\widehat{\Omega}_\varepsilon^*)}'^2 = \int_{\widehat{\Omega}_\varepsilon^*} \int_{\widehat{\Omega}_\varepsilon^*} \frac{|\phi(x) - \phi(y)|^2}{|x - y|^{n+2s}} \, dx \, dy$$

$$= \sum_{\xi_1 \in \Xi_\varepsilon} \sum_{\xi_2 \in \Xi_\varepsilon} \int_{\varepsilon(\xi_1 + Y^*)} \int_{\varepsilon(\xi_2 + Y^*)} \frac{|\phi(x) - \phi(y)|^2}{|x - y|^{n+2s}} \, dx \, dy$$

$$= \sum_{\xi_1 \in \Xi_\varepsilon} \sum_{\xi_2 \in \Xi_\varepsilon} \int_{Y^*} \int_{Y^*} \frac{|\phi(\varepsilon\xi_1 + \varepsilon q) - \phi(\varepsilon\xi_2 + \varepsilon t)|^2}{\varepsilon^{n+2s}|\xi_1 + q - \xi_2 - t|^{n+2s}} \varepsilon^{2n} \, dq \, dt$$

$$\leq \varepsilon^{n-2s} \sum_{\xi \in \Xi_\varepsilon} \int_{Y^*} \int_{Y^*} \frac{|\phi(\varepsilon\xi + \varepsilon q) - \phi(\varepsilon\xi + \varepsilon t)|^2}{|\xi + q - \xi - t|^{n+2s}} \, dq \, dt$$

$$+ 2\varepsilon^{n-2s} \sum_{\xi_1 \in \Xi_\varepsilon} \int_{Y^*} |\phi(\varepsilon\xi_1 + \varepsilon q)|^2 \sum_{\substack{\xi_2 \in \Xi_\varepsilon \\ \xi_2 \neq \xi_1}} \int_{Y^*} \frac{1}{|\xi_1 + q - \xi_2 - t|^{n+2s}} \, dt \, dq$$

$$+ 2\varepsilon^{n-2s} \sum_{\xi_2 \in \Xi_\varepsilon} \int_{Y^*} |\phi(\varepsilon\xi_2 + \varepsilon t)|^2 \sum_{\substack{\xi_1 \in \Xi_\varepsilon \\ \xi_1 \neq \xi_2}} \int_{Y^*} \frac{1}{|\xi_1 + q - \xi_2 - t|^{n+2s}} \, dq \, dt$$

$$= \varepsilon^{n-2s} \sum_{\xi \in \Xi_\varepsilon} \|\phi(\varepsilon \xi + \varepsilon \cdot)\|^{'2}_{W_2^s(Y^*)}$$

$$+ 4\varepsilon^{n-2s} \sum_{\xi_1 \in \Xi_\varepsilon} \int_{Y^*} |\phi(\varepsilon \xi_1 + \varepsilon q)|^2 \left[\sum_{\substack{\xi_2 \in \Xi_\varepsilon \\ \xi_2 \neq \xi_1}} \int_{Y^*} \frac{dt}{|\xi_1 + q - \xi_2 - t|^{n+2s}} \right] dq$$

$$\leq \varepsilon^{n-2s} \frac{1}{|\varepsilon Y|} \left(\|\mathscr{T}_\varepsilon(\phi)\|_{L^2(\Omega, W_2^s(Y^*))} \right)^2$$

$$+ 4 C_s \varepsilon^{n-2s} \sum_{\xi_1 \in \Xi_\varepsilon} \int_{Y^*} |\phi(\varepsilon \xi_1 + \varepsilon q)|^2 dist(\xi_1 + q, \partial Y_{\xi_1})^{-2s} dq.$$

Here, similar to the proof of Theorem 3.33 in [Mc00], we used the estimate

$$\sum_{\substack{\xi_2 \in \Xi_\varepsilon \\ \xi_2 \neq \xi_1}} \int_{Y^*} \frac{dt}{|\xi_1 + q - \xi_2 - t|^{n+2s}} \leq \int_{\mathbb{R}^n \setminus Y_{\xi_1}} \frac{d\tau}{|\xi_1 + q - \tau|^{n+2s}}$$

$$\leq C_s dist(\xi_1 + q, \partial Y_{\xi_1})^{-2s},$$

where C_s is a constant and $Y_{\xi_1} := \xi_1 + Y$. Applying now Lemma 3.32 from [Mc00], we obtain

$$\|\phi\|^{'2}_{W_2^s(\widehat{\Omega}_\varepsilon^*)} \leq \varepsilon^{-2s} \frac{1}{|Y|} \|\mathscr{T}_\varepsilon(\phi)\|^{'2}_{L^2(\Omega, W_2^s(Y^*))}$$

$$+ 4 C_s \varepsilon^{n-2s} \sum_{\xi_1 \in \Xi_\varepsilon} C_{Y^*} \|\phi(\varepsilon \xi_1 + \varepsilon \cdot)\|^2_{W_2^s(Y^*)}$$

$$= \varepsilon^{-2s} \frac{1}{|Y|} \|\mathscr{T}_\varepsilon(\phi)\|^{'2}_{L^2(\Omega, W_2^s(Y^*))} + 4 C_s C_{Y^*} \varepsilon^{-2s} \frac{1}{|Y|} \|\mathscr{T}_\varepsilon(\phi)\|^2_{L^2(\Omega, W_2^s(Y^*))}$$

$$\leq C_1 \varepsilon^{-2s} \frac{1}{|Y|} \|\mathscr{T}_\varepsilon(\phi)\|^2_{L^2(\Omega, W_2^s(Y^*))},$$

where C_{Y^*} and hence C_1 do not depend on ε. Since $|\mathbf{Y}| = 1$, we obtain (19.20).

Taking into account that $\|\phi\|^2_{L_2(\widehat{\Omega}_\varepsilon^*)} = \|\mathscr{T}_\varepsilon(\phi)\|^2_{L_2(\Omega \times Y^*)}$ by item (i) of Proposition 1, the definition $\|u\|^2_{W_2^s(\Omega')} := \|u\|^2_{L_2(\Omega')} + \|u\|^{'2}_{W_2^s(\Omega')}$, employed with $\Omega' = \widehat{\Omega}_\varepsilon^*$ and $\Omega' = Y^*$, implies (19.21).

(v) Let now $s \in (1, 3/2)$ and, similar to the proof of item (iii), $\mu = s - 1$. Then by (19.1), item (i) of Proposition 1 and inequality (19.20) with ϕ replaced by $\nabla \phi$ and s by μ, we obtain

$$\varepsilon^{2s}\|\phi\|^{\prime 2}_{W^s_2(\widehat{\Omega}^*_\varepsilon)} = \varepsilon^{2\mu+2}\|\nabla\phi\|^{\prime 2}_{W^\mu_2(\widehat{\Omega}^*_\varepsilon)} \le C_1\varepsilon^2\|\mathcal{T}_\varepsilon(\nabla\phi)\|^2_{L_2(\Omega,W^\mu_2(Y^*))}$$

$$= C_1\|\nabla\mathcal{T}_\varepsilon(\phi)\|^2_{L_2(\Omega,W^\mu_2(Y^*))} \le C_1\|\mathcal{T}_\varepsilon(\phi)\|^2_{L_2(\Omega,W^s_2(Y^*))}$$

which implies inequality (19.20) also for $s \in (1,3/2)$.

Definition of the Sobolev–Slobodetskii space $\|\mathcal{T}_\varepsilon(\phi)\|^2_{L_2(\Omega,W^s_2(Y^*))}$ for $1 < s < 3/2$ together with relations (19.18) and (19.20) imply (19.22).

□

19.5 Rescaling of the Trace Theorem in W^s_2

For $u \in W^s_2(\Omega^*_\varepsilon)$, $s \in (1/2,3/2)$, the trace operator (in the Gagliardo sense) $\gamma :$ $W^s_2(\Omega^*_\varepsilon) \to W^{s-1/2}_2(\partial\Omega^*_\varepsilon)$, is continuous, see, e.g., [Mc00].

Now, we can rewrite the trace theorem using the scaling estimates from Theorems 3 and 4.

Theorem 5 *Let $u \in W^s_2(\widehat{\Omega}^*_\varepsilon)$, $s \in (1/2,3/2)$, $\varepsilon > 0$.*

(i) If $s \in (1/2,1)$, then

$$\varepsilon\left(\|\gamma_{\partial T_\varepsilon}u\|^2_{L_2(\partial T_\varepsilon)} + \varepsilon^{2s-1}\|\gamma_{\partial T_\varepsilon}u\|^{\prime 2}_{W^{s-1/2}_2(\partial T_\varepsilon)}\right)$$
$$\le C\left(\|u\|^2_{L_2(\widehat{\Omega}^*_\varepsilon)} + \varepsilon^{2s}\|u\|^{\prime 2}_{W^s_2(\widehat{\Omega}^*_\varepsilon)}\right). \qquad (19.23)$$

(ii) If $s = 1$, then

$$\varepsilon\left(\|\gamma_{\partial T_\varepsilon}u\|^2_{L_2(\partial T_\varepsilon)} + \varepsilon\|\gamma_{\partial T_\varepsilon}u\|^{\prime 2}_{W^{1/2}_2(\partial T_\varepsilon)}\right) \le C\left(\|u\|^2_{L_2(\widehat{\Omega}^*_\varepsilon)} + \varepsilon^2\|\nabla u\|^2_{L_2(\widehat{\Omega}^*_\varepsilon)}\right). \qquad (19.24)$$

(iii) If $s \in (1,3/2)$, then

$$\varepsilon\left(\|\gamma_{\partial T_\varepsilon}u\|^2_{L_2(\partial T_\varepsilon)} + \varepsilon^{2s-1}\|\gamma_{\partial T_\varepsilon}u\|^{\prime 2}_{W^{s-1/2}_2(\partial T_\varepsilon)}\right)$$
$$\le C\left(\|u\|^2_{L_2(\widehat{\Omega}^*_\varepsilon)} + \varepsilon^2\|\nabla u\|^2_{L_2(\widehat{\Omega}^*_\varepsilon)} + \varepsilon^{2s}\|u\|^{\prime 2}_{W^s_2(\widehat{\Omega}^*_\varepsilon)}\right). \qquad (19.25)$$

In all three cases the constant C is independent of u and ε and, using the ϵ-dependent norms, they can be written in the same form,

$$\varepsilon\|\gamma_{\partial T_\varepsilon}u\|^2_{W^{s-1/2}_{2,\varepsilon}(\partial T_\varepsilon)} \le C\|u\|^2_{W^s_{2,\varepsilon}(\widehat{\Omega}^*_\varepsilon)}, \quad 1/2 < s < 3/2.$$

Proof If $1/2 < s < 3/2$, then by the trace theorem in Y^*, there exists a constant C independent of u and ε, such that

$$\|\gamma_{\partial T} \mathcal{T}_\varepsilon(u)(x, \cdot)\|^2_{L_2(\partial T)} + \|\gamma_{\partial T} \mathcal{T}_\varepsilon(u)(x, \cdot)\|'^2_{W_2^{s-1/2}(\partial T)}$$

$$\leq C\left(\|\mathcal{T}_\varepsilon(u)(x, \cdot)\|^2_{L_2(Y^*)} + \|\mathcal{T}_\varepsilon(u)(x, \cdot)\|'^2_{W_2^s(Y^*)}\right).$$

Integrating in x, we have

$$\|\gamma_{\partial T} \mathcal{T}_\varepsilon(u)\|^2_{L_2(\Omega, W_2^{s-1/2}(\partial T))} \leq C\|\mathcal{T}_\varepsilon(u)\|^2_{L_2(\Omega, W_2^s(Y^*))}. \tag{19.26}$$

Let first $1/2 < s < 1$. Employing inequality (19.16) in the right hand side of (19.26) and, cf. (19.14), the relation

$$\|\gamma_{\partial T} \mathcal{T}_\varepsilon(u)\|^2_{L_2(\Omega, W_2^{s-1/2}(\partial T))} = \|\mathcal{T}_\varepsilon^b(\gamma_{\partial T_\varepsilon} u)\|^2_{L_2(\Omega, W_2^{s-1/2}(\partial T))}$$

$$= \varepsilon\left(\|\gamma_{\partial T_\varepsilon} u\|^2_{L_2(\partial T_\varepsilon)} + \varepsilon^{2s-1}\|\gamma_{\partial T_\varepsilon} u\|'^2_{W_2^{s-1/2}(\partial T_\varepsilon)}\right)$$

in the left hand side, we arrive at (19.23).

Similar reasoning with relations (19.18) and (19.19) instead of (19.16) leads to (19.24) and (19.25), respectively.   □

Note that the inequality similar to (19.24), for $s = 1$, was first given in [GaEtAl14, Lem.3.1(iv)], and appears as an auxiliary result in [GrOr18].

Acknowledgements The work of the first author on this paper was supported by the Department of Mathematics at the University of Padua, during his research visits there. The work of the second author was supported by the DAAD during his stay at the Fraunhofer ITWM and by the 'Gruppo Nazionale per l'Analisi Matematica, la Probabilità e le loro Applicazioni' (GNAMPA) of the 'Istituto Nazionale di Alta Matematica' (INdAM).

References

[CiEtAl12] Cioranescu, D., Damlamian, A., Donato, P., Griso, G., and Zaki, R.: The periodic unfolding method in domains with holes. *SIAM J. Math. Anal.*, **44**, 718–760 (2012).

[Du77] Duchon, J.: Splines minimizing rotation-invariant semi-norms in Sobolev spaces. In: *Constructive Theory of Functions of Several Variables. Proceedings of a Conference Held at Oberwolfach, April 25 - May 1, 1976*, W. Schempp and K. Zeller (eds.), Springer: Berlin, Heidelberg, New York (1977), pp. 85–100.

[GaEtAl14] Gahn, M., Knabner, P., and Neuss-Radu, M.: *Homogenization of reaction-diffusion processes in a two-component porous medium with a nonlinear flux condition at the interface, and application to metabolic processes in cells*, Preprint Angew. Math., Uni Erlangen, No. 384 (2014).

[GaEtAl16] Gahn, M., Neuss-Radu, M., and Knabner, P.: Homogenization of Reaction-Diffusion Processes in a Two-Component Porous Medium with Nonlinear Flux Conditions at the Interface. *SIAM J. Appl. Math.* **76**, 1819–1843 (2016).

[GaMe18] Gaudiello, A. and Melnyk, T.: Homogenization of a Nonlinear Monotone Problem with Nonlinear Signorini Boundary Conditions in a Domain with Highly Rough Boundary. *The Oberwolfach Preprints* (2018) (OWP, ISSN 1864–7596), https://doi.org/10.14760/OWP-2018-06.

[GrEtAl12] Gómez, D., Pérez, E., and Shaposhnikova, T.A.: On homogenization of nonlinear Robin type boundary conditions for cavities along manifolds and associated spectral problems. *Asymptotic Analysis* **80**, 289–322 (2012) https://doi.org/10.3233/ASY-2012-1116.

[GrOr18] Griso, G., and Orlik, J.: *Homogenization of contact problem with Coulomb's friction on periodic cracks*, arXiv:1811.06615 (2018).

[KhEtAl17] Khruslov, E.Ya., Khilkova, L.O., and Goncharenko, M.V.: Integral Conditions for Convergence of Solutions of Non-Linear Robin's Problem in Strongly Perforated Domain. *J. Math. Phys., Analysis, Geometry*, **13**, 283–313 (2017).

[Ma11] Maz'ya, V.: *Sobolev spaces, with applications to elliptic partial differential equations*, Springer, Berlin Heidelberg (2011).

[Mc00] McLean, W.: *Strongly elliptic systems and boundary integral equations*, Cambridge University Press, Cambridge (2000).

Chapter 20
Design and Performance of a Multiphase Flow Manifold

Mobina Mohammadikharkeshi, Asad Molayari, Ramin Dabirian, Ram S. Mohan, and Ovadia Shoham

20.1 Nomenclature

A_P	Pipe cross-sectional area (m^2)
A_G	Gas cross-sectional area (m^2)
C_0	Flow distribution coefficient (–)
d	Diameter (m)
g	Gravitational acceleration (m/s^2)
h_L	Liquid level (m)
H_{F1}	Liquid holdup in film region at the front of slug (–)
H_{F2}	Liquid holdup in film region at the back of slug (–)
H_S	Liquid holdup in the slug body region (–)
L_{BFM}	Dissipation length in the main manifold (manifold length) (m)
L_s	Slug body length (m)
S_I	Gas and liquid interface length (m)
t	Time (s)
v_D	Drift velocity (m/s)
v_{F1}	Liquid film velocity at the front of slug (m/s)
v_{F2}	Liquid film velocity at the back of slug (m/s)
v_m	Mixture velocity (m/s)
v_{T1}	Translational velocity at the front of slug (m/s)
v_{T2}	Translational velocity at the back of slug (m/s)
v_{SG}	Superficial gas velocity (m/s)

M. Mohammadikharkeshi · A. Molayari · R. Dabirian · R. S. Mohan · O. Shoham (✉)
The University of Tulsa, Tulsa, OK, USA
e-mail: mobina-mohammadikharkeshi@utulsa.edu; asad-molayari@utulsa.edu;
ramin-dabirian@utulsa.edu; ram-mohan@utulsa.edu; ovadia-shoham@utulsa.edu

© Springer Nature Switzerland AG 2019   253
C. Constanda, P. Harris (eds.), *Integral Methods in Science and Engineering*,
https://doi.org/10.1007/978-3-030-16077-7_20

v_{SL}	Superficial liquid velocity (m/s)
Δt	Dissipation time (s)
θ	Inclination angle from horizontal (°)
ρ_G	Gas density (kg/m^3)
ρ_L	Liquid density (kg/m^3)

20.2 Introduction

Multiphase flow manifolds are utilized in the Petroleum Industry to gather production from various pipelines, promote separation and stratification, and redistribute the phases into downstream processing facilities for further separation, treatment and transportation. The challenge for the multiphase manifold is to handle the different flow conditions from various upstream pipelines and to ensure equal splitting of the gas and the liquid in its outlets. Maldistribution of the flow from the manifold outlets, i.e., liquid carry-over or gas carry-under, may cause operational problems in downstream separation facilities. Among the different flow patterns occurring in the pipelines upstream of manifolds, slug flow is the most challenging one due to the difficulty to dissipate it in the manifold.

Parallel pipelines are used by the Petroleum Industry as a flow splitter, or they can be combined to form manifolds [DaEtAl13, DaEtAl16]. Also, multiphase flow in parallel and looped pipelines increase the flow capacity and decrease the pressure drop in the system [AlEtAl10]. Numerous studies were published on multiphase flow in parallel lines, looped lines, as well as multiphase flow manifolds. Different geometries such as horizontal or vertical branches were investigated, with different upstream flow patterns, such as stratified (smooth and wavy) flow, slug flow, churn flow and annular flow [Co80, OsEtAl99, Bu03, TaEtAl03, ViPe04, PuEtAl06, PuEtAl10].

Slug flow has been widely studied for horizontal, vertical and inclined flow [DuHu75, Sy87, VoSo89, TaBa90, ScKo90, ZhEtAl03a, Sh06, Go08]. Slug tracking and slug dissipation models in hilly terrain pipelines, downward inclined flow, helical pipes and flow into larger diameter pipe sections were published by [Zh91, ZhEtAl94, TaBa98, TaBa00, Ra00, TaEtAl00, Di03, ZhEtAl00, ZhEtAl03b]. The difference between the slug dissipation models and steady-state slug flow models is that in the former models the slug length is not considered constant but rather it shrinks continuously until it disappears completely.

No studies have been published on slug dissipation in multiphase manifold. This is the gap that present study attempts to fill. A special multiphase flow manifold called the Balanced Feed Manifold (BFM) is utilized. In addition to the main manifold, the BFM has a secondary liquid manifold, which aims at a better separation of the gas and the liquid phases and an equal distribution of the phases into downstream facilities.

20.3 Experimental Setup

Descriptions of the multiphase flow loop, the BFM test section, test matrix and the acquired experimental results are presented in this section.

20.3.1 Flow Loop

The experimental facility utilized is shown schematically in Fig. 20.1. As shown, water is stored in a water tank and is delivered to the loop by a centrifugal pump. The water flow rate is controlled by a liquid control valve and is measured by a Coriolis mass flow meter. The gas phase, air, is supplied by a compressor and the gas flow rate is measured and controlled by a Coriolis mass flow meter and a gas control valve, respectively. The gas and liquid lines are combined upstream of the BFM test section. Downstream of the test section, 2 Gas Liquid Cylindrical Cyclones (GLCC©s) are installed, which are connected to the 2 outlets of the BFM. The GLCC©s are used to separate the gas and liquid and their respective flow rates exiting from each BFM outlet are measured using single phase meters. After measuring the gas and liquid flow rates, both lines are recombined and sent to the water tank. The water recirculates to the loop and the air is vented to the atmosphere. All the inlet liquid and gas lines are 0.05 m ID and the GLCC© diameters are 0.074 m.

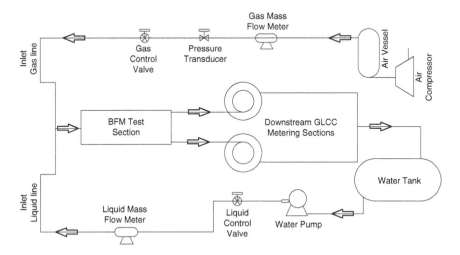

Fig. 20.1 Schematic of balanced feed manifold flow loop

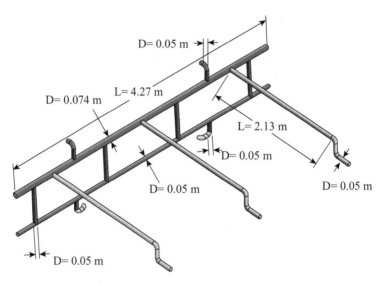

Fig. 20.2 Schematic of balanced feed manifold test section

20.3.2 BFM Test Section

A schematic of the test section is given in Fig. 20.2. As can be seen, the BFM consists of a 0.074 m ID 4.27 m long main manifold (on the top), which is connected to a secondary manifold that is 0.05 m ID and same length (at the bottom). Four sections of 0.05 m ID connect the main manifold to the secondary manifold. The BFM is fed by three inlet pipes 0.05 m ID and 2.13 m long, which are connected to the main manifold. As mentioned, the BFM has two outlets, each of which consists of a gas outlet from the top of the main manifold and a liquid outlet off the secondary liquid manifold.

20.3.3 Data Acquisition System

The control valves, mass flow meters, pressure and temperature transducers are monitored and controlled by a LabVIEW program. The experimental data are recorded and stored in an Excel spreadsheet for future analysis. Also, video cameras are used to follow the slug along the main manifold and to measure the slug dissipation length in the main manifold.

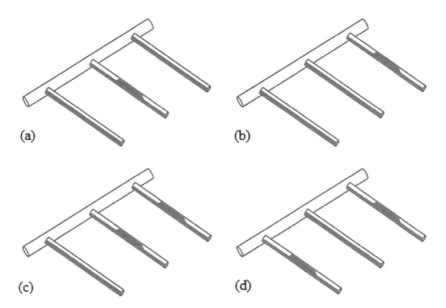

Fig. 20.3 Balanced feed manifold inlet flow configurations, (**a**) slug flow in middle pipe, (**b**) slug flow in one side pipe, (**c**) slug flow in two adjacent pipes, and (**d**) slug flow in two side pipes

20.3.4   Test Matrix

A total of 16 data points are acquired to study the slug dissipation in the BFM. Figure 20.3 shows the four different inlet flow configurations in the 3 feeding pipes used to acquire the data. As can be seen, in some of the test runs only 1 inlet pipe operates under slug flow, which occurs either in the middle inlet pipe (Fig. 20.3a) or in the side inlet pipe (Fig. 20.3b). For the other test runs, 2 inlet pipes are in slug flow, as shown in Fig. 20.3c and d. For both cases, the remaining inlet pipe(s) operates in stratified flow. Table 20.1 shows the BFM test matrix which presents the operating flow condition for different inlet flow configuration test runs.

20.3.5   Experimental Results

The BFM experimental results are presented for the 4 inlet flow conditions in Table 20.2. The results demonstrate that by increasing the mixture velocity, the experimental dissipation length increases. Also, as expected, when increasing the number of the inlet pipes that operate in slug flow the slug dissipation length in the manifold increases.

Table 20.1 Balanced feed manifold test matrix

Configuration	Run#	Property	Inlet pipe#1	Inlet pipe#2	Inlet pipe#3
Slug flow in middle pipe	1	v_{SL} (m/s)	0.06	0.43	0.06
		v_{SG} (m/s)	1.52	2.44	1.52
		Flow pattern	Stratified	Slug	Stratified
	2	v_{SL} (m/s)	0.06	0.49	0.06
		v_{SG} (m/s)	1.52	2.9	1.52
		Flow pattern	Stratified	Slug	Stratified
	3	v_{SL} (m/s)	0.06	0.46	0.06
		v_{SG} (m/s)	1.52	3.35	1.52
		Flow pattern	Stratified	Slug	Stratified
	4	v_{SL} (m/s)	0.06	0.52	0.06
		v_{SG} (m/s)	1.52	3.66	1.52
		Flow pattern	Stratified	Slug	Stratified
Slug flow in one side pipe	5	v_{SL} (m/s)	0.43	0.06	0.06
		v_{SG} (m/s)	2.44	1.52	1.52
		Flow pattern	Slug	Stratified	Stratified
	6	v_{SL} (m/s)	0.49	0.06	0.06
		v_{SG} (m/s)	2.9	1.52	1.52
		Flow pattern	Slug	Stratified	Stratified
	7	v_{SL} (m/s)	0.46	0.06	0.06
		v_{SG} (m/s)	3.35	1.52	1.52
		Flow pattern	Slug	Stratified	Stratified
	8	v_{SL} (m/s)	0.52	0.06	0.06
		v_{SG} (m/s)	3.66	1.52	1.52
		Flow pattern	Slug	Stratified	Stratified
Slug flow in two adjacent pipes	9	v_{SL} (m/s)	0.55	0.55	0.06
		v_{SG} (m/s)	3.63	3.63	1.52
		Flow pattern	Slug	Slug	Stratified
	10	v_{SL} (m/s)	0.58	0.58	0.06
		v_{SG} (m/s)	3.84	3.84	1.52
		Flow pattern	Slug	Slug	Stratified
	11	v_{SL} (m/s)	0.52	0.52	0.06
		v_{SG} (m/s)	4.27	4.27	1.52
		Flow pattern	Slug	Slug	Stratified
	12	v_{SL} (m/s)	0.61	0.61	0.06
		v_{SG} (m/s)	4.7	4.7	1.52
		Flow pattern	Slug	Slug	Stratified

(continued)

Table 20.1 (continued)

Configuration	Run#	Property	Inlet pipe#1	Inlet pipe#2	Inlet pipe#3
Slug flow in two side pipes	13	v_{SL} (m/s)	0.55	0.06	0.55
		v_{SG} (m/s)	3.63	1.52	3.63
		Flow pattern	Slug	Stratified	Slug
	14	v_{SL} (m/s)	0.58	0.06	0.58
		v_{SG} (m/s)	3.84	1.52	3.84
		Flow pattern	Slug	Stratified	Slug
	15	v_{SL} (m/s)	0.52	0.06	0.52
		v_{SG} (m/s)	4.27	1.52	4.27
		Flow pattern	Slug	Stratified	Slug
	16	v_{SL} (m/s)	0.61	0.06	0.61
		v_{SG} (m/s)	4.7	1.52	4.7
		Flow pattern	Slug	Stratified	Slug

Table 20.2 Experimental slug dissipation length results

Configuration	Run#	Slug dissipation length (m)
Slug flow in middle pipe	1	0.76
	2	0.88
	3	0.94
	4	1.13
Slug flow in one side pipe	5	0.79
	6	0.88
	7	0.97
	8	1.11
Slug flow in two adjacent pipes	9	1.40
	10	1.64
	11	1.95
	12	2.07
Slug flow in two side pipes	13	1.80
	14	1.74
	15	1.89
	16	2.04

20.4 Modeling

Slug dissipation occurs due to liquid shedding from the back of the slug body, as well as liquid drainage and backward penetration of elongated bubble into the slug body (bubble turning) that occur at the slug front. A slug dissipation model is developed based on tracking the passage of one slug body ("slug tracking") after entering the larger diameter section. In a normal and steady-state slug flow, the translational velocities at the front and at the back of the slug body, as well as the

liquid shedding and pickup at the back and at the front of the slug are equal, resulting
in a constant slug body length. However, when the slug enters the main manifold,
the translational velocities at the front and back of slug are not the same. Due to
the bubble turning phenomenon and liquid drainage from the slug body front, the
translational velocity at the slug front, v_{T1}, is lower than the translational velocity at
its back, v_{T2}. As a result, as the slug body moves along the main manifold the slug
length becomes shorter and shorter until it dissipates completely leading to stratified
flow occurs. The model developed for the prediction of the main manifold length and
diameter, which are the two parameters needed for its design, is presented next.

20.4.1 Main Manifold Diameter

The manifold diameter is determined based on [TaDu76] transition boundary from
non-stratified to stratified, as follows: it is assumed that all the inlet gas and liquid
flow axially in the manifold, whereby the minimum manifold diameter that satisfies
the transition to stratified flow under these conditions is reported as the manifold
diameter. This diameter ensures that stratified flow will occur in the main manifold.
The transition criterion is given by

$$v_G \leq \left(1 - \frac{h_L}{d}\right)\left[\frac{(\rho_L - \rho_G)g\,A_G\cos\theta}{\rho_G S_I}\right]^{0.5}. \tag{20.1}$$

Iterations are carried out varying the BFM diameter until (20.1) is satisfied.

20.4.2 Main Manifold Length

The main manifold length is determined based on the slug dissipation length in manifold. Since the main manifold consists of three Enlarged Impacting Tee-junction
(EIT) sections, the slug dissipation model developed by [MoEtA118a, MoEtA118b]
for EIT is extended for the BFM geometry. Thus, prediction of the slug dissipation
length in manifold and as a result the manifold length are determined.

Following [MoEtA118a, MoEtA118b], a liquid mass balance is applied on the
shrinking slug body in the main manifold in a coordinate system moving at v_{T2}
results in

$$\rho_L A_P H_S \frac{dL_S}{dt} - (v_{T2} - v_{F2})\rho_L A_P H_{F2} + (v_{T2} - v_{T1})\rho_L A_P H_S$$
$$+ (v_{T2} - v_{F1})\rho_L A_P H_{F1} = 0. \tag{20.2}$$

Assuming that the film velocity and liquid holdup at the front and at the back of slug are the same, (20.2) reduces to

$$dL_S + (v_{T2} - v_{T1}) \, dt = 0. \tag{20.3}$$

Integrating (20.3), the dissipation time interval, which is the time that takes for the slug body to dissipate completely, is

$$\Delta t = \frac{L_S}{(v_{T2} - v_{T1})}.$$

Finally, the dissipation length, which is the required manifold length, is given by

$$L_{BFM} = \Delta t \frac{(v_{T2} + v_{T1})}{2}.$$

In the above equations, v_{T2} is assumed to be a normal slug flow translational velocity, namely, $v_{T2} = C_0 v_m + v_D$ [Ni62]. The translational velocity at the front of the slug is lower due to the penetration of bubble turning and liquid drainage, and is given by $v_{T1} = v_m - v_D$. The drift velocity is determined by [Be84], i.e., $v_D = 0.54\sqrt{gd}\cos\theta + 0.35\sqrt{gd}\sin\theta$.

20.5   Comparison Study

The slug length in the BFM inlet feeding pipes is assumed to be $32d$ for fully developed flow. When the slug enters into the main manifold its length is adjusted owing to the larger diameter manifold, which is found using a simple mass balance over the slug body. Also, the mixture velocity in drift velocity equation is the summation of the superficial gas and liquid velocities of all the inlet pipes. Figures 20.4, 20.5, 20.6, 20.7 provide comparisons between model predictions and acquired experimental data for the 4 different inlet flow configurations. These inlet flow configurations include slug flow in the middle pipe, slug flow in one side pipe, slug flow in two adjacent pipes and slug flow in two side pipes. The model predictions for all four cases are in good agreement with the experimental data, with an average absolute relative error of 7.5%. An overall evaluation between the experimental data and model predictions is presented in Fig. 20.8, demonstrating discrepancies less than ±10% for most of the data.

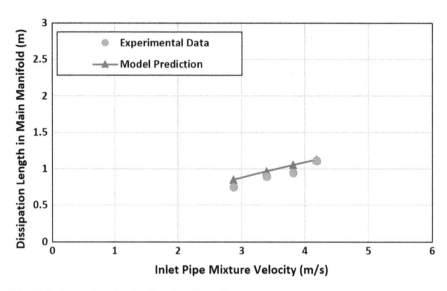

Fig. 20.4 Comparison for slug flow in middle pipe

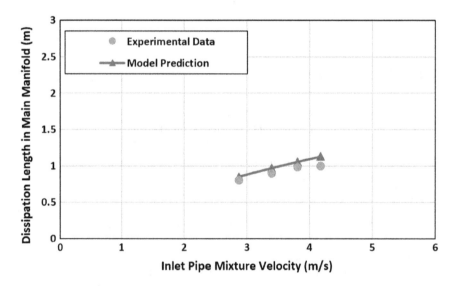

Fig. 20.5 Comparison for slug flow in one side pipe

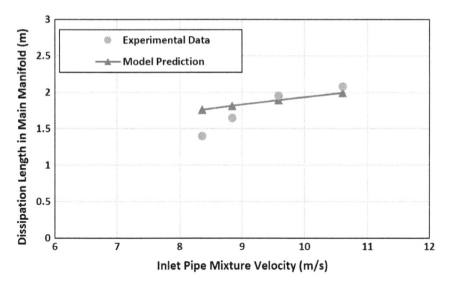

Fig. 20.6 Comparison for slug flow in two adjacent pipes

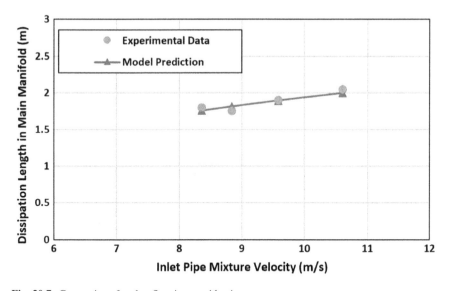

Fig. 20.7 Comparison for slug flow in two side pipes

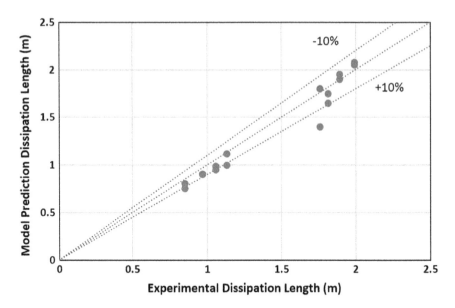

Fig. 20.8 Overall comparison between model predictions and experimental data

Acknowledgements The authors thank the Tulsa University Separation Technology Projects (TUSTP) and the Tulsa University Center of Research Excellence (TUCoRE) for the financial support.

References

[AlEtAl10] Alvarez, L. P., Mohan, R. S., Shoham, O., and Avila, C.: Multiphase flow splitting in parallel/looped pipelines. In *SPE Annual Technical Conference and Exhibition*, Society of Petroleum Engineers, Florence, Italy (2010).

[Be84] Bendiksen, K.H.: An experimental investigation of the motion of long bubbles in inclined tubes. *International Journal of Multiphase Flow*, **11**(6), 797–812 (1984).

[Bu03] Bustamante, A. R.: Design and performance of multiphase distribution manifold. M.S. Thesis, The University of Tulsa, Tulsa, OK (2003).

[Co80] Coney, M. W. E.: Two-phase flow distribution in a manifold system. In *European Two-Phase Flow Group Meeting* (1980).

[DaEtAl13] Dabirian, R., Thompson, L., Mohan, R., Shoham, O. and Avila, C.: Prediction of tow-phase flow splitting in looped lines based on energy minimization. In *SPE Annual Technical Conference and Exhibition* (2013).

[DaEtAl16] Dabirian, R., Thompson, L., Mohan, R. S., and Shoham, O.: Pressure-minimization method for prediction of two-phase-flow splitting. *Oil and Gas Facilities*, **5**(5) (2016).

[Di03] Di Matteo, C. A.: Mechanistic modeling of slug dissipation in helical pipes. M.S. Thesis, The University of Tulsa, Tulsa, OK (2003).

[DuHu75] Dukler, A. E., and Hubbard, M. G.: A model for gas-liquid slug flow in horizontal and near horizontal tubes. *Industrial & Engineering Chemistry Fundamentals*, **14**(4), 337–347 (1975).

[Go08]   Gokcal, B.: An experimental and theoretical investigation of slug flow for high oil viscosity in horizontal pipes. Ph.D. Dissertation, The University of Tulsa, Tulsa, OK (2008).

[MoEtAl18a]   Mohammadikharkeshi, M., Dabirian, R., Cole, T., Shoham, O., and Mohan, R. S.: Slug dissipation in a horizontal enlarged impacting tee-junction. In *SPE Western Regional Meeting*, Society of Petroleum Engineers, Garden Grove, California (2018a).

[MoEtAl18b]   Mohammadikharkeshi, M., Dabirian, R., Shoham, O., and Mohan, R. S.: Effect of fluid properties on slug dissipation in enlarged impacting tee. In *ASME FEDSM Montreal*, Quebec, Canada (2018b).

[Ni62]   Nicklin, D.J.: Two-Phase Bubble Flow. *Chemical Engineering Science*, **17**, 693–702 (1962).

[OsEtAl99]   Osakabe, M., Hamada, T., and Horiki, S.: Water flow distribution in horizontal header contaminated with bubbles. *International Journal of Multiphase Flow*, **25**(5), 827–840 (1999).

[PuEtAl06]   Pustylnik, L., Barnea, D., and Taitel, Y.: Prediction of two-phase flow distribution in parallel pipes using stability analysis *AIChE Journal*, **52**(10), 3345–3552 (2006).

[PuEtAl10]   Pustylnik, L., Barnea, D., and Taitel, Y.: Adiabatic flow distribution of gas and liquid in parallel pipes - Effect of additional restrictions. *Chemical Engineering Science*, **65**(8), 2552–2557 (2010).

[Ra00]   Ramirez, R.: Slug dissipation in helical pipes. M.S. Thesis, The University of Tulsa, Tulsa, OK (2000).

[ScKo90]   Scott, S. L., and Kouba, G. E.: Advances in slug flow characterization for horizontal and slightly inclined pipelines. In *SPE Annual Technical Conference and Exhibition*, Society of Petroleum Engineers (1990).

[Sh06]   Shoham, O.: *Mechanistic modeling of gas-liquid two-phase flow in pipes*, Society of Petroleum Engineers, Richardson, TX (2006).

[Sy87]   Sylvester, N. D.: A mechanistic model for two-phase vertical slug flow in pipes. *Journal of Energy Resources Technology*, **109**(4), 206–213 (1987).

[TaBa98]   Taitel, Y., and Barnea, D.: Effect of gas compressibility on a slug tracking model. *Chemical Engineering Science*, **53**(11), 2089–2097 (1998).

[TaBa00]   Taitel, Y., and Barnea, D.: Slug-tracking model for hilly terrain pipelines. *SPE Journal*, **5**(1), 102–109 (2000).

[TaBa90]   Taitel, Y., and Barnea, D.: Two-phase slug flow. *Adv. Heat Transfer*, **20**, 83–132 (1990).

[TaDu76]   Taitel, Y., and Dukler, A. E.: A model for predicting flow regime transitions in horizontal and near horizontal gas-liquid flow. *AIChE Journal*, **22**(1), 47–55 (1976).

[TaEtAl00]   Taitel, Y., Sarica, C., and Brill, J.P.: Slug flow modeling for downward inclined pipe flow: Theoretical considerations. *International Journal of Multiphase Flow*, **11**(26), 833–844 (2000).

[TaEtAl03]   Taitel, Y., Pustylnik, L., Tshuva, M., and Barnea, D.: Flow distribution of gas and liquid in parallel pipes. *International Journal of Multiphase Flow*, **29**(7), 1193–1202 (2003).

[ViPe04]   Vist, S., and Pettersen, J.: Two-phase flow distribution in compact heat exchanger manifolds. *Experimental Thermal and Fluid Science*, **28**(2–3), 209–215 (2004).

[VoSo89]   Vo, D. T., and Shoham, O.: A note on the existence of a solution for two-phase slug flow in vertical pipes. *Journal of Energy Resources Technology*, **111**(2), 64–65 (1989).

[ZhEtAl00]   Zhang, H., Redus, C. L., Brill, J. P., and Yuan, H.: Observations of slug dissipation in downward flow. *Journal of Energy Resources Technology*, **122**(3) 110–114 (2000).

[ZhEtAl03a]   Zhang, H., Wang, Q., Sarica, C., and Brill, J. P.: Unified model for gas-liquid pipe flow via slug dynamics, part 1: model development. *Journal of Energy Resources Technology*, **125**(4), 266–273 (2003a).

[ZhEtAl03b] Zhang, H., Al-Safran, E. M., Jayawardena, S. S., Redus. C. L., Sarica, C., and Brill, J. P.: Modeling of slug dissipation and generation in gas-liquid hilly-terrain pipe flow. *Journal of Energy Resources Technology*, **125**(3) 161–168 (2003b).

[Zh91] Zheng, G.: Two-phase slug flow in hilly terrain pipelines. Ph.D. Dissertation, The University of Tulsa, Tulsa, OK (1991).

[ZhEtAl94] Zheng, G., Brill, J. P., and Taitel, Y.: Slug flow behavior in a hilly terrain pipeline. *International Journal of Multiphase Flow*, **20**(1), 63–79 (1994).

Chapter 21
On the Polarization Matrix
for a Perforated Strip

Sergey A. Nazarov, Rafael Orive-Illera, and María-Eugenia Pérez-Martínez

21.1 Introduction

Let $\Pi = \{\xi = (\xi_1, \xi_2) : \xi_1 \in \mathbb{R}, \ \xi_2 \in (0, H)\}$ be a strip of width $H > 0$ and let ω be a bounded domain in the plane $\mathbb{R}^2$, with a Lipschitz boundary $\partial\omega$. Let $R > 0$ be such that

$$\overline{\omega} \subset (-R, R) \times (0, H). \tag{21.1}$$

In the unbounded domain $\varXi = \Pi \setminus \overline{\omega}$ (see Fig. 21.1), we consider the following boundary value problem:

$$-\Delta W(\xi) = 0, \ \xi \in \varXi = \Pi \setminus \overline{\omega}, \tag{21.2}$$

$$W(\xi_1, H) = W(\xi_1, 0), \ \frac{\partial W}{\partial \xi_2}(\xi_1, H) = \frac{\partial W}{\partial \xi_2}(\xi_1, 0), \ \xi_1 \in \mathbb{R}. \tag{21.3}$$

$$W(\xi) = 0, \ \xi \in \partial\omega. \tag{21.4}$$

S. A. Nazarov
St. Petersburg State University, St. Petersburg, Russia

Institute of Problems of Mechanical Engineering RAS, St. Petersburg, Russia

R. Orive-Illera
Universidad Autónoma de Madrid, Madrid, Spain
e-mail: rafael.orive@icmat.es

M.-E. Pérez-Martínez (✉)
Universidad de Cantabria, Santander, Spain
e-mail: meperez@unican.es

© Springer Nature Switzerland AG 2019
C. Constanda, P. Harris (eds.), *Integral Methods in Science and Engineering*,
https://doi.org/10.1007/978-3-030-16077-7_21

Fig. 21.1 The unbounded
domain Ξ with the hole ω

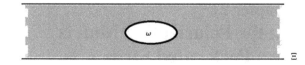

   Problem (21.2)–(21.4), with the periodic conditions (21.3), is of interest since, for
instance, it arises in homogenization processes in perforated media. We are led to
this *unit cell problem*, also the so-called *local problem*, in boundary homogenization
problems to describe boundary layer phenomena. In particular, it has been obtained
when addressing the band-gap structure of a spectral problem for the Laplace
operator in an unbounded strip periodically perforated by a family of holes, which
are also periodically distributed along a line, the so-called *perforation string*, cf.
[NaEtAl18]. That is, a double periodicity occurs: periodicity $O(1)$ in the ξ_1-
direction and $O(\varepsilon)$ in the ξ_2-direction, ε being a small parameter, $\varepsilon \ll 1$, while
each hole is homothetic to ω of ratio ε. It has been shown that the coefficients of
the so-called *polarization matrix* $p(\Xi)$ (cf. (21.6), (21.7)) play an important role
in the asymptotic determination of the endpoints of the spectral bands, as $\varepsilon \to 0$.
This matrix $p(\Xi)$ is an integral characteristic of the "Dirichlet hole" $\overline{\omega}$ in the strip
Π and is quite analogous to the classical polarization tensor in the exterior Dirichlet
problem, see Appendix G in [PoSz51].
   In particular, in [NaEtAl18], we proved the existence of holes ω such that the
anti-diagonal of the polarization matrix does not vanish (cf. (21.17), (21.18)). This
is an essential fact to determine the precise length of the above mentioned bands.
In this paper, we investigate the dependence of the coefficients of the polarization
matrix on the shape and the dimension of the hole which may provide a higher
precision in the detection of the band-gap structure of the spectrum when dealing
with a periodically perforated waveguide. To highlight the dependence of the
polarization matrix on the characteristics of the hole, in Sects. 21.3 and 21.4 we
provide two examples. The results can be applied when the holes asymptotically
collapse on a horizontal crack (cf. Theorem 2) or on a point, respectively (cf.
Theorem 3); however, a two-dimensional positive measure of the holes must be
maintained. Section 21.2 describes some general properties of coefficients of the
polarization matrix.
   It is worth recalling that, according to the general theory of elliptic problems in
domains with cylindrical outlets to infinity, cf., e.g., Ch. 5 in [NaPl94], problem
(21.2)–(21.4) has just two solutions with a polynomial growth as $\xi_1 \to \pm\infty$. We
denote these solutions as $W^{\pm}(\xi)$ by setting ± 1 for the constants accompanying ξ_1.
In order to introduce the polarization matrix associated to the perforated strip Ξ, we
consider two cut-off functions $\chi_{\pm} \in C^{\infty}(\mathbb{R})$ such that

$$\chi_{\pm}(y) = \begin{cases} 1, & \text{for } \pm y > 2R, \\ 0, & \text{for } \pm y < R, \end{cases}$$

where the subindex $\pm$ represents the support in $\pm\xi_1 \in [R, \infty)$.

Theorem 1 *There are two normalized solutions of* (21.2)–(21.4) *in the form*

$$W^{\pm}(\xi) = \pm\chi_{\pm}(\xi_1)\xi_1 + \sum_{\tau=\pm} \chi_{\tau}(\xi_1)p_{\tau\pm} + \widetilde{W}^{\pm}(\xi), \qquad \xi \in \Xi, \qquad (21.5)$$

where the remainder $\widetilde{W}^{\pm}(\xi)$ gets the exponential decay rate $O(e^{-|\xi_1|2\pi/H})$, and the coefficients $p_{\tau\pm} \equiv p_{\tau\pm}(\Xi)$, with $\tau = \pm$, compose a 2×2 polarization matrix

$$p(\Xi) = \begin{pmatrix} p_{++}(\Xi) & p_{+-}(\Xi) \\ p_{-+}(\Xi) & p_{--}(\Xi) \end{pmatrix}, \qquad (21.6)$$

where the coefficients are defined as follows:

$$p_{\tau\pm}(\Xi) = \lim_{T\to\infty} \frac{1}{H} \int_0^H (W^{\tau}(\pm T, \xi_2) - \tau\delta_{\tau,\pm}T)d\xi_2. \qquad (21.7)$$

The proof of Theorem 1 can be found in [NaEtAl18]. Here we only emphasize that the existence of two linearly independent normalized solutions $W^{\pm}$ of (21.2)–(21.4) with a linear polynomial behavior $\pm\xi_1 + p_{\pm\pm}$, as $\pm\xi_1 \to \infty$, is a consequence of the Kondratiev theory [Ko67] (cf. also Ch. 5 in [NaPl94] and Sect. 3 in [Na99]). Each solution has a polynomial growth in one direction and stabilizes towards a constant $p_{\mp\pm}$ in the other direction. In addition, it lives in an exponential weighted Sobolev space which guaranties that, after subtracting the linear part, the remaining functions have a gradient in $(L^2(\Xi))^2$.

Also it should be noted that the above results hold in the case where ω is a vertical crack, cf. (21.16).

21.2 Some General Properties of the Polarization Matrix

In this section we provide some properties of the coefficients of the polarization matrix $p(\Xi)$. Some of these properties have been proved in [NaEtAl18] and we recall them here for the sake of completeness. Also, for brevity, if no confusion arises, we avoid writing (Ξ) in the coefficients of the matrix $p(\Xi)$.

Proposition 1 *The matrix $p(\Xi) + R\,\mathbb{I}$ is symmetric and positive, where $\mathbb{I}$ stands for the 2×2 unit matrix and R given in* (21.1).

Proposition 2 *Let ω be such that its two-dimensional measure $mes_2(\omega) > 0$. Then, the coefficients of the polarization matrix $p(\Xi)$ satisfy*

$$H(2p_{+-} - p_{++} - p_{--}) > mes_2(\omega). \qquad (21.8)$$

The proof of Propositions 1 and 2 is in [NaEtAl18].

An immediate consequence of Proposition 1 are the following results:

$$p_{\pm\pm} + R \geq 0,$$ (21.9)

$$p_{++} + p_{--} + 2R \geq 2|p_{+-}|.$$ (21.10)

In addition, by (21.10), we get

$$2p_{+-} + 2R \geq -p_{++} - p_{--}.$$

Also, using this in (21.8), we obtain a lower bound

$$p_{+-} > \frac{1}{4H} \left(mes_2(\omega) - 2RH \right),$$ (21.11)

where the right-hand side is negative by the hypothesis (21.1). Obviously, (21.11) reads

$$|p_{+-}| < \frac{1}{4H} \left(2RH - mes_2(\omega) \right), \quad \text{when } p_{+-} < 0.$$

Finally, let us note that by Proposition 1, the coefficients of the anti-diagonal of p satisfy

$$p_{+-} = p_{-+}.$$ (21.12)

21.2.1 The Case of a Symmetric Hole

In this section, we consider a symmetric hole ω_s with respect to the ξ_1-axis, namely ω such that if $(\xi_1, \xi_2) \in \omega$, then $(-\xi_1, \xi_2) \in \omega$. Denoting by Ξ^s the unbounded domain with this symmetric hole, we show that the matrix $p(\Xi^s)$ becomes symmetric with respect to the anti-diagonal, namely

$$p_{++}(\Xi^s) = p_{--}(\Xi^s).$$ (21.13)

Indeed, this is due to the fact that each one of the two normalized solutions in (21.5) is related with each other by symmetry. Note that, by (21.5), $W^+(\xi)$ has a linear polynomial growth $\xi_1 + p_{++}$ as $\xi_1 \to +\infty$, while it stabilizes towards the constant p_{-+} as $\xi_1 \to -\infty$. Considering the function $W(\xi_1, \xi_2) = W^+(-\xi_1, \xi_2)$, $W(\xi)$ satisfies (21.2)–(21.4), it has a linear polynomial growth $-\xi_1 + p_{++}$ as $\xi_1 \to -\infty$ and stabilizes towards the constant p_{-+} as $\xi_1 \to +\infty$. Therefore

$$W^-(\xi_1, \xi_2) = W^+(-\xi_1, \xi_2),$$ (21.14)

and this provides

$$p_{--}(\varXi^s) = p_{++}(\varXi^s), \quad p_{+-}(\varXi^s) = p_{-+}(\varXi^s).$$

Consequently, (21.13) holds, while we observe that the symmetry with respect to the diagonal holds for any shape of the hole ω, cf. (21.12).

Moreover, in the symmetric case, the eigenvalues of the matrix p are

$$\lambda_1 = p_{++}(\varXi^s) + p_{+-}(\varXi^s), \qquad \lambda_2 = p_{++}(\varXi^s) - p_{+-}(\varXi^s), \qquad (21.15)$$

and by Proposition 2, $\lambda_2 =< -mes_2(\omega^s)(2H)^{-1} < 0$, when $mes_2(\omega^s) > 0$.

Proposition 3 *Let ω be the crack*

$$\omega = \{\xi \in \mathbb{R}^2 : \xi_1 = 0, \ \xi_2 \in (h, H - h)\}, \qquad (21.16)$$

where $h < H/2$. Then,

$$p_{+-} = p_{-+} > 0. \qquad (21.17)$$

In addition, $p_{--} = p_{++} = p_{-+} = p_{+-}$.

Proposition 4 *There are symmetric holes ω^s, with a smooth boundary and $mes_2(\omega^s) > 0$, for which*

$$p_{+-} = p_{-+} > 0. \qquad (21.18)$$

The proof of Propositions 3 and 4 has been performed in [NaEtAl18].

As a consequence of Proposition 3, when ω is a vertical crack, the inequality in Proposition 2 must be replaced by $H\,(2p_{+-} - p_{++} - p_{--}) = mes_2(\omega) = 0$. Also, in this case $\lambda_2 = 0$, cf. (21.15).

21.3   The Case of a "Big" Rectangular Hole

Let us consider the specific geometry in Fig. 21.1 with hole $\omega_\square = (-B, B) \times (h, H - h)$ with $2h < H$, see Fig. 21.2. Throughout this section, we denote by $\varXi^\square$ the perforated strip.

Theorem 2 *Let us assume that h is a small parameter, $h \ll 1$. Then, the coefficients of the polarization matrix satisfy*

$$p_{++}(\varXi^\square) = p_{--}(\varXi^\square) = -B + O(h^2), \quad p_{-+}(\varXi^\square) = p_{+-}(\varXi^\square) = O(h^2). \tag{21.19}$$

Fig. 21.2 The strips $\varXi^{\square}$ and $\varXi^{h}$ with the hole $\omega_{\square}$

Proof We use (21.7) to show (21.19). It proves useful to introduce the domain

$$\varXi^{h} = \varPi_{+}(B) \cup \varPi_{-}(B) \cup \varUpsilon^{h}(B),$$

see Fig. 21.2, where

$$\varPi_{\pm}(B) = \{\xi : \pm\xi_1 > B, |\xi_2| < H/2\}$$
$$\varUpsilon^{h}(B) = \{\xi \in \varPi : |\xi_1| \leq B, |\xi_2| < h\}.$$

Due to the symmetry of $\omega_{\square}$, the functions in (21.5) satisfy (21.14). Furthermore, we consider the natural periodic extension of W^{+} to $\varXi^{h}$,

$$W^{+}(\xi_1, \xi_2) = \begin{cases} W^{+}(\xi_1, \xi_2), & \xi_2 \in [0, H/2], \\ W^{+}(\xi_1, H + \xi_2), & \xi_2 \in [-H/2, 0). \end{cases}$$

Also, we assume that $\omega_{\square}$ is repeated by periodicity in the direction of the ξ_2-axis. Thus, W^{+} satisfies

$$-\Delta W^{+}(\xi) = 0, \ \xi \in \varXi^{h},$$

with the periodicity condition on $|\xi_1| \geq B$,

$$W^{+}(\xi_1, H/2) = W^{+}(\xi_1, -H/2), \quad \frac{\partial W^{+}}{\partial \xi_2}(\xi_1, H/2) = \frac{\partial W^{+}}{\partial \xi_2}(\xi_1, -H/2),$$

$$\tag{21.20}$$

and the Dirichlet boundary condition on $\partial\omega_{\square} \cap \partial\varXi^{h}$, namely on

$$W^{+}(\xi) = 0, \ \xi \in \{|\xi_1| = B, \text{ and } h \leq |\xi_2| < H/2\},$$
$$W^{+}(\xi) = 0, \ \xi \in \{\xi_1 \in [-B, B] \text{ and } |\xi_2| = h\}.$$

$$\tag{21.21}$$

The composite function

$$
W_0^+(\xi_1, \xi_2) = \begin{cases} W^+(\xi_1, \xi_2) - \xi_1 + B, & \xi_1 > B, \\ W^+(\xi_1, \xi_2), & \xi_1 < B, \end{cases} \tag{21.22}
$$

is harmonic in $\varXi^h$ with the exception of a line segment γ^h, namely

$$
-\Delta W_0^+(\xi) = 0, \; \xi \in \varXi^h \setminus \gamma^h, \qquad \gamma^h = \{\xi \; : \; \xi_1 = B, \; |\xi_2| < h\}. \tag{21.23}
$$

Also, it satisfies the periodicity and Dirichlet conditions (21.20)–(21.21). Moreover, (21.22) meets the following jump conditions:

$$
[W_0^+]_B(\xi_2) = 0, \qquad \left[\frac{\partial W_0^+}{\partial \xi_1}\right]_B (\xi_2) = -1, \quad |\xi_2| < h,
$$

where $[W]_B(\xi_2) = W(B + 0, \xi_2) - W(B - 0, \xi_2)$. Obviously, also W_0^+ can be extended by periodicity to a harmonic function in $\varXi^\square$.

Since the behavior at infinity of W_0^+ is given by (21.22) and the behavior of W^+ (cf. (21.5)–(21.7)), W_0^+ stabilizes towards $p_{++} + B$ when $\xi_1 \to +\infty$, while it stabilizes towards p_{-+} when $\xi_1 \to -\infty$. This also guarantees (cf. Theorem 1) $W_0^+ \in H^1((0, R) \times (0, H) \cap \varXi^h)$ for any $R > B$ and $\nabla W_0^+ \in (L^2(\varXi^h))^2$. In addition, it is simple to verify that W_0^+ satisfies

$$
(\nabla W_0^+, \nabla v)_{\varXi^h} = \int_{-h}^{h} v(B, \xi_2) d\xi_2, \qquad \forall v \in \mathscr{H}_0^1(\varXi^h), \tag{21.24}
$$

where $\mathscr{H}_0^1(\varXi^h)$ denotes the space completion of $C_{c,per}^\infty(\overline{\varXi^h} \setminus \partial\omega_\square)$ with the norm $\|\nabla v; L^2(\varXi^h)\|$. Here, $C_{c,per}^\infty(\overline{\varXi^h} \setminus \partial\omega_\square)$ is the space of the infinitely differentiable ξ_2-periodic functions, vanishing in the vicinity of $\partial\omega_\square$ (cf. (21.20)–(21.21)), with a compact support in $\overline{\varXi^h}$.

Let $B_{2h}(B, 0)$ be the disk of radius $2h$ with the center at $(B, 0)$, $\gamma^h \subset B_{2h}(B, 0)$. Due to the Dirichlet conditions at $\partial\varXi^h \cap B_{2h}(B, 0)$, the Poincare inequality after stretching the coordinates $\xi \to h^{-1}(\xi_1 - B, \xi_2)$ gives

$$
\|\nabla v; L^2(B_{2h}(B, 0) \cap \varXi^h)\|^2 \geq ch^{-2}\|v; L^2(B_{2h}(B, 0) \cap \varXi^h)\|^2
$$

and, by the trace theorem, we get

$$
\|v; L^2(\gamma^h)\|^2 \leq ch\|\nabla v; L^2(B_{2h}(B, 0) \cap \varXi^h)\|^2 + c\|v; L^2(B_{2h}(B, 0) \cap \varXi^h)\|^2
$$

$$
\leq ch\|\nabla v; L^2(\varXi^h)\|^2. \tag{21.25}
$$

The latter applied to W_0^+ means that the functional on the right-hand side of (21.24) has the norm $O(h)$ and, therefore (21.24) implies

$$\|\nabla W_0^+; L^2(\Xi^h)\| \leq ch. \tag{21.26}$$

Considering (21.7) and the definition (21.22), we write

$$p_{++}(\Xi^\square) = \frac{1}{H} \lim_{T \to \infty} \int_0^H (W^+(T, \xi_2) - T)d\xi_2 = \frac{1}{H} \lim_{T \to \infty} \int_0^H (W_0^+(T, \xi_2) - B)d\xi_2$$

and

$$p_{-+}(\Xi^\square) = \frac{1}{H} \lim_{T \to \infty} \int_0^H W^+(-T, \xi_2)d\xi_2 = \frac{1}{H} \lim_{T \to \infty} \int_0^H W_0^+(-T, \xi_2)d\xi_2,$$

which by the periodicity of W_0^+ transform into

$$p_{++}(\Xi^\square) = \frac{1}{H} \lim_{T \to \infty} \int_{-H/2}^{H/2} W_0^+(T, \xi_2)d\xi_2 - B \tag{21.27}$$

and

$$p_{-+}(\Xi^\square) = \frac{1}{H} \lim_{T \to \infty} \int_{-H/2}^{H/2} W_0^+(-T, \xi_2)d\xi_2. \tag{21.28}$$

Now, considering (21.23) and the equation $-\Delta \xi_1 = 0$, $\xi \in \Pi$, we apply the Green formula and take limits as $T \to \infty$ to show the following equalities:

$$\int_{-h}^{h} W_0^+(\pm B, \xi_2)d\xi_2 = \lim_{T \to \infty} \int_{-H/2}^{H/2} W_0^+(\pm T, \xi_2)d\xi_2, \tag{21.29}$$

and

$$(\nabla W_0^+, \nabla W_0^+)_{\Xi^h} = \int_{-h}^{h} W_0^+(B, \xi_2)d\xi_2.$$

Then, from (21.29) for $+B$ and $+T$, the second equality above, (21.26), and (21.27) give the first formula in (21.19). Similarly, using (21.29) for $-B$ and $-T$, in (21.28), we write

$$|p_{-+}(\Xi^{\square})|^2 \le 2hH^{-2} \int_{-h}^{h} |W_0^+(-B, \xi_2)|^2 d\xi_2 \le ch^2 \|\nabla W_0^+; L^2(\Xi^h)\|^2 \le ch^4.$$

Indeed, we have applied the Cauchy–Schwarz inequality, the Poincaré inequality in a ball of radius $2h$ centered in $(-B, 0)$ as in (21.25), and finally (21.26). Consequently, we get the second formula in (21.19), and the theorem is proved.

21.4   The Case of a "Small" Symmetric Hole

In this section, we deal with the strip $\Pi = \{\xi : \xi_1 \in \mathbb{R}, |\xi_2| < H/2\}$. Note that this does not imply any restriction and we can transfer the periodicity conditions on its lateral sides, in a similar way as has been performed in Sect. 21.3. Let Ξ^h be $\Xi^h = \Pi \setminus \overline{\omega^h}$, where ω^h is the hole of diameter $O(h)$,

$$\omega^h = \{\xi : \zeta := h^{-1}\xi \in \omega\},$$

with $h < H$ and ω a domain symmetric with respect to ξ_1 (cf. Sect. 21.2.1 and Fig. 21.3), $\omega \subset (-\infty, \infty) \times (-1/2, 1/2)$ such that $0 \in \omega$.

Theorem 3 *Let us assume that h is a small parameter, $h \ll 1$. Then, the coefficients of the polarization matrix satisfy*

$$p_{\pm\pm}(\Xi^h) = -\frac{H}{2\pi} \ln h + O(1).$$

Proof By the symmetry it suffices to prove the result of Theorem 3 for $p_{++}(\Xi^h)$. We divide the proof into several steps.

First Step: The Green function for (21.2)–(21.4) and the highlight of the main asymptotics for $p_{++}(\Xi^h)$.

Fig. 21.3 The strip with the small hole ω^h

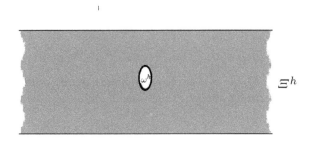

We introduce the Green function $G(\xi, \xi^0)$, that is, the periodic in $\xi_2 \in$ $(-H/2, H/2)$ solution of the equation

$$\Delta_\xi G(\xi, \xi^0) = \delta(\xi - \xi^0), \quad \xi, \xi^0 \in \Pi, \quad \xi^0 \neq \xi,$$

with $\delta(\xi - \xi^0)$ the Dirac delta at ξ^0. For $\xi^0 = 0$, this Green function satisfies

$$G(\xi) = \frac{1}{2\pi} \ln|\xi| + G_0 + \widetilde{G}(\xi), \quad \widetilde{G}(\xi) = O(|\xi|), \quad |\xi| \to 0,$$

$$G(\xi) = \frac{1}{2H}|\xi_1| + O\left(e^{-2\pi|\xi_1|/H}\right), \quad |\xi_1| \to \infty. \tag{21.30}$$

Notice that this function can be constructed explicitly with the help of a conformal mapping of a strip of width π onto the half-plane (cf., e.g., [La72] and [NaPl94] for other domains). Also, notice that the absence of a constant term in the second equation (21.30) makes G, as well as the constant G_0, to be determined uniquely.

We search for the function W^+ (21.5) in the form

$$W^+(\xi) = HG(\xi) + \frac{1}{2}\xi_1 + C_\omega(h) + Z(h^{-1}\xi)\chi(\xi) + \cdots, \tag{21.31}$$

where $C_\omega(h)$ is a constant, Z is a boundary layer term, and χ is a smooth cut-off function $\chi(\xi)$, which satisfies

$$\chi(\xi) = \begin{cases} 0, & \text{for } |\xi| > H/3, \\ 1, & \text{for } |\xi| < H/6. \end{cases}$$

Let us set $C_\omega(h)$ and Z as follows. Changing to the stretched coordinates $\zeta = h^{-1}\xi$, and computing the discrepancy in the Dirichlet condition on $\partial\omega^h$, up to the order $O(1)$, yields the following problem for the boundary layer term:

$$\begin{cases} \Delta Z(\zeta) = 0, & \zeta \in \mathbb{R}^2 \setminus \overline{\omega}, \\ Z(\zeta) = -\frac{H}{2\pi} \ln(h|\zeta|) - HG_0 - C_\omega(h), & \zeta \in \partial\omega. \end{cases} \tag{21.32}$$

Considering the logarithmic capacity potential P, cf. Chapter II in [La72], which is a harmonic function in $\mathbb{R}^2 \setminus \overline{\omega}$, vanishes on $\partial\omega$, and admits the representation

$$P(\zeta) = \frac{1}{2\pi} \ln|\zeta| - \frac{1}{2\pi} \ln c_{\log}(\omega) + \widetilde{P}, \quad \widetilde{P}(\zeta) = O(|\zeta|^{-1}), \quad |\zeta| \to \infty, \tag{21.33}$$

where $c_{\log}(\omega) > 0$ is the logarithmic capacity of the set $\overline{\omega}$ in the plane $\mathbb{R}^2$, yields the decaying solution of problem (21.32),

$$Z(\zeta) = H\widetilde{P}(\zeta), \tag{21.34}$$

provided that

$$C_\omega(h) = -\frac{H}{2\pi}\ln h - \frac{H}{2\pi}\ln c_{\log}(\omega) - HG_0. \qquad (21.35)$$

In this way, we have determined C_ω and Z (21.31), and we can write, cf. (21.7),

$$p_{+\pm}(\Xi^h) = C_\omega(h) + \tilde{p}_{+\pm}(\Xi^h), \qquad (21.36)$$

where the remainders $\tilde{p}_{+\pm}(\Xi^h)$ have to be computed from the coefficients of the polarization matrix for the remainder function arising in the ellipsis in (21.31). In the next steps we obtain the estimate

$$|\tilde{p}_{+\pm}(\Xi^h)| \leq Ch(1 + |\ln h|)^2, \qquad (21.37)$$

which in view of (21.35) allows us to rewrite the dominant part in (21.36) as

$$p_{+\pm}(\Xi^h) = -\frac{H}{2\pi}\ln h + O(1), \qquad (21.38)$$

and this ends the proof of the theorem.

Second Step: The polarization matrix for the remainder of W^+ in (21.31).

In order to estimate the remainders $\tilde{p}_{+\pm}(\Xi^h)$ in (21.36), we use the following decomposition for W^+:

$$W^+(\xi) = \chi^h(\xi)\left(HG(\xi) + \frac{1}{2}\xi_1\right) + (1 - \chi^h(\xi))H\left(\frac{1}{2\pi}\ln|\xi| + G_0\right)$$

$$+ C_\omega(h) + Z(h^{-1}\xi)\chi(\xi) + \widehat{W}^+(\xi),$$

where χ^h is a smooth cut-off function satisfying

$$\chi^h(\xi) = \begin{cases} 1, & \text{for } |\xi| > 2R_0h, \\ 0, & \text{for } |\xi| < R_0h, \end{cases}$$

with $|\nabla^p\chi^h(\xi)| \leq c_p h^{-p}$, $R_0 > 0$ is a fixed real such that $\omega \subset \{\zeta : |\zeta| < R_0\}$, and $\widehat{W}^+(\xi)$ is the solution of a problem that we set here below, cf. (21.39)–(21.40). Indeed, by the definition of χ^h and Z, for $\xi \in \partial\omega^h$, we have

$$W^+(\xi) = H\left(\frac{1}{2\pi}\ln|\xi| + G_0\right) + C_\omega(h) + Z(h^{-1}\xi)\chi(\xi) + \widehat{W}^+(\xi) = \widehat{W}^+(\xi)$$

and, hence, we have to impose the boundary condition

$$\widehat{W}^+(\xi) = 0, \quad \xi \in \partial\omega^h. \qquad (21.39)$$

The remainder $\widehat{W}^+$ is still ξ_2-periodic in $\varXi^h$ and satisfies the Poisson equation

$$-\Delta\widehat{W}^+(\xi) = \widehat{F}^+(\xi) := -[\Delta, \chi^h(\xi)]Y(\xi) - [\Delta, \chi(\xi)]Z(h^{-1}\xi), \quad \xi \in \varXi^h, \tag{21.40}$$

where $[\Delta, \chi^h]Y = 2\nabla\chi^h \cdot \nabla Y + Y\Delta\chi^h$ is the commutator of the Laplacian and the cut-off function χ^h and

$$Y(\xi) = \frac{1}{2}\xi_1 + H\left(G(\xi) - \frac{1}{2\pi}\ln|\xi| - G_0\right).$$

Next, under the basis of the following estimates for $\widehat{W}^+$:

$$\|\nabla\widehat{W}^+; L^2(\varXi^h)\| \leq Ch(1 + |\ln h|), \tag{21.41}$$

and

$$\|\widehat{W}^+(\pm T_0; .); L^2(-H/2, H/2)\| \leq C(1 + |\ln h|)h(1 + |\ln h|), \tag{21.42}$$

for a fixed $T_0 > H/3$, the Fourier representation of $\widehat{W}^+$ and application of the Green formula (cf. (21.7) and (21.29)) give

$$\widetilde{p}_{+\pm}(\varXi^h) = \lim_{T\to\infty}\frac{1}{H}\int_0^H \widehat{W}^+(\pm T, \xi_2)d\xi_2 = \frac{1}{H}\int_0^H \widehat{W}^+(\pm T_0, \xi_2)d\xi_2.$$

Finally, applying the Cauchy–Schwarz inequality, (21.41) and (21.42) lead to the estimates (21.37) which allows us to assert (21.38).

Thus, it remains to obtain (21.41) and (21.42), which is performed in the next step below.

Third Step: The estimates for $\widehat{W}^+$.

In order to show (21.41), we need to prove some estimate for the right-hand side of the integral identity

$$(\nabla\widehat{W}^+, \nabla v)_{\varXi^h} = (\widehat{F}_Y^+, v)_{\varXi^h} + (\widehat{F}_Z^+, v)_{\varXi^h}, \quad \forall v \in \mathscr{H}_0^1(\varXi^h), \tag{21.43}$$

which defines a functional in $\mathscr{H}_0^1(\varXi^h)$, $\mathscr{H}_0^1(\varXi^h)$ being the space completion of $C_{c,per}^\infty(\overline{\varXi^h})$ with the norm $\|\nabla v; L^2(\varXi^h)\|$, and the functions $\widehat{F}^+$ and $\widehat{F}_Z^+$ are defined by

$$\widehat{F}_Y^+(\xi) = -[\Delta, \chi^h(\xi)]Y(\xi) \quad \text{and} \quad \widehat{F}_Z^+(\xi) = -[\Delta, \chi(\xi)]Z(h^{-1}\xi),$$

respectively. Here and in what follows, $C^\infty_{c,per}(\overline{\Xi^h})$ is the space of the infinitely differentiable ξ_2-periodic functions, vanishing on $\partial\omega^h$ (cf. (21.39)) and with compact support in $\overline{\Xi^h}$.

The estimates to be proved read

$$\left|(\widehat{F}^+_Y, v)_{\Xi^h}\right| + \left|(\widehat{F}^+_Z, v)_{\Xi^h}\right| \le ch(1 + |\ln h|) \|v; \mathcal{H}^1_0(\Xi^h)\|, \quad \forall v \in \mathcal{H}^1_0(\Xi^h). \tag{21.44}$$

Let us show (21.44). First, we observe that, because of the cut-off functions χ^h and χ, the supports of $\widehat{F}^+_Y$ and $\widehat{F}^+_Z$ satisfy

$$\widehat{F}^+_Y(\xi) = 0 \text{ outside the annulus } a^h = \{\xi : R_0 h \le |\xi| \le 2R_0 h\},$$

$$\widehat{F}^+_Z(\xi) = 0 \text{ outside the annulus } A = \{\xi : \frac{H}{6} \le |\xi| \le \frac{H}{3}\}.$$

In addition,

$$|\widehat{F}^+_Y(\xi)| \le \frac{C}{h} \text{ for } \xi \in a^h, \tag{21.45}$$

$$|\widehat{F}^+_Z(\xi)| \le C\left((1 + h^{-1}|\xi|)^{-2} + (1 + h^{-1}|\xi|)^{-1}\right) \le Ch \text{ for } \xi \in A. \tag{21.46}$$

Indeed, we get (21.45) thanks to

$$|Y(\xi)| \le C|\xi|, \quad |\nabla Y(\xi)| \le C,$$

and we get (21.46) thanks to (cf. (21.33) and (21.34))

$$|Z(\zeta)| \le C(1 + |\zeta|)^{-1}, \quad |\nabla_\zeta Z(\zeta)| \le C(1 + |\zeta|)^{-2}.$$

Second, we note that, since the Dirichlet condition is imposed on the small contour $\partial\omega^h$ only, it proves useful to use the Hardy inequality

$$\int_\delta^d \left|\ln\frac{\tau}{\delta}\right|^{-2} |V(\tau)|^2 \frac{d\tau}{\tau} \le 4 \int_\delta^d \left|\frac{dV}{d\tau}(\tau)\right|^2 \tau d\tau, \quad \forall V \in C^\infty_c(\delta, d], \tag{21.47}$$

for any $d > \delta$ (in particular, $V(\delta) = 0$) considering $\delta = hR_0$ and $d = H/3$. Furthermore, the Poincare inequality in the coordinates ζ gives

$$\|\nabla v; L^2(B_{2R_0 h} \setminus \omega^h)\|^2 \ge ch^{-2}\|v; L^2(B_{2R_0 h} \setminus \omega^h)\|^2 \tag{21.48}$$

due to the Dirichlet condition on $\partial\omega^h$.

Then, we apply (21.47) to the product $V = v\chi^h$ so that we have

$$\|\nabla v; L^2(B_{H/3} \setminus \omega^h)\|^2 \geq c\|\tau^{-1}(1 + |\ln \tau| + |\ln h|)^{-1}v; L^2(B_{H/3} \setminus \omega^h)\|^2,$$
(21.49)

which is due to

$$\left|\ln \frac{\tau}{hR_0}\right|^{-1} \geq C(1 + |\ln h| + |\ln \tau|)^{-1},$$

and (21.48). Hence, by (21.45), we obtain the estimate

$$\left|(\widehat{F}_Y^+, v)_{\Xi^h}\right| \leq c \left(\int_{a^h} \tau^2(1 + |\ln \tau| + |\ln h|)^2|\widehat{F}_Y^+(\xi)|^2 d\xi\right)^{\frac{1}{2}} \|\nabla v; L^2(\Xi^h)\|$$

$$\leq c\left(h^2(1 + |\ln h|)^2 \frac{1}{h^2} mes_2(a^h)\right)^{\frac{1}{2}} \|v; \mathscr{H}_0^1(\Xi^h)\|$$

$$\leq ch(1 + |\ln h|) \|v; \mathscr{H}_0^1(\Xi^h)\|.$$

Besides, from (21.49) and (21.46), we get

$$\left|(\widehat{F}_Z^+, v)_{\Xi^h}\right| \leq ch(1 + |\ln h|) \|v; \mathscr{H}_0^1(\Xi^h)\|.$$

Consequently, we have proved (21.44).

Now, from (21.44) we conclude that the norm of the functional on the right-hand side of (21.43) is $O(h(1 + |\ln h|))$ and, therefore, the solution $\widehat{W}^+$ meets the estimate (21.41). Furthermore, the above weighted estimates prove that, for each fixed $T_0 > H/3$, the estimate (21.42) holds true. Indeed, the factor $(1 + |\ln h|)$ in the estimate above comes from the inequality

$$\|\widehat{W}^+; L^2((-T_0, T_0) \times (-H/2, H/2) \setminus \omega^h)\|^2$$

$$\leq C\left(\|\nabla \widehat{W}^+; L^2((-T_0, T_0) \times (-H/2, H/2) \setminus \omega^h)\|^2\right.$$

$$\left. + \|\widehat{W}^+; L^2(B_{H/3} \setminus \omega^h)\|^2\right),$$

and using (21.49) for the second term on the right-hand side in the above inequality. The other factor, namely $h(1 + |\ln h|)$, comes from (21.41). Therefore, the theorem is proved.

Acknowledgements The research of the first author has been partially supported by the Russian Foundation on Basic Research grant 18-01-00325. The research of the second author has been partially supported by MINECO, through the Severo Ochoa Programme for Centres of Excellence in RaD (SEV-2015-0554) and MTM2017-89976-P. The research of the third author has been partially supported by MINECO, MTM2013-44883-P.

References

[Ko67] Kondratyev, V.A.: Boundary value problems for elliptic equations in domains with conic and corner points. *Transactions Moscow Matem. Soc.* 16, 227–313 (1967).

[La72] Landkof, N.S.: *Foundations of modern potential theory*. Springer-Verlag (1972).

[Na99] Nazarov, S.A.: The polynomial property of self-adjoint elliptic boundary-value problems and the algebraic description of their attributes. *Uspehi mat. nauk.* 54, n. 5, 77–142 (1999). English transl.: *Russ. Math. Surveys.* 54, n. 5, 947–1014 (1999).

[NaEtAl18] Nazarov, S.A., Orive-Illera, R., and Pérez-Martínez, M.-E.: Asymptotic structure of the spectrum in a Dirichlet-strip with double periodic perforations. Submitted (2018).

[NaPl94] Nazarov, S.A., and Plamenevskii, B.A.: *Elliptic Problems in Domains with Piecewise Smooth Boundaries*, Walter de Gruyter (1994).

[PoSz51] Polya, G., and Szego, G.: *Isoperimetric Inequalities in Mathematical Physics*, Annals of Mathematics Studies, no. 27, Princeton University Press (1951).

Chapter 22
Operator Perturbation Approach for Fourth Order Elliptic Equations with Variable Coefficients

Julia Orlik, Heiko Andrä, and Sarah Staub

22.1 Periodic Boundary Value Problem

Let $Y' \in \mathbb{R}^d$ be a bounded domain with a heterogeneous structure, such that Y' is a representative periodic cell of the structure. For the periodic functions φ on Y' with a zero mean value $\langle \varphi \rangle = 0$ we introduce the Sobolev space

$$H^2_{per[0]}(Y') = \{\varphi \in H^2_{per}(Y') : \langle \varphi \rangle = 0\}.$$

Furthermore, we use the abstract index notation for tensors in $\mathbb{R}^d$. We introduce Latin indices $ijkl = 1, \ldots, d$ and unit vectors e_i, $i = 1, \ldots, d$. We consider the periodic boundary value problem (PBVP) to find $w \in H^2_{per[0]}(Y')$ from

$$\frac{\partial^2}{\partial y_k \partial y_l}\left(C_{ijkl}(y)\left(E_{0\,ij} + \frac{\partial^2 w(y)}{\partial y_i \partial y_j}\right)\right) = 0, \quad \forall y \in Y'. \tag{22.1}$$

Here, the tensor $E_0 \in \mathbb{R}^{d \times d}$ denotes the given mean value of rotational deformations.

The coefficients of the fourth-order tensor $C \in L^\infty(Y', \mathbb{R}^{d \times d \times d \times d})$ are bounded and satisfy the symmetry conditions

$$C_{ijkl} = C_{klij}, \; C_{ijkl} = C_{jikl} = C_{ijlk}, \; \exists \alpha > 0 : \alpha\xi : \xi \leq \xi : C(y) : \xi \leq \alpha^{-1}\xi : \xi \tag{22.2}$$

for any symmetric non-zero tensor $\xi \in \mathbb{R}^{d \times d}$ and $y \in Y'$.

J. Orlik (✉) · H. Andrä · S. Staub
Fraunhofer ITWM, Kaiserslautern, Germany
e-mail: julia.orlik@itwm.fraunhofer.de; heiko.andrae@itwm.fraunhofer.de;
sarah.staub@itwm.fraunhofer.de

© Springer Nature Switzerland AG 2019   283
C. Constanda, P. Harris (eds.), *Integral Methods in Science and Engineering*,
https://doi.org/10.1007/978-3-030-16077-7_22

Remark 1 In the two-dimensional case $Y' \subset \mathbb{R}^2$, one important application is the bending of plates made of heterogeneous materials. According to [FrEtAl02], for thin three-dimensional plate-like structures, where the ratio between the thickness and the representative lateral size tends to zero, the theoretical limits lead to averaging in the thickness and then solving the plate bending equation with variable coefficients in 2D-domains.

22.1.1 Problem in the Weak and Operator Form

For a scalar function w and a tensor-valued function g we introduce the following operators to shorten the notation:

$$Dw(y) := \nabla \otimes \nabla w(y) = \frac{\partial w}{\partial y_k \partial y_l}(y)e_k \otimes e_l, \tag{22.3}$$

$$D^*g(y) := \nabla \cdot \nabla \cdot g(y) = \text{div div } g(y) = \frac{\partial^2 g_{kl}}{\partial y_k \partial y_l}(y), \quad \Delta^2 = D^*D, \tag{22.4}$$

$$[Aw](y) := D^*(C(y)Dw), \tag{22.5}$$

where Einstein's summation convention is used on the last line. So Eq. (22.1) can be written in the short form

$$D^*(C(y)(E_0 + Dw(y))) = 0, \quad y \in Y', \quad \text{or} \quad Aw = -D^*g, \quad g := C(y) : E_0. \tag{22.6}$$

The weak formulation of the problem (22.1) is given by

$$\langle C(Dw + E_0) : D\varphi \rangle = 0, \quad \forall \varphi \in H^2_{per[0]}(Y'). \tag{22.7}$$

where $C = C(y)$

Now, we introduce the bilinear form

$$c(w, v) := \int_{Y'} C_{ijkl}(y)[\nabla \otimes \nabla w]_{ij}[\nabla \otimes \nabla v]_{kl}(y)dy, \quad \forall w, v \in H^2_{per[0]}(Y'), \tag{22.8}$$

which is symmetric, because the tensor of coefficients is symmetric, continuous and coercive.

We also introduce an operator $L : H^2(Y') \to H^{-2}(Y')$ and the right-hand side functional

$$Lw := c(w, \cdot), \quad l_y(v) = \int_{Y'} D^* : gv(y)dy. \tag{22.9}$$

Our problem (22.1) can be reformulated in the operator form: Find $u \in H^2_{per[0]}(Y')$ for $l_y \in H^{-2}(Y')$ and $L \in \mathscr{L}(H^2_{per}(Y'), H^{-2})$, satisfying

$$Lw = l_y. \tag{22.10}$$

According to [Pa16], the norm in $H^2_{per[0]}(Y')$ can be introduced in one of the equivalent ways:

$$\|\varphi\|^2_{H^2_{per[0]}(Y)} =< |\nabla^2\varphi|^2 + |\nabla\varphi|^2 + |\varphi|^2 >, \quad \|\varphi\|^2_{H^2_{per[0]}(Y)} =< |\nabla^2\varphi|^2 > \tag{22.11}$$

The equivalence of the norms is ensured by the Poincare inequality

$$< |\varphi|^2 > \le C_P < |\nabla\varphi|^2 > \quad \forall \varphi \in H^1(Y') \text{ with } < \varphi >= 0.$$

For any periodic functions $w, v \in C^4(\bar{Y}')$ we have the second Green identity,

$$\int_{Y'} \left[v\,Aw - Av\,w \right] dx = \left\langle T^+w\,,\,\gamma^+v \right\rangle_{\partial Y'} - \left\langle T^+v\,,\,\gamma^+w \right\rangle_{\partial Y'}.$$

Take $v = G$, where $G\star \equiv (L)^{-1}$ and $G(y, x)$ is the Green's function for the periodic problem $Av = \delta(x)$ and the second expression gives w. The first and second boundary integrals disappear, since the Green-function, DG, w and Dw satisfy periodic conditions.

$$-w(x) + \int_{Y'} G(x, y)\,Aw(y)\,dy = 0, \quad x \in Y'.$$

Replacing Aw by the right-hand side of Eq. $(22.6)_2$, we obtain

$$w(x) = \left(G \star D^*g\right)(x) = \int_{Y'} G(x, y)D^*g(y)dy.$$

Note that

$$Dw(x) = D\left(G \star D^*g\right)(x) = \int_{Y'} D_x G(x, y) \cdot D^*_y g(y)\,dy$$

$$= \int_{Y'} D_y(D_x G(x, y)) : g(y)\,dy.$$

Lemma 1 *Let $C_0 \in \mathbb{R}^{d \times d \times d \times d}$ be a constant symmetric tensor and $w \in H^2(Y')$ a scalar function, then $D^*(C_0 w) = C_0 : Dw$.*

Proof Here we use the following rule for a constant matrix C_0 and a scalar w:

$$D^*(C_0 w) = w D^* C_0 + C_0 : Dw + 2(\text{div } C_0) \cdot \nabla w;$$

$$D^*(C_0 : Dw) = D^* D^*(C_0 w) = C_0 :: (DDw) = -D^*(g). \quad (22.12)$$

Remark 2 Note that a fundamental solution for an anisotropic plate exists and is known. For example, a periodic fundamental solution can be found in a series. Anyway, it is possible to write a Newton-type potential, G, which will be inverse to the operator $C_0 :: DD[\cdot]$.

Let us check that $G\star : H^{-2}(Y') \to H^2_{per[0]}(Y')$, i.e., that $\delta \in H^{-2}(\mathbb{R}^2)$.

Definition 1 We define the norm $\|u\|^2_{H^s(\mathbb{R}^d)}$ for $u \in H^s(\mathbb{R}^d)$ as

$$\|u\|^2_{H^s(\mathbb{R}^d)} = \int_{\mathbb{R}^d} (|\xi|^2 + 1)^s |\tilde{u}|^2 d\xi, \quad (22.13)$$

where $\tilde{u}$ is the Fourier transform.

This definition can also be found on [Mc00, p. 76].

Lemma 2 *The Dirac function δ belongs to $H^s(\mathbb{R}^d)$ for any $s < \frac{-d}{2}$.*

Proof From the transformation formula for integrals it follows

$$\|\delta\|^2_{H^s(\mathbb{R}^d)} = \int_{\mathbb{R}^d} \frac{(1 + |\xi|^2)^s}{(2\pi)^d} d\xi \sim \frac{1}{(2\pi)^n} \int_1^\infty \rho^{2s} \rho^{n-1} d\rho$$

$$= \frac{1}{(2\pi)^n (2s + n)} < \infty, \quad \text{if} \quad s < \frac{-d}{2}. \quad (22.14)$$

Corollary 1 $\delta(x) \in H^{-2}(\mathbb{R}^2)$.

Remark 3 Following the first Green's Identity

$$\langle T^+ w, \gamma^+ v \rangle_{\partial Y'} = \int_{Y'} \left[v \, Aw + c(w, v) \right] dy$$

and accounting for the fact that $\langle T^+ w, \gamma^+ v \rangle_{\partial Y'}$ vanishes because of the periodicity, we can conclude that $v Aw = -Lwv$. Because of Lemma 1, $G\star \equiv (L)^{-1}$ exists for variable coefficients as well. We need to find a way to construct this Newton-type potential.

Potential Polarization Field

We refer to [EyMi99] and repeat the same approach applied to the second order elliptic PDE there to the higher order elliptic PDE with a symmetric coefficient matrix.

The idea is to modify the problem, given in the variational formulation to the integral equation of the second kind, with the integral operator mapping in the same space. To solve the desired equations, we consider a related problem with a reference operator L_0, corresponding to the constant reference tensor C_0.

$$D^*(C_0 : E(y)) + D^*((C(y) - C_0) : E(y)) = 0. \tag{22.15}$$

Denoting $C(y) - C_0 = \delta C(y)$ yields

$$- D^*(C_0 : (E(y) - E_0) = D^*(\delta C(y) : E(y)), \quad \text{or} \tag{22.16}$$

$$D^*(C_0 : Dw(y)) + D^*(\delta C(y) : Dw(y)) = -D^*(C_0 : E_0). \tag{22.17}$$

If the polarization tensor $P = \delta C(y) : E(y)$ were known,

$$D^*[C_0 : Dw] = -D^* P, \tag{22.18}$$

the related problem could be solved using the periodic Green's operator.

Lemma 3 *Problem (22.18) has a periodic unique solution $w \in H^2_{per[0]}(Y')$, which can be represented by (22.18) through an integral operator $\Gamma \star : L^2(Y', \mathbb{R}^{d \times d}) \to L^2(Y', \mathbb{R}^{d \times d})$ given by the formula*

$$[\Gamma \star P](y) \equiv -DL_0^{-1} D^* P(y), \quad \forall P \in L^2_{per}(Y', \mathbb{R}^{d \times d}) \tag{22.19}$$

$$[\Gamma \star P](y) = \int_{Y'} D_y [D_x G(x, y)] : P(y) dy,$$

$G(x, y)$ is the Green's function for the periodic 4th-order problem

$$C_0 :: (D_y D_y G) = \delta(y - x),$$

where $\delta(y - x)$ is the Dirac delta distribution.

Now we can rewrite (22.17) as the following

$$E(y) - E_0 = -\Gamma \star P(E), \quad \text{or} \quad E(y) + \Gamma \star \delta C(y) : E(y) = E_0 \tag{22.20}$$

and

$$E(y) = (I - \Gamma \star \delta C(y))^{-1} E_0 =: R \star E_0. \tag{22.21}$$

with

$$\Gamma \star \delta C := -DL_0^{-1}D^*\delta C, \quad [\Gamma \delta C](x,y) := D_y[D_x G(x,y)]\delta C(y) \in L^1(Y' \times Y')$$
$$(22.22)$$

22.1.2 Orthogonal Decomposition of the 4th-Order Differential Operator on Ker and Im

Our problem reduces to (22.18) and finding

$$P \equiv (C(y) - C_0) : E(y) \in V_{pot[0]}(Y', \mathbb{R}^{d\times d}).$$

Remark 4 If C_0 would be the effective coefficients (see (3.11) in [Pa16]), then

$$< C(y) : E(y) >= C_0 :< E > .$$

Following ideas from [GaMi98], [Pa16], [JiEtAl94], we decompose the L^2-space into the solenoidal and potential in the sense of higher-order div^n-free and $curl^n$-free, or higher-order potential tensor spaces:
  $V_{sol}(Y', \mathbb{R}^{d\times d}) = \{P = \{g_{sh}\}_{s,h} \in L^2(Y', \mathbb{R}^{d\times d})$ be symmetric, s. t. $D^*P = 0.\}$
  $V_{pot}(Y', \mathbb{R}^{d\times d}) = \left\{D\Gamma, \ \Gamma \in H^2_{per}(Y')\right\}.$
  According to (22.7), one has the orthogonality property:
  $V_{sol}(Y', \mathbb{R}^{d\times d}) = V_{pot}(Y', \mathbb{R}^{d\times d})^{\perp}$
  $= \{P \in L^2(Y', \mathbb{R}^{d\times d}) :< P : e >= 0 \quad \forall e \in V_{pot}(Y', \mathbb{R}^{d\times d})\}.$
  Define another potential field
  $V^*_{pot[0]}(Y', \mathbb{R}^{d\times d}) = \left\{P \in V_{sol}(Y', \mathbb{R}^{d\times d}) \text{ and } \langle P \rangle = 0. \ \exists \ \Gamma \in H^2_{per}\right.$
$(Y', \mathbb{R}^{d\times d\times d\times d}),$
  s.t. $D^*\Gamma = P, \ \Gamma^{sh}_{ij} = \Gamma^{sh}_{ji}$ (sym), $\Gamma^{sh}_{ij} = -\Gamma^{ij}_{sh}$ (skew-sym)$\Big\}.$
  $V_{sol[0]}(Y', \mathbb{R}^{d\times d}) = \{P \in L^2(Y', \mathbb{R}^{d\times d}) : \ D^*P = 0, \quad < P >= 0\}.$
  $V_{sol}(Y', \mathbb{R}^{d\times d}) = V_{sol[0]}(Y', \mathbb{R}^{d\times d}) \otimes \mathbb{R}^{d\times d}$
  $V_{pot}(Y', \mathbb{R}^{d\times d}) = V_{pot[0]}(Y', \mathbb{R}^{d\times d}) \otimes \mathbb{R}^{d\times d}$

Lemma 4 *For* $g \in V_{sol,[0]}(Y')$ $\langle f, g \rangle \equiv \int_{Y'} f : g \, dy = 0, \forall f \in V^*_{pot}(Y').$

Proof Let Γ be a solution of the periodic boundary value problems for the equations $D^*\Gamma^{ij}_{sh} = g_{sh}, D^*g = 0.$
Then

$$\int_{Y'} D^*_{ij} \Gamma^{ij}_{sh} f_{sh} \, dy = \int_{Y'} \Gamma^{ij}_{sh} D_{ij} f_{sh} \, dy \overset{\Gamma^{ij}_{sh} \text{ skew sym.}}{=} 0.$$

That is, the following Weyl's decomposition (see [JiEtAl94]), known for the theory of second order elliptic operators, should be valid also for the higher-order operators:

$$L^2(Y') = V_{sol[0]}(Y', \mathbb{R}^{d\times d}) \otimes V_{pot[0]}(Y', \mathbb{R}^{d\times d}) \otimes \mathbb{R}^{d\times d}$$
$$= V_{pot[0]}(Y', \mathbb{R}^{d\times d}) \otimes V_{sol}(Y', \mathbb{R}^{d\times d}).$$

We describe the way to construct a Newton potential. Let us take the Fourier series expansion

$$g_{ij}(y) = \sum_{0\neq n\in\mathbb{Z}^d} g_n^{ij} e^{2\pi n\cdot y\sqrt{-1}}.$$

Then the condition $D^*P = 0$ means that

$$g_n^{ij} n_i n_j = 0 \quad \text{for all} \quad n. \tag{22.23}$$

We introduce the matrix $\Gamma^{sh} = \{\Gamma_{ij}^{sh}\}_{i,j}$ with the Fourier series coefficients

$$\Gamma_{ij,\,n}^{sh} = (-g_n^{ij} n_s n_h + g_n^{sh} n_i n_j)|n|^{-4}(-4\pi^2)^{-1}.$$

The n-th term of the Fourier series satisfies

$$D^*[\Gamma_n^{sh} e^{2\pi n\cdot y\sqrt{-1}}] = \frac{\partial^2}{\partial y_i \partial y_j} \Gamma_{ij,n}^{sh} e^{2\pi n\cdot y\sqrt{-1}}$$
$$= e^{2\pi n\cdot y\sqrt{-1}}|n|^{-4}(-g_n^{ij} n_s n_h n_i n_j + g_n^{sh} n_i n_j n_i n_j) = e^{2\pi n\cdot y\sqrt{-1}} g_n^{sh}. \tag{22.24}$$

Here relation (22.23) and the identity $n_i n_j n_i n_j = |n|^4$ were taken into account.

22.1.3   Periodic Fundamental Solution of the Biharmonic Equation

Assume now that the tensor C_0 acts on a matrix ξ as $C_0\xi = \lambda_0 C_0(\mathrm{Tr}\,\xi)I$, where $\mathrm{Tr}\,\xi = \xi_{ii}$ is the trace of the matrix ξ, I is the identity matrix, and λ_0 is a scalar. Obviously, $C_0\xi \cdot \xi = \lambda_0(\mathrm{Tr}\,\xi)^2$, and the matrix $\xi = D\varphi$ satisfies $\mathrm{Tr}\,D\varphi = \Delta\varphi$, $C_0 D\varphi \cdot D\varphi = \lambda_0\Delta\varphi\Delta\varphi$.

Let $E = E_0 + Dw$ be the bending deformations of the plate, or the curvature.

$$E_{n+1}(y) = \Gamma \star [(I - \lambda_0^{-1}C(y)) : E_n(y)], \quad n = 0, 1, 2, 3, \ldots,$$

where

$$[\Gamma \star E](x) = \int_{Y'} D_y(D_x G(x, y)) : E(y)dy,$$

$G(x, y)$ is the Green's function for the periodic biharmonic problem, i.e. for each fixed $x \in Y'$ the function $G(x, y)$ satisfies the equation

$$\Delta_y^2 G = \delta(y - x), \qquad (22.25)$$

where $\delta(y - x)$ denotes the Dirac measure.

We redefine here $g = -\lambda_0^{-1} g$. Let us denote by $(\Delta^2)^{-1} D^* g$ for $g \in (L^2(Y'))^{d \times d}$ a solution of the periodic problem

$$\Delta^2 w = -D^* g, \qquad v \in H_{per[0]}^2(Y'), \qquad (22.26)$$

which exists due to Lemma 1.

We can construct a periodic fundamental solution in the following way. We take the Fourier series expansion $G(y) = \sum_{n \in \mathbb{Z}^d} c_n(G) e^{2\pi n \cdot y \sqrt{-1}}$, where the Fourier coefficients

$$c_n(G) = \int_{Y'} G(x) e^{2\pi n \cdot y \sqrt{-1}} dy, \quad n \in \mathbb{Z}^d.$$

The Fourier series for the Dirac sequence has coefficients $c_n(\delta) = 1 : \delta(y) = \sum_{0 \neq n \in \mathbb{Z}^d} e^{2\pi n \cdot y \sqrt{-1}}$. The n-th term of the Fourier series satisfies

$$\Delta^2[c_n(G) e^{2\pi n \cdot y \sqrt{-1}}] \equiv D^* D[c_n(G) e^{2\pi n \cdot y \sqrt{-1}}]$$

$$= \sum_{i=1}^{d} \sum_{j=1}^{d} \frac{\partial^2}{\partial y_i \partial y_j} \left(\frac{\partial^2}{\partial y_i \partial y_j} c_n(G) e^{2\pi n \cdot y \sqrt{-1}} \vec{e}_i \otimes \vec{e}_j \right)$$

$$= -e^{2\pi n \cdot y \sqrt{-1}} (2\pi)^4 c_n(G)(n_i n_j n_i n_j). \qquad (22.27)$$

Here the identity $n_i n_j n_i n_j = |n|^4$ is taken into account.

$$c_n(G) = -(2\pi)^{-4} |n|^{-4}$$

$$G(y) = -(2\pi)^{-4} \sum_{0 \neq n \in \mathbb{Z}^d} |n|^{-4} e^{2\pi n \cdot y \sqrt{-1}}. \qquad (22.28)$$

Remark 5 According to [Wi77] (see Appendix), if take Y' such that it will contain many periodicity cells, it is possible to take the fundamental solution and do not look for the periodic Green's function. Since we require that the right-hand side of the equation has a zero mean-value over Y', the boundary values oscillate around zero and, according to the St. Veinant hypothesis, the corresponding boundary layer decays exponentially. Its thickness is proportional to the period of the structure (or characteristic size of inclusions related to the distance between them) (see, e.g., [To65]).

22.2   Neumann Series and Its Convergence Estimate by Spectral Properties

Definition 2 Let $\mathscr{A}$ be a Banach-algebra with unit, $\Gamma \in \mathscr{A}$ and $\sigma(\Gamma\star)$ be the spectrum of $\Gamma\star$. $\rho(\Gamma\star) = \sup\{|\lambda| \ |\lambda \in \sigma(\Gamma\star)\}$ is the spectral radius of $\Gamma\star$.

The class of kernels of integral operators on the compact set forms a Banach algebra with product $\star$ without a unit. We can add the identity as a unit to our algebra $\mathscr{A}$.

Definition 3 Let $Y' \subset \mathbb{R}^d$ be a compact set, and $\Gamma \star \delta C \in \mathscr{L}(L^2(Y'))$. Then there exists a resolvent $\underline{\tilde{R}}(\lambda)$ of $\Gamma\star$ in $\mathscr{L}(L^2(Y')))$, given by

$$\underline{\tilde{R}}(\lambda) := (\lambda I - \Gamma \star \delta C)^{-1} = \sum_{j=0}^{\infty} \frac{(\Gamma \star \delta C)^j}{\lambda^{j+1}}, \quad \forall \lambda \in \mathscr{C} \setminus \sigma(\Gamma \star \delta C). \quad (22.29)$$

In order for solution to (22.20) to exist, i.e., (22.29) the Neumann series to converge for $\lambda = 1$, the spectral radius should be $\rho(\Gamma \star \delta C) < 1$. And this depends on the choice of the constant coefficient matrix C_0. We postpone this question to the next section and discuss properties of the resolvent kernel now.

We recall further properties of the resolvent and refer to [SaVa02, Corol. 5.5.1.].

Lemma 5 *Let $R\star$ be an integral operator and R be its kernel and $c_n(R)$ ($n \in \mathbb{Z}^d$) its Fourier coefficients. If $|c_n(R)| \leq c|n|^\alpha$ with some $\alpha \in \mathbb{R}$, then $R\star \in \mathscr{L}(H^\lambda, H^{\lambda-\alpha})$ for any $\lambda \in \mathbb{R}$. Moreover, if $c_1|n|^\alpha \leq |c_n(R)| \leq c_2|n|^\alpha$ ($n \in \mathbb{Z}^d$) with some positive constants c_1 and c_2, then $R\star$ builds an isomorphism between H^λ and $H^{\lambda-\alpha}$ for any $\lambda \in \mathbb{R}$.*

Lemma 6 *Let R be a resolvent kernel to G and $c_n(R)$ ($n \in \mathbb{Z}^d$) its Fourier coefficients. $|c_n(R)| \leq c|n|^{-4}$ and $R \in L^1(Y', \mathscr{L}(H^{-2}, H^2))$. That means, the resolvent kernel is weakly singular and has the same leading singularity as G (the fundamental solution of the 4th order PDE with constant coefficients). The last statement is true for the kernels' pare $\tilde{R}$ and Γ also.*

Proof Owing to [Gretal90, Definition 9.3.1], the resolvent kernel R of kernel G satisfies the following equality:

$$R + G \star R = R + R \star G = G. \quad (22.30)$$

In order to check that $|c_n(R)| \leq c|n|^{-4}$, we again apply the Fourier series technique, replacing the Fourier transform for the periodic functions. According to [SaVa02], the coefficients of the series, corresponding to a product of two functions correspond to the product of their coefficients. Then $(I + c_n(G))c(R) = c_n(G)$, equivalently (see [Gretal90, Th.2.8]) $(I + c_n(G))(I - c(R)) = I$. Recall that $c_n(G) = -(2\pi)^{-4}|n|^{-4}$. Hence, we can see by the developing in the series

$$|c(R)| = \sum_{l=1}^{\infty}((2\pi)^{-4}|n|^{-4})^l.$$

Furthermore, the expression (22.29) of the resolvent by the multiple multiplication of $G\star$ maps into the same algebra and the same space.

Theorem 1 *If for $\Gamma \star \delta C$ defined by (22.22) $\rho(\Gamma \star \delta C) < 1$, the unique solution of problem (22.17), $w \in H^2_{per[0]}(Y')$, exists and satisfies the estimate*

$$\|w\|_{H^2(Y')} \leq \sup_{j \in \mathbb{N}} \left\| (D_y D_x G\star)^j \right\|_{L^1(Y' \times Y')} \sum_{j=0}^{\infty} \left\| \frac{\delta C}{\alpha(C_0)} \right\|^j_{(L^\infty(Y'))^{d \times d}} \|E_0\|_{L^2(Y')},$$

$$(22.31)$$

where G is the periodic fundamental solution of the 4th-order PDE with constant coefficients C_0.

Proof This estimate is based on (22.11), estimate of the resolvent, given by the Neumann series (22.29) and the Lax-Milgram theorem.

In recent works [Su18], [NiXu17], [NiEtAl18], also the leading term asymptotics for the resolvent were discussed. We note that our results and techniques can also be generalized for high-order PDEs.

22.3 Bounds on $C_0(x)$

As we mentioned above, the Neumann series for problem (22.21) converges for $\lambda = 1$, if the spectral radius $\rho(\Gamma \star \delta C) < 1$. And this depends on the choice of the constant coefficient matrix C_0.

Let λ and E_λ be an eigenvalue and a corresponding eigenvector of

$$\lambda E_\lambda = -\Gamma \star ((C(x) - C_0) : E_\lambda), \quad (\lambda C_0 + (C(x) - C_0)) : E_\lambda = 0,$$
$$C'(x) : E_\lambda = 0,$$

where $C'(x) = C(x) - (1 - \lambda)C_0$. We need that for $|\lambda| \geq 1$ the last problem has only trivial solution, i.e.

$$\xi : C' : \xi > 0, \quad \text{or} \quad \xi : C' : \xi < 0, \quad \forall \xi \in \mathbb{R}^{d \times d}$$

In order for λ to be a non-trivial eigenvalue, C' cannot be either positive definite or negative definite. One must have, for the eigenvalues of the matrix

$$\exists x_1; \; \mu'_i(x_1) < 0, \quad \text{and} \quad \exists x_2; \; \mu'i(x_2) > 0, \; \forall i = 1, \dots, d.$$

In other words:

$$\min_i \left\{ \min_x \left(1 - \frac{\mu_i(x)}{\mu_{i0}} \right) \right\} \leq \lambda \leq \max_i \left\{ \max_x \left(1 - \frac{\mu_i(x)}{\mu_{i0}} \right) \right\}. \qquad (22.32)$$

A sufficient condition for the scheme to converge is $\rho < 1$, i.e., $|\lambda| < 1$, hence $\mu_{i0} > \mu_i(x)/2 > 0$, $i = 1, \dots d$. Under this constraint, we look for μ_{i0} giving bounds in (22.32) with opposite sign and equal absolute value. This is ensured by

$$\mu_{i0} = \frac{1}{2} (\min_x \mu_i(x) + \max_x \mu_i(x)). \qquad (22.33)$$

With these parameters, the spectral radius of $\Gamma \star \delta C$ is bounded from above by

$$\rho \leq \max_i \left\{ \frac{\max_x \mu_i(x) - \min_x \mu_i(x)}{\max_x \mu_i(x) + \min_x \mu_i(x)} \right\}. \qquad (22.34)$$

Furthermore, owing to [MiEtAl01], for the high contrast coefficients, it is better to choose

$$\mu_{i0} = -\sqrt{\min_x(\mu_i(x)) \cdot \max_x(\mu_i(x))}. \qquad (22.35)$$

22.3.1   Voigt-Reuss Bounds for the Effective Coefficients

Let L^{hom} be the effective operator with tensor C^{hom}, defined (see (3.6) in [Pa16]) by

$$J_0 = C^{hom} E_0 = < C(x)E(x) > = < C(y)(Dw(y) + E_0) >, \qquad (22.36)$$

where $J_0 = < J >$ the mean value of the field J. The approach to find the effective tensor for the fourth order PDE is presented in [Pa16]. The tensor C_{hom} inherits the properties of symmetry and positive definiteness.

The following Voigt-Reuss estimates for the homogenized tensor are known, e.g. form[JiEtAl94] for the homogenization of the second-order elliptic operators, and were generalized in [Pa16] for higher order PDEs:

$$\left\langle C^{-1}(x) \right\rangle^{-1} \leq C_{hom} \leq \langle C(x) \rangle, \tag{22.37}$$

called the Voigt-Reuss bracketing. Here C^{-1} is the inverse tensor for C and $< \cdot >$ is an averaging over the periodicity cell. The following examples can be found, e.g., in [JiEtAl94]

Example 1

(i) For a laminate with two isotropic faces along the faces with conductivities α and β the effective homogenized coefficient is the mean value of both and across the layers the harmonic mean.

(ii) For the chess-board structure, assuming $C_1 = \alpha I$, $C_2 = \beta I$, $C_{hom} = \sqrt{\alpha\beta} I$ and for a monocrystal C_1 with eigenvalues λ_1, λ_2, and $C_2 = R^T C_1 R$, R is an orthogonal matrix. The homogenized tensor will be $C_{hom} = \sqrt{\lambda_1\lambda_2} I$

References

[FrEtAl02] Friesecke, G., James, R.D. and Müller, S. A theorem on geometric rigidity and the derivation of nonlinear plate theory from three-dimensional elasticity. *Comm. Pure Appl. Math.*, 55(11):1461–1506, 2002.

[Pa16] Pastukhova, S. E. Operator error estimates for homogenization of fourth order elliptic equations, *St. Petersburg Math. 2*, Vol. 28 (2017), No. 2, Pages 273–289, 2016.

[MiEtAl01] J. C. Michel J. C., Moulinec, H., and Suquet, P. A computational scheme for linear and non-linear composites with arbitrary phase contrast, *Int. J. Numer. Meth. Engng*, 52:139–160 (DOI: 10.1002/nme.275), 2001.

[GaMi98] Grabovsky, Y. and Milton G. W. Rank one plus a null-Lagrangian is an inherited property of two-dimensional compliance tensors under homogenisation, *Proceedings of the Royal Society of Edinburgh*, 128A, 283–299, 1998.

[EyMi99] Eyre D. J. and Milton, G. W. A fast numerical scheme for computing the response of composites using grid refinement, *Eur. Phys. J. AP*, 6, 41–47, 1999.

[JiEtAl94] Jikov, V.V., Kozlov, S. M., and Oleinik, O. A. *Homogenization of Differential Operators and Integral Functionals*, Springer, 1994.

[Mc00] McLean W., *Strongly elliptic systems and boundary integral equations*, Cambridge University Press, 2000.

[Gretal90] Gripenberg, G., Londen, S.-O. and Staffan O. *Volterra Integral and Functional Equations* , Cambridge University Press, 1990.

[Wi77] J.R. Willis, J. R. Bounds and Self-Consistent Estimates for the Overall Properties of Anisotropic Composites. *Journal of the Mechanics and Physics of Solids*. 25. 185–202. https://doi.org/10.1016/0022-5096(77)90022-9 (1977)

[To65] Toupin, R. A. Saint-Venant's Principle, *Archive for Rational Mechanics and Analysis*, 1965 - Springer

[Su18] Suslina, T. A. Homogenization of the Dirichlet problem for higher-order elliptic equations with periodic coefficients, *St. Petersburg Math. J.,* Vol. 29, No. 2, Pages 325–362 http://dx.doi.org/10.1090/spmj/1496, 2018.

[NiXu17] Niu, W. and Xu, Y. Uniform boundary estimates in homogenization of higher-order elliptic systems, *Annali di Matematica Pura ed Applicata (1923-)*, 2017.

[NiEtAl18] Niu, W., Shen Z. and Xu, Y. Convergence rates and interior estimates in homogenization of higher order elliptic systems, *Journal of Functional Analysis*, 2018.

[SaVa02] Saranen, J. and Vainikko, G. *Periodic Integral and Pseudodifferential Equations with Numerical Approximation*, Springer, 2002.

Chapter 23
Extension of the Fully Lagrangian Approach for the Integration of the Droplet Number Density on Caustic Formations

Andreas Papoutsakis

23.1 Introduction

Caustic formations in dispersed two-phase flows appear as accumulation regions, and result to zones of high concentration of the dispersed phase. These formations play a significant role in a great variety of engineering, environmental and biological applications. Accordingly, the vortical structures of the in-cylinder carrier phase flow field during the fuel injection in internal combustion engines induce accumulation regions of the spray droplets [Sa14] and affect the local stoichiometry of the mixture [Pa18]. Sea currents and weather systems result to the local accumulation of pollutants, plastic debris and aerosols [Kn12]. In biological systems the accumulation of nutrients plays a significant role in the behaviour of microorganisms [Vo81] and affects the immigration of sea life and coral colonies [SeEtAl16]. The statistical identification of caustic formations with traditional Lagrangian approaches demands a computationally prohibitive number of representative particles [HeYo05]. The Fully Lagrangian Approach (Osiptsov method) [Os84] overcomes this problem by incorporating the solution of the droplet number conservation equation in Lagrangian form along a single trajectory.

FLA is based on the Lagrangian form of the continuity equation for the particulate phase, treated as a continuum, and incorporates a method for the calculation of the components of the Jacobi matrix of the transformation from the Eulerian to the Lagrangian coordinates along a single trajectory. This is essentially a method of characteristics for the solution of the continuity equation on the Lagrangian trajectories. This approach can deal with such complex cases as the regions of intersecting droplet trajectories and caustics and presents unique characteristics

A. Papoutsakis (✉)
City University of London, London, UK
e-mail: andreas.papoutsakis@city.ac.uk

© Springer Nature Switzerland AG 2019
C. Constanda, P. Harris (eds.), *Integral Methods in Science and Engineering*,
https://doi.org/10.1007/978-3-030-16077-7_23

in capturing the occurrence of the fine structure of the accumulation regions. In [IjEtAl09], the efficiency of the FLA for the calculation of the droplet number density and its modelling capability in identifying the spatial structure of caustics was demonstrated. The introduction of the FLA into the study of turbulent flows ([MeRe11] and [PiEtAl05]) resulted in the identification and analysis of spatial structures of the dispersed phase distribution using the moments of concentration.

In this paper the second order extension of the FLA which relates the point-wise number density as defined in the standard FLA to a number density defined at a finite given length-scale is presented. The second order FLA addresses the singularities of the standard FLA and demonstrates the integrability of the point-wise number density of the standard FLA on caustics. Furthermore, with the approach presented here the FLA number density can be related to a finite length scale inferred from the LES filter width Δ needed for the introduction of the FLA to turbulent flows [PaEtAl18].

In the next section the calculation of the number density, assuming a second order structure of the dispersed continuum by the introduction of the Hessian of the Lagrangian transformation (i.e. the curvature of the deformed dispersed continuum) is presented. In the third section a method for the calculation of the Hessian is derived which is similar to the initial value problem used for the calculation of the Jacobian for the standard FLA. In the fourth section an approach for the calculation of the Hessian magnitude across a caustic formation for multiple dimensions is presented. In the fifth section the method is implemented for the two-dimensional problem of the dispersion of inertia droplets in an array of Taylor vortices.

23.2   The Number Density in a Finite Volume

In Fig. 23.1 the temporal evolution of inertia droplets with Stokes number $St = 0.1$ dispersing in a field of homogeneous and isotropic turbulence taken from [PaEtAl16] is presented. It can be observed that the droplets accumulate creating caustic filaments. In the analysis presented in this work it is assumed that the caustic formations are characterised by this one-dimensional structure. Thus, one primary direction perpendicular to the filament of accumulated droplets can be assumed.

The one-dimensional distribution of droplets as a Stokesian droplet continuum along the primary axis x is considered at time t which corresponds to an initial Lagrangian distribution at the location x_0 at time $t = 0$ which spans from $x_0 = 0$ to $x_0 = 1$, as shown in Fig. 23.2. This continuum disperses and deforms as shown by the solid curve in Fig. 23.2. Function $x(x_0)$ is uniquely defined along the x_0 axis. The number density at any point C averaged over a finite length scale Δ is obtained by spatially averaging the number of droplets within this filtering volume of size Δ on the x axis. For each point of interest C the local Lagrangian coordinate system $\delta = x_0 - x_0^C$ and the local Eulerian coordinate system $\epsilon = x - x^C$ for the distribution

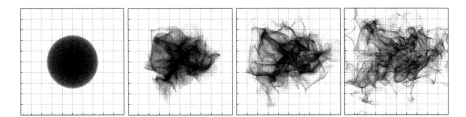

Fig. 23.1 The evolution of a subset of individual inertia droplets with St=0.1 initially located within a spherical volume. The results of calculation were taken from [PaEtAl16] and correspond to the simulation of droplet dispersion in homogeneous and isotropic turbulence, for dimensionless times $t^* = 0, 1, 1.5$ and 2.0

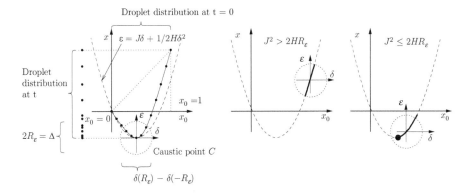

Fig. 23.2 Left: Schematic representation of the deformation of the dispersed continuum at the vicinity of a fold. The horizontal axis represents the initial Lagrangian coordinate x_0. The vertical axis corresponds to the current Eulerian coordinate x at time t. Straight dotted line: the distribution $x(x_0)$ for $t = 0$. Solid curve: the distribution $x(x_0)$ at time t. Dashed curve: the approximation of the continuum distribution in the second order FLA. Axes δ and ϵ correspond to the local coordinate system at C. Middle: The position of the filtering interval $\Delta = 2R_\epsilon$ around C when $J^2 > 2R_\epsilon H$. Right: The position of the filtering interval $\Delta = 2R_\epsilon$ around C when $J^2 < 2R_\epsilon H$. Straight dashed line: the approximation of the continuum distribution in the first order FLA

of the deformed continuum are defined. The point-wise number density n_d at a point ϵ_0 in the local coordinate system is calculated as:

$$n_d(\epsilon_0) = \left| \lim_{\epsilon \to \epsilon_0} \frac{\delta(\epsilon) - \delta(-\epsilon)}{2\epsilon} \right| = \left| \frac{\partial \delta}{\partial \epsilon} \right|_{\epsilon=\epsilon_0}, \tag{23.1}$$

where $\frac{\partial \epsilon}{\partial \delta}$ is equal to the Jacobian J and $n_d = 1/|\mathbf{J}|$. If C is located at a caustic point, where $\frac{\partial \epsilon}{\partial \delta}$ is zero, then the number density at C is infinitely large. It will be shown later in this section that the number density distribution can be integrable over a finite volume, and a filtered number density $\hat{n}_d$ can be defined and predicted. Assuming that the number density $\frac{\partial \delta}{\partial \epsilon}$ is integrable in the interval $\epsilon_0 \in [-R_\epsilon, R_\epsilon]$,

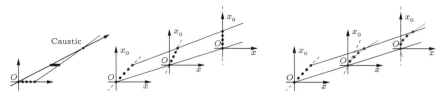

Fig. 23.3 Left: Schematic representation of the formation of a caustic point as inertia droplets converge to the same point by the effect of the carrier phase flow-field. Middle: The equivalent deformation of the dispersed continuum $x(x_0)$ for the same scenario assuming the first order FLA structure $x = Jx_0$. The linear approximation of the local continuum space results to an infinite number density where all droplets fall on the same point. Right: The equivalent deformation of the dispersed continuum $x(x_0)$ for the same scenario assuming a second order structure $x = Jx_0 + \frac{1}{2}Hx_0^2$. Accounting for the curvature of the local continuum space results to a finite large density on the caustic

the spatially averaged or filtered number density $\hat{n}_d$ for an interval $\Delta = 2R_\epsilon$ on the Eulerian space ϵ can be estimated as:

$$\hat{n}_d = \frac{1}{2R_\epsilon} \int_{-R_\epsilon}^{R_\epsilon} n_d d\epsilon = \left| \frac{\delta(\epsilon = R_\epsilon) - \delta(\epsilon = -R_\epsilon)}{2R_\epsilon} \right|, \tag{23.2}$$

where n_d is evaluated from Eq. (23.1).

For the first order FLA, a simple linear expression for the dispersed continuum distribution $\epsilon = J\delta$ is assumed, shown in Fig. 23.3 (Middle). In this case, Eq. (23.2) leads to the standard first order expression for the filtered number density $\hat{n}_d = 1/J$. As shown in Fig. 23.2, in the vicinity of a fold the linear approximation for $\epsilon(\delta)$ cannot represent the topology of the fold. The second order Taylor approximation for $\epsilon(\delta)$ (see Fig. 23.3) (Right) is introduced as:

$$\epsilon(\delta) = \left. \frac{\partial \epsilon}{\partial \delta} \right|_{\delta=0} \delta + \frac{1}{2} \left. \frac{\partial^2 \epsilon}{\partial \delta^2} \right|_{\delta=0} \delta^2. \tag{23.3}$$

The first derivative $\frac{\partial \epsilon}{\partial \delta} = \frac{\partial x}{\partial x_0}$ in Expression (23.3) is equal to the Jacobian J, while the second derivative $\frac{\partial^2 \epsilon}{\partial \delta^2} = \frac{\partial^2 x}{\partial x_0^2}$ is equal to the Hessian H of the transformation from the Eulerian x to the Lagrangian coordinate x_0 at the point $\epsilon = 0$. Thus, the approximation of the transformation $\epsilon(\delta)$ (dashed parabola in Fig. 23.2 **Left**) can be presented as:

$$\epsilon(\delta) = J\delta + \frac{1}{2}H\delta^2. \tag{23.4}$$

The orientation of the local coordinate system $\delta = \pm(x_0 - x_0^C)$ and $\epsilon = \pm(x - x^C)$ can be chosen in order to ensure that both H and J are positive. The solution to (23.4) can be presented as:

$$\delta(\epsilon) = \frac{-J + \sqrt{J^2 + 2H\epsilon}}{H} , \qquad (23.5)$$

where the root closest to the Taylor expansion reference point is chosen. Thus Expression (23.2) for the filtered number density can be evaluated as:

$$\hat{n}_d = \frac{\frac{-J+\sqrt{J^2+2HR_\epsilon}}{H} - \frac{-J+\sqrt{J^2-2HR_\epsilon}}{H}}{2R_\epsilon} = \frac{2}{\sqrt{J^2 + 2HR_\epsilon} + \sqrt{J^2 - 2HR_\epsilon}} , \qquad (23.6)$$

when $J^2 > 2HR_\epsilon$. For $J^2 < 2HR_\epsilon$, Solution (23.5) is defined only for $\epsilon = R_\epsilon$ as for $\epsilon = -R_\epsilon$ the filtering interval extends outside the fold of the dispersed continuum. In this case the integral in Eq. (23.2) is evaluated only in the interval $[\epsilon_{min}, R_\epsilon]$. ϵ_{min} corresponds to the minimum limit of the droplet distribution that occurs at $\delta_{min} = -J/H$. Thus, for the case when Jacobian J is small relative to the curvature of the fold, the number density inferred from Eq. (23.2) can be estimated as:

$$\hat{n}_d = \frac{\frac{-J+\sqrt{J^2+2HR_\epsilon}}{H} + \frac{J}{H}}{2R_\epsilon} = \frac{\sqrt{J^2 + 2HR_\epsilon}}{2R_\epsilon H} , \qquad (23.7)$$

when $J^2 - 2HR_\epsilon < 0$. As follows from Eq. (23.7), the number density for a caustic point $J = 0$ becomes:

$$\hat{n}_d = \frac{1}{\sqrt{2R_\epsilon H}} .$$

Both Eqs. (23.5) and (23.6) reduce to the classical FLA expression $n_d = 1/J$ for $R_\epsilon \to 0$. Thus, even if the Jacobian is zero at C the number density is integrable in the vicinity of C if Hessian H is non-zero (Morse critical point). In the general case, the integral of the number density averaged within a volume with radius R_ϵ can be calculated as:

$$\hat{n}_d = \begin{cases} \dfrac{2}{\sqrt{J^2+2HR_\epsilon}+\sqrt{J^2-2HR_\epsilon}} & \text{if } J^2 - 2HR_\epsilon > 0 \\[4mm] \dfrac{\sqrt{J^2+2HR_\epsilon}}{2R_\epsilon H} & \text{if } J^2 - 2HR_\epsilon < 0. \end{cases} \qquad (23.8)$$

When $J^2 = 2HR_\epsilon$ then $\hat{n}_d = \sqrt{2}/J$ for both the first and second branches of Eq. (23.8). For $J^2 \gg 2HR_\epsilon$, where the dispersed phase is diluted, the above expression simplifies to the classical FLA expression $\hat{n}_d = 1/J$.

23.3 The Calculation of the Hessian in the Second Order FLA for Multiple Dimensions

In this section a method for the calculation of the Hessian matrix along a trajectory is presented. This method is similar to the approach used in the classical FLA for the calculation of the Jacobian and is not bound to the one-dimensional assumption for the structure of the dispersed phase. An initial value problem for the time derivative of the Hessian, represented by the auxiliary variable ψ_{ijk}, can be formulated by differentiating the equivalent expression for ω_{ij} over the Lagrangian coordinate x_k^0. The Hessian matrix of the transformation from the Lagrangian to the Eulerian coordinates of the dispersed continuum is defined as $H_{ijk} = \frac{\partial^2 x_i}{\partial x_j x_k}$.

In the classical FLA, the Jacobian $J_{ij} = \frac{\partial x_i}{\partial x_j}$ is calculated by introducing the auxiliary variable ω_{ij} which is the time derivative of the Jacobian:

$$\omega_{ij} = \frac{\partial J_{ij}}{\partial t} = \frac{\partial}{\partial t}\left(\frac{\partial x_i}{\partial x_j^0}\right) = \frac{\partial V_i}{\partial x_j^0}. \tag{23.9}$$

As shown in [PaEtAl18] the calculation of the Jacobian can be extended for non-Stokesian droplets. Here, however, the Stokesian case is considered. In this case the droplet acceleration $\frac{\partial V_i}{\partial t} = f_i(S_i)$, $S_i = U_i - V_i$ is governed by the Stokes law $f_i = \frac{1}{\tau_d} S_i$ for the drag force f_i. Following [Os84] the initial value problem for ω_{ij} is obtained as:

$$\frac{\partial \omega_{ij}}{\partial t} = \frac{\partial}{\partial x_j^0}\left(\frac{\partial V_i}{\partial t}\right) = \frac{1}{\tau_d}\left(\frac{\partial x_m}{\partial x_j^0}\frac{\partial U_i}{\partial x_m} - \frac{\partial V_i}{\partial x_j}\right) = \frac{1}{\tau_d}\left(\frac{\partial U_i}{\partial x_m}J_{mj} - \omega_{ij}\right),$$

$$\tag{23.10}$$

where addition is assumed among the terms of index m. Equation (23.10) is integrated using the initial condition $\omega_{ij} = \frac{\partial V_i}{\partial x_j}\big|_{t=0}$ stemming from Eq. (23.9).

For the derivation of an expression of the Hessian H_{ijk}, a similar approach is followed by introducing the auxiliary variable ψ_{ijk}, defined as the time derivative of the Hessian:

$$\psi_{ijk} = \frac{\partial H_{ijk}}{\partial t}.$$

A simple way to calculate the Hessian time rate is to obtain the spatial derivative of the rate of the Jacobian ω_{ij} over the Lagrangian coordinate x_k^0 as:

$$\frac{\partial \psi_{ijk}}{\partial t} = \frac{\partial}{\partial t}\left(\frac{\partial \omega_{ij}}{\partial x_k^0}\right) = \frac{\partial}{\partial x_k^0}\left(\frac{\partial \omega_{ij}}{\partial t}\right).$$

The time derivative of ψ can be presented using the expression of the rate of the Jacobian in Eq. (23.10) as:

$$\frac{\partial \psi_{ijk}}{\partial t} = \frac{\partial}{\partial x_k^0}\left(\frac{1}{\tau_d}\left(\frac{\partial U_i}{\partial x_m}J_{mj} - \omega_{ij}\right)\right).$$

Using the chain rule as in (23.10), an expression for the time rate of the Hessian is obtained:

$$\frac{\partial \psi_{ijk}}{\partial t} = \frac{1}{\tau_d}\left(H_{mj}^k\frac{\partial U_i}{\partial x_m} + J_{mj}J_{nk}\frac{\partial^2 U_i}{\partial x_m \partial x_n} - \psi_{ijk}\right),$$

where addition is assumed among the terms with indices m and n. For $t = 0$ the Lagrangian derivative coincides with the Eulerian derivative thus the initial condition for $\psi_{ijk}(t = 0)$ is expressed as:

$$\psi_{ijk}(t = 0) = \frac{\partial^2 V_i}{\partial x_j \partial x_k}.$$

Assuming that at $t = 0$ the dispersed continuum is not deformed, $H(t = 0) = 0$ is used as the initial value for H_{ijk}. The initial value problem for the calculation of the Hessian and the Jacobian under the second order FLA concept is finally described by the non-linear first order differential system summarised by the following initial value problem:

$$\frac{\partial}{\partial t}\begin{bmatrix} J_{ij} \\ \omega_{ij} \\ H_{ijk} \\ \psi_{ijk} \end{bmatrix} = \begin{bmatrix} \omega_{ij} \\ \frac{1}{\tau_d}\left(\frac{\partial U_i}{\partial x_m}J_{mj} - \omega_{ij}\right) \\ \psi_{ijk} \\ \frac{\partial \psi_{ijk}}{\partial t} = \frac{1}{\tau_d}\left(H_{mj}^k\frac{\partial U_i}{\partial x_m} + J_{mj}J_{nk}\frac{\partial^2 U_i}{\partial x_m \partial x_n} - \psi_{ijk}\right) \end{bmatrix},$$

with initial conditions:

$$\begin{bmatrix} J_{ij} \\ \omega_{ij} \\ H_{ijk} \\ \psi_{ijk} \end{bmatrix}_{t=0} = \begin{bmatrix} 1 \\ \frac{\partial V_i}{\partial x_j} \\ 0 \\ \frac{\partial^2 V_i}{\partial x_{jk}} \end{bmatrix} \tag{23.11}$$

For the results of the second order FLA presented in the fifth section, the differential system (23.11) is numerically integrated over time using a fourth order Runge-Kutta method.

23.4 Calculation of the Hessian Magnitude H Across the Caustic Formation

In this section the calculation of the Hessian magnitude H is presented as a function of the entries of the Hessian matrices H_i. For the calculation of the Hessian magnitude H along the primary direction of a caustic filament the angle θ between the local coordinate systems (η, ζ) and (η_0, ζ_0) (attached to the fold) in relation to the global coordinate systems (x, y) and (x_0, y_0) is assumed. The relations between the two systems are expressed by the transformations:

$$\eta = x \cos(\theta) - y \sin(\theta) , \quad \zeta = x \cos(\theta) + y \sin(\theta)$$

$$\eta_0 = x_0 \cos(\theta) - y_0 \sin(\theta) , \quad \zeta_0 = x_0 \cos(\theta) + y_0 \sin(\theta) . \tag{23.12}$$

The above relations describe the rotation of the coordinate system x, y to the local coordinate system η, ζ, which is used in Eq. (23.1). Taking the second derivative of the relation (23.4), the Hessian H is obtained as the second derivative (curvature) of the dispersed continuum distribution $H = \frac{\partial^2 \eta}{\partial \eta_0^2}$. In order to calculate H in terms of the entries of the global Hessian matrix H_{ijk} one can start from the first derivative $\frac{\partial \eta}{\partial \eta_0}$ which can be written as:

$$\frac{\partial \eta}{\partial \eta_0} = \frac{\partial \eta}{\partial x_0} \frac{\partial x_0}{\partial \eta_0} + \frac{\partial \eta}{\partial y_0} \frac{\partial y_0}{\partial \eta_0} .$$

Using the transformation Eq. (23.12) for the expression of η and ζ in terms of x and y and taking into account that the partial derivatives of the Lagrangian coordinates can be deducted from (23.12) (e.g. $\frac{\partial x_0}{\partial \eta_0} = cos(\theta)$ and $\frac{\partial y_0}{\partial \eta_0} = -sin(\theta)$) the following expression for the first derivative is obtained:

$$\frac{\partial \eta}{\partial \eta_0} = \frac{\partial x}{\partial x_0} \cos(\theta)^2 - \frac{\partial y}{\partial x_0} \sin(\theta) \cos(\theta) + \frac{\partial x}{\partial y_0} \sin(\theta) \cos(\theta) - \frac{\partial y}{\partial y_0} \sin(\theta)^2 .$$

$$\tag{23.13}$$

The second derivative $H = \frac{\partial}{\partial \eta_0} \left(\frac{\partial \eta}{\partial \eta_0} \right)$ is obtained by differentiating (23.13) over η_0 using the partial derivatives of the Lagrangian coordinates deducted from (23.12) (e.g. $\frac{\partial x_0}{\partial \eta_0} = cos(\theta)$ and $\frac{\partial y_0}{\partial \eta_0} = -sin(\theta)$), and also the following evaluations of the chain rule:

$$\frac{\partial}{\partial \eta_0}\left(\frac{\partial x}{\partial x_0}\right) = \frac{\partial^2 x}{\partial x_0 \partial x_0}\frac{\partial x_0}{\partial \eta_0} + \frac{\partial^2 x}{\partial y_0 \partial x_0}\frac{\partial y_0}{\partial \eta_0}$$

$$\frac{\partial}{\partial \eta_0}\left(\frac{\partial y}{\partial x_0}\right) = \frac{\partial^2 y}{\partial x_0 \partial x_0}\frac{\partial x_0}{\partial \eta_0} + \frac{\partial^2 y}{\partial y_0 \partial x_0}\frac{\partial y_0}{\partial \eta_0}$$

$$\frac{\partial}{\partial \eta_0}\left(\frac{\partial x}{\partial y_0}\right) = \frac{\partial^2 x}{\partial x_0 \partial y_0}\frac{\partial x_0}{\partial \eta_0} + \frac{\partial^2 x}{\partial y_0 \partial y_0}\frac{\partial y_0}{\partial \eta_0}$$

$$\frac{\partial}{\partial \eta_0}\left(\frac{\partial y}{\partial y_0}\right) = \frac{\partial^2 y}{\partial x_0 \partial y_0}\frac{\partial x_0}{\partial \eta_0} + \frac{\partial^2 y}{\partial y_0 \partial y_0}\frac{\partial y_0}{\partial \eta_0}$$

Thus, the second derivative $H = \frac{\partial^2 \eta}{\partial \eta_0^2}$ is obtained from (23.12) as:

$$\frac{\partial^2 \eta}{\partial \eta_0^2} = \frac{\partial^2 x}{\partial x_0 \partial x_0}\cos(\theta)^3 - \frac{\partial^2 x}{\partial y_0 \partial y_0}\cos(\theta)^2 \sin(\theta) \tag{23.14}$$

$$-\frac{\partial^2 y}{\partial x_0 \partial x_0}\cos(\theta)\sin(\theta)^2 + \frac{\partial^2 y}{\partial y_0 \partial y_0}\sin(\theta)^3,$$

In the final derivation for the expression of H shown in Eq. (23.14) the symmetry of the sub-matrices H_i (i.e. $H_{112} = H_{121}$ and $H_{212} = H_{221}$) was taken into account which results to the elimination of the non-diagonal terms during the derivation. The value of the magnitude of the Hessian H across a direction η at an angle θ in relation to x, in terms of the entries of the Hessian matrices H_{ijk}, becomes:

$$H = H_{111}\cos(\theta)^3 - H_{122}\cos(\theta)^2 \sin(\theta) - H_{211}\cos(\theta)\sin(\theta)^2 + H_{222}\sin(\theta)^3 \tag{23.15}$$

In the analysis presented in here and in the implementation presented in the next section, a NR method is used for finding the maximum absolute value of H within the interval $[0 : \pi]$ as dictated by Eq. (23.15), which reveals the primary direction of the fold. This maximum value for H is used for the evaluation of the number density for a finite length-scale R_ϵ using the model Eq. (23.8).

23.5   Droplets in a Periodic Two-Dimensional Array of Taylor Vortices

In this section the calculation number density n_d using the second order extension of the FLA for a set of different length scales $\Delta = 2R_\epsilon$ is demonstrated. For the dimensionless distance x, normalised by L, the droplets are initially distributed

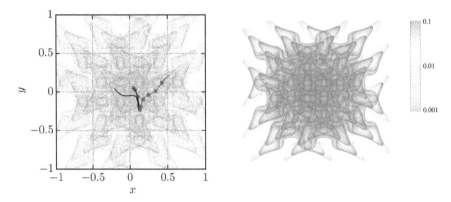

Fig. 23.4 Left: Distribution of droplets at $t^* = 0.4$. Solid red line: Trajectory of the reference droplet from $t^* = 0$ to $t^* = 1$. Red circles: Position of the reference droplet at $\Delta t^* = 0.1$ intervals, the size of the red circles corresponds to the length-scale $R_\epsilon = 0.01$. Solid black line: The trajectory of a neighbouring droplet, initially close to the reference droplet. Right: The same instance represented as a deformed dispersed continuum. The surface is coloured by the number density value $\hat{n}_d$ as predicted by the second order FLA using $R_\epsilon = 0.001$. The red areas correspond to high number density and appear at the vicinity of the folds of the surface

within a square of unity size in the interval $[-1, -1 : 1, 1]$. A periodic two-dimensional array of Taylor vortices is assumed to represent the flow field of the carrier phase:

$$u_x = -5\sin(4\pi x)\cos(4\pi y), u_y = 5\sin(4\pi x)\sin(4\pi y) .$$

The problem is modelled using a standard Lagrangian approach with 400×400 individual droplets dispersing under the influence of the carrier phase for Stokes droplets. The Stokes number for this case is assumed equal to St= 0.5. The number density for each FLA droplet is calculated using $n_d = 1/\mathbf{J}$ for the first order FLA and Eq. (23.8) for the second order FLA. The Jacobian and the Hessian are calculated using the fourth order Runge-Kutta method for the integration of (23.11), as described in the previous section. In Fig. 23.4 (Left) the distribution of the droplets at $t^* = 0.4$ is shown. In the same figure the trajectory of a reference droplet located at $x_0 = y_0 = L/32$ from $t^* = 0$ to $t^* = 1$ is shown. In Fig. 23.4 (Right) the same instance is presented as a deformed surface of the dispersed continuum. Caustics are located at the edges of the surface where $\frac{\partial \epsilon}{\partial \delta} = 0$, from our analysis $\frac{\partial^2 \epsilon}{\partial \delta^2}$ is not necessarily zero, thus the number density at these points shown with red colour is finite. The colour map of this figure corresponds to the number density inferred from the second order FLA for a length-scale $R_\epsilon = 0.001$. This value is finite and reaches a converged finite maximum values $\hat{n}_d \approx 4$ for this specific length scale at this instance.

Figure 23.5 shows the comparison of the number densities for the reference droplet of the simulation $x_0 = y_0 = L/32$ calculated using both the standard FLA

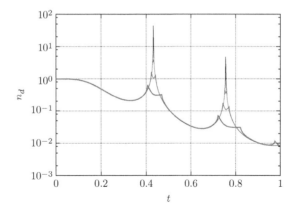

Fig. 23.5 Number density $\hat{n}_d$ along the reference droplet trajectory assuming a length-scales $R_\epsilon = 0.0001, 0.001$ and 0.01. Solid black curve: standard FLA. Thin solid red curve: result inferred from the second order FLA with $R_\epsilon = 0.0001$. Solid red curve: result inferred from the second order FLA with $R_\epsilon = 0.001$. Thick red curve: result inferred from the second order FLA with $R_\epsilon = 0.01$

and the second order FLA for different length scales R_ϵ, using the model Eq. (23.8). The Hessian arrays and the magnitude H are calculated as described in the sections three and four for this two-dimensional problem. As it can be inferred from the Fig. 23.5 the droplet goes through two caustic points at $t^* \approx 0.4$ (identified also by the red filament at $y = -0.25$ in Fig. 23.4 (Right)) and a second one at $t^* \approx 0.8$. As it is expected the first order FLA identifies the caustic as a singularity point. For the second order FLA the larger the length-scale on which the number density is defined the more diffused the caustic formation appears. Thus, with the second order FLA not only the occurrence of the caustic is identified by a single Lagrangian particle but also the intensity of the caustic is calculated.

23.6 Conclusion

In this paper the second order extension of the FLA which relates the point-wise number density as defined in the standard FLA to a number density defined at a finite given length-scale was presented. The second order FLA addresses the singularities of the standard FLA and demonstrates the integrability of the point-wise number density of the standard FLA on caustics. The FLA accounts for the history of the droplet trajectories by integrating an initial value problem, in the second order FLA the second derivative of the carrier phase flow field is also accounted for, which enters in the calculation of the Hessian, thus, affecting the number density (intensity) of the caustic formations. Furthermore, with the approach presented here the FLA number density can be related to a finite length scale inferred from the LES filter width Δ needed for the introduction of the FLA to turbulent flows.

References

[Sa14] Sazhin, S.S.: *Droplets and Sprays*, Springer, 2014. https://doi.org/10.1007/978-1-4471-6386-2

[Pa18] Papoutsakis, A., Sazhin, S.S., Begg S., Danaila I., and Luddens, F.: An efficient Adaptive Mesh Refinement (AMR) algorithm for the Discontinuous Galerkin method: Applications for the computation of compressible two-phase flows. *Journal of Computational Physics*, **363**, 399–427 (2018).

[Kn12] Knight, G.D.: *Plastic Pollution*, Heinemann Library, 2012.

[Vo81] Vogel, S.: *Life in Moving Fluids. The Physical Biology of Flow.*, Princeton University Press, 1981. https://doi.org/10.2307/1352661

[SeEtAl16] Serrano, X. M. Baums, I.B., Smith, T.B., Jones, R.J., Shearer, T.L, and Baker, A.C.: Long distance dispersal and vertical gene flow in the Caribbean brooding coral Porites astreoides. *Sci. Rep*, **6**, 21619; https://doi.org/10.1038/srep21619 (2016).

[HeYo05] Healy, D.P., and Young, J.B.: Full Lagrangian methods for calculating particle concentration fields in dilute gas-particle flows. *Proceedings of the Royal Society of London A: Mathematical, Physical and Engineering Sciences*, **461**, 2197–2225 (2005).

[Os84] Osiptsov, A.N.: Investigation of regions of unbounded growth of the particle concentration in disperse flows. *Fluid Dynamics*, **19**, 3, 378–385 (1984).

[IjEtAl09] Ijzermans, R., Reeks, M., Meneguz, E., Picciotto, M., and Soldati: Measuring segregation of inertial particles in turbulence by a full Lagrangian approach. *Phys. Rev. E*, **80**, 015302 (2009).

[MeRe11] Meneguz, E., and Reeks, M.: Statistical properties of particle segregation in homogeneous isotropic turbulence. *J. Fluid Mech.*, **686**, 338–351 (2011).

[PiEtAl05] Picciotto, M., Marchioli, C., Reeks, M., and Soldati, A.: Statistics of velocity and preferential accumulation of micro-particles in boundary layer turbulence. *Nuclear Eng. Design*, **235**, 1239–1249 (2005).

[PaEtAl18] Papoutsakis, A., Rybdylova, O., Zaripov, T., Danaila, L., Osiptsov, A., and Sazhin, S.: Modelling of the evolution of a droplet cloud in a turbulent flow. *International Journal of Multiphase Flows*, **104**, 233–257 (2018).

[PaEtAl16] Papoutsakis, A., Rybdylova, O., Zaripov, T., Danaila, L., Osiptsov, A., and Sazhin, S.: Modelling of the evolution of a droplet cloud in a turbulent flow. In *Proceedings of the 27th European Conference on Liquid Atomization and Spray Systems ILASS-Europe.*, G. de Sercey and S.S. Sazhin (eds.). Brighton, UK (2016), MF-07 (2018).

Chapter 24
The Nodal LTS_N Solution in a Rectangular Domain: A New Method to Determine the Outgoing Angular Flux at the Boundary

Aline R. Parigi, Cynthia F. Segatto, and Bardo E. J. Bodmann

24.1 Introduction

A variety of solutions may be found in the literature, that solve S_N nodal problems for neutron transport in rectangular domains (for instance, see [HaEtAl02, SeEtAl99b] and the references therein). This procedure results in a set of coupled one-dimensional S_N problems with unknown terms representing the outgoing angular flux at the respective boundary. To the best of our knowledge, there are no reports based either on theoretical reasoning or experimental evidence, which allow to estimate these fluxes. It is noteworthy that only ad hoc hypothesis is encountered. Hence, in this work we present an approach to determine these angular fluxes approximating the rectangular domain by a set of one-dimensional problems with known solution at the boundary, which allows to estimate the outgoing fluxes for the two-dimensional problem as, for instance, by constant or exponential expressions, where the latter is used in the present work.

A. R. Parigi
Instituto Federal Farroupilha, Santa Maria, RS, Brazil

C. F. Segatto · B. E. J. Bodmann (✉)
Universidade Federal do Rio Grande do Sul, Porto Alegre, RS, Brazil
e-mail: bardo.bodmann@ufrgs.br

© Springer Nature Switzerland AG 2019
C. Constanda, P. Harris (eds.), *Integral Methods in Science and Engineering*,
https://doi.org/10.1007/978-3-030-16077-7_24

24.2   The LTS_N Transport Equations in 2D

We consider a neutron transport problem in a rectangular domain $(x, y) \in [0, a] \times [0, b]$ for discrete ordinates and isotropic scattering represented by the following equation

$$\mu_m \frac{\partial \Psi_m}{\partial x}(x, y) + \eta_m \frac{\partial \Psi_m}{\partial y}(x, y) + \sigma_t \Psi_m(x, y) = Q_m(x, y) + \frac{\sigma_s}{4} \sum_{n=1}^{M} \omega_n \Psi_m(x, y)$$

(24.1)

and associated boundary conditions,

$$\Psi_m(0, y, \Omega_m(\mu_m, \eta_m)) = \Psi_m(0, y, \Omega_m(-\mu_m, \eta_m)), \quad \mu_m > 0$$

$$\Psi_m(x, 0, \Omega_m(\mu_m, \eta_m)) = \Psi_m(x, 0, \Omega_m(\mu_m, -\eta_m)), \quad \eta_m > 0$$

$$\Psi_m(a, y, \Omega_m(\mu_m, \eta_m)) = 0, \quad \mu_m < 0$$

$$\Psi_m(x, b, \Omega_m(\mu_m, \eta_m)) = 0, \quad \eta_m < 0.$$

Here, $m = 1, \ldots, \frac{M}{2}$, with $M = \frac{N(N+2)}{2}$, where $\Omega_m = (\mu_m, \eta_m)$ represents a discrete direction and ω_m is the weight associated with direction m. Further, N is the order of ordinates, M is the total number of discrete directions according to the level-symmetric quadrature scheme of reference [LeEtAl93], σ_s and σ_t are the macroscopic total and scattering cross sections and $Q(x, y)$ is an isotropic neutron source term. The nodal method consists in the transverse integration of the 2D transport equation, which results in a set of one-dimensional problems. Thus, integrating Eq. (24.1) in x yields the one-dimensional S_N equation in y,

$$\eta_m \frac{d\tilde{\Psi}_{xm}}{dy}(y) + \frac{\mu_m}{a}[\Psi_m(a, y) - \Psi_m(0, y)] + \sigma_t \tilde{\Psi}_{xm}(y) = Q_{xm}(y)$$

$$+ \frac{\sigma_s}{4} \sum_{n=1}^{M} \omega_n \tilde{\Psi}_{x_n}(y),$$

(24.2)

where

$$\tilde{\Psi}_{xm}(y) \equiv \frac{1}{a} \int_0^a \Psi_m(x, y)dx, \qquad Q_{xm}(y) \equiv \frac{1}{a} \int_0^a Q_m(x, y)dx,$$

and by an analogue procedure one obtains the one-dimensional S_N equation in x.

$$\mu_m \frac{d\hat{\Psi}_{ym}}{dx}(x) + \frac{\eta_m}{b}[\Psi_m(x, b) - \Psi_m(x, 0)] + \sigma_t \hat{\Psi}_{ym}(x) = Q_{ym}(x)$$

$$+ \frac{\sigma_s}{4} \sum_{n=1}^{M} \omega_n \hat{\Psi}_{y_n}(x)$$

(24.3)

where

$$\hat{\Psi}_{y_m}(x) \equiv \frac{1}{b}\int_0^b \Psi_m(x,y)dy, \qquad\qquad Q_{y_m}(x) \equiv \frac{1}{b}\int_0^b Q_m(x,y)dy$$

Equations (24.2) and (24.3) are the average fluxes for each direction m in x and y, respectively, and may be cast in matrix form.

$$\frac{d\tilde{\Psi}_x}{dy}(y) - \mathbf{A}_y\tilde{\Psi}_x(y) = Q_x(y)\mathbf{N}^{-1}\mathbf{1} - \frac{1}{a}\mathbf{M}\mathbf{N}^{-1}[\Psi(a,y) - \Psi(0,y)] \qquad (24.4)$$

$$\frac{d\hat{\Psi}_y}{dx}(x) - \mathbf{A}_x\hat{\Psi}_y(x) = Q_y(x)\mathbf{M}^{-1}\mathbf{1} - \frac{1}{b}\mathbf{N}\mathbf{M}^{-1}[\Psi(x,b) - \Psi(x,0)] \qquad (24.5)$$

where

$$\mathbf{A}_x(i,j) = \begin{cases} \frac{-\sigma_t}{\mu_i} + \frac{\sigma_s\omega_j}{4\mu_i}, & \text{if } i=j \\ \frac{\sigma_s\omega_j}{4\mu_i}, & \text{if } i\neq j \end{cases} \qquad \mathbf{A}_y(i,j) = \begin{cases} \frac{-\sigma_t}{\eta_i} + \frac{\sigma_s\omega_j}{4\eta_i}, & \text{if } i=j \\ \frac{\sigma_s\omega_j}{4\eta_i}, & \text{if } i\neq j \end{cases}$$

$\mathbf{M}$ and $\mathbf{N}$ are diagonal matrices of order M with components μ_m and η_m, respectively, where $m = 1, \ldots, M$ and $\mathbf{1}$ is a vector of order M containing components all equal to 1. The solution obtained, for (24.4) and (24.5), by the LTS_N method for $\tilde{\Psi}_x$ and $\hat{\Psi}_y$ is given below.

$$\tilde{\Psi}_x(y) = \mathbf{Y}e^{\mathbf{E}^*(y)}\mathbf{Y}^{-1}\tilde{\Psi}_x(0) \qquad\qquad (24.6)$$

$$+\mathbf{Y}e^{\mathbf{E}y}\mathbf{Y}^{-1}\star\left[Q_x(y)\mathbf{N}^{-1}\mathbf{1} - \frac{1}{a}\mathbf{M}\mathbf{N}^{-1}\left(\Psi(a,y) - \Psi(0,y)\right)\right]$$

$$\hat{\Psi}_y(x) = \mathbf{X}e^{\mathbf{D}^*(x)}\mathbf{X}^{-1}\hat{\Psi}_y(0) \qquad\qquad (24.7)$$

$$+\mathbf{X}e^{\mathbf{D}x}\mathbf{X}^{-1}\star\left[Q_y(x)\mathbf{M}^{-1}\mathbf{1} - \frac{1}{b}\mathbf{M}^{-1}\mathbf{N}\left(\Psi(x,b) - \Psi(x,0)\right)\right]$$

Here, $\mathbf{D} = \text{diag}\{d_1, d_2, \ldots, d_M\}$ with M eigenvalues of $\mathbf{A}_x$, $(\mathbf{D}^*(x))_i = d_i x$ for $d_i < 0$ and $d_i(x-a)$ for $d_i > 0$. In analogy $\mathbf{E} = \text{diag}\{e_1, e_2, \ldots, e_M\}$, with M eigenvalues of $\mathbf{A}_y$, $\mathbf{E}^*(y) = e_i y$ for $e_i < 0$ and $e_i(y-b)$ for $e_i > 0$. Further, $\mathbf{X}$ is the matrix of eigenvectors of $\mathbf{A}_x$, $\mathbf{Y}$ the matrix of eigenvectors of $\mathbf{A}_y$ and "$\star$" signifies the convolution operation.

For the outgoing flux hypothesis we follow reference [HaEtAl02].

$$\Psi(x,0) = e^{-\text{sign}(\mathbf{M})\lambda x}\mathbf{V}^3 \qquad\qquad \Psi(0,y) = e^{-\text{sign}(\mathbf{N})\lambda y}\mathbf{V}^4$$

$$\Psi(x,b) = e^{-\text{sign}(\mathbf{M})\lambda x}\mathbf{V}^5 \qquad\qquad \Psi(a,y) = e^{-\text{sign}(\mathbf{N})\lambda y}\mathbf{V}^6$$

Here $\lambda = \sigma_a$, $\mathbf{V}^3$, $\mathbf{V}^4$, $\mathbf{V}^5$, $\mathbf{V}^6$ are vectors of order M to be determined and sign$(\mu) = 1$ for $\mu > 0$ and -1 for $\mu < 0$ denotes the usual signal function. Finally, to determine the two-dimensional LTS_N solution one solves a compatible homogeneous linear system of $4M$ equations obtained from the evaluation of the nodal LTS_N solution, (24.6), (24.7) together with the definition of the angular fluxes at the boundary.

24.3 Numerical Results for Case 1

The solution derived in the previous section is applied to two cases. The rectangular domain is specified by $[0, a] \times [0, b]$ with $a = b = 1$ cm, with a neutron source strength $Q(x, y) = 1\,\text{cm}^{-3}\text{s}^{-1}$ active in a subdomain $0 \le x \le a_s = 0.52$ cm and $0 \le y \le b_s = 0.52$ cm, using $\sigma_t = 1.0\,\text{cm}^{-1}$ and for σ_s we consider three scenarios $\sigma_s = 0.5\,\text{cm}^{-1}$, $\sigma_s = 0.1\,\text{cm}^{-1}$ and $\sigma_s = 0.05\,\text{cm}^{-1}$. In case 1 the boundary conditions are reflective at $x = 0$ and $y = 0$ and vacuum for $x = a$ and $y = b$ as shown in Fig. 24.1 (left). The results for the mean scalar flux are presented in Table 24.1.

A second case was considered with different boundary conditions in comparison with case 1, namely reflective at $x = 0$, $y = 0$, $y = b$ and vacuum for $x = a$, as shown in Fig. 24.1 (right). The results for the mean scalar flux for case 2 are presented in Table 24.3.

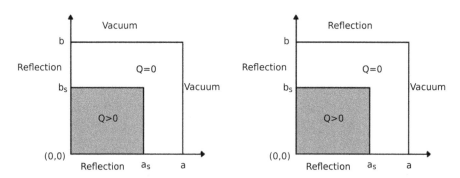

Fig. 24.1 Domains of Case 1 (left) and Case 2 (right)

Table 24.1 Scalar flux $\hat{\Phi}_y(x)$ for case 1

σ_s	N	LTS_N 2D-DiagExp		
		$x = 0.5$ cm	$x = 0.7$ cm	$x = 0.98$ cm
0.5	2	0.295	0.201	0.115
	4	0.312	0.196	0.097
	6	0.313	0.187	0.090
	8	0.313	0.180	0.087
	12	0.313	0.174	0.086
	16	0.314	0.172	0.085
0.1	2	0.211	0.141	0.077
	4	0.219	0.128	0.061
	6	0.219	0.120	0.057
	8	0.219	0.115	0.055
	12	0.220	0.110	0.054
	16	0.220	0.110	0.054
0.05	2	0.204	0.135	0.074
	4	0.211	0.122	0.059
	6	0.211	0.114	0.054
	8	0.211	0.110	0.052
	12	0.212	0.105	0.051
	16	0.213	0.103	0.051

Table 24.2 $\hat{\Psi}_y(x)$ for $\sigma_s = 0.5$ by LTS_N 2D-DiagExp, case 1

$x = 0$	$x = 0.1$	$x = 0.2$	$x = 0.3$
0.569687129496	0.614877062408	0.650497164397	0.677591017408
0.310024200332	0.313021061165	0.317230044072	0.321781441371
0.310024200332	0.256110804576	0.195104433932	0.126354488012
0.569687129496	0.514347269906	0.449053881649	0.372482572288
$x = 0.4$	$x = 0.5$	$x = 0.6$	$x = 0.7$
0.696994419105	0.709360623569	0.648008169440	0.576738717673
0.325911725868	0.328938387865	0.263065398601	0.191165442668
0.049137821924	**−0.037345405124**	**−0.056811139696**	**−0.052907616350**
0.283102173626	0.179144856658	0.135736103675	0.107842098996
$x = 0.8$	$x = 0.9$	$x = 1.0$	
0.513027275830	0.456101737373	0.405263777723	
0.129940220322	0.077974843231	0.034037604579	
−0.042714125904	**−0.057913528976**	0.000000000000	
0.076727625178	0.041225485705	0.000000000000	

It is noteworthy that some of the numerical values for the angular flux close to the boundary are negative as shown in Tables 24.2 and 24.4 (bold values), which from a physical point of view is absurd. This artefact is due to the adopted hypothesis for the outgoing flux at the boundary.

Table 24.3 Scalar flux, $\hat{\Phi}_y(x)$ and $\tilde{\Phi}_x(y)$, for case 2

| | | LTS_N 2D-DiagExp | | | | | |
| | | $x = y = 0.5\,\text{cm}$ | | $x = y = 0.7\,\text{cm}$ | | $x = y = 0.98\,\text{cm}$ | |
σ_s	N	$\hat{\Phi}_y(x)$	$\tilde{\Phi}_x(y)$	$\hat{\Phi}_y(x)$	$\tilde{\Phi}_x(y)$	$\hat{\Phi}_y(x)$	$\tilde{\Phi}_x(y)$
0.5	2	0.359	0.348	0.272	0.282	0.176	0.248
	4	0.389	0.373	0.270	0.277	0.157	0.225
	6	0.392	0.378	0.263	0.269	0.149	0.214
	8	0.394	0.379	0.258	0.264	0.146	0.210
	12	0.394	0.380	0.253	0.260	0.144	0.210
	16	0.395	0.380	0.251	0.258	0.144	0.210
0.1	2	0.243	0.232	0.174	0.178	0.110	0.150
	4	0.256	0.246	0.164	0.165	0.090	0.125
	6	0.257	0.248	0.156	0.158	0.084	0.117
	8	0.258	0.249	0.152	0.153	0.081	0.113
	12	0.258	0.250	0.148	0.150	0.080	0.112
	16	0.259	0.250	0.145	0.148	0.080	0.113
0.05	2	0.233	0.223	0.166	0.169	0.103	0.142
	4	0.245	0.236	0.155	0.156	0.084	0.117
	6	0.246	0.237	0.148	0.149	0.079	0.109
	8	0.247	0.238	0.144	0.145	0.077	0.106
	12	0.248	0.239	0.139	0.141	0.076	0.105
	16	0.248	0.240	0.137	0.139	0.076	0.106

Table 24.4 $\hat{\Psi}_y(x)$ for $\sigma_s = 0.5$ by LTS_N 2D-DiagExp, case 2

$x = 0$	$x = 0.1$	$x = 0.2$	$x = 0.3$
0.583353595524	0.627691490813	0.663445012458	0.691326493333
0.583353595524	0.528886229007	0.463780627122	0.386791134657
0.355096375283	0.301882944669	0.240545854240	0.170431747042
0.355096375283	0.401294613751	0.440292302752	0.472499779737
$x = 0.4$	$x = 0.5$	$x = 0.6$	$x = 0.7$
0.711912150378	0.725653095690	0.665711461591	0.595768620264
0.296457947689	0.191079212411	0.145843042519	0.115777075700
0.090791113663	0.000773012280	-0.023415999111	-0.025530284952
0.498243289270	0.517767525785	0.464064646591	0.400676519550
$x = 0.8$	$x = 0.9$	$x = 0.98$	$x = 1.0$
0.533207506962	0.477185668675	0.436568184330	0.426950824604
0.082208722172	0.044037201476	0.009343605563	0.000000000000
-0.022786789395	-0.014547683054	-0.003459832538	0.000000000000
0.344872032671	0.295715934688	0.260618633477	0.252380864452

24.4   An Alternative to Determine the Unknown Angular Fluxes at the Boundaries

Due to the shortcoming of the method outlined in the previous section a new approach was developed. It is noteworthy that the problem of negative angular fluxes does not appear in the one-dimensional problem. This is basically a consequence of having only two boundary points at the extreme of the domain, whereas the two-dimensional problem is characterized by a closed line boundary. The question that arises is whether it is possible to construct a solution for the two-dimensional problem starting from the solution of the one-dimensional problem with known angular fluxes at the endpoints, which opens pathways to avoid the otherwise non-physical behaviour of the fluxes close to the boundaries. To this end, the two-dimensional domain is covered with a finite set of sufficiently narrow sub-domains so that to an approximation each strip may be considered a one-dimensional problem. A sketch of this procedure is given in Fig. 24.2 for the horizontal version of one-dimensional problems $r_k \in [0, b]$ with $k = 1, \ldots, J$ and continuous $x \in [0, a]$ and the analogue for the vertical implementation ($r_k \in [0, a]$, $k = 1, \ldots, J$ and continuous $y \in [0, b]$). Thus, along each line the heterogeneous one-dimensional transport problem was considered, with the domain divided in two regions that are distinguished by the presence or absence of a source term. Each one-dimensional problem is represented by the following equation,

$$\gamma_n \frac{d\phi_{k_n}^{i}}{dx}(x) + \sigma_t \phi_{k_n}^{i}(x) = \frac{\sigma_s}{2} \sum_{n=1}^{\mathcal{N}} \phi_{k_n}^{i}(x)\omega_n + q_k^{i}(x) , \qquad (24.8)$$

Fig. 24.2 Construction of the $2DLTS_N$ solution by the solution of the $1D$ problem (here is shown a horizontal version)

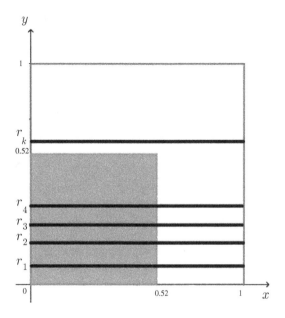

subject to boundary the conditions

$$\phi_k^1(0, \gamma_n) = \phi_k^1(0, -\gamma_n) \quad \gamma_n > 0 \quad \text{with} \quad n = \{1, .., \frac{\mathcal{N}}{2}\},$$

$$\phi_k^1(l_s, \gamma_n) = \phi_k^2(l_s, \gamma_n) \quad \text{with} \quad n = \{1, .., \mathcal{N}\}, \quad \text{where } L_s = a_s \text{ or } L_s = b_s,$$

$$\phi_k^2(L, \gamma_n) = 0, \quad \gamma_n < 0 \quad \text{with} \quad n = \{\frac{\mathcal{N}}{2} + 1, \ldots, \mathcal{N}\} \text{ and } L = a \text{ or } L = b,$$

where $\phi_{k_n}^i$ represents the angular flux in region i, γ_n and ω_n for $n = 1, \ldots, \mathcal{N}$ are the directions and weights, $\mathcal{N}$ is the quadrature order. The source terms in region 1 are

$$q_k^1(x) = \begin{cases} 1 & \forall x \in [0, 0.52] \text{ for } y_k \le 0.52 \\ e^{-\beta(y_k - 0.52)} & \forall x \in [0, 0.52] \text{ for } y_k > 0.52 \end{cases},$$

and for region 2 is $q_k^2(x) = 0$ for $x \in (0.52, 1]$.

For the one-dimensional neutron transport problem in the region originally without neutron source ($y_k \in (b_s, b]$, we consider a weak source of the type $q(x)_k^1 = e^{-\beta(y_k - b_s)}$, where β is a constant to be determined in a way to maintain this source term $q(x)_k^1$ close to zero. This modification has to be introduced into the method in order to avoid an otherwise trivial solution for sub-domains without source term contributions. Moreover, this scheme preserves the condition of a reflective contour for the one-dimensional problem, where the structure of the solution is known.

Upon application of the LTS_N method [SeEtAl99b], one obtains the solution for (24.8).

$$\phi^1(x) = B^1(x)\xi^1 + H^1(x) \quad \text{for } x \in [0, L_s]$$
$$\phi^2(x) = B^2(x)\xi^2 \quad \text{for } x \in [L_s, L]$$

where

$$B^i(x) = \begin{cases} Xe^{Dx} & \text{for } D < 0 \\ Xe^{D(x - L_i)} & \text{for } D > 0 \end{cases} \quad \text{for } i = 1, 2.$$

$$H^1(x) = \begin{cases} X \int_0^x e^{D(x-\zeta)} q^1(\zeta) \, d\zeta X^{-1} & \text{for } D < 0 \\ X \int_x^{L_s} e^{D(x-\zeta)} q^1(\zeta) \, d\zeta X^{-1} & \text{for } D > 0 \end{cases}$$

Here X and D are the eigenvector and eigenvalue matrices of the one-dimensional problem.

To estimate the two-dimensional angular fluxes at the boundaries through the one-dimensional angular fluxes we use the LTS_N method associated with the so-called dummy-nodes inclusion technique (DNI) [ChEtAl00]. In order to interpolate the directions of the two-dimensional problem using the one-dimensional directions, we include the cosine of the directions Ω in the quadrature scheme associating a zero weight for each contribution pointing into the domain. Details of the scheme are shown next.

- $\gamma_i = \cos(\widehat{\Omega}_i), \longrightarrow w_i = 0 \quad$ for $\quad i = 1, \frac{M}{4}$
- $\gamma_{i+\frac{M}{4}} = \lambda_i \longrightarrow w_{i+\frac{M}{4}} = \omega_i \quad$ for $\quad i = 1, \mathcal{N}$
- $\lambda_{i+\frac{M}{2}+\mathcal{N}} = \cos(\widehat{\Omega}_i), \longrightarrow w_{i+\frac{M}{2}+\mathcal{N}} = 0 \quad$ for $\quad i = \frac{M}{4}+1, \frac{M}{2}$

In order for the nodal LTS_N solution to be completely determined, it is sufficient to solve a compatible homogeneous linear system of M equations obtained from the evaluation of the nodal LTS_N solution in the contours and the definition of the angular fluxes at the boundary from the one-dimensional LTS_N solution.

24.5 Numerical Results for Case 2

In this section we present the numerical results obtained from the new methodology presented in the previous section considering problem 1, described in Sect. 24.3. The results are presented in Tables 24.5, 24.6 and 24.7. We emphasize that the results obtained by the new methodology proposed are comparable to those presented in the

Table 24.5 Scalar flux, $\hat{\Phi}_y(x)$, for $x = 0.5\,\text{cm}$, problem 1

				LTS_N 2D
			LTS_N 2D-DiagExp	LTS_N 1D
σ_s	$N = 2, 4, 6, 8, 12, 16$	$N = 2, 4, 6, 8, 12, 16$	$N = 2, 4, 6, 8, 12, 16$	$N = 2, 4, 6, 8, 12, 16$
0.5	0.304	0.326	0.295	0.280
	0.320	0.337	0.312	0.319
	0.320	0.336	0.313	0.325
	0.320	0.335	0.313	0.328
	0.320	0.335	0.313	0.330
	0.321	0.335	0.314	0.330
0.1	0.219	0.236	0.211	0.211
	0.226	0.236	0.219	0.229
	0.226	0.235	0.219	0.231
	0.225	0.234	0.219	0.232
	0.226	0.234	0.220	0.233
	0.227	0.234	0.220	0.234
0.05	0.212	0.228	0.204	0.204
	0.218	0.228	0.211	0.220
	0.217	0.226	0.211	0.223
	0.217	0.226	0.211	0.224
	0.218	0.226	0.212	0.225
	0.218	0.226	0.213	0.225

Table 24.6 Scalar flux, $\hat{\Phi}_y(x)$, for $x = 0.7$, problem 1

σ_s	$N = 2, 4, 6, 8, 12, 16$	$N = 2, 4, 6, 8, 12, 16$	LTS_N 2D-DiagExp $N = 2, 4, 6, 8, 12, 16$	LTS_N 2D LTS_N 1D $N = 2, 4, 6, 8, 12, 16$
0.5	0.210	0.225	0.201	0.211
	0.199	0.207	0.196	0.221
	0.190	0.196	0.187	0.218
	0.184	0.189	0.180	0.214
	0.178	0.182	0.174	0.211
	0.175	0.179	0.172	0.209
0.1	0.144	0.156	0.141	0.151
	0.130	0.135	0.128	0.146
	0.122	0.126	0.120	0.140
	0.117	0.120	0.115	0.137
	0.112	0.115	0.110	0.133
	0.110	0.112	0.110	0.131
0.05	0.138	0.150	0.135	0.145
	0.124	0.129	0.122	0.139
	0.116	0.120	0.114	0.134
	0.112	0.114	0.110	0.130
	0.107	0.109	0.105	0.126
	0.105	0.107	0.103	0.124

Table 24.7 Scalar flux, $\hat{\Phi}_y(x)$, for $x = 0.98$, problem 1

σ_s	$N = 2, 4, 6, 8, 12, 16$	$N = 2, 4, 6, 8, 12, 16$	LTS_N 2D-DiagExp $N = 2, 4, 6, 8, 12, 16$	LTS_N 2D LTS_N 1D $N = 2, 4, 6, 8, 12, 16$
0.5	0.112	0.107	0.115	0.137
	0.092	0.084	0.097	0.128
	0.085	0.076	0.090	0.123
	0.082	0.072	0.087	0.121
	0.081	0.070	0.086	0.119
	0.080	0.070	0.085	0.118
0.1	0.072	0.066	0.077	0.094
	0.055	0.048	0.061	0.080
	0.050	0.043	0.057	0.075
	0.048	0.040	0.055	0.073
	0.047	0.040	0.054	0.072
	0.047	0.040	0.054	0.071
0.05	0.069	0.062	0.074	0.090
	0.052	0.045	0.059	0.076
	0.047	0.040	0.054	0.071
	0.045	0.038	0.052	0.069
	0.045	0.037	0.051	0.068
	0.045	0.037	0.051	0.067

Table 24.8 $\hat{\psi}_y(0.5)$, for $\sigma_s = 0.5$, by $LT S_N$ 2D- $LT S_N$ 1D, problem 1

$x = 0$	$x = 0.1$	$x = 0.2$	$x = 0.3$
0.149896183201	0.165049980315	0.177235384763	0.186365596207
0.588051865155	0.645823493082	0.693849007552	0.733119502107
0.588051865155	0.519345631960	0.438291404659	0.343218196198
0.149896183201	0.131867235935	0.111073872692	0.087657456232
$x = 0.4$	$x = 0.5$	$x = 0.6$	$x = 0.7$
0.192350838296	0.195090186284	0.161580039899	0.150540269463
0.764451671578	0.788506901878	0.720126426291	0.625456295802
0.232158681506	0.102803426912	0.050185885741	0.035875650064
0.061805744652	0.033771935078	0.042944426436	0.033739679842
$x = 0.8$	$x = 0.9$	$x = 1.0$	
0.141999980870	0.132019916756	0.122727165227	
0.541694841578	0.469431542035	0.406028716183	
0.023946972290	0.010782946491	0.000000000000	
0.021645425005	0.012706342622	0.000000000000	

literature, in addition, no negative angular fluxes were found through this approach. Results for the mean angular flux, with $N = 2$ are shown in Table 24.8.

24.6   Conclusion

The present work is the first approach to determine the outgoing angular fluxes at the boundaries for the S_N approximation of the neutron transport problem in a rectangular domain from the known solutions of the one-dimensional problem. Note that the undetermined fluxes at the boundary of the two-dimensional problem is an inherent feature of the S_N nodal problem. Besides the fact that the results obtained are comparable to the one in the literature, one may affirm that by the new technique the commonly observed negative values for the angular fluxes near the boundaries are mitigated. Further, the presented procedure allows to construct an approximate solution in semi-analytical representation for the nodal S_N equations in a rectangular domain. In future work we continue the outlined reasoning and develop corrections to the obtained solution together with error estimates.

References

[ChEtAl00] Chalhoub, E. S., and Garcia, R. D. M.: In *The equivalence between two techniques of angular interpolation for the discrete-ordinates method*, Journal of Quantitative Spectroscopy & Radiative Transfer, **64**, 517–535 (2000).

[HaEtAl02] Hauser, E. B., Panta Pazos, R., Barros, R.C., and Vilhena, M. T.: In *Solution and study of Nodal Neutron Transport Equation applying the LT S_N DiagExp Method*, Proceedings of the 18th International Conference on Transport Theory, Rio de Janeiro, Brazil, 303–307 (2002).

[LeEtAl93] Lewis, E., and Mille, W.: In *Computational Methods of Neutron Transport*, John Wiley and Sons, New York (1993).

[SeEtAl99b] Segatto, C. F., Vilhena, M. T., and Gomes, M.: In *The One-Dimensional LTSN Solution in a Slab with High Degree of Quadrature*, Annals of Nuclear Energy, **26**, 925–934 (1999b).

Chapter 25
Image Processing for UAV Autonomous Navigation Applying Self-configuring Neural Network

Gerson da Penha Neto, Haroldo F. de Campos Velho, Elcio H. Shiguemori, and José Renato G. Braga

25.1   Introduction

Application and development of Unmanned Aerial Vehicles (UAV) or *drone* have had a rapid growth for various purposes [VaVa14], such as monitoring in agriculture and livestock [BiEtAl17], search for missing persons and rescue operations [GoRe15], land mapping [JaEtAl17], and environmental and forest fire monitoring [DaEtAl17].

There is an expectation for a continuous increasing of the UAV in the future, mainly due to lower cost for operation and manufacturing when compared to traditional aircraft driven by a pilot [FiEtAl17].

The flight control of *drones* can be performed remotely or autonomously. There are different strategies for the UAV autonomous navigation, [CoDo09, SiEtAl15, BrEtAl15, BrEtAl16]. The positioning estimation can be done by using inertial sensor and General Navigation Satellite Systems (GNSS) signal [Gr15]. The use of the GNSS signal can present some difficulties due to natural [CoDo08] or not natural [FaEtAl18] interference. One approach to solve the mentioned troubles is to apply a computer vision system combining with the inertial (INS) sensor to estimate the position of the UAV [CoDo09, BrEtAl16].

A computer vision process can be done by image edge extraction from geo-referenced satellite [DaEtAl17] and caught by the UAV. A correlation between the segmented images is computed to estimate the UAV position [SiEtAl15]. The vision-based architecture developed by [CoDo09] is composed of a Kalman filter

G. da Penha Neto · H. F. de Campos Velho (✉) · J. R. G. Braga
Instituto Nacional de Pesquisas Espaciais (INPE), São José dos Campos, SP, Brazil
e-mail: gerson.penha@inpe.br; haroldo.camposvelho@inpe.br

E. H. Shiguemori
Departamento de Ciência e Tecnologia Aeroespacial (DCTA), São José dos Campos, SP, Brazil
e-mail: elcio@ieav.cta.br

© Springer Nature Switzerland AG 2019   321
C. Constanda, P. Harris (eds.), *Integral Methods in Science and Engineering*,
https://doi.org/10.1007/978-3-030-16077-7_25

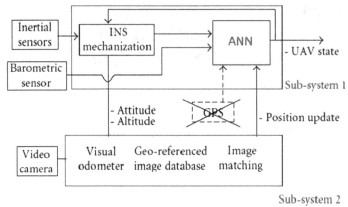

Fig. 25.1 Sensor fusion architecture

(KF), combining the INS signal and the UAV positioning by image processing mentioned before.

The application of a KF has a relative high computational effort. A supervised artificial network (ANN) can be trained to emulate the Kalman filter for reducing the computational effort with similar operation performance. The new system replacing KF by ANN is shown in Fig. 25.1. The use of ANN requires the adjustment of several parameters [Ha07, AnEtAl13]. The process of obtaining an adequate ANN architecture to solve a specific problem is a complex task. Such task usually requires a help from an expert on neural network and a prior knowledge about the problem to determine the ANN topology to produce good results [SaEtAl12a, SaEtAl12b]. There are proposals in the literature for obtaining an adequate ANN architecture [BeVo07, LoEtAl12]. An automatic strategy to identify the best topology for the neural network is obtained by minimizing a functional by a new meta-heuristic called Multi-Particle Collision Algorithm (MPCA) [SaEtAl16].

Our results show similar accuracy between the ANN and the Kalman filter, with better processing performance to the neural network.

25.2 Applied Model

A fixed coordinate system is used on the body of the aircraft to model the UAV movement. The orientation of the coordinate axes of the body is fixed in:

- The x axis points from the middle of the aircraft to its nose.
- The y axis points to the right of the x axis, facing the direction of the pilot, perpendicular to the x axis.
- The z axis points down from the aircraft, perpendicular to the xy plane.

Fig. 25.2 Coordinate system
for the body of a UAV

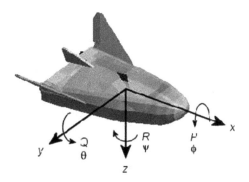

Figure 25.2 describes the coordinate system applied to the body of the UAV. Navigation is done by moving along these axes from origin. The rotations for aircraft movement are defined by the Euler angles (Φ, Θ, Ψ). Φ (Roll) represents the ability to roll on the x axis, Θ (Pitch) represents the ability to roll on the y axis, and Ψ (Yaw) represents the ability to roll on the z axis. The non-inertial speed coordinate system is placed on the aircraft at the initial position. The orientation of the coordinate system is defined in relation to the aircraft speed v—it is similar to the orientation of body coordinates.

In addition to the UAV body and velocity coordinates, the trajectory modeling requires positioning and orientation of the aircraft in relation to the rotation of the Earth. The navigation coordinates are defined in relation to the center and surface of the Earth. For this case, geocentric and geodetic latitudes, the descending-northeaster system, and the Earth-centered inertial system are used.

25.2.1 Platform Used

The framework proposed by [CoDo09] was tested in real flight by the group of the Department of Computer Science and Information Science of Linköping University, Sweden. A stand-alone helicopter shown in Fig. 25.3—the commercial model Yamaha Rmax. The total length of the helicopter is 3.6 m, with capacity to load up to 96 kg.

The developed positioning system was capable of flying fully automatic from takeoff to landing. Sensors used in the helicopter were three accelerometers and three gyroscopes along the three axes of the aircraft body, providing the acceleration and angular rate, a barometric altitude sensor, and a monocular video camera.

Fig. 25.3 UAV helicopter
Yamaha Rmax

25.2.2 Kalman Filter Applied to Autonomous Navigation

In [CoDo09], the first part of the proposed architecture (subsystem 1) is represented by a KF to estimate the UAV: position, velocity, and altitude; the second part is represented by a Bayesian filter called point-mass filter (PMF).

The implemented KF [Ma02] is a component of the navigation system. It is used to combine the UAV estimated position from image processing with data from an inertial sensor. The KF uses a dynamic model to find the linear error in space. The model is derived from a perturbation analysis of the motion equations [MiEtAl02]. The equation below represents the system model:

$$\begin{bmatrix} \delta^* \vec{r}^n \\ \delta^* \vec{v}^n \\ \delta^* \vec{\epsilon}^n \end{bmatrix} = \begin{bmatrix} F_{rr} & F_{rv} & 0- \\ F_{vr} & F_{vv} & A \\ F_{\epsilon r} & F_{\epsilon v} & -\vec{\omega}_{in}^n \end{bmatrix} \begin{bmatrix} \delta \vec{r}^n \\ \delta \vec{v}^n \\ \delta \vec{\epsilon}^n \end{bmatrix} + \begin{bmatrix} 0 & 0 \\ R_b^n & 0 \\ 0 & -R_b^n \end{bmatrix} \vec{u} \tag{25.1}$$

and the estimation model equation is given by:

$$\begin{bmatrix} \varphi_i - \varphi_v \\ \lambda_i - \lambda_v \\ h_i - (\Delta h_b - h_0) \end{bmatrix} = \begin{bmatrix} I & 0 & 0 \end{bmatrix} \begin{bmatrix} \delta \vec{r}^n \\ \delta \vec{v}^n \\ \delta \vec{\epsilon}^n \end{bmatrix} + \vec{\epsilon} \tag{25.2}$$

being:

$$A = \begin{bmatrix} 0 & -a_d & a_\epsilon \\ a_d & 0 & a_n \\ -a_\epsilon & a_n & 0 \end{bmatrix}; \ \delta \vec{r}^n = \begin{bmatrix} \delta \varphi \\ \delta \lambda \\ \delta h \end{bmatrix}; \ \delta \vec{v}^n = \begin{bmatrix} \delta v_n \\ \delta v_\epsilon \\ \delta v_d \end{bmatrix}; \ \vec{u} = \begin{bmatrix} \delta \vec{f}_{acc}^T \\ \vec{\omega}_{gyro}^T \end{bmatrix} \tag{25.3}$$

where $\delta \vec{r}^n$ represent latitude, longitude and altitude errors; $\delta \vec{v}^n$ denotes the errors of north, east, and descending velocity; $\vec{u}$ is the accelerometers and gyroscopes noises; $\vec{\epsilon}$ represents the noise measurements; $\vec{\omega}_{in}^n$ is the rotation rate; a_n, a_e, a_d are the north, east, and vertical accelerations; the vectors F_{xx} are the matrix entries of the dynamical system.

Estimation from subsystem 2—(latitude, longitude)—is used to update the UAV position by the KF. The updating of the altitude measurement is done using the barometer information. For calculating the altitude in the WGS84 reference system, an absolute reference value is necessary for h_0, the initial altitude. For example, if h_0 taken in the take-off position, the barometric altitude variation δh_{baro} related to h_0 can be obtained from a pressure sensor, and the altitude of the UAV can finally be calculated. This technique works if the ambient static pressure remains constant.

The system presented in Eqs. (25.1) and (25.2) is fully observable. The matrix entries F_{xx} and $\vec{\omega}_{in}^n$ have small values, because they depend on the Earth rotation rate and the navigation rotation rate relative to the curvature of the Earth. These elements are influential in high-speed flight conditions. Such conditions are not representative for the standard flight to small UAVs.

25.3   Neural Network Applied to Autonomous Navigation

Artificial neural network (ANN) is a key branch from the artificial intelligence. Many problems are solved by using ANN. Two classes of neural networks are described as *supervised* and *unsupervised*. For the supervised NN, a training set (*target*) is used to identify the connections weights—the *learning phase*. The unsupervised NN has no target or training set: the connections weights are updated during the ANN application.

Here, the supervised multi-layer perceptron (MLP) neural network is trained to emulate the Kalman filter. This ANN has an input layer, one or more hidden layers (processing units), and one output layer [Ha07]. Each neuron is fed by a weighted combination from the inputs plus a bias, being this combination the element for the non-linear activation function, representing the neuron output. The back-propagation algorithm to identify the connection weights is employed here during the learning phase [Ha07].

Some parameters need to be determined to find the MLP-NN architecture: number of hidden layers, number of neurons for each layer, the type of activation function, and the learning rate and momentum to the back-propagation algorithm. Due to the strong link between the ANN architecture definition and the quality of the obtained solution, the search for the optimal architecture becomes one important issue. Therefore, the ANN topology is an important topic in the context of the autonomous navigation.

25.3.1 MPCA Metaheuristic for ANN Optimal Architecture

Empirical approach is the most widely used method to design the ANN architecture. But, this strategy is a time-consuming process, requiring an expertise on ANN and on the problem to be addressed, with a degree of uncertainty [Ha07]. However, there is no guarantee that an optimal architecture for the neural network will be determined.

There are approaches to find the ANN architecture by using statistical algorithms, constructive and evolutionary approaches [BeVo07, LoEtAl12].

In our scheme, the optimal topology for the ANN is formulated as a solution of an optimization problem. The metaheuristic called Multiple Particle Collision Algorithm (MPCA) is employed as optimizer [AnEtAl13, SaEtAl16]. The MPCA is an extension derived from the standard Particle Collision Algorithm (PCA) [Sa05]. The MPCA is a stochastic optimization algorithm emulating the neutron traveling inside of a nuclear reactor, where two phenomena take place: scattering and absorption.

The cost function ϕ to be minimized is a combination of two factors: a metric for the neural network complexity, and the quadratic difference between the desired values and the ANN output:

$$\phi = \left[\theta_1 e^{x^2} + \theta_2 y + 1\right]\left(\frac{\rho_1 E_t + \rho_2 E_g}{\rho_1 + \rho_2}\right) \tag{25.4}$$

where x is the number of neurons, y corresponds to the number of epochs up to the convergence during the training phase for each MPCA algorithm iteration, θ_1 and θ_2 are adjustment parameters to calculate the complexity quantification of the neural network; ρ_1 and ρ_2 taking into account the balance between training (E_t) and generalization (E_g) errors.

25.4 Experiment Results

The methodology proposed by [CoDo09] to the UAV autonomous navigation is capable of flying fully automatic from takeoff to landing—see in Fig. 25.4 the red line. The set of sensors during the tests are: three accelerometers and three gyroscopes, providing the acceleration and angular rate of the helicopter, along the three axes of the body; a barometric altitude sensor; and a monocular video camera.

The video camera sensor is a standard CCD. The capture rate of the camera is 25 Hz, and the images resolution are reduced to half from the original image in order to have a smaller computational effort. During the experiment, the video camera was facing down and fixed to the body of the helicopter. For each sensor, the reading of the data is stored in a matrix composed of two columns. The first column is the instant of time in seconds and the second column is the value measured by

Fig. 25.4 Orthorectified aerial image

Table 25.1 Available features of sensors used in the autonomous navigation algorithm

Sensor	Acquisition rate	Resolution
Accelerometers	66 Hz	1 mG
Gyroscopes	200 Hz	0.1 deg/s
Barometer	40 Hz	0.1 m
Camera	25 Hz	384 × 288 pixels

the sensor. Table 25.1 presents a summary of the characteristics of the sensors. The ANN merges information from the inertial sensors with data from subsystem 2 and provides the estimation of the UAV position.

The sensor data and image were recorded during the flight. Subsystem 2, based on computer vision, the UAV initial position, before the takeoff, is provided by the inertial sensor and GPS (Global Positioning System) signal. After initialization, the GPS signal is no longer used for position estimation. The UAV positioning is computed by subsystem 2. Indeed, the goal is to do the navigation without using GPS. The data are processed at 50 Hz rate, meaning, every 50 Hz the network is activated with a sample coming from the inertial sensors and the processing result from subsystem 2.

For computing the optimal ANN architecture by the MPCA algorithm, a total of 16,249 patterns were used. The data set was split into three subsets: 70% to be used during the training phase, 20% for generalization, and 10% for validation. This amount is a result of the information from each sensor. The input vector has 12 entries. Table 25.2 presents a summary of the architecture found by the MPCA. The ANN architecture contains 32 neurons in the hidden layer, and the activation

Table 25.2 ANN architecture to emulate KF

Parameter	Value computed
Neurons in the input layer	32
Number of hidden layers	1
Neurons in the hidden layer	16
Neurons in the output layer	2
Learning rate	0.7
Rate of momentum	0.52
Activation Function	Sigmoid

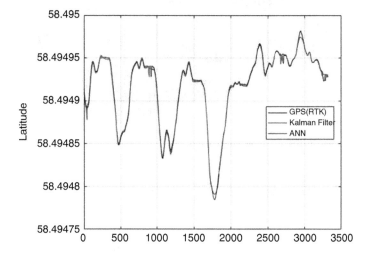

Fig. 25.5 Comparison latitudes estimated by GPS (RTK), KF and ANN

function Sigmoid presented the best performance. The mean square error obtained from the network training was 8.62×10^{-5}.

Figures 25.5 and 25.6 present the comparison of the estimation of latitudes together with their respective errors. Figures 25.7 and 25.8 present the comparison of the estimation of latitudes together with their respective errors. The errors presented in Figs. 25.6 and 25.8 are the difference between the GPS and the KF or ANN positioning.

The experimental results of this work confirm the validity of the proposed approach. The fusion of sensors based on computer vision can guarantee navigation, when the GPS signal is not available.

A next step to be evaluated is the implementation of ANN in FPGA [SaEtAl16]. The goal is to improve system performance by using dedicated hardware.

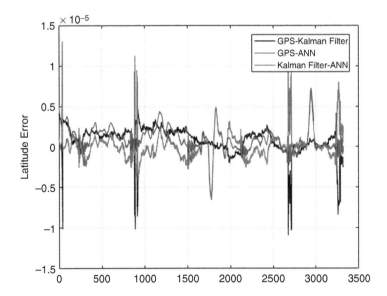

Fig. 25.6 Error in estimation of latitude

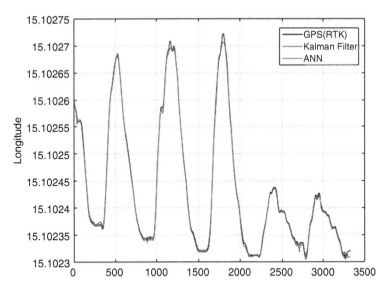

Fig. 25.7 Comparison latitudes estimated by GPS (RTK), KF and ANN

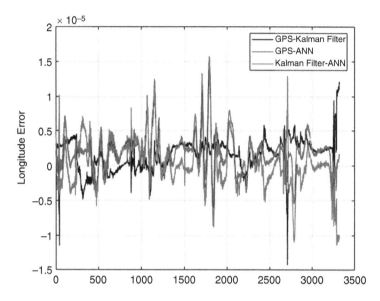

Fig. 25.8 Error in the estimation of the longitude

25.5 Final Remarks

The multi-layer perceptron artificial neural network was trained to emulate a Kalman filter for data fusion employed to estimate the UAV position. The difference between UAV positioning by GPS and by image processing was less than 2×10^{-5} using KF or MLP-NN.

The optimal architecture for the MLP-NN was found by employing the MPCA optimizer. Providing an efficient neural network to substitute the KLF with reduction of the computational effort, implying a lower energy demand, for the designed UAV positioning system.

Acknowledgements The authors would like to thank the FAPESP and CNPq, Brazilian agencies for research support.

References

[AnEtAl13] Anochi, J. A., Sambatti, S. B., Luz, E. F. P. d., and Campos Velho, H. F. d.: New learning strategy for supervised neural network: Mpca meta-heuristic approach. In Anais..., Congresso Brasileiro de Inteligência Computacional, (CBIC)., Sociedade Brasileira de Inteligência Computacional, **01**, pp. 01–06 (2013).

[BeVo07] Benardos, P. G. and Vosniakos, G. C.: Optimizing feedforward artificial neural network architecture. Eng. Appl. Artif. Intell, **20**, pp. 365–382 (2007).

[BiEtAl17] Birdal, A. C., Avdan, U., and Târk, T.: Estimating tree heights with images from an unmanned aerial vehicle. Geomatics, Natural Hazards and Risk, **2**, pp. 1144–1156 (2017)

[BrEtAl15] Braga, J. R. G., Campos Velho, H. F., and Shiguemori, E. H.: Estimation of UAV position using LiDAR images for autonomous navigation over the ocean. In 9th International Conference on Sensing Technology (ICST), pp. 811–816 (2015).

[BrEtAl16] Braga, J. R. G., Campos Velho, H. F., Conte, G., Doherty, P., and Shiguemori, E. H.: An image matching system for autonomous UAV navigation based on neural network. In 14th International Conference on Control, Automation, Robotics and Vision (ICARCV), pp. 1–6 (2016).

[CoDo08] Conte, G. and Doherty, P. An integrated UAV navigation system based on aerial image matching. In IEEE Aerospace Conference, pp. 1–10 (2008).

[CoDo09] Conte, G. and Doherty, P. Vision-based unmanned aerial vehicle navigation using geo-referenced information. EURASIP Journal on Advances in Signal Processing, **1**, 387–308 (2009).

[SiEtAl15] Silva, W., Shiguemori, E. H., Vijaykumar, N. L., and Campos Velho, H. F.: Estimation of UAV position with use of thermal infrared images. In 9th International Conference on Sensing Technology (ICST), pp. 828–833 (2015).

[DaEtAl17] Dash, J. P., Watt, M. S., Pearse, G. D., Heaphy, M., and Dungey, H. S.: Assessing very high resolution UAV imagery for monitoring forest health during a simulated disease outbreak. ISPRS Journal of Photogrammetry and Remote Sensing, **131**, pp. 1–14 (2017).

[FaEtAl18] Faria, L., Augusto, M. S., Correia, A. F., M., and Roso, N.: Susceptibility of GPS-dependent complex systems to spoofing, In Journal of Aerospace Technology and Management, Sâo José dos Campos - SP (2018), vol. 10.

[FiEtAl17] Fiori, L., Doshi, A., Martinez, E., Orams, M. B., and Bollard-Breen, B.: The use of unmanned aerial systems in marine mammal research. Remote Sensing, **9** (2017).

[GoRe15] Goncalves, J. and Renato, H.: UAV photogrammetry for topographic monitoring of coastal areas. In ISPRS Journal of Photogrammetry and Remote Sensing, vol. 104, pp. 101–111 (2015).

[Gr15] Groves, P. D.: Principles of GNSS, inertial, and multisensor integrated navigation systems, 2nd Ed. IEEE Aerospace and Electronic Systems Magazine, vol. 30, pp. 26–27 (2015).

[Ha07] Haykin, S.: Neural networks: principles and practical. Artmed (2007).

[JaEtAl17] James, M. R., Robson, S., d'Oleire-Oltmanns, S., and Niethammer, U.: Optimising UAV topographic surveys processed with structure-from-motion: Ground control quality, quantity and bundle adjustment. Geomorphology, vol. 280, pp.51–66 (2017).

[LoEtAl12] Loghmanian, S. M. R., Jamaluddin, H., Ahmad, R., Yusof, R., and Khalid, M.: Structure optimization of neural network for dynamic system modeling using multiobjective genetic algorithm. Neural Computing and Applications, vol. 21, pp. 1281–1295 (2012).

[Ma02] Maybeck, P.: Stochastic Models, Estimation, and Control - vol. 1, Academic Press (2002).

[MiEtAl02] Michaelsen, E., Kirchhof, M., and Stilla, U. Stilla.: Sensor pose inference from airborne videos by decomposing homography estimates. In Accepted for ISPRS (2004).

[Sa05] Sacco, W.: A new stochastic optimization algorithm based on a particle collision metaheuristic (2005).

[SaEtAl12a] Sambatti, S. B. M., Anochi, J. A., Luz, E. F. P. d., Carvalho, A. R., Shiguemori, E. H., and Campos Velho, H. F.: Mpca meta-heuristics for automatic architecture optimization of a supervised artificial neural network. In Proceedings... World Congress on Computational Mechanics, **10** (2012).

[SaEtA112b] Sambatti, S. B. M., Furtado, H. C. M., Anochi, J. A., Luz, E. F. P. d., and Campos Velho, H. F.: Automatic configuration of an artificial neural network with application to data assimilation. In Proceedings...International Conference on Integral Methods in Science and Engineering, **12** (2012).

[SaEtA116] Sambatti, S. B. M., Campos Velho, H. F., Furtado, H. C. M., Gomes, V. C., and Charão, A. S.: Self-configured neural network for data assimilation using FPGA for ocean circulation. Conference of Computational Interdisciplinary Sciences, **4** (2016).

[VaVa14] Valavanis, K. P. and Vachtsevanos, G. J.: Handbook of Unmanned Aerial Vehicles. Springer Publishing Company, Incorporated (2014).

Chapter 26
Towards the Super-Massive Black Hole Seeds

Eduardo S. Pereira, Pedro A. Santos, and Haroldo F. de Campos Velho

26.1 Introduction

In science, there are different problems to be solved. One of them is the evolution of the Universe. This paper discusses a little piece of the cited problem, which is the evolution of the black holes.

From nowadays, observations show super-massive black holes with mass about $10^9 M_\odot$ are ubiquitous in galaxy centers. However, the origin of this kind of object is not completely understood yet. A possibility rely in the fact the super-massive black holes (SMBH) seeds had their origin in the death of the Population III stars, during the called Dark Cosmological Ages – at redshift in the interval $z = [10, 20]$. In order to verify this hypothesis, the first step is to try to reconstruct the SMBH mass function at high redshift. The next step is to do a comparison between the computed distribution with the reminiscent of the first stars.

An inverse problem methodology is applied to identify the initial distribution of the SMBH seeds. The regularized inverse backward solution is obtained by minimizing a functional with square difference between the forward problem and observations, associated with the Tikhonov zeroth order regularization operator. The forward problem is the mass conservation law, representing the time variation of the black hole mass distribution n_{bh} in balance with the gradient of the function given by the product between the n_{bh} and the average of the mass accretion rate $\dot{m}_{bh}$. The time integration of the forward problem is performed by the Lax-Wendroff method. The formulation of the forward problem is described in terms of mass accretion instead of space coordinate.

E. S. Pereira · P. A. Santos · H. F. de Campos Velho (✉)

Instituto Nacional de Pesquisas Espaciais (INPE), São José dos Campos, SP, Brazil

e-mail: haroldo.camposvelho@inpe.br

C. Constanda, P. Harris (eds.), *Integral Methods in Science and Engineering*,
https://doi.org/10.1007/978-3-030-16077-7_26

A hybrid optimization method, combining a genetic algorithm with quasi-Newton approaches, is applied to compute the inverse solution. Synthetic observations are used to evaluate the inverse technique. Good inverse solutions are obtained even considering high level of noise in the data.

26.2 The Forward Problem: Conservation Law to the Ancient Black Holes

To determine the super-massive black hole (SMBH) seeds, a conservation law is assumed for the numerical mass density function of these objects, $n_{bh}(m_{bh}, t)$. This function represents the number of black holes per unit of comoving volume and per unit of solar mass at look-back time t.

The SMBH mass function (BHMF) must obey the following conservation law [YuTr02, SmBl92]:

$$\frac{\partial n_{bh}}{\partial t} + \frac{\partial \langle \dot{m}_{bh} \rangle n_{bh}}{\partial m_{bh}} = S(m_{bh}, t) \qquad (26.1)$$

where $S(t, m_{bh})$ represents the source that accounts for black hole mergers and, according to [YuTr02], [SmBl92] and [Li12], it is considered $S(t, m_{bh}) = 0$. The accretion is the dominating process of growth of SMBH, as stressed by [Li12], [ShEtAl10] and [ShEtAl9], with respect to the merges among this objects. This fact is assumed to justify the consideration of the source term to be null.

The $\langle \dot{m}_{bh} \rangle$ is the mean accretion rate weighted by the fraction of active SMBH [RaFa09] and [Li12]:

$$\langle \dot{m}_{bh} \rangle \equiv \delta(m_{bh}, z) \dot{m}_{bh} . \qquad (26.2)$$

The $\delta(m_{bh}, z)$ is the duty cycle of quasar, that takes into account the activity lifetime of quasars. On the other hand, the mean accretion rate can be written as [PeMi14]:

$$\langle \dot{m}_{bh} \rangle = \frac{1}{c^2} \frac{1}{f} \bar{L}_b(z, m_{bh}), \qquad (26.3)$$

where $\bar{L}_b(z, m_{bh})$ is the mean bolometric luminosity of quasars (MBLQ) for objects with central SMBHs of mass m_{bh} at a given redshift z, c is the speed of light, $f = \bar{\eta}/(1 - \bar{\eta})$, and $\bar{\eta}$ is the mean radiative efficiency, assumed as 0.1.

The MBLQ can be fitted by a parametric function—see [PeMi14]:

$$\bar{L}_b(z, m_{bh}) = \bar{L}_b^* \left(\frac{m_{bh}}{m_{bh}^*} \right)^\alpha \left(\frac{\tau_*}{t_u(z)} \right) \exp \left(-\frac{t_u(z)}{\tau_*} \right), \qquad (26.4)$$

here $\bar{L}_b^*$, m_{bh}^*, and τ_* are free parameters, being $t_u(z)$ the look-back time. Observe that m_{bh}^* represents a typical SMBH mass, in cosmic scale, τ_* and $\bar{L}_b^*$ represent, respectively, a lifetime scale and the dominant bolometric luminosity of Activity Galaxy Nuclei (AGN).

The relation between the cosmic time and the redshift, for Cold Dark Matter (CDM) plus cosmological constant (Λ), called ΛCDM model, is given by [PeMi10]:

$$\left| \frac{dt_u}{dz} \right| = \frac{9.787h^{-1}\mathrm{Gyr}}{(1+z)\sqrt{\Omega_\Lambda + \Omega_m(1+z)^3}} \; . \tag{26.5}$$

In the above equation, the Hubble's factor is defined as: $H_0/h \equiv 100 \; km \; s^{-1}$, being H_0 the Hubble's constant; the parameters $(\Omega_b, \Omega_m, \Omega_\Lambda)$ are density parameters for barionic matter, dark matter, and dark energy, respectively; and "Gyr" represents *giga years*. The numerical values for there parameters are: $h = 0.73$, $\Omega_b = 0.04$, $\Omega_m = 0.24$, $\Omega_\Lambda = 0.76$. The duty cycle of quasar can be computed from t_u by integrating the above expression: $t_u = \int_z^\infty g(z) \, dz$, where the function $g(z)$ is the right hand side of equation (26.5).

26.3   Mathematical Framework for the Inverse Solution

Before discussing the mathematical framework necessary to treat the problem, the inverse problem can be illustrated with a simple example. Consider $p(x)$ a polynomial of degree n. The forward problem is to find the roots $x_1, x_2, \ldots, x_n$ of this polynomial: $p(x_k) = 0$. An inverse problem can be expressed as: knowing the set of roots, find the polynomial. This example introduces some interesting aspects of the inverse problems: sometimes the forward and the inverse problem can be changed. The inverse problem exampled here is a polynomial regression, which in other contexts can be seen as the forward problem. This example has a very simple inverse solution $p(x) = c \prod_{k=1}^n (x - x_k)$. However this is not the case for the majority of inverse problems.

One of the best ways to understand the nature of the inverse problem is attributed to the Oleg Alifanov: *The solution of an inverse problem consists to determine unknown causes from the observed effects*. So, the forward problem can be understood as the evaluation of the effects considering the causes. The mathematical equations describing the forward model have the role of the evaluation. The *causes* will be a set of parameters or functions containing material properties of the system (parameters from the constitutive equations), source or sink terms, the initial and/or boundary conditions. In the case of ambiguity, the forward problem will be the one which was studied at first.

Hadamard had studied differential equations, and he has defined three conditions to characterize a *well-posed problem*: (1) existence of the solution, (2) the solution must be unique, (3) solution continuously depends on the input data [Ha52]. Inverse problems are generally ill-posed, because one cannot guarantee existence, uniqueness, and/or continuous dependence of the data.

Some techniques have been developed to deal with these badly posed problems. The truncated singular value decomposition (TSVD) is a technique to deal with such problem, where the smaller singular values are dropped. But, the TSVD is appropriate to linear problems. In addition, for large system, TSVD can be very expansive. Tikhonov has provided an alternative formulation with the regularization approach [Ti77]. The regularized inverse solution is obtained by minimizing the functional:

$$\arg \min_{u \in U} \|A(u) - f^\delta\|^2 + \alpha \Omega(u) \tag{26.6}$$

where U is the normed space of the possible solutions, A is a linear or nonlinear mapping from the normed space U to the normed space F, f^δ is the observed data with δ level of noise, α is the regularization parameter, and Ω is the regularization operator. If the observations have second statistical moment defined, the parameter α can be determined by the Morozov's discrepancy criterion [Mo84], finding the root of the following relation:

$$\|A(u) - f^\delta\| \approx N\sigma^2 \tag{26.7}$$

where N is the number of parameters to be estimated, and σ^2 is the uncertainty in the measurements. The generalized Morosov's discrepancy approach can be applied [ShEtAl04] if the statistics for the measurements has not its moments defined.

The forward problem can be given by a partial differential equation with initial and boundary conditions. Our inverse problem is to identify an unknown initial condition.

26.3.1 Regularization

There are different regularization operators. A general expression for the Tikhonov regularization is given by [Ti77]

$$\Omega(u) = \sum_{k=0}^{P} \mu_k \|u^{(k)}\|^2 \tag{26.8}$$

where $u^{(k)}$ denote the k-th derivative (or difference) and $\mu_k > 0$. Generally, it is not used combinations of several terms with derivative of different orders, instead just one. In this case, the expression becomes

$$\Omega(u) = \|u^{(k)}\| \tag{26.9}$$

and the technique is called Tikhonov regularization of order k. The zero order regularization looks for a function $u \approx 0$, implying in a reduction of oscillations in the inverse solution u. In the case of regularization of order 1, token $u^{(1)} \approx 0$, so u is approximately constant. Basically, the regularization acts in the amplitude of u, and the regularization parameter α has an important rule to compute the inverse solution. If the parameter $\alpha \to 0$ spurious oscillations will be present in the inverse solution, and if $\alpha \to \infty$ the solution u goes to zero.

26.3.2 Optimization

There are many methods to solve optimization problems presented in the literature. They can be split into two major groups: deterministic methods and stochastic methods. Examples of deterministic method are gradient descent, Newton method, quasi-Newton, conjugate gradient, and many others. Some stochastic methods are simulated annealing, genetic algorithms, ant colony optimization, swarm particle optimization, and others.

Some authors have proposed hybrid methods, combining the strategy of global search of the stochastic methods with the local search of the deterministic methods. That is the adopted strategy here, applying a genetic algorithm with the linear Newton-CG method.

Genetic Algorithm

A genetic algorithm (GA) is an optimization method inspired by the process of natural selection, belonging to the class of evolutionary algorithms [BaEtal00]. This metaheuristic is conceived using the modern version of evolution theory. In the context of optimization, the objective function evaluates the fitness of a population. The population individual is a candidate solution for the problem, and belongs to a set of possible solutions. Using the elitism procedure, individuals with the best fitness will be able to generate the new individuals. The new generation has distinctions introduced by mutations.

In the classical version of the algorithm, there are three fundamental operators: selection, crossover, and mutation. The Python package DEAP was used, based on evolutionary computation framework for rapid prototyping [FoEtal12]. This framework has some basic elements, a toolbox, where the operators act on the individuals to be selected: population, algorithms, and operators.

The algorithm **eaSimple**—*Algorithm 1*—was adopted, where population is a list of individuals. The toolbox contains the evolution operators **cxpb**, the probability of crossover, **mutpb**, the probability of mutation, and **varAnd** [FoEtal12], the smaller part of the algorithm. The **varAnd** works firstly duplicating the parental population, applying the probability of crossover selecting two individuals, and two offsprings generated will take the place of the parents. Finally, a mutation operator is applied to every individual with a probability. New individuals can be generated by only crossover, only mutation, and by crossover and mutation.

Algorithm 1 eaSimple

1: evaluate(population)   ▷ Calculate the fitness of every individual of the population
2: **for** g in number of generations **do**
3:       population ← select(population)
4:       offspring ← varAnd(population, toolbox, cxpb, mutpb)
5:       evaluate(offspring)
6:       population ← offspring

The Tournament selection was employed, selecting the best individual in a random list in the population; mutation was performed by a Gaussian random choice; and a Blend crossover, which makes a linear combination of the parents, using a random number in the interval [0, 1].

Linear Newton Conjugate Gradient (CG) Method
The traditional Newton method has the iteration procedure given by

$$x_{k+1} = x_k + [\nabla^2 f_k]^{-1} \nabla f_k \qquad (26.10)$$

where $f_k = f(x_k)$. Computationally, the inverse of the Hessian matrix $[\nabla^2 f(x_k)]^{-1}$ is not explicitly calculated, instead the linear system is solved:

$$[\nabla^2 f_k] p_k = -\nabla f_k \qquad (26.11)$$

where $p_k \equiv x_{k+1} - x_k$ is the basic Newton's method step. Most rules for terminating the iterative solver for (26.10) are based on the residual, given by [CoBo80] (see Chapter 4.6):

$$r_k = [\nabla^2 f_k] p_k + \nabla f_k , \qquad (26.12)$$

where p_k is the inexact Newton step. The idea of Linear Newton-CG method is to solve Eq. (26.11) using the conjugate gradient method [NoWr06] to find p_k, then the conjugate gradient method stops when

$$\|r_k\| \leq \eta_k \| \nabla f_k \| \qquad (26.13)$$

where the sequence $\{\eta_k\}_{k=1,2,...}$, with $0 < \eta_k < 1 \ \forall k$, is called the forcing sequence. The choice of η_k changes the convergence of the method. There exists two theorems implying the local convergence [NoWr06]:

Theorem 1 *Suppose that $\nabla^2 f(x)$ exists and is continuous in a neighborhood of a minimizer x^*, with $\nabla^2 f(x^*)$ is positive definite. Consider the iteration $x_{k+1} = x_k + p_k$ where p_k satisfies (26.13), and assume that $\eta_k \leq \eta$ for some constant $\eta \in [0, 1)$. Then, if the start point x_0 is sufficiently near x^*, the sequence $\{x_k\}$ converges to x^* and satisfies*

$$\|\nabla^2 f(x^*)(x_{k+1} - x^*)\| \leq \hat{\eta} \|\nabla^2 f(x^*)(x_k - x^*) , \qquad (26.14)$$

for some constant $\hat{\eta}$ with $\eta < \hat{\eta} < 1$.

Algorithm 2 Linear search Newton-CG

1: Given initial point x_0
2: **for** k = 0, 1, 2, ... **do**
3:     Define tolerance $\varepsilon_k = \min(0.5, \sqrt{\|\nabla f_k\|}) \|\nabla f_k\|$
4:     $z_0 \leftarrow 0$
5:     $r_0 \leftarrow \nabla f_k$
6:     $d_0 \leftarrow -r_0$
7:     **for** j=0, 1, 2, ... **do**
8:         **if** $(d_j, B_k d_j) \leq 0$ **then**
9:             **return** $p_k = -\nabla f_k$
10:         **else**
11:             **return** $p_k = -z_j$
12:         $\alpha_j = \|r_j\|^2/(d, B_k d_j)$
13:         $z_{j+1} = z_j + \alpha_j d_j$
14:         $r_{j+1} = r_j + \alpha_j B_k d_j$
15:         **if** $\|r_j + 1\| < \varepsilon_k$ **then**
16:             **return** $p_k = z_{j+1}$
17:         $\beta_{j+1} = \|r_{j+1}\|^2/\|r_j\|^2$
18:         $d_{j+1} = -r_{j+1} + \beta_{j_1} d_j$
19:     $x_{k+1} = x_k + \alpha_k p_k$

Theorem 2 *Suppose that the conditions of the Theorem 1 hold, and assume that the iterates $\{x_k\}$ generated by the inexact Newton method converge to x^*. Then the rate of convergence is superlinear if $\eta_k \to 0$. If addition, $\nabla^2 f(x)$ is Lipschitz continuous for x near x^* and if $\eta_k = O(\|\nabla f_k\|)$, then the convergence is quadratic.*

So the choice $\min\{0.5, \sqrt{\|\nabla f_k\|}\}$ [NoWr06] gives a superlinear convergence. Some variables are defined to the algorithm description. In the conjugate gradient iteration, d_j will represent the search directions, z_j will be the variable that in the conjugate gradient loop will converge to p_k, the solution of equation (26.10), which is the k-th Newton step in the outside loop, and the (a, b) denotes the intern product, which also induces the norm of the space as $\|a\| = \sqrt{(a, a)}$. The iteration follows up to reach a small value ε. The algorithm is applied to solve the linear system (26.11).

There are two stop conditions for Algorithm 2: for the inner loop and for the conjugate gradient step. First $\alpha_j = \|r_j\|^2/(d_j, B_k d_j)$—with: $B_k \equiv \nabla^2 f_k$, if the Hessian is not positive definite, where the conjugate gradient method cannot be applied. However the return condition implemented will yet generate descent direction, and the whole algorithm works. Second, $\|r_{j+1}\| < \varepsilon_k$, the standard stopping condition for conjugate gradient method. The algorithm was implemented with the Python language, using the **scipy** package, module *Optimize*, where the objective function, including the forward model and regularization operator, is called as a function to be minimized.

26.4 Identifying Black Hole Initial Distribution

As already mentioned, the inverse problem is expressed as an optimization problem. The objective function to be minimizing is given by—see also (26.6):

$$J(n_{bh}) = \sum_{j=1}^{N} \|A\left[(n_{bh})_j\right] - f_j^{\delta}\|_2^2 + \alpha \|\Omega^{(1)}(n_{bh})_j\|_2^2 \qquad (26.15)$$

where N is the number of observations, $\| \cdot \|_2$ is the L_2 norm, f^{δ} represents the observation with δ level of noise, and the first order Tikhonov regularization was employed—see above. The parameter α was determined by numerical experimentation.

The forward problem is solved by central finite difference, where the time variable is replaced by the redshift, denoted by z [Ma96]. The discretization parameters for the forward model were $\Delta m = 0.1$, $\Delta z = -0.1$, with the $z_0 = 20$ (initial time). The discretization parameters were selected to represent the observations (Δm) and numerical stability (Δz). The numerical values for the free parameters in Eq. (26.4) are given in Table 26.1.

The inverse methodology was evaluated with synthetic observations:

$$f^{\delta} \equiv n_{bh}^{\text{Exp}} = \left(1 + \sigma^2 \mu\right) n_{bh}^{\text{Mod}} \qquad (26.16)$$

being n_{bh}^{Mod} the result obtained by integration of the forward model from $z = z_0$ up to $z = 0$, the level of the noise is denoted by σ^2, and μ is a random number with Gaussian distribution with zero mean.

Some numerical experiments were carried out, with noiseless data, and observations with 10% of noise. For the noise data, the regularization parameter α was 70.0. The genetic algorithm was executed with a population of 1000 elements and 200 generations. Figure 26.1 shows inverse solution (blue curve) and the true answer (green curve), using only GA for solving the optimization problem (Fig. 26.1a), and using the hybrid method (GA + quasi-Newton)—see Fig. 26.1b.

The inversion with noisy data is shown in Fig. 26.2. Clearly, the GA alone was not able to find an acceptable solution—Fig. 26.2a (blue curve). However, applying the hybrid optimizer, combining the GA metaheuristic and quasi-Newton scheme, a good inverse backward distribution to the ancient Black Hole seeds can be computed—blue curve in Fig. 26.2b.

Table 26.1 Numerical values for free parameters in Eq. (26.4)

Parameter	Value
L_m^*	1.0
m_{bh}^*	2.90×10^{11}
α^*	2.71×10^{-1}
η	0.5
τ_*	4.81×10^9

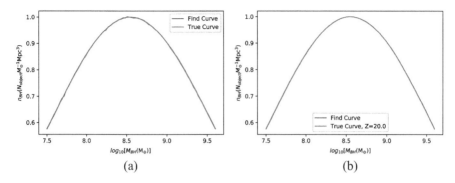

Fig. 26.1 Black Hole initial distribution (noiseless observation): (**a**) only GA, (**b**) GA + quasi-Newton

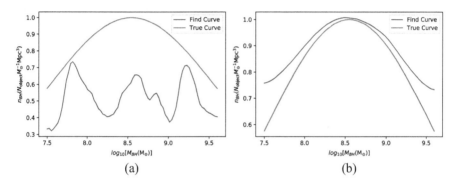

Fig. 26.2 Black Hole initial distribution (noisy observation): (**a**) only GA, (**b**) GA + quasi-Newton

26.5 Final Remarks

A procedure to identify the distribution for a large redshift to the super-massive black hole seeds was presented and tested with synthetic observation. A perfect inversion was obtained with noiseless data. However, the GA metaheuristic alone was not able to calculate the inversion with noisy observation. Only the hybrid optimization approach can compute a good inverse solution.

The inverse problem worked here is a hard one. Other inverse approaches can be employed to identify the Black Hole initial distribution [ChEtAl03], including the use of other regularization operators, such as higher order Tikhonov regularization [Ti77] and entropic regularization [CaEtAl97, RaEtAl99].

Acknowledgements The authors would like to thank the FAPESP and CNPq, Brazilian agencies for research support.

References

[BaEtal00] Baeck, T., Fogel, D.B., Michalewicz, Z.: *Evolutionary Computation 1: Basic Algorithms and Operators*, Taylor & Francis (2000).

[CaEtAl97] Campos Velho, H. F., Ramos, F. M.: Numerical Inversion of Two-Dimensional Geoelectric Conductivity Distributions from Electromagnetic Ground Data, *Braz. J. Geophys.*, **15**, 133–143 (1997).

[ChEtAl03] Chiwiacowsky, L.D., Campos Velho, H. F.: Different Approaches for the Solution of a Backward Heat Conduction Problem, *Inverse Problems in Engineering*, **11**, 471–494 (2003).

[CoBo80] Conte, S. D., de Boor, C.: *Elementary Numerical Analysis: An Algorithmic Approach*, McGraw Hill, 3rd Edition (1980).

[FoEtal12] Fortin, F.-A., De Rainville, F.-M., Gardner, M.-A., Parizeau, M., Gagné, C.: DEAP: Evolutionary Algorithms Made Easy. *Journal of Machine Learning Research*, **13**, 2171–2175 (1992).

[Ha52] Hadarmard, J.: *Lectures on Cauchy's Problem in Linear Partial Differential Equations*, Dover (1952).

[Li12] Li Y., Wang J., Ho L. C.: Cosmological Evolution of Supermassive Black Holes II. Evidence for Downsizing of Spin Evolution, *The Astrophysical Journal*, **749**, 187–198 (2012).

[Ma96] Madsen, M. S.: *The Dynamic Cosmos – Exploring the Physical Evolution of the Universe*, Chapman & Hall, New York (NY), USA (1996).

[Mo84] Morosov, V. A.: *Methods for Solving Incorrectly Posed Problems*, Springer Verlag (1984).

[NoWr06] Nocedal, J., Wright, S. J.: *Numerical Optimization*, Springer (2006).

[PeMi14] Pereira, E. S., Miranda, O. D., Accretion history of active black holes from type 1 AGN. *Astrophys Space Sci*, **352**, 801–807 (2014).

[PeMi10] Pereira, E. S., Miranda, O. D., Stochastic background of gravitational waves generated by pre-galactic black holes *Monthly Notices Of Royal Astronomic Society*, **401**, 1924–1932 (2010).

[RaFa09] Raimundo, S. I., Fabian, A. C. Eddington ratio and accretion efficiency in active galactic nuclei evolution, *Monthly Notes of Royal Astronomic Society*, **396**, 12–17, (2009).

[RaEtAl99] Ramos, F. M., Campos Velho, H. F.,Carvalho, J. C., Ferreira, N. J.: Novel Approaches on Entropic Regularization, *Inverse Problems*, bf 15, 1139–1148 (1999).

[ShEtAl04] Shiguemori, E. H., Campos Velho, H. F., Silva, J. D. S.: Generalized Morosov's Principle. In: *Inverse Problems, Design and Optimization Symposium* (IPDO), Angra dos Reis (RJ), Brazil, 290–298 (2004).

[SmBl92] Small, T. A., Blandford R. D.: Quasar evolution and the growth of black holes. *Monthly Notices of the Royal Astronomical Society*, **259**, 4, 725–737 (1992).

[ShEtAl10] Shankar F., et al.: On the Radiative Efficiencies, Eddington Ratios, and Duty Cycles of Luminous High-Redshift Quasars, *The Astrophysical Journal*, **718**, 213–250 (2010).

[ShEtAl9] Shankar F., Weinberg D.H., Miralda-Escude J.: Self-Consistent Models of the AGN and Black Hole Populations: Duty Cycles, Accretion Rates, and the Mean Radiative Efficiency *The Astrophysical Journal*, **690**, 20–41 (2009).

[Ti77] Tikhonov, A. N., Arsenin, V. Y.: *Solution of Ill-posed Problems*, John Wiley & Sons (1977).

[YuTr02] Yu, Q., Tremaine, S.: Observational constraints on growth of massive black holes, *Monthly Notices of the Royal Astronomical Society*, **335**, 965–976 (2002).

Chapter 27
Decomposition of Solutions of the Wave Equation into Poincaré Wavelets

Maria V. Perel and Evgeny A. Gorodnitskiy

27.1 Introduction

Derivation of formulas that decompose solutions of the wave equation in terms of local ones with known properties is justified by its applications. Fourier analysis yields the decomposition of solutions into plane waves if the medium is homogeneous. If the medium is nonhomogeneous, asymptotic formulas are known [Po07] for decomposition of solutions into Gaussian beams, which are particular solutions that are local in the sense of Gauss near rays. These formulas are valid in a medium with a smooth inhomogeneity, and are applied in problems of seismics [PoEtAl10] or various problems of wave propagation and diffraction, see [LeHe17], [ShEtAl04] and the references therein. An exact decomposition of solutions in the homogeneous medium in terms of very special ones, which decrease in a power law, was given in [Ka94] by the techniques from the continuous wavelet analysis (CWA). After the obtaining of a particle-like exact solution in a homogeneous medium [KiPe99], [KiPe00], the question arose whether any solution can be decomposed into such solutions. Decompositions in terms of these particle-like solutions, called Gaussian wave packets, were found by means of techniques from CWA, see [AnEtAl94], in [PeSi03], [PeSi06], [PeSi07], [PeSi09] and [PeEtAl11]. A particle-like solution is just an example of elementary local solution. It turned out that such formulas can be used for solutions of a wider class. The CWA enables us to decompose a given function in terms of wavelets, which are obtained from a single function called a mother wavelet by means of some group of transformations. In the papers mentioned above, the similarity group was used. In [Pe09], [PeGo12], [GoPe11], we found and studied decompositions of solutions constructed by the affine Poincaré CWA; that

M. V. Perel (✉) · E. A. Gorodnitskiy
Saint Petersburg University, Saint Petersburg, Russia
e-mail: m.perel@spbu.ru

C. Constanda, P. Harris (eds.), *Integral Methods in Science and Engineering*,
https://doi.org/10.1007/978-3-030-16077-7_27

is, by a CWA based on the affine Poincaré group. This chapter contains a brief review of the main results of [Pe09], [PeGo12], [GoPe11], [GoPe17], and of papers concerning decompositions in an inhomogeneous medium and their applications [GoEtAl16], where asymptotic elementary solutions named quasiphotons [BaUl81], [Ka84], [Ba07] were applied.

27.2 Statement of the Problem

Below, we show an integral representation of solutions of the boundary value problem for the wave equation in the half plane

$$\Delta u(t, x, z) - \frac{1}{c^2} u_{tt}(t, x, z) = 0,$$

$$u(x, z, t) \mid_{z=0} = f(t, x).$$

(27.1)

The solution of this problem is not unique. An additional condition similar to causality will be given later.

A specific feature of the problem is the multiscale structure of the boundary data $f(t, x)$. An example of boundary data for problems of seismics is shown in Fig. 27.1. The lighter the color, the larger the value of the function.

Our aim is to give an integral representation of the solution that is convenient for boundary data processing. The mathematical tool for obtaining the necessary formulas is the affine Poincaré continuous wavelet analysis (CWA). The outline of the chapter is as follows. We decompose the boundary data into a specially chosen affine Poincaré wavelets. Each wavelet is the boundary datum for some local solution. The reconstruction formula for wavelets yields the integral representation of solutions in terms of elementary local solutions with known properties.

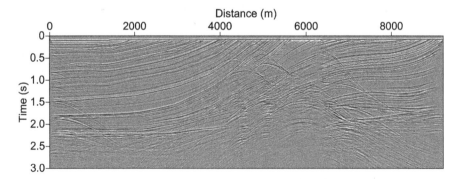

Fig. 27.1 Marmousi model [Ver94] seismograms as boundary value function $f(t, x)$. See details in [GoEtAl16]

27.3 Affine Poincaré CWA

We list here the results from the affine Poincaré CWA, see [AnEtAl94], which constitute the basis of the constructions of the present paper.

First, we determine the space of functions, which will be decomposed in affine Poincaré wavelets. The space $\mathbb{L}_2(\mathbb{R}^2)$ can be decomposed as a direct sum of four subspaces $\mathscr{D}_j$ as follows:

$$\mathbb{L}_2(\mathbb{R}^2) = \mathscr{D}_1 \oplus \mathscr{D}_2 \oplus \mathscr{D}_3 \oplus \mathscr{D}_4. \tag{27.2}$$

This means that any function $f(t, x) \in \mathbb{L}_2(\mathbb{R}^2)$ can be expanded as a sum:

$$f(t, x) = \sum_{j=1}^{4} f_j(t, x), \tag{27.3}$$

where $f_j(t, x) \in \mathscr{D}_j$ and is defined by means of the support of its Fourier transform

$$f_j(t, x) = \iint_{D_j} d\omega \, dk_x \, \hat{f}_j(\omega, k_x) \exp(i(-\omega t + k_x x)).$$

The domain of integration is determined by the inequalities

$$D_1 : |\omega| > c|k_x|, \quad \omega > 0, \qquad D_2 : |\omega| > c|k_x|, \quad \omega < 0, \tag{27.4}$$

$$D_3 : |\omega| < c|k_x|, \quad k_x > 0, \qquad D_4 : |\omega| < c|k_x|, \quad k_x < 0. \tag{27.5}$$

The affine Poincaré CWA is developed for each of the spaces $\mathscr{D}_j$, $j = 1, 2, 3, 4$. We give here results for $\mathscr{D}_1$. For other spaces all the formulas are obtained analogously.

Secondly, we determine a pair of mother wavelets. Let us choose two functions $\zeta, \psi \in \mathscr{D}_1$, which fulfill an additional condition:

$$0 < C_{\zeta\psi} \equiv \iint_{\mathscr{D}_1} dk_x d\omega \, \frac{\overline{\hat{\zeta}(\omega, k_x)}\hat{\psi}(\omega, k_x)}{\omega^2/c^2 - k_x^2} < \infty \tag{27.6}$$

named the admissibility condition. Such functions form a pair of mother wavelets. These functions can coincide if the condition (27.6) is met.

Thirdly, we select a group of transformation, with which families of wavelets are constructed. We apply the affine Poincaré group. It comprises shifts by x_s, t_s : $x \to x + x_s$, $t \to t + t_s$, scaling by a scale $a : x \to ax$, $z \to az$, $t \to at$ and the

Lorentz transform with the rapidity ϕ :

$$\begin{pmatrix} ct \\ x \end{pmatrix} \to \Lambda_\phi \begin{pmatrix} ct \\ x \end{pmatrix}, \quad \Lambda_\phi = \begin{pmatrix} \cosh\phi & \sinh\phi \\ \sinh\phi & \cosh\phi \end{pmatrix}. \tag{27.7}$$

Families of wavelets obtained from $\zeta(t,x)$ and $\psi(t,x)$ read

$$\zeta_{\phi a}(t,x) = \frac{1}{a}\zeta(t',x'), \quad \psi_{\phi a}(t,x) = \frac{1}{a}\psi(t',x'), \quad \begin{pmatrix} ct' \\ x' \end{pmatrix} = \frac{1}{a}\Lambda_{-\phi}\begin{pmatrix} ct \\ x \end{pmatrix}.$$

The affine Poincaré wavelet transform is determined as an integral

$$\mathcal{F}(\phi,a,t_s,x_s) = \iint_{\mathbb{R}^2} dt dx f(t,x)\overline{\zeta_{\phi a}(t-t_s,x-x_s)}. \tag{27.8}$$

If the affine Poincaré wavelet transform is known, the function $f(t,x)$ can be reconstructed by the formula

$$f(t,x) = \frac{1}{C_{\zeta\psi}} \int_{\mathbb{R}} d\phi \int_0^\infty \frac{da}{a^3} \iint_{\mathbb{R}^2} dt_s dx_s \mathcal{F}(\phi,a,t_s,x_s)\psi_{\phi a}(t-t_s,x-x_s), \tag{27.9}$$

see [AnEtAl94].

27.4 Wavelet Analysis for Solutions in Homogeneous Medium

We present here an integral representation of solutions of the problem (27.1) in terms of elementary solutions. At the beginning we give an additional condition, which is analogous to the radiation condition. We define the space, to which the solution should belong. Let $f(t,x) \in \mathcal{D}_1$, i.e., $\operatorname{supp}\hat{f}(\omega,k_x) \subset D_1$, $D_1 = \{(\omega,k_x) : \omega > c|k_x|\}$.

Define spaces of solutions $\mathcal{H}_1^+$ and $\mathcal{H}_1^-$ by their Fourier expansions as follows: $u^+ \in \mathcal{H}_1^+$, $u^- \in \mathcal{H}_1^-$ if

$$u^+(t,x,z) = \frac{1}{(2\pi)^2} \iint_{D_1} d\omega dk_x\, e^{i(-\omega t+k_x x+k_z z)}\hat{f}(\omega,k_x),$$

$$u^-(t,x,z) = \frac{1}{(2\pi)^2} \iint_{D_1} d\omega dk_x\, e^{i(-\omega t+k_x x-k_z z)}\hat{f}(\omega,k_x),$$

$$k_z = \sqrt{\omega^2/c^2 - k_x^2}.$$

Functions $u^{\pm}(t, x, z)$ satisfy the wave equation in the sense of distributions, see details in [GoPe17]. The Fourier expansion of solutions from the space $\mathscr{H}_1^+$ contains the plane waves with a positive phase speed, i.e., such plane waves propagate away from the boundary $z = 0$. The space $\mathscr{H}_1^-$ consists of the plane waves moving from the infinity to the boundary $z = 0$.

If $f(t, x) \in \mathscr{D}_1$, there is a unique solution, which satisfies (27.1) and belongs to the class $\mathscr{H}_1^+$.

The second step is to define the mother solution and a family of solutions. Let choose a solution $\Psi(t, x, z) \in \mathscr{H}_1^+$. It is a mother solution. Then $\psi(t, x) = \Psi(t, x, 0) \in \mathscr{D}_1$ and can be regarded as a mother wavelet.

A family of solutions parameterized by the rapidity ϕ and the scale a is constructed as follows:

$$\Psi_{\phi a}(t - t_s, x - x_s, z) = \frac{1}{a}\Psi\left(\frac{t'}{a}, \frac{x'}{a}, \frac{z}{a}\right), \quad \begin{pmatrix} ct' \\ x' \end{pmatrix} = \Lambda_{-\phi}\begin{pmatrix} c(t - t_s) \\ x - x_s \end{pmatrix}.$$

$$(27.10)$$

It is important that such transformations as shifts, scaling in all the coordinates and time, and the Lorentz transformations map solutions to another ones.

The traces of these solutions on the boundary can be regarded as wavelets from the family obtained by transformations of the affine Poincaré group applied to the mother wavelet, i.e.,

$$\Psi_{\phi a}(t - t_s, x - x_s, 0) = \psi_{\phi a}(t - t_s, x - x_s). \qquad (27.11)$$

We assume that the continuous affine Poincaré transform of the function $f(t, x)$ is calculated by means of (27.8) with a mother wavelet $\zeta(t, x)$. The mother wavelets $\psi(t, x)$ and $\zeta(t, x)$ should satisfy the admissibility condition (27.6).

The reconstruction formula for the boundary data $f(t, x)$ is given by (27.9).

The solution $u(t, x, z) \in \mathscr{H}_1^+$ can be decomposed in terms of elementary ones $\Psi(t, x, x) \in \mathscr{H}_1^+$ as follows:

$$u(t, x, z) = \frac{1}{C_{\zeta\psi}} \int_{\mathbb{R}} d\phi \int_0^{\infty} \frac{da}{a^3} \iint_{\mathbb{R}^2} dt_s dx_s \mathscr{F}(\phi, a, t_s, x_s)\Psi_{\phi a}(t - t_s, x - x_s, z).$$

$$(27.12)$$

The condition (27.11) and the reconstruction formula (27.9) yield the fulfillment of the boundary condition (27.1). The formula (27.12) is the main result for decomposition of solutions in homogeneous medium. This formula is an integral superposition of solutions. Therefore we expect that an integral is a solution also. To justify the formula (27.12), we proved the following theorem in [GoPe17].

Theorem 1 *Let $u(t, x, z) \in \mathscr{H}_1^{+}$.*

Let wavelets $\zeta, \psi \in \mathscr{D}_1$ be a pair of admissibility wavelets, and $\Psi(t, x, z) \in \mathscr{H}_1^{+}$ is the solution corresponding to $\psi(t, x) \in \mathscr{D}_1$.

Then

1. $\tilde{u}(t, x, z) \in \mathscr{H}_1^{+}$, where

$$\tilde{u}(t, x, z) = \frac{1}{C_{\psi\zeta}} \int_{A_1}^{A_2} \frac{da}{a^3} \int_{\Phi_1}^{\Phi_2} d\phi \iint dt_s dx_s \mathscr{F}(\phi, a, t_s, x_s) \Psi_{\phi a}(t - t_s, x, x_s, z),$$

where $0 < A_1 < A_2 < \infty, -\infty < \Phi_1 < \Phi_2 < \infty, 0 < c^2 t_s^2 + x_s^2 < \rho < \infty,$

2.

$$\lim_{\substack{A_1 \to 0, A_2 \to \infty \\ \Phi_1 \to -\infty, \Phi_2 \to \infty \\ \rho \to \infty}} \|u(t, x, z) - \tilde{u}(t, x, z)\| = 0.$$

We give now an example of an exact solution of the wave equation, which may be taken as a mother solution. It is a solution named the Gaussian Wave Packet, which was found in [KiPe99], [KiPe00] and studied from the point of view of continuous wavelet analysis in [PeSi07]. It is determined by an explicit formula

$$\Psi(t, x, z) = \sqrt{\frac{2}{\pi}} \frac{\exp(-ps)}{\sqrt{z + ct - i\varepsilon}}, \qquad s = \sqrt{1 - i\theta/\gamma},$$

$$\theta = z - ct + \frac{x^2}{z + ct - i\varepsilon},$$

where $\varepsilon, \gamma,$ and p are positive parameters, and satisfies the wave equation

$$\Psi_{tt} - c^2(\Psi_{xx} + \Psi_{zz}) = 0. \tag{27.13}$$

It is shown in [KiPe00] that for large p and moderate t it has a maximum in the point running with a speed c along a straight line. Near the maximum it has a Gaussian asymptotics

$$\Psi(t, x, z) \sim \exp\left[i\frac{p}{2\gamma}(z - ct) - \frac{(z - ct)^2}{2\sigma_z^2} - \frac{x^2}{2\sigma_x^2}\right], \tag{27.14}$$

$$\sigma_z^2 = \frac{4\gamma^2}{p}, \quad \sigma_x^2 = \frac{\gamma\varepsilon}{2p}.$$

If $z = 0$, it is the Morlet wavelet in the 2D space.

The Fourier transform of this solution is found in [PeSi07]. The solution belongs to the $\mathscr{H}_1^+$ class.

27.5   Decomposition of Solutions for an Inhomogeneous Medium

We suggest an integral formula for the solution decomposition in an inhomogeneous medium

$$u(t, x, z) = \frac{1}{C_{\zeta \psi}} \int_{\mathbb{R}} d\phi \int_0^{\infty} \frac{da}{a^3} \iint_{\mathbb{R}^2} dt_s dx_s \mathscr{F}(\phi, a, t_s, x_s) \Psi_{\phi a}(t - t_s, x, x_s, z).$$

$$(27.15)$$

Decomposition formulas in cases of homogeneous and inhomogeneous media differ in the construction of a family of solutions. For an inhomogeneous medium, these solutions are determined as solutions of the wave equation (27.13) with the boundary condition

$$\Psi_{\phi a}(t - t_s, x, x_s, 0) = \psi_{\phi a}(t - t_s, x - x_s).$$

Such solution in the general case cannot be described by a formula like (27.10) and should be found numerically. A condition similar to radiation condition is not found here for a wide class of the mother solutions $\Psi(t, x, z)$. For particular case such as an asymptotic solution like (27.14), we can determine the direction of propagation of the maximum of the solution. This maximum should move away from the boundary $z = 0$. A strict statement of the problem for Ψ and the proof of convergence of fourfold integrals in (27.15) will be done in our further paper.

The calculation of the fourfold integrals is a separate problem. We show in [GoEtAl16] that these integrals can be effectively calculated in applied problem of seismics. The solution of the problem of the depth migration is reduced in particular to the problem (27.1), where $f(t, x)$ is a seismograms, i.e., an acoustic field on the boundary of the Earth dependent on time t and the coordinate on the Earth x for some sources, see details in [GoEtAl16]. The boundary data for the Marmousi model [Ver94] is shown in Fig. 27.1.

The excitation coefficient $\mathscr{F}(\phi, a, t_s, x_s)$ of elementary solutions can be a sparse function of ϕ, a, t_s, x_s. Particularly, in case of the function $f(t, x)$ shown in Fig. 27.1 most of the wavelet coefficients are smaller than the 5% of the maxima, see Fig. 27.2. In the figure the sparsity map is shown: the white color corresponds to the coefficients less then 5%. The sign at the top means the shift x_s in meters, the numbers along vertical sides are time shift t_s counts from 0 to 3 s, the numbers

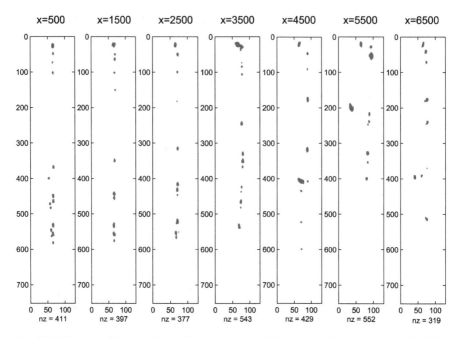

Fig. 27.2 Sparsity of the wavelet coefficients in case of the function $f(t, x)$ shown in Fig. 27.1. The number nz is a number of points taken into account from more than 90000 points. See discussion in the text

along horizontal sides counts the rapidity ϕ. The rapidity ϕ rules the direction θ of the propagation of the solution via the relation

$$\tan \theta = \sinh \phi,$$

see [GoEtAl16].

27.6   Conclusions

We have obtained an integral representation of solutions of the wave equation in the half plane based on the wavelet decomposition of the boundary data. In a homogeneous medium, we present and study accurately this representation. In an inhomogeneous medium, we give a formal result, apply it to a problem of seismics and show its effectiveness. The advantage is that the coefficients of elementary solutions in obtained decomposition have a meaning of the wavelet transform of the boundary data. The wavelet transform is often applied to data processing and can be a sparse function. Therefore our results contain built-in filtering ability and selectivity for valuable, or the most interesting, portions of the boundary data.

References

[Ka94] Kaiser, G.: *A Friendly Guide to Wavelets*, Birkhäuser (1994).

[AnEtAl94] Antoine, J.-P., Murenzi, R., Vandergheynst, P., and Ali, S. T.: *Two-Dimensional Wavelets and their Relatives*, Cambridge University Press (1994).

[BaUl81] Babich, V. M., and Ulin, V. V.: The complex space-time ray method and "quasi-photons". (Russian). In *Mathematical questions in the theory of wave propagation. 12 Zap. Nauchn. Sem. Leningrad. Otdel. Mat. Inst. Steklov. (LOMI)*, **117**, 5–12 (1981). Transl. *J. Soviet Math.*, **24**, 269–274 (1984).

[Ka84] Kachalov, A. P.: A coordinate system for describing the "quasiphoton". In *Mathematical questions in the theory of wave propagation. 14 Zap. Nauchn. Sem. Leningrad. Otdel. Mat. Inst. Steklov. (LOMI)*, **140**, 73–76 (1984). Transl. *J. Soviet Math.*, **32**, 151–153 (1986).

[Po07] Popov, M. M.: A new method for calculating wave fields in high-frequency approximation. (Russian). In *Mat. Vopr. Teor. Rasprostr. Voln.. 36 Zap. Nauchn. Sem. S.-Peterburg. Otdel. Mat. Inst. Steklov. (POMI)*, **342**, 5–13 (2007). Transl. *J. Math. Sci.*, N.Y. **148**, 633–638 (2008).

[ShEtAl04] Shlivinski, A., Heyman, E., Boag, A., and Letrou, C.: A phase-space beam summation formulation for ultrawide-band radiation. *IEEE Transactions on Antennas and Propagation*, **52**, 2042–2056 (2004).

[PoEtAl10] Popov, M. M., Semtchenok, N. M., Popov, P. M., and Verdel, A. R.: Depth migration by the Gaussian beam summation method. *Geophysics*, **75**, S81–S93 (2010).

[LeHe17] Leibovich, M., and Heyman, E.: Beam Summation Theory for Waves in Fluctuating Media. Part I: The Beam Frame and the Beam-Domain Scattering Matrix. *IEEE Transactions on Antennas and Propagation*, **65**, 5431–5442 (2017).

[Ba07] Babich, V. M.: Quasiphotons and the space-time ray method. (Russian) In *Mathematical questions in the theory of wave propagation. 36 Zap. Nauchn. Sem. S.-Peterburg. Otdel. Mat. Inst. Steklov.*, **342**, 5–13 (2007). Transl. *J. Math. Sci. (N.Y.)*, **148**, 633–638 (2008).

[KiPe99] Kiselev, A.P., and Perel, M. V.: Gaussian wave packets. *Optics and Spectroscopy*, **86**, pp. 307–309 (1999).

[KiPe00] Kiselev, A.P., and Perel, M. V.: Highly localized solutions of the wave equation. *J. Math. Phys.*, **41**, pp. 1034–1955 (2000).

[PeSi03] Perel, M. V., and Sidorenko, M. S.: Wavelet analysis in solving the Cauchy problem for the wave equation in three-dimensional space. In *Mathematical and Numerical Aspects of Wave Propagation: Waves 2003*, G. C. Cohen (ed), Springer, Berlin (2003) pp.794–798.

[PeSi06] Perel, M. V., and Sidorenko, M. S.: Wavelet analysis for the solutions of the wave equation. In *Proceedings of the International Conference Days on Diffraction*, Saint-Petersburg (2006), pp. 208–217.

[PeSi07] Perel, M. V., and Sidorenko, M. S.: New physical wavelet 'Gaussian wave packet'. *J. of Phys. A: Math. and Theor.*, **40**, 3441 (2007).

[PeSi09] Perel, M. V., and Sidorenko, M. S.: Wavelet-based integral representation for solutions of the wave equation. *J. of Phys. A: Math. and Theor.*, **42**, 375211 (2009).

[Pe09] Perel, M. V.: Integral representation of solutions of the wave equation based on Poincaré wavelets. In *Proceedings of the International Conference Days on Diffraction* , Saint-Petersburg (2009), pp. 159–161.

[GoPe11] Gorodnitskiy, E. A., and Perel, M. V.: The Poincaré wavelet transform: implementation and interpretation. In *Proceedings of the International Conference Days on Diffraction* , Saint-Petersburg (2011), pp. 72–77.

[PeEtAl11] Perel, M., Sidorenko, M., and Gorodnitskiy, E.: Multiscale Investigation of Solutions of the Wave Equation. In *Integral Methods in Science and Engineering*, C. Constanda and M. E. Perez (eds.), Birkhäuser, Boston (2011), pp. 291–300.

[PeGo12] Perel, M., and Gorodnitskiy, E.: Integral representations of solutions of the wave equation based on relativistic wavelets. *J. of Phys. A: Math. and Theor.*, **45**, 385203 (2012).

[GoEtAl16] Gorodnitskiy, E., Perel, M. V., Geng, Yu, and Wu, R.-S.: Depth migration with Gaussian wave packets based on Poincaré wavelets. *Geophys. J. Int.*, **205**, pp. 314–331 (2016).

[GoPe17] Gorodnitskiy, E. A., Perel, M. V.: Justification of a Wavelet-Based Integral Formula for Solutions of the Wave Equation. (Russian) In *Mathematical questions in the theory of wave propagation. Part 47 Zap. Nauchn. Sem. S.-Peterburg. Otdel. Mat. Inst. Steklov. (POMI)*, **461**, 107–123 (2017). Transl. *J. Math. Sci. (N.Y.)*, 238, pp. 630–640 (2019).

[Ver94] Versteeg, R.: The Marmousi experience: Velocity model determination on a synthetic complex data set. *The Leading Edge*, **13**, 927–936 (1994).

Chapter 28
The Method of Fundamental Solutions for Computing Interior Transmission Eigenvalues of Inhomogeneous Media

Lukas Pieronek and Andreas Kleefeld

28.1 Introduction

The interior transmission eigenvalue problem (ITEP) arises in the study of inverse scattering as a precursor to justify the feasibility of quantitative reconstruction methods, see [CaHa12]. The corresponding eigenvalues (ITEs) are associated with certain critical wave numbers which allow for incident test waves with arbitrary small scattering responses that would hardly be detected in experiments. One can prove in a mathematically rigorous way, see [KiGr08], that this loss of information due to the scatterer's practical invisibility indeed complicates its recovery process on the basis of sampling methods, for example.

While it has been known for quite a long time that ITEs fortunately only form an at most discrete set, their accurate computation for a given scatterer is still rather challenging since the underlying ITEP is both non-elliptic and non-self-adjoint. In this paper, we will present a relatively easy algorithm for the efficient approximation of ITEs using a robust version of the method of fundamental solutions (MFS). Having thus been positively tested for perfectly homogeneous and possibly anisotropic scatterers so far, see [KlPi18a] and [KlPi18b], we will now analyze and reinvestigate the numerical merits of the method in the generalized context of inhomogeneous media that decompose into several homogeneities stuck in a surrounding bulk. For competitive alternative methods in this context, see the solution of the direct problem in [GiPa13], for example.

L. Pieronek (✉) · A. Kleefeld
Forschungszentrum Jülich GmbH, Jülich, Germany
e-mail: l.pieronek@fz-juelich.de; a.kleefeld@fz-juelich.de

© Springer Nature Switzerland AG 2019

C. Constanda, P. Harris (eds.), *Integral Methods in Science and Engineering*,
https://doi.org/10.1007/978-3-030-16077-7_28

28.2   The ITEP and the Modified MFS

Let $D \subset \mathbb{R}^2$ be a bounded and simply connected domain with smooth boundary representing the contour of some scatterer. According to our modeling assumptions, there exist an upper component number $1 \leq N_c \in \mathbb{N}$ and a disjoint decomposition $\bigcup_{i=1}^{N_c} D_i \subset D \subset \overline{D} = \bigcup_{i=1}^{N_c} \overline{D}_i$ of homogeneous composites such that the D_i are each open, connected and fulfill for $i_1 \neq i_2$ either $\partial D_{i_1} \cap \partial D_{i_1} = \emptyset$ or $\partial D_{i_1} \cap \partial D_{i_1}$ being a smooth and non-intersecting closed contour. Besides, we assume the existence of some bulk which will be emphasized by D_1 and which distinguishes from the other components through its surrounding property $\partial D \subset \partial D_1$ unlike $\partial D \cap \partial D_i = \emptyset$ for $2 \leq i \leq N_c$, cf. Fig. 28.1. As such, D may be considered as an inner pollution of the original material D_1 by the components D_i for $i > 2$. These different material parameters will be reflected via piecewise-constant functions on D with jumps only across ∂D_i and encompass the index of refraction $n \in L^\infty(D, \mathbb{R}_{>0})$, as a local measure for the propagation speed of the penetrable wave under consideration, and the diffusivity tensor $A \in L^\infty(D, \mathbb{R}_{sym}^{2 \times 2})$ to reflect the material's anisotropic structure, if necessary (otherwise set $A = I$, where I denotes the identity matrix in $\mathbb{R}^2$). Depending on whether $A = I$ on D or not, there are also mathematical constraints for n and the eigenvalues of A that we will specify later. When focusing on the analysis of each homogeneous component, we also write $f_i := f_{|D_i}$ for any function f on D.

In contrast to inverse problems, we assume all these material data to be known a priori and want to compute ITEs on that basis in the following. An ITE is a wave number $k > 0$ for which the ITEP

$$\operatorname{div}(A\nabla w) + k^2 n w = 0 \quad \text{in } D ,$$

$$\Delta v + k^2 v = 0 \quad \text{in } D ,$$

$$w = v \quad \text{on } \partial D ,$$

$$\partial_{\nu_A} w = \partial_\nu v \quad \text{on } \partial D ,$$

$$(28.1)$$

Fig. 28.1 Exemplary scatterer D with bulk D_1 that contains three inner components D_2, D_3, D_4. For demonstration purposes, some admissible source boundary $\Gamma_4 = \Gamma_{4,1} \cup \Gamma_{4,2}$ for D_4 is added as dotted contours in the component's outside and inside, respectively

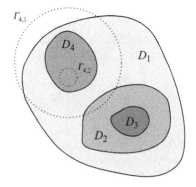

where ν_A denotes the (co-)normal derivative, can be solved appropriately in a non-trivial way. The penultimate condition refers to the regularity assumptions of admissible eigenfunctions being $(v, w) \in L^2(D)$ with $(v - w) \in H_0^2(D)$ for the isotropic case and $(v, w) \in H^1(D)$ otherwise. The origin for such a differentiation is that the usual Fredholm setting approved for eigenvalue investigations can at most be guaranteed then.

Having settled a proper abstract framework for such a non-linear eigenvalue problem, approximations of ITEs can be obtained as a by-product of finite dimensional approximation of the corresponding eigenfunctions. The standard MFS achieves this for homogeneous media by looking for superposed trial functions that are translations of certain fundamental solutions from the governing PDE. This ansatz is somewhat converse to finite element methods as each solution candidate automatically fulfills the interior conditions of (28.1) exactly, but not necessarily the prescribed boundary conditions along ∂D, cf. [GiPa13]. However, since we actually work with scatterers that are piecewise-homogeneous, we want to apply this method at least locally for each composite and properly supported trial functions. This then requires additional control of function transitions across adjacent components D_j and D_i according to our globally imposed regularity assumptions of exact eigenfunctions. Fortunately, these extra costs only affect the approximation of w since v always obeys a pure Helmholtz equation with constant wave number so that its trial functions can still be defined throughout D. These observations finally make us consider the following MFS-based approximation spaces of order m. For v, we set

$$V_m := \text{span}\{\phi_1^v, \ldots, \phi_m^v\},$$

where

$$\phi_r^v(x) := \frac{i}{4} H_0^{(1)}(k|x - s_r^v|),$$

$H_0^{(1)}$ is the first Hankel function of order zero, $x \in \overline{D}$ is the actual argument whereas $\{s_1^v, \ldots, s_m^v\}$ are the so-called source points for v which have to lie on some closed exterior contour Γ^v disjoint to D. For w, we need to take into account the refined treatment of the domain and define an artificial boundary for each composite in its individual exterior. Since the D_i are in general multiply-connected, each of the $N_e(i)$ enclaves E_{i,j_i} of $\mathbb{R}^2 \setminus \overline{D}_i$ is labeled by j_i, always starting with the unbounded component by convention. Then we endow each of them with their own closed contour $\Gamma_{i,j_i}^w \Subset E_{i,j_i}$ and wish to choose $\Gamma_{1,j_1}^w = \Gamma^v$. We may thus drop the superscripts w or v from now and abbreviate the union over j_i for all assigned source contours Γ_{i,j_i} of D_i by Γ_i. We can finally set up the approximation space W_m as the span of

$$\phi_{i,j_i,r}^w(x) := \frac{i}{4 \det A_i^{\frac{1}{2}}} H_0^{(1)}\left(\sqrt{n_i}k \left|A_i^{-\frac{1}{2}}\left(x - s_{i,j_i,r}^w\right)\right|\right) \mathbb{1}_{\overline{D}_i},$$

where $s^w_{i,j_i,r}$ are now the source points for w on Γ^w_{i,j_i} with $1 \leq r \leq m$. Thus the dimension of W_m is actually a multiple of m depending on the number of composites N and of their actual voids.

The optimal MFS solution pair $(v_m, w_m) \in V_m \times W_m$ then tries to fulfill both the boundary conditions of the ITEP (28.1) and the transitional regularity criteria for global eigenfunctions best in the sense of a minimal collocation error. Therefore we select representative points $\{x_{i_1,i_2,1}, \ldots, x_{i_1,i_2,m}\}$ along the interface of $\partial D_{i_1} \cap \partial D_{i_2}$ where $0 \leq i_1 < i_2 \leq N_c$ and $D_0 := \mathbb{R}^2 \backslash \overline{D}$ is introduced for simplicity. We implicitly omit the definition of those $x_{i_1,i_2,\ell} \in D$ whose pair (i_1, i_2) would be associated with an empty intersection. The same convention will affect all the x_{0,i_2} with $i_2 > 1$ due to the separating bulk. Again, the total number of collocation points is a multiple of m, but this time depending on the cumulated number of boundaries of all the D_i.

Instead of performing the collocation procedure now in a straightforward way that would result in a numerically ill-conditioned problem, we follow the stabilization improvement from Betcke and Trefethen, see [BeTr05], that also gives rise to call our corresponding MFS update the *modified MFS* henceforth. For that we pick m-independent auxiliary points $\{a_1, \ldots, a_{N_a}\}$ from $\bigcup_{i=1}^{N_c} D_i$ which will later ensure that our approximations of eigenfunctions are sufficiently large in the interior. Thus the utterly critical zero function will never be detected as a theoretically admissible (with respect to (28.1)), but practically undesirable solution candidate.

Let us structure this amount of introduced data in matrix form to see more easily how they emerge in the modified MFS collocation procedure. If (i_1, i_2) is a feasible pair such that $\partial D_{i_1} \cap \partial D_{i_2} \neq \emptyset$, we define $W_{i_1,i_2,i,j_i} \in \mathbb{C}^{2m \times m}$ parametrized by $1 \leq i \leq N_c$ and $0 \leq i_1 < i_2 \leq N_c$ for the varyingly supported trial functions of W_m via

$$(W_{i_1,i_2,i,j_i})_{\ell,r} := \phi^w_{i,j_i,r}(x_{i_1,i_2,\ell}) ,$$

$$(W_{i_1,i_2,i,j_i})_{m+\ell,r} := \partial_\nu \phi^w_{i,j_i,r}(x_{i_1,i_2,\ell}) ,$$

whereas for V_m it suffices to consider $V \in \mathbb{C}^{2m \times m}$ given by

$$(V)_{\ell,r} := \phi^v_r(x_{0,1,\ell}) ,$$

$$(V)_{m+\ell,r} := \partial_\nu \phi^v_r(x_{0,1,\ell}) .$$

Here, $1 \leq \ell, r \leq m$ and ν is a unit normal vector along $\partial D_{i_1} \cap \partial D_{i_2} \neq \emptyset$ pointing in the same direction for both components D_{i_1} and D_{i_2}. The evaluations of trial functions at interior points are summarized in an analogical fashion to matrices in $\mathbb{C}^{N_a \times m}$ and read

$$(\widetilde{W}_{i,j_i})_{\ell,r} := \phi^w_{i,j_i,r}(a_\ell)$$

as well as

$$(\widetilde{V})_{\ell,r} := \phi_r^v(a_\ell) \,,$$

respectively. In the anisotropic case, $\widetilde{W}_{i,j}$ and $\widetilde{V}$ may even be extended to matrices in $\mathbb{C}^{3N_a \times m}$ by attaching corresponding partial derivative evaluations of $\phi_{i,j_i,r}^w$ and ϕ_r^v, respectively, for being more consistent with the norms introduced in the abstract setting later. Through this matrix reformulation, the m-independent indices i_1, i_2, i, j_i for the different scattering composites are separated from the m-dependent point discretizations labeled by ℓ, r. Since their matrices are implicitly parametrized by the wave number k, we finally define the block-type system $T(k)$ by

$$T(k) := \begin{pmatrix} V & W_{0,1,1,1} & \cdots & W_{0,1,1,N_e(1)} & 0 & \cdots\cdots & 0 \\ 0 & W_{1,2,1,1} & \cdots & W_{1,2,1,N_e(1)} & W_{1,2,2,1} & \cdots\cdots & W_{1,2,N_c,N_e(N_c)} \\ \vdots & \vdots & & \vdots & \vdots & & \vdots \\ \vdots & \vdots & & \vdots & \vdots & & \vdots \\ 0 & W_{N_c-1,N_c,1,1} & \cdots & W_{N_c-1,N_c,1,N_e(1)} & W_{N_c-1,N_c,2,1} & \cdots\cdots & W_{N_c-1,N_c,N_c,N_e(N_c)} \\ \widetilde{V} & 0 & \cdots & \cdots & \cdots & \cdots\cdots & 0 \\ 0 & \widetilde{W}_{1,1} & \cdots & \widetilde{W}_{1,N_e(1)} & \widetilde{W}_{2,1} & \cdots\cdots & \widetilde{W}_{N_c,N_e(N_c)} \end{pmatrix}.$$

Recall that the majority of the W matrices are zero since not all components D_{i_1} and D_{i_2} will be adjacent, i.e. $W_{i_1,i_2,i,j_i} = 0 \in \mathbb{C}^{m \times m}$ if $i \notin \{i_1, i_2\}$. This might motivate to treat T in parallel from a programming perspective, especially because the last technical thing left to do is performing a QR factorization of $T(k) = Q_T(k)R_T(k)$. Extracting its unitary part, we may write

$$Q_T(k) = \begin{pmatrix} Q(k) \\ \widetilde{Q}(k) \end{pmatrix} \,.$$

Similar as above, $\widetilde{Q}(k)$ corresponds to the lower $2N_a$ scalar rows whereas $Q(k)$ comprises the remaining upper part of $Q_T(k)$. Such a decomposition of $Q_T(k)$ and thus of $T(k)$ will implicitly help us later to distinguish numerically between real and spurious eigenvalue approximations in an effective way.

Now we have everything together to formulate our approximate ITEP based on the modified MFS in a very compact form: Find those k for which $k \mapsto Q(k)$ is almost singular. For this purpose, the smallest singular value $\sigma_1(k)$ of $Q(k)$ will serve as a convenient measure for the degeneracy of $Q(k)$ whenever its magnitude is close to zero. Corresponding wave numbers will be called approximate ITEs and will be denoted by k_m to relate its dimensional origin to the underlying eigenfunction approximation. In the next section we will discuss our derived approach from a more abstract perspective and show its feasibility in practice.

28.3   Approximation Analysis

The major difference for the application of the modified MFS between purely and
piecewise homogeneous media is apparently the additional treatment of interior
composite transitions for w. So far we take them numerically into account by
pointwise collocation, but it seems more natural to formulate precise assumptions
with respect to their original Sobolev spaces. Therefore we focus on $N_c > 1$ in
the sequel and assume first that $A = I$, i.e. the scatterer is entirely isotropic.
Then the corresponding ITEP eigenfunctions are only in $L^2(D)$, in particular the
w-dependent PDE has to be fulfilled in a distributional sense and reads

$$\int_D (\Delta\psi + k^2 n\psi)w \ dx = 0$$

for any $\psi \in C_c^\infty(D)$. If we now approximate w by some proper $w_m \in W_m$, the
integral above does not vanish any more for all test functions in general. However,
choosing $\psi \in C_c^\infty(D\backslash\bigcup_{i\neq i_1,i_2} \overline{D}_i) \subset C_c^\infty(D)$, where $1 \leq i_1 < i_2 \leq N_c$
are some adjacent component indices, the following reformulation shows that the
resulting deviations are completely due to certain integral misfits over $\partial D_{i_1} \cap \partial D_{i_2}$.
This is because the trial functions in W_m, although they solve the corresponding
interior ITEP condition pointwise almost everywhere, are a priori discontinuous
along composite transitions:

$$\int_D (\Delta\psi + k^2 n\psi)w_m \ dx$$

$$= \int_{D_{i_1}} (\Delta\psi + k^2 n_{i_1}\psi)w_{m,i_1} \ dx + \int_{D_{i_2}} (\Delta\psi + k^2 n_{i_2}\psi)w_{m,i_2} \ dx$$

$$= \int_{\partial D_{i_1} \cap \partial D_{i_2}} \partial_\nu\psi(w_{m,i_1} - w_{m,i_2}) - \psi\partial_\nu(w_{m,i_1} - w_{m,i_2}) \ ds$$

$$+ \int_{D_{i_1}} \underbrace{(\Delta w_{m,i_1} + k^2 n_{i_1} w_{m,i_1})}_{=0} \psi \ dx + \int_{D_{i_2}} \underbrace{(\Delta w_{m,i_2} + k^2 n_{i_2} w_{m,i_2})}_{=0} \psi \ dx$$

$$= \int_{\partial D_{i_1} \cap \partial D_{i_2}} \partial_\nu\psi(w_{m,i_1} - w_{m,i_2}) - \psi\partial_\nu(w_{m,i_1} - w_{m,i_2}) \ ds \ . \tag{28.2}$$

Conversely, we hope to recover some exact eigenfunction w with $w, \Delta w \in L^2(D)$
in the limit $m \to \infty$, so we expect $(w_{m,i_1} - w_{m,i_2})$ to be evanescent at least with
respect to $H^{-\frac{1}{2}}(\partial D_{i_1} \cap \partial D_{i_2})$ and likewise $\partial_\nu(w_{m,i_1} - w_{m,i_2})$ should be controlled
as $H^{-\frac{3}{2}}(\partial D_{i_1} \cap \partial D_{i_2})$-traces. However, for technical reasons that will become clear
later, we assume the convergence to hold more strongly in $H^{\frac{3}{2}}(\partial D_{i_1} \cap \partial D_{i_2})$ and
$H^{\frac{1}{2}}(\partial D_{i_1} \cap \partial D_{i_2})$, respectively. In the anisotropic case, however, we can stay with
the natural approximation assumptions $(w_{m,i_1} - w_{m,i_2}) \to 0$ in $H^{\frac{1}{2}}(\partial D)$ and
$(\partial_{\nu_{A_{i_1}}} w_{m,i_1} - \partial_{\nu_{A_{i_2}}} w_{m,i_2}) \to 0$ in $H^{-\frac{1}{2}}(\partial D)$ that originate from $w \in H^1(D)$.

In order to finally adapt the prescribed boundary conditions on ∂D of the eigenfunctions from (28.1) in our approximation procedure, the explicit regularity conditions on $(v - w)$ need to be taken into account. Accordingly, the isotropic case asserts $(v - w) \in H_0^2(D)$ which suggests imposing $(v_m - w_m) \to 0$ in $H^{\frac{3}{2}}(\partial D)$ as well as $\partial_v(v_m - w_m) \to 0$ in $H^{\frac{1}{2}}(\partial D)$ by continuity of the corresponding trace operators. In the anisotropic case we similarly arrive at $(v_m - w_m) \to 0$ in $H^{\frac{1}{2}}(\partial D)$ and $(\partial_v v_m - \partial_{v_A} w_m) \to 0$ in $H^{-\frac{1}{2}}(\partial D)$.

Altogether, these preliminaries are to show that the analysis of the ITEP based on the modified MFS (embedded into the abstract setting) is conceptually the same for undiluted and piecewise homogeneous media. Therefore we expect most of the results from [KlPi18a] and [KlPi18b] to be extendable to our specially inhomogeneous setting such as the following theorem. It can be considered as the justification for the applicability of the modified MFS to ITE approximations.

Theorem 1 *Consider a sequence $\{(v_m, w_m, k_m)\}_{m \in \mathbb{N}} \subset V_m \times W_m \times \mathbb{R}_{>0}$ for either $A = I$ with restriction $n > 1$ or $n < 1$ throughout D, $0 < A < I$ or $I < A$ (where the matrix order is understood with respect to positive definiteness) which fulfills the following properties, respectively:*

In the isotropic case, we assume

1. *eigenvalue convergence: $k_m \to k \neq 0$,*
2. *uniform interior bounds: $C^{-1} < \left(\|v_m\|_{L^2(D)}^2 + \|w_m\|_{L^2(D)}^2 \right) < C$ for some $C > 1$ and for all m large enough,*
3. *vanishing boundary data: $\left(\|v_m - w_m\|_{H^{\frac{3}{2}}(\partial D)} + \|\partial_v(v_m - w_m)\|_{H^{\frac{1}{2}}(\partial D)} \right) \to 0$ and $\left(\|w_{m,i_1} - w_{m,i_2}\|_{H^{\frac{3}{2}}(\partial D_{i_1} \cap \partial D_{i_2})} + \|\partial_v(w_{m,i_1} - w_{m,i_2})\|_{H^{\frac{1}{2}}(\partial D_{i_1} \cap \partial D_{i_2})} \right) \to 0$ for adjacent components $D_{i_1}, D_{i_2} \subset D$,*

whereas in the anisotropic case, the corresponding assumptions read

1′. *eigenvalue convergence: $k_m \to k \neq 0$,*
2′. *uniform interior bounds: $C^{-1} < \left(\|v_m\|_{H^2(D)}^2 + \|w_m\|_{H^2(D)}^2 \right) < C$ for some $C > 1$ and for all m large enough,*
3′. *vanishing boundary data: $\left(\|v_m - w_m\|_{H^{\frac{1}{2}}(\partial D)} + \|\partial_v v_m - \partial_{v_A} w_m\|_{H^{-\frac{1}{2}}(\partial D)} \right) \to 0$ and*

$$\left(\|w_{m,i_1} - w_{m,i_2}\|_{H^{\frac{1}{2}}(\partial D_{i_1} \cap \partial D_{i_2})} + \|\partial_{v_{A_{i_1}}} w_{m,i_1} - \partial_{v_{A_{i_2}}} w_{m,i_2}\|_{H^{-\frac{1}{2}}(\partial D_{i_1} \cap \partial D_{i_2})} \right) \to 0$$

for adjacent components $D_{i_1}, D_{i_2} \subset D$.

In either case, the limit k of the approximate eigenvalues k_m is an ITE.

Proof Since our proof to be presented works structurally similar for the anisotropic case based on the corresponding techniques from the homogeneous scenario, see [KlPi18b], but much easier due to more consistent control assumptions of (v_m, w_m) throughout $\overline{D}$, see condition 3′, we only focus on the isotropic case in the following.

We aim to construct an eigenfunction candidate (v, w) and show that it fulfills the required properties for k being a real ITE. We take the weak $L^2(D)$-limit of our approximate pairs (v_m, w_m) which exists (actually only for a subsequence which we will, however, not explicitly restate in the sequel) by weak compactness and the uniform bounds provided in assumption 2. The fact that v is then a distributional solution of the Helmholtz equation on D is quite trivial because it can be shown exactly as in the homogeneous case. The corresponding result for w relies on an additional treatment of (28.2) which is then also straightforward by condition 3. So far, we therefore know that $(v, w) \in L^2(D) \times L^2(D)$ fulfills the interior conditions of (28.1) and we want to prove next that $u := (v - w) \in H_0^2(D)$, i.e. u has zero boundary data and is twice weakly differentiable in the interior (the latter criterion would be redundant for the anisotropic demonstration).

We modify the piecewise smooth but generally discontinuous difference functions $u_m := (v_m - w_m) \in L^2(D)$ to $\tilde{u}_m \in H^2(D)$ that will be uniformly bounded with respect to m. As potential jumps of u_m go back to those of w_m across ∂D_i for $i > 1$, we want to fill these discontinuity gaps by adding certain lifting functions $\theta_{m,i} \in H^2(D_i)$ to u_m. More precisely, we set for $i > 1$

$$\Delta^2 \theta_{m,i} = 0 \quad \text{in } D_i ,$$

$$\theta_{m,i} = (w_{m,i^*} - w_{m,i}) \mathbb{1}_{\partial D_i \cap \partial E_{i,1}} \quad \text{on } \partial D_i ,$$

$$\partial_\nu \theta_{m,i} = \partial_\nu (w_{m,i^*} - w_{m,i}) \mathbb{1}_{\partial D_i \cap \partial E_{i,1}} \quad \text{on } \partial D_i ,$$

where $1 \leq i^* \leq N_c$ is determined uniquely by $D_{i^*} \subset \partial E_{i,1}$ and $\partial D_i \cap \partial D_{i^*} \neq \emptyset$. Then we extend $\theta_{m,i}$ by zero in $D \backslash D_i$. Standard a priori estimates ensure that our lifting functions can be bounded within their support D_i by

$$\|\theta_{m,i}\|_{H^2(D_i)}$$
$$\leq C \left(\|w_{m,i} - w_{m,i^*}\|_{H^{\frac{3}{2}}(\partial D_i \cap \partial D_{i^*})} + \|\partial_\nu (w_{m,i} - w_{m,i^*})\|_{H^{\frac{1}{2}}(\partial D_i \cap \partial D_{i^*})} \right)$$

and globally they cumulate by definition to

$$\tilde{u}_m := u_m + \sum_{i=2}^{N_c} \theta_{m,i} \in H^2(D) .$$

Therefore, $\tilde{u}_m$ solves

$$\Delta \tilde{u}_m = -k_m(v_m - n w_m) + \sum_{i=2}^{N_c} \Delta \theta_{m,i} \quad \text{in } D_i ,$$

$$\tilde{u}_m = v_m - w_m \quad \text{on } \partial D_i ,$$

$$\partial_\nu \tilde{u}_m = \partial_\nu (v_m - w_m) \quad \text{on } \partial D_i ,$$

and is bounded by

$$\|\widetilde{u}_m\|_{H^2(D)} \leq C \left(\|\Delta \widetilde{u}_m\|_{L^2(D)} + \|v_m - w_m\|_{H^{\frac{3}{2}}(\partial D)} + \|\partial_\nu (v_m - w_m)\|_{H^{\frac{1}{2}}(\partial D)} \right).$$

In particular, the $\widetilde{u}_m$ converge both weakly in $H^2(D)$ and strongly in $L^2(D)$. Since $\theta_{m,i} \to 0$, the strong $L^2(D)$-limit of u_m and $\widetilde{u}_m$ coincide which then implies that $u \in H^2(D)$. The fact that even $u \in H_0^2(D)$ finally follows by assumption 3 and the continuity of the trace operators from $H^2(D)$ to $H^{\frac{3}{2}}(\partial D)$ and to $H^{\frac{1}{2}}(\partial D)$, respectively.

It remains to prove that $u \neq 0$. We will contrarily assume that $u = 0$ which would then imply $u_m \to 0$ in $L^2(D)$ according to our previous derivations. Expanding the $L^2(D)$-norm in its scalar product representation, we may conclude, including assumption 2

$$\liminf_{m \to \infty} \mathrm{Re} \int_D v_m \overline{w}_m \, dx = \liminf_{m \to \infty} \frac{1}{2} \left(\|v_m\|_{L^2(D)}^2 + \|w_m\|_{L^2(D)}^2 - \|v_m - w_m\|_{L^2(D)}^2 \right)$$

$$= \liminf_{m \to \infty} \frac{1}{2} \left(\|v_m\|_{L^2(D)}^2 + \|w_m\|_{L^2(D)}^2 \right) \geq \frac{1}{2C} > 0.$$

Determined by keeping positive signs above, we multiply the latter inequality either with $(1 - n_i)$ or with $(n_i - 1)$, assuming the latter without loss of generality. Since $\min_{1 \leq i \leq N_c} (n_i - 1) > 0$ and $k_m \to k > 0$, we thus obtain

$$0 < \liminf_{m \to \infty} \mathrm{Re} \int_D k_m^2 (n - 1) v_m \overline{w}_m \, dx$$

$$= \liminf_{m \to \infty} \mathrm{Re} \int_D \overline{w}_m \Delta v_m - v_m \Delta \overline{w}_m \, dx$$

$$= \liminf_{m \to \infty} \sum_{i=1}^{N_c} \mathrm{Re} \int_{D_i} \overline{w}_m \Delta v_m - v_m \Delta \overline{w}_m \, dx$$

$$= \liminf_{m \to \infty} \sum_{0 \leq i_1 < i_2 \leq N_c} \mathrm{Re} \int_{\partial D_{i_1} \cap \partial D_{i_2}} \overline{w}_m \partial_\nu v_m - v_m \partial_\nu \overline{w}_m \, ds$$

$$= \liminf_{m \to \infty} \sum_{0 \leq i_1 < i_2 \leq N_c} \mathrm{Re} \int_{\partial D_{i_1} \cap \partial D_{i_2}} \overline{w}_m \partial_\nu v_m - \overline{v}_m \partial_\nu v_m + v_m \partial_\nu \overline{v}_m - v_m \partial_\nu \overline{w}_m \, ds$$

$$= \liminf_{m \to \infty} \sum_{0 \leq i_1 < i_2 \leq N_c} \mathrm{Re} \int_{\partial D_{i_1} \cap \partial D_{i_2}} (\overline{w}_m - \overline{v}_m) \partial_\nu v_m + v_m \partial_\nu (\overline{v}_m - \overline{w}_m) \, ds$$

$$\leq \liminf_{m\to\infty} \sum_{0\leq i_1 < i_2 \leq N_c} \left(\|v_m - w_m\|_{H^{\frac{3}{2}}(\partial D_{i_1}\cap\partial D_{i_2})} \|v_m\|_{H^{-\frac{3}{2}}(\partial D_{i_1}\cap\partial D_{i_2})} \right.$$

$$\left. + \|\partial_\nu v_m\|_{H^{-\frac{1}{2}}(\partial D_{i_1}\cap\partial D_{i_2})} \|\partial_\nu(v_m - w_m)\|_{H^{\frac{1}{2}}(\partial D_{i_1}\cap\partial D_{i_2})} \right)$$

$$= 0 . \tag{28.3}$$

The last equality follows by assumption 3 and by some uniform upper bound on the negative dual norms as inherited from our interior control of v_m, cf. assumption 2. Obviously, (28.3) gives a contradiction and manifests that $u \neq 0$ which thus completes the proof.

Remark The proof above indicates why our initial modeling assumptions for D restrict to material components D_i facing never more than one another at each transitional point. Otherwise, the traces of different w_m parts might be incompatible in any crossing point which would then lock the possibility to find sufficiently regular lifting functions in its vicinity.

28.4 Numerical Examples

In this section we use the modified MFS to compute the first four real ITEs for the unit disc D varied in two different inner configurations. One of the two, let us say $D^{\circ\circ}$, will have separated disc-shaped components, while for the other we choose a corresponding concentric composition and denote it by D°, cf. Fig. 28.2. More concretely, we analyzed three-components-scatterers whose set representation decomposes into

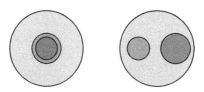

A	D	ITE 1	ITE 2	ITE 3	ITE 4
$=I$	$D^{\circledcirc}$	3.3472649097009	3.5339744459219	3.8215531039714	4.0276794096285
	$D^{\circ\circ}$	2.9695607637622	3.8151728473562	4.2620616635742	4.3612725527356
$\neq I$	$D^{\circledcirc}$	1.143183893	1.64497010	2.443821	3.273054
	$D^{\circ\circ}$	1.2372795	1.569690	2.261430	3.1939135

Fig. 28.2 Visualization of the scatterers $D^{\circledcirc}$ (left) and $D^{\circ\circ}$ (right) and listing of their first four real ITEs both for the isotropic ($= I$) and anisotropic case ($\neq I$) with common index of refraction given by $n_1 = 4, n_2 = 3, n_2 = 2$

$$D_2^{\circ\circ} := B_{0.4}\big((0.5,0)\big) , \quad D_3^{\circ\circ} := B_{0.3}\big((-0.5,0)\big) ,$$
$$D_1^{\circ\circ} := B_1\big((0,0)\big)\backslash(D_2^{\circ\circ} \cup D_3^{\circ\circ})$$

and

$$D_2^{\circledcirc} := B_{0.4}\big((0,0)\big) , \quad D_3^{\circledcirc} := B_{0.3}\big((0,0)\big) , \quad D_1^{\circledcirc} := B_1\big((0,0)\big)\backslash(D_2^{\circledcirc} \cup D_3^{\circledcirc}) ,$$

respectively. Here, $B_r(c)$ denotes the disc of radius r centered at $c \in \mathbb{R}^2$. We equipped $D^{\circledcirc}$ and $D^{\circ\circ}$ each with some identical sample of refractive indices for both the isotropic and anisotropic case which read $n_1 = 4, n_2 = 3, n_3 = 2$ and set additionally for $A \neq I$

$$A_1 = \begin{pmatrix} 2 & 0 \\ 0 & 4 \end{pmatrix} , \quad A_2 = \begin{pmatrix} 3 & 0 \\ 0 & 2 \end{pmatrix} , \quad A_3 = \begin{pmatrix} 2 & 0 \\ 0 & 3 \end{pmatrix} .$$

Having thus fixed our material parameters, there is now much freedom in choosing the computational points for the modified MFS procedure. First of all, we associated source boundaries $\Gamma_i^{\circledcirc}$ and $\Gamma_i^{\circ\circ}$ to the components of $D_i^{\circledcirc}$ and $D_i^{\circ\circ}$, respectively, by scaling the underlying circles with respect to their individual center by some factor $S = 1.5 > 1$ for the outer part of the source boundary, cf. $\Gamma_{i,1}^{\circledcirc}$ and $\Gamma_{i,1}^{\circ\circ}$, and similarly by $s = 0.5 < 1$ for the remaining ones, if required. Therefore we arrive at

$$\Gamma_{1,1}^{\circ\circ} := \partial B_S\big((0,0)\big) , \quad \Gamma_{1,2}^{\circ\circ} := \partial B_{0.4s}\big((0.5,0)\big) , \quad \Gamma_{1,3}^{\circ\circ} := \partial B_{0.3s}\big((-0.5,0)\big) ,$$
$$\Gamma_{2,1}^{\circ\circ} := \partial B_{0.4S}\big((0.5,0)\big) ,$$
$$\Gamma_{3,1}^{\circ\circ} := \partial B_{0.3S}\big((-0.5,0)\big) ,$$

for $D^{\circ\circ}$ and

$$\Gamma_{1,1}^{\circledcirc} := \partial B_S\big((0,0)\big) , \quad \Gamma_{1,2}^{\circledcirc} := \partial B_{0.4s}\big((0,0)\big) ,$$
$$\Gamma_{2,1}^{\circledcirc} := \partial B_{0.4S}\big((0,0)\big) , \quad \Gamma_{2,2}^{\circledcirc} := \partial B_{0.3s}\big((0,0)\big) ,$$
$$\Gamma_{3,1}^{\circledcirc} := \partial B_{0.3S}\big((0,0)\big) ,$$

for $D^{\circledcirc}$. Since the location of interior points turns out not to contribute significantly to the output, we placed them also on a circle with identical center as $D_3^{\circledcirc}$ and $D_3^{\circledcirc}$, respectively, but half its radius. Conveniently, all the computational points introduced in the second section of this paper could thus be distributed equidistantly on their corresponding circles.

In the process of our numerical experiments, we fixed $N_a = 10$ and varied m to improve the accuracy of our approximate ITEs. Minimizing the first singular value, the optimal results from Fig. 28.2 were achieved for $60 \leq m \leq 100$ and thus at most 500 collocation points were needed altogether. Exceeding this regime leads more and more to the emergence of ill-conditioning effects and thus to unreliable results. However, approaching the admissible threshold for m from below, the cut-off mantissa of our ITEs approximations tend to converge with increasing m, so we believe all the listed digits to be correct (modulo round-off-errors). Note that due to the rotational-invariant structure of $D^{\odot}$ at least in the case $A = I$ all ITEs can easily be computed analytically using a Fourier Bessel ansatz, cf. [KlPi18a] with a correspondingly extended matrix system for the component transitions of v and w. The first four of them were found to be 3.347264909700945, 3.533974445921942, 3.821553103971393, 4.027679409628525 and thus confirm that the approximations obtained by the modified MFS are indeed correct up to machine precision here.

Generally, our current observations are mostly consistent with those made for purely homogeneous scatterers, see the references [KlPi18a] and [KlPi18b]. In particular, the computational results from the isotropic case are still significantly better than for $I \neq A$ which goes back to the more advanced body of trial functions for w. Novelties affect the irregular behavior of the smallest singular value function whose minimal dips were sometimes extremely steep and thus hard to detect (such as the largest eigenvalue given above that was recomputed with the Fourier Bessel ansatz). As an explanation for that, the optimal number of collocation points necessary for the transitional boundaries to preserve a comparable quality of eigenvalue approximations turned out to be surprisingly large. While the undiluted isotropic disc required only around $m = 20$ collocation points altogether to recover ITEs almost up to machine precision, our latest experiments necessitate almost about its fivefold per boundary component (500 in total) and thus seem to scale quadratically.

28.5   Conclusion

In this paper, the recovery of interior transmission eigenvalues for inhomogeneous media in two dimensions was investigated on the basis of the method of fundamental solutions. Although best suited for homogeneous scatterers to benefit most from the lower dimensional boundary description, our numerical examples show that highly accurate results can still be obtained for scatterers which consist of a moderate number of homogeneous components. Conversely, the more complex the inner structure of D is, including anisotropic behavior, the more collocation points are generally needed and in correlation with that the less precise the eigenvalue approximation becomes. Our theoretical studies additionally show that our method will, under appropriate assumptions of the output, never detect spurious eigenvalues in the limiting process and thus proves its practical reliability.

References

[BeTr05] Betcke, T. and Trefethen, L. N.: Reviving the method of particular solutions. *SIAM Review*, **47**, 469–491 (2005).

[CaHa12] Cakoni, F. and Haddar, H.: Transmission eigenvalues in inverse scattering theory. *Inside Out II*, **60**, 527–578 (2012).

[GiPa13] Cakoni, F., and Haddar, H.: Gintides, D. and Pallikaraki, N. A computational method for the inverse transmission eigenvalue problem. *Inverse Problems*, **29**, 104010 (2013).

[KiGr08] Kirsch A. and Grinberg N.: The Factorization Method for Inverse Problems. *Oxford University Press*, (2008).

[KlPi18a] Kleefeld, A. and Pieronek, L.: The method of fundamental solutions for computing acoustic interior transmission eigenvalues. *Inverse Problems*, **34**, 035007 (2018).

[KlPi18b] Kleefeld, A. and Pieronek, L.: Computing interior transmission eigenvalues for homogeneous and anisotropic media. *Inverse Problems*, **34**, 105007 (2018).

Chapter 29
Tensor Product Approach to Quantum Control

Diego Quiñones-Valles, Sergey Dolgov, and Dmitry Savostyanov

29.1 Introduction

Quantum control plays a central part in both established technologies such as nuclear magnetic resonance (NMR) and emerging technologies such as quantum computing [GlEtAl15]. The 'hardware' of a quantum computer is a quantum system (e.g. a spin system), which we manipulate with time-dependent fields (e.g. laser pulses). Our goal is to design a pulse sequence that steers the system from an initial state $|\psi_0\rangle$ to a desired state $|\psi_T\rangle$ during time T—a 'firmware', so to say. The main challenge for numerical methods is the notoriously known *curse of dimensionality*—the dimension of the state space growing exponentially with the number of particles d. Even for the simplest spin-$\frac{1}{2}$ systems, the wavefunction $|\psi\rangle$, the density matrix ρ, and the Hamiltonian $\hat{H}$ require at least $O(2^d)$ storage. To find the optimal trajectory we need to compute $\rho(t)$ in 100s to 1000s points of time. This leads to gigabyte-scale storage for $d \simeq 20$ spins and becomes unfeasible for $d \gtrsim 30$.

The main approaches used currently are the sparse format for $\hat{H}$, $|\psi\rangle$'s, ρ's (only large entries are stored, zeroes and small elements are discarded), and dimensionality reduction—the equations of system dynamics are projected into a subspace of low-lying eigenstates [Wó6EtAl05] or its heuristically chosen equivalent [KuEtAl07, CaEtAl11].

D. Quiñones-Valles · D. Savostyanov (✉)
University of Brighton, Brighton, UK
e-mail: d.quinonesvalles@brighton.ac.uk; d.savostyanov@brighton.ac.uk

S. Dolgov
University of Bath, Bath, UK
e-mail: s.dolgov@bath.ac.uk

© Springer Nature Switzerland AG 2019   367
C. Constanda, P. Harris (eds.), *Integral Methods in Science and Engineering*,
https://doi.org/10.1007/978-3-030-16077-7_29

Tensor product algorithms are a novel approach to high-dimensional problems. Based on the idea of separation of variables, they approximate high-dimensional data with a tensor product of low-dimensional factors. For example, when interaction of particles is a superposition of a few near-neighbour interactions, the corresponding quantum state is a superposition of a few unentangled states, i.e. it has a low Schmidt rank. The corresponding tensor format also has a low rank, providing good compression and low storage costs. By maintaining the compressed format for all operations we can compute the system's trajectory and optimise it without any uncontrollable truncations or heuristic assumptions. Model examples of such systems include Ising and Heisenberg spin chains [Sc11], more realistic applications include quantum wires [BiEtAl17] and backbones of simple protein molecules [SaEtAl14].

In quantum physics a tensor product format was first proposed in the 1970s as the renormalisation group formalism [Wi75], followed in the 1990s by MPS/MPO [FaEtAl92, KlEtAl93] and DMRG [Wh92] algorithms for finding the ground state of a quantum spin chain. Similar formats were applied in statistics and can be traced back to works of Hitchcock in the 1930s [Hi27b, Hi27a] and Tucker in the 1960s [Tu66]. Tensor formats in three dimensions were rigorously studied in the numerical linear algebra community as a generalisation of the low-rank decomposition of matrices [OsEtAl08, FlEtAl08, GoEtAl09, Sa10]. Generalisations to higher dimensions eventually led to re-discovery of the MPS/DMRG framework under the name of the tensor train (TT) format [OsTy09, Os11]. In the last decade the TT format and the more general Hierarchical Tucker format [Gr10] have been successfully applied to a variety of problems, such as finding several low-lying states of a quantum system [DoEtAl14], performing superfast Fourier transform [DoEtAl12], solving high-dimensional linear systems [DoSa15] and first-order differential equations [Do18]. First applications of tensor product methods to optimal control problems are considered in [DoEtAl16, BeEtAl16].

The recently proposed tAMEn algorithm [Do18] allows to compute evolution of a quantum system under a given (time-dependent) Hamiltonian with high accuracy. Although the trajectory lies in extremely high-dimensional state space, tAMEn keeps the computation costs feasible by performing all calculations in the TT format and adapting the TT ranks to meet the required accuracy. In this paper we employ the tAMEn algorithm as a building block to develop a version of a classical GRAPE algorithm [KhEtAl-5] and use it to control a spin chain with $d = 41$ spins. This shows that tensor product algorithms can be used to design control sequences for quantum computers with 50 to 100 qubits, which can be expected within a decade [Go16].

29.2 Optimal Quantum Control

29.2.1 Dynamic Optimisation Problem

The dynamics of a quantum system with Hamiltonian $\hat{H}$ is described by Liouville–von Neumann equation in superoperator form

$$\dot{\rho} = -i\hat{\hat{H}}(t)\rho + \hat{\hat{R}}\rho,$$

where ρ is the quantum density matrix stretched into a vector form (referred to later as quantum state), and

$$\hat{\hat{H}}(t) = I \otimes \hat{H}(t) - \hat{H}(t) \otimes I.$$

Traditionally $\hat{H}(t)$ is split into a constant drift and a time-dependent control term:

$$\hat{H}(t) = \hat{H}_0 + \sum_{k=1}^{K} c_k(t)\hat{H}_k.$$

Control operators $\hat{H}_k$ are usually defined by model or instrument used; we manipulate the system by changing control pulse functions $c_k(t)$. Now choosing a suitable basis set $\{\varphi_n(t)\}_{n=1}^{N}$ and dropping the relaxation part $\hat{\hat{R}}$ for simplicity, we obtain an ODE for system dynamics:

$$\dot{\rho} = -i\left(\hat{\hat{H}}_0 + \sum_{k=1}^{K}\sum_{n=1}^{N} c_{n,k}\hat{\hat{H}}_k\varphi_n(t)\right)\rho, \qquad \rho(0) = \rho_0. \tag{29.1}$$

Our goal is to choose control parameters $c = [c_{n,k}]$ to maximise *fidelity* (or *overlap*)

$$F(c) = \Re\langle\rho_T \mid \rho(T)\rangle. \tag{29.2}$$

29.2.2 First-Order Optimisation Framework

Classical optimisation methods require a gradient of F w.r.t. $c_{n,k}$ which reads

$$\frac{\partial F}{\partial c_{n,k}} = \Re\left\langle\rho_T \mid \frac{\partial\rho(T)}{\partial c_{n,k}}\right\rangle. \tag{29.3}$$

To find the gradient, we differentiate (29.1) w.r.t. control parameters and obtain

$$\frac{\partial}{\partial c_{n,k}} \frac{\partial \rho}{\partial t} = -i \left(\varphi_n(t) \hat{\hat{H}}_k \rho + \hat{\hat{H}}(t) \frac{\partial \rho}{\partial c_{n,k}} \right).$$

Changing the order of derivatives in the left-hand side, we obtain a system of coupled ODEs for the density matrix and its gradient:

$$\frac{\partial}{\partial t} \begin{bmatrix} \frac{\partial \rho}{\partial c_{n,k}} \\ \rho \end{bmatrix} = -i \begin{bmatrix} \hat{\hat{H}}(t) & \varphi_n(t) \hat{\hat{H}}_k \\ 0 & \hat{\hat{H}}(t) \end{bmatrix} \begin{bmatrix} \frac{\partial \rho}{\partial c_{n,k}} \\ \rho \end{bmatrix}, \qquad \begin{bmatrix} \frac{\partial \rho}{\partial c_{n,k}} \\ \rho \end{bmatrix}_{t=0} = \begin{bmatrix} 0 \\ \rho_0 \end{bmatrix}. \tag{29.4}$$

To justify the initial condition for the gradient, write a first-order approximation for $\rho(\delta t)$ with an infinitesimal time δt:

$$\rho(\delta t) = \rho_0 - i \delta t \left(\hat{\hat{H}}_0 + \sum_{k=1}^{K} \sum_{n=1}^{N} c_{n,k} \hat{\hat{H}}_k \varphi_n(t) \right) \rho_0 + O(\delta t^2)$$

and differentiate it w.r.t. $c_{n,k}$ to see $\frac{\partial \rho}{\partial c_{n,k}}(\delta t) = O(\delta t) \to 0$.

The system (29.4) can be integrated on $[0, T]$ using a suitable method (e.g. Runge–Kutta scheme [Ru1895]) for all k, n to produce both the fidelity (29.2) and its gradient (29.3).

29.2.3 GRAPE Algorithm

A classical gradient ascend pulse engineering (GRAPE) method [KhEtAl-5] is a simple version of the approach described above. Control pulses $c_k(t)$ are assumed piecewise-constant on intervals $(t_{n-1}, t_n]$ between the nodes of the grid $\{t_n\}_{n=0}^{N}$ with $t_0 = 0$ and $t_N = T$. We will use a uniform grid $t_k = \tau k$ with step-size $\tau = T/N$. The basis functions $\varphi_n(t)$ can be taken as Heaviside functions

$$\theta_n(t) = \begin{cases} 1, & \text{if } t_{n-1} < t \leq t_n, \\ 0, & \text{otherwise,} \end{cases}$$

thus making $c_{n,k}$ the amplitude of control pulse $c_k(t)$ on $(t_{n-1}, t_n]$. The time-locality of basis functions $\theta_n(t)$ allows us to compute (29.3) more efficiently. Note that $\theta_n(t) = 0$ for $t \leq t_{n-1}$ hence ODE (29.4) for the gradient component simplifies to the homogeneous ODE

$$\frac{\partial}{\partial t} \frac{\partial \rho}{\partial c_{n,k}} = -i \hat{\hat{H}} \frac{\partial \rho}{\partial c_{n,k}} \quad \text{for} \quad t \in (0, t_{n-1}], \quad \text{s.t.} \quad \frac{\partial \rho}{\partial c_{n,k}}(0) = 0, \tag{29.5}$$

which of course gives $\frac{\partial \rho}{\partial c_{n,k}}(t_{n-1}) = 0$. Therefore, to compute (29.3) we can propagate the state $\rho(t)$ on $(0, t_{n-1}]$ and start solving the coupled system (29.4) only from t_{n-1}. When we reach t_n and compute $\frac{\partial \rho}{\partial c_{n,k}}(t_n)$, we can note that $\theta_n(t) = 0$ for $t > t_n$ hence (29.5) holds on $t \in (t_n, T]$ as well. Since $\hat{H}(t)$ is piecewise-constant,

$$\frac{\partial \rho}{\partial c_{n,k}}(T) = \exp\left[-i\int_{t_n}^{T} \hat{H}(s)\mathrm{d}s\right] \frac{\partial \rho}{\partial c_{n,k}}(t_n) = \prod_{m=n+1}^{N} \underbrace{\exp\left[-i\tau\hat{H}(t_m)\right]}_{Q_m} \frac{\partial \rho}{\partial c_{n,k}}(t_n),$$

and plugging this in (29.3) gives

$$\frac{\partial F}{\partial c_{n,k}} = \Re\left\langle \rho_T \middle| \prod_{m=n+1}^{N} Q_m \frac{\partial \rho}{\partial c_{n,k}}(t_n) \right\rangle = \Re\left\langle \prod_{m=n+1}^{N} Q_m^{\dagger}\rho_T \middle| \frac{\partial \rho}{\partial c_{n,k}}(t_n) \right\rangle.$$

$$(29.6)$$

The state

$$\lambda(t_n) = \prod_{m=n+1}^{N} Q_m^{\dagger}\rho_T$$

is the solution of ODE with the boundary condition at the right end, colloquially referred to as 'back-propagation':

$$\frac{\partial \lambda}{\partial t} = -i\hat{H}(t)\lambda \quad \text{for} \quad t \in [t_n, T), \quad \text{s.t.} \quad \lambda(T) = \rho_T. \qquad (29.7)$$

Hence the coupled system (29.4) needs to be solved only on $[t_{n-1}, t_n]$ to produce the n-th components of the gradient. By pre-computing all $\lambda(t_n)$ for $n = 1, \ldots, N$, we can implement the GRAPE iteration with complexity scaling linearly in N.

1. For $n = N, \ldots, 1$, propagate (29.7) backward: $\qquad \lambda(t_n) \longrightarrow \lambda(t_{n-1})$

2. For $n = 1, \ldots, N$, propagate (29.4) forward: $\begin{bmatrix} 0 \\ \rho(t_{n-1}) \end{bmatrix} \longrightarrow \begin{bmatrix} \frac{\partial \rho}{\partial c_{n,k}}(t_n) \\ \rho(t_n) \end{bmatrix}$

3. Compute (29.6) and update the control amplitudes: $\quad c_{n,k} \longrightarrow c_{n,k} - \epsilon\frac{\partial F}{\partial c_{n,k}}$.

Here ϵ is the step size for the gradient ascend. It should be carefully selected, since a too small value will make the algorithm converging very slowly while a very big one will make the algorithm divergent. In our implementation the step size is chosen adaptively: it is increased each time the step is successful (i.e. results in better fidelity) and decreased each time we have an overstep (i.e. fidelity decreases and the step has to be rejected). Note that line-search strategies for this optimisation problem are rather pointless, because the gradient (29.6) and the fidelity function (29.2) are computed simultaneously from (29.4), and have therefore the same complexity.

In the classic GRAPE algorithm, the propagators Q_m in (29.6) are calculated as matrix exponentials [MoVa03] and applied to the states; in more advanced versions of the algorithm, the action of matrix exponentials on the states is computed on-fly maintaining the sparse format for λ's and ρ's [Si98].

Instead of computing propagators, we use the tAMEn algorithm to solve the first-order ODEs (29.4) and (29.7) with Hamiltonians and states represented in the TT format. The next section focuses on mathematical details of this approach.

29.3   Tensor Train Format and the tAMEn Algorithm

Introducing individual state indices for each site, we can treat ρ as a multi-index (high-dimensional) array which is referred to as *tensor* in numerical linear algebra:

$$\rho = [\rho(i_1, \ldots, i_d; \ j_1, \ldots, j_d)].$$

We should admit that our use of the term *tensor* does not imply proper differentiation of upper and lower indices or *Einstein summation convention* [Ei1916]. We use the Tensor Train (TT) decomposition [Os11] to separate *pairs* of i_k, j_k, belonging to different sites, by a product approximation of the form

$$\rho(i_1, \ldots, i_d; \ j_1, \ldots, j_d) \approx \sum_{\alpha_0, \ldots, \alpha_d = 1}^{r_0, \ldots, r_d} \rho^{(1)}_{\alpha_0, \alpha_1}(i_1, j_1) \cdots \rho^{(d)}_{\alpha_{d-1}, \alpha_d}(i_d, j_d). \qquad (29.8)$$

Here, $\rho^{(k)}$, $k = 1, \ldots, d$, are called *TT blocks*, and the summation ranges $r_1, \ldots, r_d$ are called *TT ranks*. The grouping of i_k and j_k is such that unentangled (e.g. pure) states are represented by (29.8) exactly with elementary TT ranks $r_0 = \cdots = r_d = 1$. The TT representation is equivalent to matrix product states (MPS) [FaEtAl92, KlEtAl93] with open boundary conditions $r_0 = r_d = 1$. Entangled states require $r_k > 1$ for at least one intermediate $k = 1, \ldots, d - 1$. However, we aim at representing states with *weak* correlations, so that $r_1, \ldots, r_{d-1}$ can be kept bounded by a moderate constant $r_k \leq r \ll 2^d$. This yields an $O(dr^2)$ storage cost of the TT format, *linear* in the system size. For example, ground states of non-critical one-dimensional spin chains admit such bounded TT ranks due to the area law [EiEtAl10].

We use a linear piecewise Chebyshëv scheme [Do18] to discretise ODEs (29.1), (29.4) and (29.7) in time. Recall that we solve ODEs on time intervals $t \in (t_{n-1}, t_n]$, where control pulses $c_k(t)$ and hence $\hat{H}(t)$ are constant and only $\rho(t)$ depends on time. Without loss of generality we can assume that $t \in (0, \tau]$ and drop the time index n for the rest of this section. We choose a basis of Lagrange polynomials $\{L_m(t)\}_{m=1}^M$ centred at Chebyshëv nodes $\{\tau_m\}_{m=1}^M$ and represent the state as

$$\rho(t) \approx \sum_{m=1}^{M} \rho(\tau_m) L_m(t), \qquad t \in (0, \tau], \qquad 0 < \tau_1 < \tau_2 < \cdots < \tau_M = \tau.$$

Since the length of the interval τ is usually quite small to provide better resolution of control pulses, the basis size M can be very moderate. We collect the nodal values $\rho(\tau_m)$ into a vector $\overline{\rho} = [\rho(\tau_m)]_{m=1}^{M}$ of length $4^d M$, which we aim to find. The time derivative operator is discretised as a *differentiation matrix* $S = [L'_\ell(\tau_m)]_{m,\ell=1}^{M}$ in accordance with the spectral approximation theory [Tr00]. Applying this discretisation scheme to ODE (29.1) we obtain the following linear system

$$\underbrace{(S \otimes I + I \otimes i\hat{\hat{H}})\overline{\rho}}_{A} = \underbrace{(S\, 1_M) \otimes \rho(0)}_{f}. \tag{29.9}$$

Here $\overline{\rho}$ is the unknown vector of $\rho(\tau_m)$'s, 1_M is the vector of size M full of 1's, and I denotes identity matrices of appropriate size. Note that the Kronecker products are implicitly realised by the TT representation (29.10). Therefore, we never actually compute them in (29.9), but just collect the corresponding factors into TT blocks. The same applies to the states. In particular, we consider unentangled states ρ_0 and ρ_T, and construct their TT blocks explicitly. Simple mixed states can be made by concatenation of TT blocks instead of a direct summation of a large number of density matrices [SaEtAl14]. The TT ranks can be truncated down to optimal values for a desired accuracy using the singular value decomposition (SVD) [Os11].

Algebraic equations (29.9) can be solved in the TT approximation by an alternating iteration, such as the Alternating Minimal Energy (AMEn) algorithm [DoSa15], which is an enhanced version of Alternating Least Squares (ALS) [HoEtAl12]. We expand the TT representation (29.8) by an extra block which carries the time index m and write:

$$\overline{\rho}(i_1, \ldots, i_d; j_1, \ldots, j_d, m) = \sum_{\alpha} \rho_{\alpha_0,\alpha_1}^{(1)}(i_1, j_1) \cdots \rho_{\alpha_{d-1},\alpha_d}^{(d)}(i_d, j_d) \rho_{\alpha_d,\alpha_{d+1}}^{(d+1)}(m),$$

$$\overline{\rho} = \sum_{\alpha} \rho_{\alpha_0,\alpha_1}^{(1)} \otimes \cdots \otimes \rho_{\alpha_{d-1},\alpha_d}^{(d)} \otimes \rho_{\alpha_d,\alpha_{d+1}}^{(d+1)}.$$

$$\tag{29.10}$$

Now the last block $\rho^{(d+1)} = [\rho_{\alpha_d,\alpha_{d+1}}^{(d+1)}(m)]$ carries degrees of freedom in time, the rightmost TT rank is $r_{d+1} = 1$, and r_d can now be larger than 1.

The basic ALS algorithm would solve (29.9) by sweeping over TT blocks, i.e. representing $\overline{\rho}$ by (29.10) and reducing the equation to only one block $\rho^{(k)}$ in each step. This can be written using a *frame* matrix

$$P_{\neq k} = \left(\sum_{\alpha_0 \ldots \alpha_{k-2}} \rho^{(1)}_{\alpha_0,\alpha_1} \otimes \cdots \otimes \rho^{(k-1)}_{\alpha_{k-2}} \right) \otimes I \otimes \left(\sum_{\alpha_{k+1} \ldots \alpha_{d+1}} \rho^{(k+1)}_{\alpha_{k+1}} \otimes \cdots \otimes \rho^{(d+1)}_{\alpha_d,\alpha_{d+1}} \right),$$

which is of size $4^d M \times 4 r_{k-1} r_k$ for $k \leq d$, and of size $4^d M \times r_{k-1} M r_k$ for $k = d+1$. The frame matrix realises a linear map from the elements of a single TT block into the elements of the whole vector given (29.10), that is $\overline{\rho} = P_{\neq k} \rho^{(k)}$, assuming that elements of $\rho^{(k)}$ are stretched in a vector. The ALS method solves the reduced Galerkin systems $(P^\dagger_{\neq k} A P_{\neq k}) \rho^{(k)} = P^\dagger_{\neq k} f$ subsequently for $k = 1, \cdots, d+1$. This system can be assembled efficiently [HoEtAl12] due to the TT formats of $P_{\neq k}$, A and f and solved using standard methods.

The AMEn method [DoSa15] improves the convergence of ALS by expanding $\rho^{(k)}$ (and hence all $P_{\neq q}$ for $q > k$) with a TT approximation of the *residual* $f - A\overline{\rho}$, where $\overline{\rho}$ is formed by (29.10) with an updated block $\rho^{(k)}$ plugged in. This also allows to adapt the TT ranks to ensure a desired accuracy of the TT approximation. The tAMEn (time-dependent AMEn) algorithm [Do18] utilises the special meaning of the last TT block, carrying the time variable m, in order to preserve conservation laws (e.g. the Frobenius norm), and to estimate the time discretisation error.

For the interval lengths τ chosen in our numerical experiments, we have found that $M = 8$ Chebyshëv nodes in each interval are sufficient to resolve the time derivative with the relative accuracy of 10^{-5} or better. In each step the tAMEn algorithm produces a set of TT blocks $\rho^{(1)}, \ldots, \rho^{(d+1)}$, comprising the discrete-time solution $\overline{\rho}$ on the interval $(0, \tau]$. After $\overline{\rho}$ is obtained, we compute $\rho(\tau)$ in the TT format by taking the slice of the last TT block $\rho^{(d+1)}(M)$ and hence removing the time index m from consideration. Since $t = \tau$ represents the end point of the interval $(t_{n-1}, t_n]$ the obtained state represents $\rho(t_n)$ and can be used now as an initial state in the next GRAPE step over the time interval $(t_n, t_{n+1}]$.

The tAMEn algorithm is agnostic to a particular form of $\hat{H}$ (the only assumption is that it must admit a TT decomposition), and can be applied to any tensor-structured differential equation $\dot{x} = Ax$ such as the auxiliary matrix formalism (29.4) and the back-propagation (29.7).

29.4 Numerical Experiments

We consider a Heisenberg chain of spin-$\frac{1}{2}$ particles, for which a Hamiltonian is a simple sum of nearest neighbour interactions

$$\hat{H} = \sum_{k=1}^{d-1} J_x \hat{\sigma}^{(k)}_x \hat{\sigma}^{(k+1)}_x + J_y \hat{\sigma}^{(k)}_y \hat{\sigma}^{(k+1)}_y + J_z \hat{\sigma}^{(k)}_z \hat{\sigma}^{(k+1)}_z, \tag{29.11}$$

where the Pauli matrices $\hat{\sigma}^{(k)}_{\{x,y,z\}}$ [Sl18] act only on spin in position k of the chain,

$$\hat{\sigma}^{(k)}_{\{x,y,z\}} = I \otimes \cdots \otimes I \otimes \underbrace{\hat{\sigma}_{\{x,y,z\}}}_{k\text{-th place}} \otimes I \otimes \ldots \otimes I.$$

When $J_x = J_y = J_z$, this system is called the XXX Heisenberg model, when $J_x = J_y \neq J_z$ it is called the XXZ model, and for $J_x \neq J_y \neq J_z$ it is called the XYZ model. It is commonly mentioned that linear Heisenberg chains can be diagonalised exactly using the Bethe ansatz [Be31], however the eigenstates can only be written via roots of a system of algebraic equations of degree d [Sl18], that are not computable in closed form neither using analytic methods nor numerically with reasonable accuracy for $d \gtrsim 20$. For simpler XX models ($J_x = J_y$, $J_z = 0$) eigenvectors are available in closed form and can be used for dimensionality reduction, allowing chains with $d \simeq 200$ spins to be controlled [WóEtAl05]. We could not find examples of optimal control pulse computed numerically for XXX, XXZ and XYZ Heisenberg models with the number of spins $d \gtrsim 20$. We decided therefore to test our tensor product approach for XXX and XXZ Heisenberg spin-$\frac{1}{2}$ chains with $d = 11$, $d = 21$ and $d = 41$.

The initial and target states are taken as

$$\psi_0 = |\uparrow\downarrow\downarrow \cdots \downarrow\downarrow\rangle, \qquad \psi_T = |\downarrow\downarrow \cdots \downarrow\downarrow\uparrow\rangle,$$

so the task is to move the $|\uparrow\rangle$ state from the first to the last position in the chain. The control operator $H_c(t) = c(t)\hat{\sigma}^{(1)}_z$ is the magnetic field acting on the first spin only.

In theory, Heisenberg chains are fully controllable [WaEtAl16], which means the fidelity (29.2) can be made infinitely close to 1 as $T \to \infty$. In practice, however, the available time T for the pulse is limited, and the final state $\rho(T)$ will not be fully focused, leaving some *infidelity* $1 - |\langle \rho_T | \rho(T)\rangle| > 0$. We optimise the pulse sequence to reduce the infidelity as much as possible.

The results are shown in Fig. 29.1. Computing a short high-fidelity pulse for large spin system appears to be a challenging task. Using four cores on a Xeon E5-2650 CPU and 10 GB of memory, we reduced infidelity to 10^{-3} for $d = 21$ and to 10^{-2} for $d = 41$ within a few days of computation. This seems fast compared to high-fidelity calculations reported in [SpEtAl18], where infidelity 10^{-4} was reached after weeks/month of optimisation for a system with $d = 4$ qubits with 4 levels each.

The efficiency of the method depends primarily on reasonable choice of the pulse length T, which is usually of the order of the inverse of the natural frequency of the system [KhEtAl01]. The time T typically should increase with the number of spins in the chain due to the finite rate of the information exchange among them. We see in our experiments that larger intervals T are required to reduce infidelity for larger chains; also, XXZ chains demand larger T than XXX chains. We keep resolution $\tau = 10^{-2}$ for the control pulse constant for all experiments, which is the main reason why computational time per iteration grows with T. The TT ranks grow only

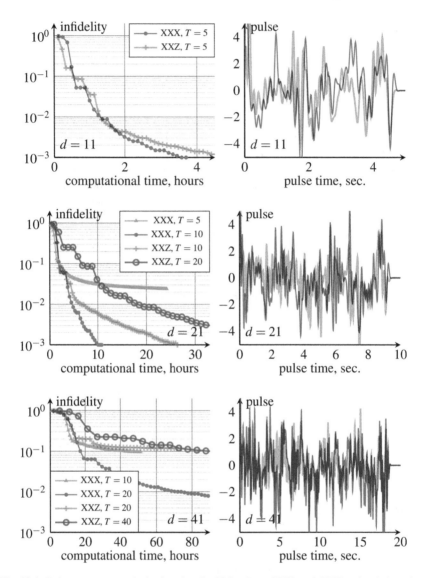

Fig. 29.1 Pulse sequence optimisation for the Heisenberg XXX and XXZ spin chains given by (29.11) with $J_x = J_y = J_z = 2\pi$ and $J_x = J_y = 2\pi$, $J_z = 2.2\pi$, respectively. Left: convergence of the GRAPE algorithm with TT compression of all states and tAMEn algorithm for time evolution. Right: the optimised pulse sequences for XXX and XXZ chains. Top, middle, bottom row: chains with $d = 11$, $d = 21$ and $d = 41$ spins, respectively

mildly with d and remain moderate $r_k \lesssim 30$ for $d = 41$. This resulted in about $70 \cdot 10^3$ unknowns in the TT decomposition of each $\rho(t_k)$ for $d = 21$ and $140 \cdot 10^3$ unknowns for $d = 41$. For comparison, a single diagonal of $\rho(t_k)$ has $2^{21} \approx 2 \cdot 10^6$ entries for $d = 21$ and $2^{41} = 2 \cdot 10^{12}$ entries for $d = 41$.

29.5   Conclusions and Future Work

Tensor product algorithms open new possibilities to control long spin wires and multi-qubit gates in quantum computers. We explored a proof-of-concept example, in which we considered a simple XXX or XXZ Heisenberg chain, optimised control pulse using a classical GRAPE method [KhEtAl-5], while representing all states in the TT format [Os11] and propagating them using the tAMEn algorithm [Do18]. This combination of relatively simple algorithms allowed us to reach fidelity of 99% for a chain of $d = 41$ spins using a single workstation for calculations. We used only controllable approximation techniques and avoided any heuristic or random truncations of the state space. Our algorithm is deterministic and flexible, i.e. can be applied to any linear or quasi-linear quantum system, for which we can expect the states to have moderate entanglement (as measured by Schmidt ranks).

We have not yet reached the fidelity of $\sim 99.99\%$ required for topological error correction [SpEtAl18], mostly because of slow first-order convergence of GRAPE, but we hope to make it possible using a second-order optimisation algorithm, such as Newton–Raphson [GoKu16] or a quasi-Newton, e.g. BFGS [FoEtAl11, EiEtAl11]. We have not imposed any constraints on pulse shape, which resulted in a few outbreaks seen in Fig. 29.1, but implementation of the box constraints $|c(t)| < C$ is a straightforward step which we postpone for future work.

Acknowledgements The authors appreciate the financial support of EPSRC which makes this work possible—grants EP/P033954/1 (D.Q. and D.S.) and EP/M019004/1 (S.D.) Numerical calculations for this paper were performed on a dedicated server, provided to D.S. by the University of Brighton (Rising Stars grant).

References

[BeEtAl16] Benner, P., Dolgov, S., Onwunta, A., Stoll, M.: Low-rank solvers for unsteady Stokes-Brinkman optimal control problem with random data. *Computer Methods in Applied Mechanics and Engineering* **304**, 26–54 (2016).

[Be31] Bethe, H.: Zur Theorie der Metalle. I. Eigenwerte und Eigenfunktionen der linearen Atomkette. *Zeitschrift für Physik* **71**, 205–226 (1931).

[BiEtAl17] Bischoff, J.M., Jeckelmann, E.: Density–matrix renormalization group method for the conductance of one-dimensional correlated systems using the Kubo formula. *Phys Rev B* **96**, 195111 (2017).

[CaEtAl11] Caneva, T., Calarco, T., Montangero, S.: Chopped random–basis quantum optimization. *Phys. Rev. A* **84**(2), 022326 (2011).

[DoEtAl16] Dolgov, S., Pearson, J.W., Savostyanov, D.V., Stoll, M.: Fast tensor product solvers for optimization problems with fractional differential equations as constraints. *Applied Mathematics and Computation* **273**, 604–623 (2016).

[Do18] Dolgov, S.V.: A tensor decomposition algorithm for large ODEs with conservation laws. *Computational Methods in Applied Mathematics* (2018).

[DoEtAl14] Dolgov, S.V., Khoromskij, B.N., Oseledets, I.V., Savostyanov, D.V.: Computation of extreme eigenvalues in higher dimensions using block tensor train format. *Computer Phys. Comm.* **185**(4), 1207–1216 (2014).

[DoEtAl12] Dolgov, S.V., Khoromskij, B.N., Savostyanov, D.V.: Superfast Fourier transform using QTT approximation. *J. Fourier Anal. Appl.* **18**(5), 915–953 (2012).

[DoSa15] Dolgov, S.V., Savostyanov, D.V.: Alternating minimal energy methods for linear systems in higher dimensions. *SIAM J. Sci. Comput.* **36**(5), A2248–A2271 (2014).

[Ei1916] Einstein, A.: Die Grundlage der allgemeinen Relativitätstheorie. *Annalen der Physik* **354**(7), 769–822 (1916).

[EiEtAl10] Eisert, J., Cramer, M., Plenio, M.B.: Colloquium: Area laws for the entanglement entropy. *Rev. Mod. Phys.* **82**, 277–306 (2010).

[EiEtAl11] Eitan, R., Mundt, M., Tannor, D.J.: Optimal control with accelerated convergence: Combining the Krotov and quasi–Newton methods. *Phys. Rev. A* **83**, 053426 (2011).

[FaEtAl92] Fannes, M., Nachtergaele, B., Werner, R.: Finitely correlated states on quantum spin chains. *Comm. Math. Phys.* **144**(3), 443–490 (1992).

[FlEtAl08] Flad, H.J., Khoromskij, B.N., Savostyanov, D.V., Tyrtyshnikov, E.E.: Verification of the cross 3D algorithm on quantum chemistry data. *Rus. J. Numer. Anal. Math. Model.* **23**(4), 329–344 (2008).

[FoEtAl11] de Fouquieres, P., Schirmer, S.G., Glaser, S.J., Kuprov, I.: Second order gradient ascent pulse engineering. *J. Magnetic Reson.* **212**(2), 412–417 (2011).

[GlEtAl15] Glaser, S.J., Boscain, U., Calarco, T., Koch, C.P., Köckenberger, W., Kosloff, R., Kuprov, I., Luy, B., Schirmer, S., Schulte-Herbrüggen, T., et al.: Training Schrödinger's cat: Quantum optimal control. *The European Phys. J. D* **69**(12), 1–24 (2015).

[GoKu16] Goodwin, D.L., Kuprov, I.: Modified Newton–Raphson GRAPE methods for optimal control of spin systems. *J. Chem. Phys.* **144**, 204107 (2016).

[GoEtAl09] Goreinov, S.A., Savostyanov, D.V., Tyrtyshnikov, E.E.: Tensor and Toeplitz structures applied to direct and inverse 3D–electromagnetic problems. In: *Proc. PIERS*, pp. 1896–1900 (2009).

[Go16] Government Office for Science: The quantum age: technological opportunities (2016). URL www.gov.uk/government/publications/quantum-technologies-blackett-review

[Gr10] Grasedyck, L.: Hierarchical singular value decomposition of tensors. *SIAM J. Matrix Anal. Appl.* **31**(4), 2029–2054 (2010).

[Hi27a] Hitchcock, F.L.: The expression of a tensor or a polyadic as a sum of products. *J. Math. Phys* **6**(1), 164–189 (1927).

[Hi27b] Hitchcock, F.L.: Multiple invariants and generalized rank of a p-way matrix or tensor. *J. Math. Phys* **7**(1), 39–79 (1927).

[HoEtAl12] Holtz, S., Rohwedder, T., Schneider, R.: The alternating linear scheme for tensor optimization in the tensor train format. *SIAM J. Sci. Comput.* **34**(2), A683–A713 (2012).

[KhEtAl01] Khaneja, N., Brockett, R., Glaser, S.J.: Time optimal control in spin systems. *Phys. Rev. A* **63**, 032308 (2001).

[KhEtAl-5] Khaneja, N., Reiss, T., Kehlet, C., Schulte-Herbrüggen, T., Glaser, S.J.: Optimal control of coupled spin dynamics: design of NMR pulse sequences by gradient ascent algorithms. *J. Magnetic Reson.* **172**(2), 296–305 (2005).

[KlEtAl93] Klümper, A., Schadschneider, A., Zittartz, J.: Matrix product ground states for one-dimensional spin-1 quantum antiferromagnets. *Europhys. Lett.* **24**(4), 293–297 (1993).

[KuEtAl07] Kuprov, I., Wagner-Rundell, N., Hore, P.J.: Polynomially scaling spin dynamics simulation algorithm based on adaptive state-space restriction. *J Magn. Reson.* **189**(2), 241–250 (2007).

[MoVa03] Moler, C., Van Loan, C.: Nineteen dubious ways to compute the exponential of a matrix, twenty-five years later. *SIAM Review* **45**(1), 3–49 (2003).

[Os11] Oseledets, I.V.: Tensor-train decomposition. *SIAM J. Sci. Comput.* **33**(5), 2295–2317 (2011).

[OsEtAl08] Oseledets, I.V., Savostianov, D.V., Tyrtyshnikov, E.E.: Tucker dimensionality reduction of three-dimensional arrays in linear time. *SIAM J. Matrix Anal. Appl.* **30**(3), 939–956 (2008).

[OsTy09] Oseledets, I.V., Tyrtyshnikov, E.E.: Breaking the curse of dimensionality, or how to use SVD in many dimensions. *SIAM J. Sci. Comput.* **31**(5), 3744–3759 (2009).

[Ru1895] Runge, C.: Ueber die numerische Auflösung von Differentialgleichungen. *Math. Ann.* **46**(2), 167–178 (1895).

[Sa10] Savostyanov, D.V.: Tensor algorithms of blind separation of electromagnetic signals. *Rus. J. Numer. Anal. Math. Model.* **25**(4), 375–393 (2010).

[SaEtAl14] Savostyanov, D.V., Dolgov, S.V., Werner, J.M., Kuprov, I.: Exact NMR simulation of protein-size spin systems using tensor train formalism. *Phys. Rev. B* **90**, 085139 (2014).

[Sc11] Schollwöck, U.: The density-matrix renormalization group in the age of matrix product states. *Annals of Physics* **326**(1), 96–192 (2011).

[Si98] Sidje, R.B.: Expokit: a software package for computing matrix exponentials. *ACM Transactions on Mathematical Software (TOMS)* **24**(1), 130–156 (1998).

[Sl18] Slavnov, N.A.: Algebraic Bethe ansatz. arXiv preprint 1804.07350 (2018).

[SpEtAl18] Spiteri, R., Schmidt, M., Ghosh, J., Zahedinejad, E., Sanders, B.C.: Quantum control for high–fidelity multi–qubit gates. *New J. Phys.* **20**, 113009 (2018). https://doi.org/10.1088/1367-2630/aae79a

[Tr00] Trefethen, L.N.: *Spectral methods in MATLAB*. SIAM, Philadelphia (2000).

[Tu66] Tucker, L.R.: Some mathematical notes on three-mode factor analysis. *Psychometrika* **31**, 279–311 (1966).

[WaEtAl16] Wang, X., Burgarth, D., Schirmer, S.: Subspace controllability of spin-$\frac{1}{2}$ chains with symmetries. *Phys. Rev. A* **94**, 052319 (2016).

[Wh92] White, S.R.: Density matrix formulation for quantum renormalization groups. *Phys. Rev. Lett.* **69**(19), 2863–2866 (1992).

[Wi75] Wilson, K.G.: The renormalization group: Critical phenomena and the Kondo problem. *Rev. Mod. Phys.* **47**(4), 773–840 (1975).

[WóEtAl05] Wójcik, A., Łuczak, T., Kurzyński, P., Grudka, A., Gdala, T., Bednarska, M.: Unmodulated spin chains as universal quantum wires. *Phys. Rev. A* **72**, 034303 (2005).

Chapter 30
Epidemic Genetic Algorithm for Solving Inverse Problems: Parallel Algorithms

Sabrina B. M. Sambatti, Haroldo F. de Campos Velho, and Leonardo D. Chiwiacowsky

30.1 Introduction

Genetic Algorithms (GA) are methods of optimization based on the evolutionary concepts proposed by Charles Darwin and Alfred Russel Wallace. Darwin and Wallace collected evidence from travels on the South America. Darwin was the invited naturalist during the Beagle's trip for mapping the South America coast, and Wallace was a naturalist who had lived in the Amazon region about 4 years. Both have concluded about natural selection process after reading the Thomas Mathus's book: *An Essay on the Principle of Population*, where it is argued that the human population grows faster than food production.

The Darwin-Wallace's theory is the main guideline, but there were many troubles with the evolution theory. For example, it was not clear how the speciation works, and how little changes appear in new generations. A central name for the evolutionary synthesis is Ernst Mayr. He achieved the *modern synthesis* that integrated Mendel's theory of heredity with Darwin-Wallace's theory of evolution and natural selection. The evolutionary theory is under development yet. The punctuated equilibrium and endosymbiotic proposal are two examples of new ideas in the field—see [Da96].

S. B. M. Sambatti
Climatempo, São José dos Campos, SP, Brazil
e-mail: sabrina.sambati@climatempo.com.br

H. F. de Campos Velho (✉)
Instituto Nacional de Pesquisas Espaciais (INPE), São José dos Campos, SP, Brazil
e-mail: haroldo.camposvelho@inpe.br

L. D. Chiwiacowsky
Universidade de Caxias do Sul (UCS), Caxias do Sul, RS, Brazil
e-mail: ldchiwiacowsky@ucs.br

© Springer Nature Switzerland AG 2019   381
C. Constanda, P. Harris (eds.), *Integral Methods in Science and Engineering*,
https://doi.org/10.1007/978-3-030-16077-7_30

The GA strategies are outlined from the modern version of the evolution theory (*great synthesis*). In the optimization context, the objective function works as the environment (acting as the evolutionary pressure). Only the most apt individuals in the population will be selected to be parents for the new generation (our *natural selection*). For the evolution works, many features have important role to produce an appropriate answer to the environment conditions: the time must be long enough, and the variability in the population, for example. Results from variations on these parameters are investigated here, for verifying if our implementation can emulate some characteristics of the biological evolutionary theory.

The application of the GA on optimization problems is an iterative process, where successive evaluations on the objective function are required. However, this method can escape local minimum, and, in addition, one of the main genetic algorithm aspects is its parallelization capacity. The parallel genetic algorithms (PGA) is made up of the distribution of the tasks of a sequential GA by different processors, providing a much better performance than the sequential GA. Depending on the parallel strategy adopted, the PGA is effectively a new algorithm, with a chance of avoiding the premature convergence which could exist in the sequential GAs.

Different strategies are investigated for the parallel implementation of a genetic algorithm (GA). The PGA is employed to solve the inverse heat conduction problem of determining the initial temperature from the noisy measurements at a given time. The parallel code is generated using calls to the message passing communication library MPI (Message Passing Interface).

In the parallel code, each processor executes the GA in its own population and migration of best-fitness individuals occurs periodically among processors. An epidemic operator purges each population whenever there is no fitness improvement. Different migration strategies are tested, as the island model (individuals may migrate to all the other processors), the stepping-stone model (migration may occur only between consecutive processors of a logical ring in an alternate manner). Performance results, quality of the solutions, and convergence are discussed, comparing the different migration strategies.

30.2 Inverse Problem

The direct problem consists of a transient heat conduction problem in a slab with adiabatic boundary condition, and initial condition $T(x, t = 0) = f(x)$. The mathematical formulation of the problem is given by:

$$\frac{\partial^2(x,t)}{\partial x^2} = \frac{\partial T(x,t)}{\partial t}, \qquad x \in (0,\ 1), \qquad t > 0, \tag{30.1}$$

$$\frac{\partial T(x,t)}{\partial x} = 0, \qquad x = 0, \qquad x = 1, \quad t > 0, \tag{30.2}$$

$$T(x,t) = f(x), \qquad x \in [0,\ 1], \qquad t = 0, \tag{30.3}$$

where $T(x, t)$ denotes temperature at a given point and time, x and t the spatial and time variables. These variables are treated as nondimensional quantities. The solution of the direct problem for $x \in (0, 1)$ and $t > 0$ for a given initial condition $f(x)$ is given by

$$T(x, t) = \sum_{m=0}^{\infty} e^{-\beta_m^2 t} \frac{X(\beta_m, x)}{N(\beta_m)} \int_0^1 X(\beta_m, x') f(x') dx' \qquad (30.4)$$

where $X(\beta_m, x)$ are the eigenfunctions associated with the problem, β_m and $N(\beta_m)$ are, respectively, the corresponding eigenvalues and *norms* [Ozi80]. The function $f(x)$ is assumed to be bounded satisfying Dirichlet's conditions in the interval $[0, 1]$. In this transient conduction problem, the goal is to estimate the unknown initial temperature distribution $f(x)$, from the knowledge of the measured temperature T_i at the time $t = \tau > 0$, for a finite number (N_x) of different locations in the domain. This inverse problem is formulated as an optimization problem, as follows. The present inverse problem admits an analytical solution, obtained from the orthogonality property of $X(\beta_m, x)$. For a given temperature profile at time τ, the initial condition can be expressed as [MuEtAll99, MuEtAll00]—with $x \in (0, 1)$:

$$f(x) = \sum_{m=0}^{\infty} e^{\beta_m^2 \tau} \frac{X(\beta_m, x)}{N(\beta_m)} \int_0^1 X(\beta_m, x') T(x', \tau) dx' . \qquad (30.5)$$

However, this *solution* does not produce a smooth solution, nor is it close to the true solution when some noise is presented in the experimental data $T(x, \tau)$—see Fig. 30.1. Since measurement errors are permanent features in the experimental data, Eq. (30.5) is not useful.

For determining the initial condition, a regularized solution is obtained by choosing the function $f(x)$ that minimizes the following functional:

$$J(f(x), \xi) = \|T^{\text{Mod}}(f) - T^{\text{Exp}}\|_2^2 + \xi \sum_{k=1}^{N_{sol}} f_k^2 , \qquad (30.6)$$

where $T^{\text{Exp}} = T^{\text{Exp}}(x_i, \tau)$ is the experimental data $(t = \tau > 0)$ and T^{Mod} is the temperature obtained using the candidate solution $f(x)$ at time $t = \tau$ using Eq. (30.4), and $\|.\|_2$ is the norm 2. The last term is the zeroth-order Tikhonov regularization term [Tik77], weighted by ξ, the regularization parameter. Each candidate solution $f(x)$ is sampled by a set of N_x discrete points: f_k, with $k = 1, 2, \ldots, N_x$. In the GA approach, each individual is a particular instance f_k and its fitness is given by this functional.

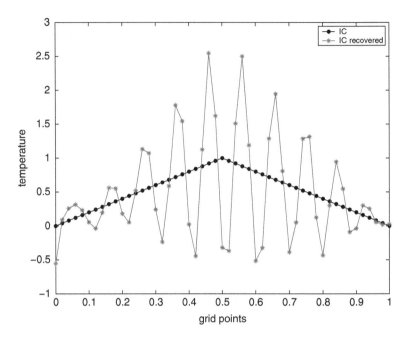

Fig. 30.1 Exact initial temperature profile and solution obtained using analytical solution Eq. (30.5)

30.3 Parallel Genetic Algorithm with Epidemic Operator

In a GA-based optimizer, an initial population is generated, composed by a group of random individuals, each one associated with a possible solution. Every individual is evaluated, being assigned a score or fitness, according to the numerical value of the objective function. In every new generation, a new population of individuals is generated from the former one, by means of combining *parents* individuals using selection, crossing-over and mutation operators. Some schemes preserve best-fitness individuals from one generation to another (elitism) [Ca95]. It is expected that, after many generations, the population will evolve and better-fitted individuals will appear. An epidemic genetic operator [Me03, ChCa03] is also used in this work. If the fitness does not improve after some number of generations, the population is renewed, preserving only best-fitness individuals.

Tournament Selection [Mic96]

This operator uses a random number *rand* from the interval $[0,1)$ with uniform distribution. *bigger* is the best fitness individual and *smaller* is the worst fitness individual:

    bigger:=rand; smaller:=rand; val:=0.75;

  if (rand < val) then
      position:=bigger;
  else
      position:=smaller;
  endif

Geometrical Crossover [Mic96]

The crossover operator breeds only one offspring from two parents. From the parents x_i and y_i, the offspring is represented by:

$$z_i = x_i^{\mu}\, y_i^{1-\mu}, \tag{30.7}$$

where μ is a number between $[0, 1]$. A typical value is $\mu = 1/2$, where the same weight is given to both parents.

Non-uniform Mutation [Mic96]

The mutation operator is defined as:

$$x_i' = \begin{cases} x_i + \triangle(t, l_{sup} - x_i) \text{ if a random binary digit is 0,} \\ x_i - \triangle(t, x_i - l_{inf}) \text{ if a random binary digit is 1,} \end{cases} \tag{30.8}$$

and

$$\triangle(t, y) \equiv y\left[1 - \text{rand}^{(1-t/T)^b}\right],$$

where *rand* is a random number from the interval $[0,1)$ with uniform distribution, T is the maximal generation number, t is the generation number, and b is a system parameter determining the degree of non-uniformity.

Epidemical Strategy [Me03, ChCa03]

This innovative operator is activated when a specific number of generations is reached without improvement of the best individual. Then, all the individuals are *affected by a plague*, and only those that have the best fit (e.g., first 2% with the best fit in the population) *survive*. The remaining individuals *die* and are substituted by new individuals with new genetic variability, such as immigrants arriving, in order to evolve the population. Two parameters need to be chosen: one determines when the strategy will be activated, i.e. the number of generations without improvement of the best individual fit, while the other parameter determines the amount of individuals that will survive the *plague*.

30.3.1 Parallel Strategies for Epidemic-GA

An important feature of GAs is their suitability for parallelization. Nowadays GAs have been widely employed, but parallel implementations are recent. According to [Ca95], GA parallelization techniques are divided into: (a) Global Parallelization: in this approach all genetic operators and the evaluation of all individuals are explicitly parallelized; (b) Coarse Grained: such approach requires a division of population into some number of demes (subpopulations) that are separated one from another ("geographic isolation"). The individuals compete only within a deme. This approach introduces an additional operator called *migration* that is used to send some individuals from one deme to another; (c) Fine Grained: such approach requires a large number of processors because the population is divided into a large number of small demes, and each one evolves separately, but subject to migration.

Many GA researchers believe that a PGA, with its multiple distributed sub-populations and local rules and interactions, is a more realistic model for the evolution of species in nature than a single large population [Le94]. This work follows the coarse grained approach and implements some migration strategies (PGA-Epidemic): island model and stepping-stone.

In the island model, best-fitness individuals may migrate to all other processors. Island-1 denotes the sending of each processor best solution to a master processor that selects and broadcasts "the best-of-the-bests" to all others. Island-2 denotes multiple broadcasts in which each processor sends its particular best solution to all others. Figure 30.2 outlines the two versions of the island PGA-Epidemic.

In the stepping-stone (SS) model, a logical ring of processors is defined and communication occurs in steps as each processor sends its best individual to the left and right-side neighbour (SS-1 model). After a finite number of steps, all processors have the best global solution. In the proposed implementation using a logical ring is

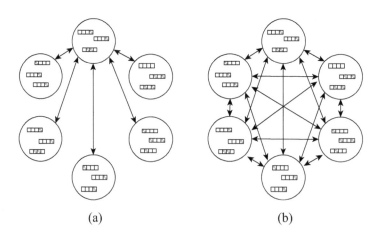

(a)   (b)

Fig. 30.2 PGA-Epidemic models: (**a**) island-1, (**b**) island-2

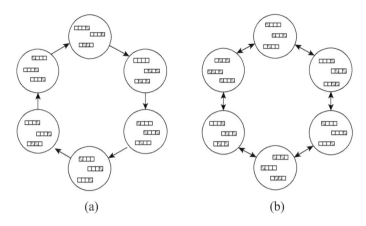

(a)   (b)

Fig. 30.3 Epidemic-PGA models: (**a**) stepping-stone-1, (**b**) stepping-stone-2

also defined and migration is also restricted to neighbour: one-directional migration
(SS-2 model). The sketch for SS-models is shown in Fig. 30.3.

The algorithm for Epidemic-PGA is presented below in a pseudo-language.

Algorithm – Epidemic-PGA

    Random initialization *Pop*(*t*)

    Population distribution in *p*-processors

    for *processor*=1 to *p* **do**

        Evaluation *Pop*(*t*)

        while stopping criterion is not satisfy

            $t \leftarrow t + 1$

            Selection *Pop*(*t*) from *Pop*(*t* − 1)

            Apply crossover on *Pop*(*t*)

            Apply mutation on *Pop*(*t*)

            Evaluated *Pop*(*t*)

        end while

        If epidemic **then**

            Apply epidemic

        end if

        If migration **then**

            Apply migration

        end if

    end for

30.4   Numerical Results

A PGA is used to solve the inverse backward heat conduction problem for a one-dimensional slab. The chosen test case assumed the following initial temperature profile, given by the triangular test function

$$f(x) = \begin{cases} 2x, & 0 \le x \le 0.5 \\ 2(1-x), & 0.5 \le x \le 1; \end{cases}$$

Experimental data, corresponding to the measured temperatures at a time $\tau > 0$, are obtained from the forward model by adding a Gaussian noise with 5% of level. It is adopted $\tau = 0.01$ and an experimental data grid of 101 points (N_x).

Many parameters must be to set up for the PAG-Epidemic: (a) population size: 336 and 1008 individuals; (b) geometrical crossover operator $\mu = 1/2$; (c) non-uniform mutation operator $b = 5$; (d) mutation probability 5%; (e) epidemic operator (the best five individuals are kept); and (f) maximal generation number: 10,000 and 50,000. The action of the epidemic operator can be realized from Fig. 30.4, where a smoother inverse solution is obtained when the epidemic operator is applied. Due to the randomness of the GA, several solutions were computed and only one *average* answer is shown in the graphical representation.

In order to check if our implementation of the GA could reproduce some features of the biological evolution, some preliminary tests are carried out. Firstly, time is a crucial parameter for the evolution, i.e., a minimum time period is necessary to produce a *good* evolutionary answer for a given condition. Secondly, a greater variability can also be important to obtain a good answer. Indeed, looking at Fig. 30.5 it is easy to recognize that a very bad answer is obtained starting with 336 individuals in the population at 10,000 generations. However, the GA-epidemic inverse solution for the same population considering 50,000 generations produces good estimation. In addition, dealing with a bigger population (that means, a greater variability) a good estimation is also obtained.

One remarkable feature in the evolutionary process is the phenomenon of the stasis (the evolution is stopped). Two factors can determine this process: intense genic flux, and/or the conditions do not change (environment and mutagenic factors). This process can be observed during the searching of the inverse solution with GA. Figure 30.6 shows the inverse solution by epidemic-PGA with 336 individuals for 10,000 generations (we already saw that under these conditions was not possible to obtain a good inverse solution—see Fig. 30.5). Using one or two processors, it is not possible to obtain an acceptable inverse solution, but considering four processors (or more), good solutions are obtained. This means that partial isolation of the sub-populations, with occasional migration, is a mechanism that can break the stasis process. The new genetic operator epidemic is employed, and it is not possible to predict when the operator will be activated. Table 30.1 shows many times this operator is used in the parallel implementation.

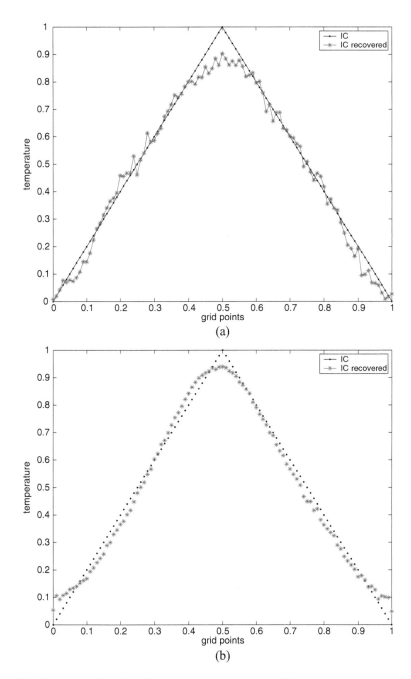

Fig. 30.4 Inverse solution: (**a**) epidemic operator non-activated; (**b**) epidemic operator activated

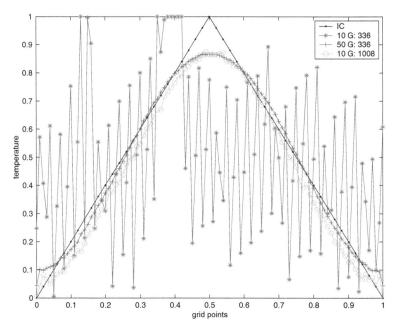

Fig. 30.5 Inverse solution with GA-Epidemic: 336 individuals with 10,000 generations (bad inverse solution: "∗" 10-G 336); 336 individuals with 50,000 generations ("+" 50-G 336); 1008 individuals with 10,000 generations ("○" 10-G 1008)

Table 30.1 Average number of epidemic operator used

Processors	Island-1	Island-2	SStone-1	SStone-2
1	1.4	1.0	1.5	3.8
2	10.5	0.0	0.0	1.0
4	9.7	11.1	10.9	9.2
6	10.6	13.2	13.5	10.6
8	18.4	22.6	17.8	20.5

From the previous considerations, performance results for the epidemic-PGA are obtained running 50,000 generations, and a population with 1008 individuals. Inversions obtained with island and stepping-stone models ranging to one up to eight processors are all similar to those in Fig. 30.6. Usually, the performance of parallel implementation can be roughly evaluated by the *speed-up* $S_p = T_1/T_p$, being T_1 the sequential time and T_p the parallel time for p-processors. Another definition is the *efficiency* $E_p = S_p/p$. All parallel strategies present good speed-up—see Fig. 30.7a, having island-2 and stepping-stone-2 the best performance, similar behaviour is noted for the efficiency (Fig. 30.7b). The stepping-stone-1 has the lowest efficiency for two processors, due to the migration time spent as shown in Table 30.2.

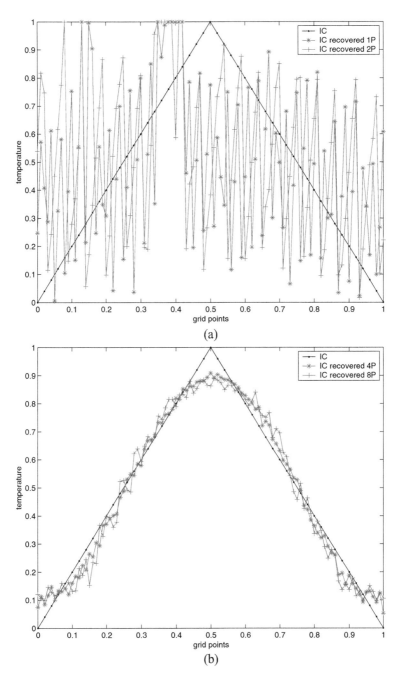

Fig. 30.6 Inverse solution with Epidemic-GA: (**a**) one and two processors; (**b**) four and eight processors

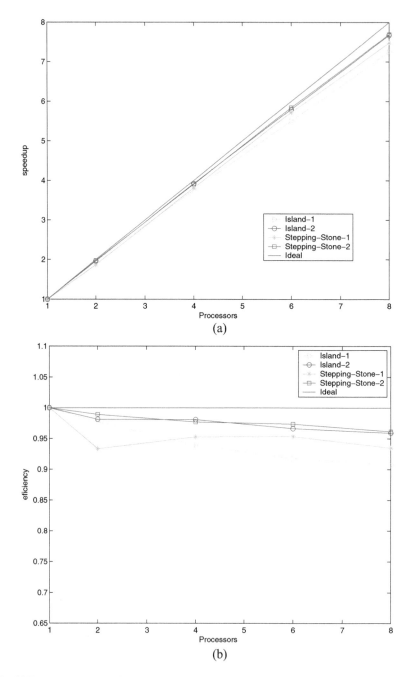

Fig. 30.7 Population with 1008-individuals: (**a**) speed-up (1008 individuals), (**b**) efficiency (1008 individuals)

Table 30.2 Performance for the PGA-epidemic stepping-stone-1

Proc.	Migration (s)	Total time (s)	Speed-up	Effic.
1	0.003	2829.104	1.000	1.000
2	66.355	1515.014	1.866	0.933
4	12.844	742.013	3.817	0.952
6	8.812	494.003	5.714	0.953
8	14.268	377.692	7.489	0.935

30.5 Conclusion

The inverse problem of estimating the unknown initial condition of heat conduction transfer in a one-dimensional slab was solved using a parallel genetic algorithm that uses a new evolutionary operator called *epidemic*. The PGA was implemented using some migration topologies, two versions of island model, and two stepping-stone models. Different from the global parallel strategy—where only the fitness evaluation is parallelized, the PGA with migration consists in a new algorithm—see Fig. 30.6.

The Epidemic-PGA performance is good, with speed-ups close to the linear. However, a study of the algorithm complexity for the strategies employed deserves to be investigated in a future work. Any way, one goal is to provide a stochastic optimization tool for the National Institute for Space Research (Brazil), which is a rich source for application of the inverse problem methodology in space science (such as image reconstruction in astronomy and astrophysics, determining the maps of the cosmic radiation background in microwaves), space technology (damage detection in aerospace structures, and inverse design for satellite thermal analysis), and space applications (atmospheric temperature retrieval from satellite data, and multispectral estimation of the optical properties for natural waters). Indeed, the epidemic-GA is being used for damage detection in a hybrid optimization procedure, combining epidemic-GA with conjugate gradient [ChCa03, ChEtAl08, ChEtAl04a, ChEtAl04b].

The island-2 strategy was applied to the inverse problem of stiffness estimation of a mass-spring system [Ca06], which is an inverse problem in vibration. The mathematical model in vibrations deals with hyperbolic equations (the mass-spring system is a prototype of partial hyperbolic differential equations). The speedup of the island-2 strategy was very similar in the two types of inverse problems: heat transfer and mechanical vibration, a close linear performance curve (ideal parallelism).

Acknowledgements The authors would like to thank the FAPESP and CNPq, Brazilian agencies for research support.

References

[Ca06] Campos Velho,H. F., Chiwiacowsky, L. D., Sambatti, S. B. M.: Structural damage identification by a hybrid approach: variational method associated with parallel epidemic genetic algorithm. Scientia, (2006), pp. 10–18.

[Ca95] Cantú-Paz, E.: *A summary of research on parallel genetic algorithms*, Genetic Algorithms Laboratory - University of Illinois, Urbana-Chapaign, (1995).

[ChCa03] Chiwiacowsky, L. D., Campos Velho, H. F: Different approaches for the solution of a backward heat conduction problem. In *Inverse problem in engineering*, 11(3), 471–494, (2003).

[ChEtAl04a] Chiwiacowsky, L. D., Gabarri, P., Campos Velho, H. F: Determining stiffness matrix by the adjoint method. In *International Conference on Computational and Experimental Engineering and Sciences*, Ilha da Madeira, Portugal, (2004), pp. 26–29.

[ChEtAl04b] Chiwiacowsky, L. D., Gabarri, P., Campos Velho, H. F: Damage assessment of large space structures through the variational approach. In *International Astronautical Congress* Vancouver, Canada, (2004), pp. 04–08.

[ChEtAl08] Chiwiacowsky, L. D., Gabarri, P., Campos Velho, H. F: Damage assessment of large space structures through the variational approach. *Acta Astronautica*, 60(10–11), 592–604, (2008).

[Da96] Dawkins, R.: The Blind Watchmaker: Why the Evidence of Evolution Reveals a Universe without Design. W. W. Norton and Company, (1996).

[Le94] Levine, D.: A parallel genetic algorithm for the set partitioning problem. Master's thesis, Mathematics and Computer Science Division - Argonne National Laboratory. Argonne, USA, (1994).

[Me03] Medeiros, F. L.: Algoritmo genético híbrido como um método de busca de estados estacionários de sistemas dinâmicos. Master's thesis, INPE-CAP. São José dos Campos, Brazil (2003).

[Mic96] Michalewicz, Z.: Genetic algorithms + data structures = evolution programs. (1996).

[MuEtAll99] Muniz, W. B., Ramos, F. M., Campos Velho, H. F.: A comparison of some methods for estimating the initial condition of the heat equation. Journal of Computational and Applied Mathematics, (1999).

[MuEtAll00] Muniz, W. B., Ramos, F. M., Campos Velho, H. F. Entropy and a Tikhonov regularization techniques applied to the backwards heat equation. Computers and mathematics with application, (2000).

[Ozi80] Özisik M. N.: Heat conduction. Wiley Interscience, USA, (1980).

[Tik77] Tikhonov, A. V., Arsenin V. Y.: Solutions of Ill-posed Problems. Winston and Sons, New York, (1977).

Chapter 31
A Chemical Kinetics Extension to the Advection-Diffusion Equation by NO_x and SO_2

Juliana Schramm and Bardo E. J. Bodmann

31.1 Introduction

Pollution emission and dispersion goes hand in hand with technological progress and its consequences of an expansive productive sector. Also research in the field of air quality has evolved considerably, so that it is possible to simulate and evaluate scenarios of emission of certain substances and their impact on the environment. Studies of this kind are useful, on the one hand, to analyse existing situations and, on the other hand, stimulate the elaboration of new strategies followed by further technological advances.

The present contribution reports on progress in pollution dispersion simulation with emphasis on the role of chemical reactions, more specifically the influence of reaction kinetics on the concentration distributions and their time evolution. Chemical reactions of pollutants after their release from power or industrial plants represent either a sink or a source term, where the first one lowers the concentration of a specific substance when the original molecules are transformed into other harmless substances. However, there also exists the possibility that the transformation by a chemical reaction gives rise to new substances, which might present an attack on health. One example is the production of ozone triggered by the presence of nitrogen oxides. In order to gain insight into the relevance or importance of chemical reactions in the dispersion process we consider in this study the time dependence of some chemical compositions in the atmosphere.

To this end, in this work the chemical kinetics of nitrogen oxides (NO_x) and sulphur dioxide (SO_2) are considered as an extension to the advection-diffusion equation and validated using experimental data from the Hanford experiment, for

J. Schramm · B. E. J. Bodmann (✉)
Federal University of Rio Grande do Sul, Porto Alegre, RS, Brazil
e-mail: bardo.bodmann@ufrgs.br

© Springer Nature Switzerland AG 2019
C. Constanda, P. Harris (eds.), *Integral Methods in Science and Engineering*,
https://doi.org/10.1007/978-3-030-16077-7_31

a stable planetary boundary layer, and from the Prairie Grass experiment, for an unstable boundary layer. One of the frequently used simulators by environmental agencies is the CALPUFF program, which we use as a complementary comparison to the aforementioned validation by observational data.

31.2 Tropospheric Chemistry

In the further we consider only substances in the emission of exhaust fumes, that arise typically in electric power production. Thus, reaction rates are added to the advection-diffusion equation, once dominant emitted substances and their associated reactions in the troposphere are identified. More specifically, in the further we consider nitrogen oxides (NO_x) and sulphur dioxide (SO_2) as a first step to extend the advection-diffusion model. One of the characteristics of the chemistry of NO_x compounds is that they undergo photolysis and therefore the reactions for these species are divided according to the possible reactions during daylight and corresponding transformations during night. In the presence of sunlight, the nitrogen oxides NO and NO_2 follow the photochemical cycle shown below.

$$NO_2 + h\nu \rightarrow NO + O$$

$$O + O_2 + M \rightarrow O_3 + M$$

$$NO + O_3 \rightarrow NO_2 + O_2$$

$$OH + NO_2 + M \rightarrow HNO_3 + M$$

Here $h\nu$ is the photon energy with h Planck's constant and frequency ν and M is the molecule that provides collision energy so that the reaction can occur. In this cycle the reaction rate of NO_2 is given by

$$\frac{d[NO_2]}{dt} = -j_{NO_2}[NO_2] + k_{O_3+NO}[O_3][NO] - k_{OH+NO_2+M}[OH][NO_2][M] ,$$

and the NO reaction rate is

$$\frac{d[NO]}{dt} = j_{NO_2}[NO2] - k_{O_3+NO}[O_3][NO] ,$$

where k is a chemical reaction rate coefficient, j is the reaction rate coefficient when photolysis occurs and $[X]$ represents the concentration of the species X in units of moles. Summing up the contributions and writing the total reaction balance as a pseudo-second-order reaction rate yields

$$\frac{d[NO_x]}{dt} = -k(T, z)_{ps, OH+NO_2}[OH][NO_2] \, ,$$

where $k(T, z)_{ps}$ is the pseudo-second-order rate coefficient.

During night, the presence of NO leads to reactions with ozone so that almost all NO_x is converted to NO_2. The NO_x night reactions are

$$NO_2 + O_3 \rightarrow NO_3 + O_2$$

$$NO_2 + NO_3 + M \leftrightarrow N_2O_5 + M$$

$$N_2O_5 + H_2O(s) \rightarrow 2HNO_3(s)$$

and the final reaction rate for NO_x is

$$\frac{d[NO_x]}{dt} = -k_{NO_2+O_3}[NO_2][O_3]$$

$$-k(T, z)_{ps, NO_2+NO_3} \left([NO_2][NO_3] - \frac{1}{K_{eq}}[N_2O_5] \right)$$

where K_{eq} is an equilibrium constant.

The homogeneous gas phase reactions of SO_2 are

$$SO_2 + OH + M \rightarrow HOSO_2 + M \, ,$$

$$HOSO_2 + O_2 \rightarrow HO_2 + SO_3$$

and

$$SO_3 + H_2O + M \rightarrow H_2SO_4 + M \, .$$

From the reactions above the final reaction rate is derived,

$$\frac{d[SO_2]}{dt} = -k_{ps, SO_2+OH}(T, z)[SO_2][OH] \, .$$

Note that the considerations made above represent only kinetic effects (time evolution) that are independent on the spatial coordinate. In the next section these reactions are inserted into a space-time model, i.e. the advection-diffusion equation.

31.3   The Extended Advection-Diffusion Equation

The advection-diffusion equation, which contains as presented in the previous section the effects of the chemical kinetics in the source term $\bar{S}$, is given below.

$$\frac{\partial \bar{c}}{\partial t} + \bar{u}\frac{\partial \bar{c}}{\partial x} + \bar{v}\frac{\partial \bar{c}}{\partial y} + \bar{w}\frac{\partial \bar{c}}{\partial z} = \frac{\partial}{\partial x}\left(K_x\frac{\partial \bar{c}}{\partial x}\right) + \frac{\partial}{\partial y}\left(K_y\frac{\partial \bar{c}}{\partial y}\right) + \frac{\partial}{\partial z}\left(K_z\frac{\partial \bar{c}}{\partial z}\right) + \bar{S}$$

The source term is considered as composed by a point source contribution and the transformations by chemical reactions.

$$S = S_F + \left(\frac{\partial c}{\partial t}\right)_{CR}$$

Considering the instantaneous point source with strength S_F, K_x and K_y are turbulent diffusion constants and K_z is considered locally constant but varies slowly with increasing height. The problem to be solved is then

$$\frac{\partial \bar{c}}{\partial t} + \bar{u}\frac{\partial \bar{c}}{\partial x} + \bar{v}\frac{\partial \bar{c}}{\partial y} + \bar{w}\frac{\partial \bar{c}}{\partial z} = K_x\frac{\partial^2 \bar{c}}{\partial x^2} + K_y\frac{\partial^2 \bar{c}}{\partial y^2} + K_z\frac{\partial^2 \bar{c}}{\partial z^2} + \left(\frac{\partial \bar{c}}{\partial t}\right)_{CR},$$

where we tacitly absorbed the contribution of a locally fixed source in the initial condition. The solution of this equation is subject to the following initial condition,

$$\bar{c}(x, y, z, 0) = Q\delta(x - x_0)\delta(y - y_0)\delta(z - H_s)$$

where x_0 and y_0 are the location of the fixed point source and H_s is its height.

Let $\bar{c}(x, y, z, t) = \bar{c}_x(x, t)\bar{c}_y(y, t)\bar{c}_z(z, t)$, then the problem may be decoupled into equations of $1 \oplus 1$ space-time dimensions.

$$\frac{\partial \bar{c}_x}{\partial t} + \bar{u}\frac{\partial \bar{c}_x}{\partial x} = K_x\frac{\partial^2 \bar{c}_x}{\partial x^2} + \left(\frac{\partial \bar{c}_x}{\partial t}\right)_{CR} \tag{31.1}$$

$$\frac{\partial \bar{c}_y}{\partial t} + \bar{v}\frac{\partial \bar{c}_y}{\partial y} = K_y\frac{\partial^2 \bar{c}_y}{\partial y^2} + \left(\frac{\partial \bar{c}_y}{\partial t}\right)_{CR}$$

$$\frac{\partial \bar{c}_z}{\partial t} + \bar{w}\frac{\partial \bar{c}_z}{\partial z} = K_z\frac{\partial^2 \bar{c}_z}{\partial z^2} + \left(\frac{\partial \bar{c}_z}{\partial t}\right)_{CR}$$

Each problem is now subject to the corresponding decoupled initial condition.

$$\bar{c}_x(x, 0) = Q^{1/3}\delta(x - x_0) \tag{31.2}$$

$$\bar{c}_y(y, 0) = Q^{1/3}\delta(y - y_0)$$

$$\bar{c}_z(z, 0) = Q^{1/3}\delta(z - H_s)$$

Due to different chemical reactions of each considered species, it is necessary to solve the Fourier transform separately for NO_x and SO_2. The NO_x daylight reaction rate in mass basis is then given by

$$\left(\frac{d\bar{c}_{NO_x,day}}{dt}\right)_{CR} = -\lambda_{NO_x,day}\bar{c}_{NO_x} ,$$

where

$$\lambda_{NO_x,day} = \frac{k(T, s)_{ps,OH+NO_2}M_{NO_x}}{M_{OH}M_{NO_2}}c_{OH}$$

and c_X is the concentration of the species X in units of (g/m^3) and M_X is the molar mass of species X in units of (g/mol).

Assuming that the chemical reactions occur in an isotropic fashion, $1/3$ may be attributed to each direction x, y and z, thus

$$\left(\frac{d\bar{c}_{x,NO_x,day}}{dt}\right)_{CR} = -\frac{\lambda_{NO_x,day}}{3}\bar{c}_{x,NO_x} ,$$

$$\left(\frac{d\bar{c}_{yNO_x,day}}{dt}\right)_{CR} = -\frac{\lambda_{NO_x,day}}{3}\bar{c}_{y,NO_x} ,$$

$$\left(\frac{d\bar{c}_{z,NO_x,day}}{dt}\right)_{CR} = -\frac{\lambda_{NO_x,day}}{3}\bar{c}_{z,NO_x} ,$$

and the Fourier transform for $\bar{c}_x(x, t)$ is given by

$$C(\alpha, t) = \frac{1}{\sqrt{2\pi}} \int_{-\infty}^{\infty} \bar{c}_x(x, t)e^{-i\alpha x}dx .$$

Upon application of the Fourier transform in (31.1) with respect to the spatial coordinate

$$\frac{\partial C_{NO_x}}{\partial t} + i\alpha\bar{u}C_{NO_x} = -\alpha^2 K_x C_{NO_x} - \frac{\lambda_{NO_x,day}}{3}C_{NO_x} ,$$

with initial condition (31.2) for the NO_x daylight reactions

$$C_{NO_x}(\alpha, 0) = \frac{Q^{1/3}}{\sqrt{2\pi}} e^{-i\alpha x_0}$$

one obtains the solution

$$C_{NO_x}(\alpha, t) = D_1 e^{-\left(\alpha^2 K_x + i\alpha\bar{u} + \frac{\lambda_{NO_x,day}}{3}\right)t} ,$$

where D_1 is an integration constant. Consequently, the solution of the transformed problem is

$$C_{NO_x}(\alpha, t) = \frac{Q^{1/3}}{\sqrt{2\pi}} e^{-\left[\alpha^2 K_x t + i\alpha(x_0 + \bar{u}t) + \frac{\lambda_{NO_x,day}}{3}t\right]} . \tag{31.3}$$

The inverse transform of (31.3) is given by

$$\bar{c}_{x,NO_x,day}(x, t) = \frac{Q^{1/3}}{2\pi} \int_{-\infty}^{\infty} e^{-\left[\alpha^2 K_x t - i\alpha(x - x_0 - \bar{u}t) + \frac{\lambda_{NO_x,day}}{3}t\right]} d\alpha .$$

Completing the square of the exponent

$$\alpha^2 K_x t - i\alpha(x - x_0 - \bar{u}t) + \frac{\lambda_{NO_x,day}}{3}t =$$

$$\left\{\alpha(K_x t)^{1/2} - \frac{i(x - x_0 - \bar{u}t)}{2(K_x t)^{1/2}}\right\}^2 + \frac{(x - x_0 - \bar{u}t)^2}{4K_x t} + \frac{\lambda_{NO_x,day}}{3}t$$

and substituting $\eta = \alpha(K_x t)^{1/2} - i(x - x_0 - \bar{u}t)/2(K_x t)^{1/2}$ then

$$\bar{c}_{x,NO_x,day}(x, t) = \frac{Q^{1/3} e^{-\frac{(x - x_0 - \bar{u}t)^2}{4K_x t}} e^{-\frac{\lambda_{NO_x,day}}{3}t}}{2\pi(K_x t)^{1/2}} \int_{-\infty}^{\infty} e^{-\eta^2} d\eta ,$$

where the integral is known to be $\sqrt{\pi}$. Thus

$$\bar{c}_{x,NO_x,day}(x, t) = \frac{Q^{1/3}}{2(\pi K_x t)^{1/2}} \exp\left(-\frac{(x - x_0 - \bar{u}t)^2}{4K_x t} - \frac{\lambda_{NO_x,day}}{3}t\right)$$

and the same procedure is used to solve for $\bar{c}_{y,NO_y,day}(y,t)$ and $\bar{c}_{z,NO_x,day}(z,t)$. The final solution for NO_x during day light is then

$$
\bar{c}_{NO_x,day}(x,y,z,t) = \frac{Q}{\sqrt{64\pi^3 K_x K_y K_z t^3}} \exp\left(-\frac{(x-x_0-\bar{u}t)^2}{4K_x t}\right.
$$

$$
\left. -\frac{(y-y_0-\bar{v}t)^2}{4K_y t} - \frac{(z-H_s-\bar{w}t)^2}{4K_z t} - \lambda_{NO_x,day}t\right).
$$

The analogue steps are used to solve NO_x during night and SO_2, where the latter has no contributions due to photo-chemical reactions. The solutions for these species are

$$
\bar{c}_{NO_x,night}(x,y,z,t)
$$

$$
= \frac{Q}{\sqrt{64\pi^3 K_x K_y K_z t^3}}
$$

$$
\times \exp\left(-\frac{(x-x_0-\bar{u}t)^2}{4K_x t} - \frac{(y-y_0-\bar{v}t)^2}{4K_y t}\right.
$$

$$
\left. -\frac{(z-H_s-\bar{w}t)^2}{4K_z t} - \lambda_{1,NO_x,night}t + \lambda_{2,NO_x,night}t\right)
$$

with

$$
\lambda_{1,NO_x,night} = 2\frac{k_{NO_2+O_3}}{M_{O_3}}c_{O_3}
$$

$$
\lambda_{2,NO_x,night} = \frac{k(T,z)_{ps,NO_2+NO_3}k_{NO_2+O_3}}{M_{O_3}K_{eq}k_{N_2O_5+H_2O(s)}}c_{O_3}
$$

and for SO_2

$$
\bar{c}_{SO_2}(x,y,z,t) = \frac{Q}{\sqrt{64\pi^3 K_x K_y K_z t^3}} \exp\left(-\frac{(x-x_0-\bar{u}t)^2}{4K_x t}\right.
$$

$$
\left. -\frac{(y-y_0-\bar{v}t)^2}{4K_y t} - \frac{(z-H_s-\bar{w}t)^2}{4K_z t} - \lambda_{SO_2}t\right)
$$

with

$$
\lambda_{SO_2} = \frac{k(T,z)_{ps,SO_2+OH}}{M_{OH}}c_{OH}
$$

Assuming that a continuous emission is a superposition of instantaneous emissions during a small time interval $d\tau$, then

$$\bar{C}(x, y, z, t) = \int_0^t \bar{c}(x, y, z, t - \tau) d\tau .$$

Using a period of 6.26 h/day with sunlight (based on data from INMET - National Institute of Meteorology, Brazil) then photolysis occurs in that period of time, so that the final solution for a mean concentration of NO_x and SO_2 from a continuous source is

$$\bar{C}_{NO_x}(x, y, z, t) = \frac{Q}{\sqrt{64\pi^3 K_x K_y K_z}} \int_0^t \left\{ \frac{1}{\sqrt{(t - \tau)^3}} \exp\left[-\frac{(x - x_0 - \bar{u}(t - \tau))^2}{4K_x(t - \tau)} \right. \right.$$

$$-\frac{(y - y_0 - \bar{v}(t - \tau))^2}{4K_y(t - \tau)} - \frac{(z - H_s - \bar{w}(t - \tau))^2}{4K_z(t - \tau)}$$

$$\left. \left. -0.26\lambda_{NO_x,day}(t - \tau) + 0.74(-\lambda_{1,NO_x,night}(t - \tau) + \lambda_{2,NO_x,night}(t - \tau)) \right] \right\}$$

and

$$\bar{C}_{SO_2}(x, y, z, t) = \frac{Q}{\sqrt{64\pi^3 K_x K_y K_z}} \int_0^t \left\{ \frac{1}{\sqrt{(t - \tau)^3}} \exp\left[-\frac{(x - x_0 - \bar{u}(t - \tau))^2}{4K_x(t - \tau)} \right. \right.$$

$$\left. \left. -\frac{(y - y_0 - \bar{v}(t - \tau))^2}{4K_y(t - \tau)} - \frac{(z - H_s - \bar{w}(t - \tau))^2}{4K_z(t - \tau)} - \lambda_{SO_2}(t - \tau) \right] \right\} .$$

31.4 Model Validation and Effects Due to Chemical Reactions

The dispersion mechanism by advection-diffusion was validated using the Hanford experiment with a stable boundary layer. In this experiment a tracer substance (SF_6) was released during 30 min from a tower of two meter height. The measurements were conducted in five distances between 100 and 3200 m, where in this work the 100 m distance was not considered. For more details about the experiment, see reference [DoHo85]. The eddy diffusivities used in this work are given in reference [DeEtAl96], for a stable planetary boundary layer.

The contribution of the extension by chemical reactions was evaluated by the Prairie Grass experiment, where the boundary layer was unstable. From a total of 61 experiments, only 20 were used in this work, those that were under unstable boundary layer conditions. In this experiment a SO_2 tracer was released from a position 46 cm above ground and during 10 min. Samples were taken in five distances between 50 and 800 m, for more details, see [Ba58]. In this case, the vertical eddy diffusivity from reference [DeEtAl97] was employed.

The wind speed at height z used in the current simulations follows a logarithmic profile.

$$u(z) = \frac{u_*}{\kappa}\left[\ln\left(\frac{z+z_0}{z_0}\right) - \psi_m\left(\frac{z}{L}\right) + \psi_m\left(\frac{z_0}{L}\right)\right], z \le z_B$$

$$u(z) = u(z_B), z > z_B$$

Here $z_B = \max(0.1z_i; |L|)$, z_0 is the roughness length and ψ_m is the stability function, with $\psi_m = 4.7(z/L)$ for stable conditions and for unstable conditions ψ_m is given by [BaBa75] as shown below.

$$\psi_m = -\ln\left[\left(\frac{1+(1-\frac{15z}{L})^{\frac{1}{2}}}{2}\right)\left(\frac{1+(1-\frac{15z}{L})^{\frac{1}{4}}}{2}\right)^2\right] + 2\arctan\left(1-\frac{15z}{L}\right)^{\frac{1}{4}} - \frac{\pi}{2}$$

In order to calculate the effects of the chemical reactions it is necessary to obtain the reaction constants (λ). Both the chemical kinetics coefficients and photochemical data were taken from reference [BuEtAl15]. The temperature was $295 K$ and the atmospheric pressure was 1 atm. Further, the maximum natural tropospheric ozone concentration according to [Ja99] was 10×10^{-6} mol of O_3 per mol of air. Using these data the reaction constants used in this work are $\lambda_{NO_x,day} = 9.701 \times 10^{-6}s^{-1}$, $\lambda_{1,NO_x,night} = 0.015s^{-1}$, $\lambda_{2,NO_x,night} = 4.858 \times 10^{-22}s^{-1}$ and $\lambda_{SO_2} = 1.189 \times 10^{-6}s^{-1}$.

Although there were no measurements of NO_x and SO_2 in the Hanford experiment and only SO_2 measurements, the findings of these projects were used to show the effects of chemical reactions in the results, comparing the results for pure advection-diffusion simulation to the ones where the chemical reactions were "switched on" in the simulations. The results are shown in Fig. 31.1, where observed

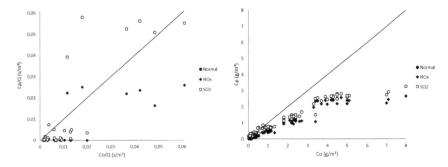

Fig. 31.1 Scatter plot for observed (C_o) and predicted (C_p) concentration for the Hanford (left) and for the Prairie Grass experiment (right). The attribute "Normal" indicates dispersion by pure advection-diffusion, whereas NO_x and SO_2 are the cases where chemical reactions were taken into account

versus simulated concentrations are shown for the Hanford experiment (left) and the Prairie Grass experiment (right). Here the attribute "Normal" indicates that only advection-diffusion is active without any chemical reactions and NO_x and SO_2 are the results for dispersion considering also chemical reactions, as introduced in the previous section.

In both figures the concentrations in "Normal" and SO_2 simulation mode show coincidence so that the "Normal" dots are hidden behind the SO_2 ones. This is different for the mean concentration of NO_x taking into account the chemical reactions, where a reduction of 87.87% under Hanford conditions and 41.08% under Prairie Grass experiment conditions occurs. Considering the chemical reactions of SO_2 the mean concentration reduction was 0.06% in the Hanford case and 0.008% in Prairie Grass case. These findings allow to conclude that for the SO_2 dispersion the simulation by only advection-diffusion is sufficient, the inclusion of chemical reactions does not alter significantly the results for short time intervals.

An additional comparison was made using the results from simulations by the CALPUFF program, where the micrometeorological conditions were handled by the CALMET/CALPUFF implementation [Sc16]. Again, runs with pure advection-diffusion and considering also chemical reactions were performed and compared. As a domain of the simulation we considered the environment around a thermoelectric power station close to the city of Linhares (Espirito Santo state) in the southeast of Brazil. This location was chosen since the terrain is considerably flat and thus provides almost laboratory like conditions. For the CALPUFF simulations a grid cell 15 km × 15 km was chosen with a 1 km resolution and the duration of the simulation was set to 90 h. The average values of the wind speed u, v, w, the friction velocities u_*, w_*, the Obukhov length L and the planetary boundary layer height z_i were taken from CALPUFF, and since $L > 0$ the turbulent diffusion parameterization from [DeEtAl96] for stable conditions was used. Again, a significant concentration reduction was observed for NO_x, with 93.04% using CALPUFF and 99.99% using the proposed model, whereas for SO_2 the reduction was 1.31% and 1.85% for the CALPUFF and for the proposed model, respectively. In order to show the chemical reaction effects on the dispersion process, Fig. 31.2 shows the concentration isolines obtained by the proposed model for a purely advection-diffusion process and for the simulation including chemical reactions (Fig. 31.3). The x and y axes are the geographical coordinates in units of km.

31.5 Conclusion

In the present contribution we discussed the influence of some chemical reactions, here for NO_x and SO_2 on the dispersion process that was implemented by the advection-diffusion equation together with the equations that represent the reaction kinetics of the chemical process. A general comment is in order here, the dispersion process in nature follows a stochastic process, whereas the advection-diffusion model calculates only mean values, so that one does not expect a perfect alignment

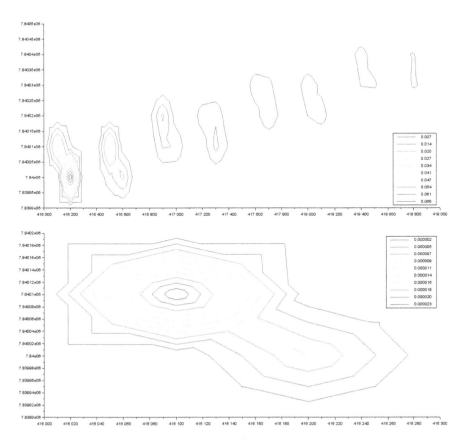

Fig. 31.2 Concentration isolines (in units of $\mu g/m^3$) for NO_x without (above) and with chemical reactions (below)

between observed and simulated concentrations but observes a distribution of points around the bi-sector. For cases where the mean value is sufficient to characterize the dispersion observed and simulated concentrations should coincide provided the advection-diffusion equation is an appropriate model.

In our analysis of simulations and experiments, despite the fact that the Hanford and Prairie Grass experiments are only short duration observations (30 and 10 min, respectively) the chemical reactions contributed to a reduction in the mean concentration values, which was significant for NO_x, where the reduction was two orders in magnitude and spurious for SO_2. This tendency was also confirmed by the simulations using the CALPUFF program, where the concentration reduction (in %) was fairly close to the findings of the present model. However, there is a difference in both simulations, CALPUFF uses constant reaction rates, which in our model have an explicit temperature dependence, this feature is especially important, since the vertical temperature profile is not a constant, so that the reaction rates vary also with height.

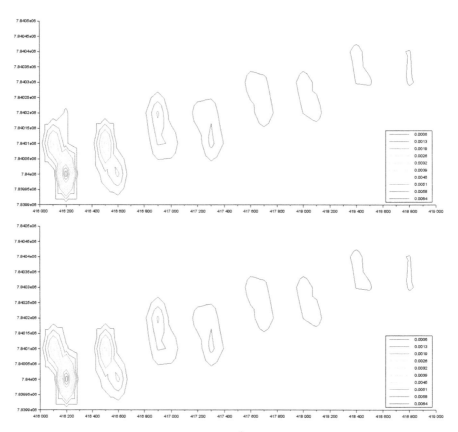

Fig. 31.3 Concentration isolines (in units of $\mu g/m^3$) for SO_2 without (above) and with chemical reactions (below)

Our results clearly show that the chemical reactions have to be considered in pollution dispersion models in the atmospheric boundary layer since they have contributions to changes also in the time dependence of pollutant concentrations. The authors of this work are completely aware of the fact that so far the experimental validation is not complete due to the fact that only one experiment made use of one of the chemical compositions of interest, namely, SO_2. Nevertheless, in the close future we will continue our analysis and include from the theoretical point of view more substances such as CO and CO_2 and from the experimental part using measurements of all these compounds around the thermoelectric power station LORM close to the city Linhares in the Espirito Santo state in Brazil.

References

[Ba58] Barad, M. L.: Project Prairie Grass, a field program in diffusion. *AFCRL-TR-58-235*, **I–III** (1958).

[BaBa75] Barker, E. H., Baxter, T. L.: A Note on the Computation of Atmospheric Surface Layer Fluxes for Use in Numerical Modeling. *Journal of Applied Meteorology*, **14**, 620–622 (1975).

[BuEtAl15] Burkholder, J. B., Sander, S. P., Abbatt, J., Barker, J. R., Huie, R. E., Kolb, C. E., Kurylo, M. J., Orkin, V. L., Wilmouth, D. M., Wine, P. H: Chemical Kinetics and Photochemical Data for Use in Atmospheric Studies, Evaluation No. 18. *JPL Publication 15–10*, Jet Propulsion Laboratory, Pasadena (2015).

[DeEtAl97] Degrazia, G. A., Rizza, U., Mangia, C., Tirabassi, T.: Validation of a new turbulent parameterization for dispersion models in convective conditions. *Boundary-Layer Meteorology*, **85**, 243–254 (1997).

[DeEtAl96] Degrazia, G. A., Vilhena, M. T., Moraes, O. L. L.: An algebraic expression for the eddy diffusivities in the stable boundary layer: a description of near-source diffusion. *Il Nuovo Cimento*, **19C**, 399–403 (1996).

[DoHo85] Doran, J. C., Horst, T. W.: An evaluation of Gaussian plume-depletion model with dual-tracer field measurements. *Atmospheric Environment*, **19**, 939–951 (1985).

[Ja99] Jacob, D. J.: Introduction to atmospheric chemistry. Princeton University Press, Princeton (1999).

[Sc16] Schramm, J.: Estudo da dispersao de poluentes em uma usina termeletrica localizada em Linhares utilizando o modelo CALPUFF. Master thesis. Universidade Federal do Rio Grande do Sul, Porto Alegre, Brazil (2016).

Chapter 32
On the Development of an Alternative Proposition of Cross Wavelet Analysis for Transient Discrimination Problems

Adalberto Schuck Jr. and Bardo E. J. Bodmann

32.1   Introduction

The wavelet transform in both continuous (CWT) and discrete (DWT) form is a well-established tool in many application fields for one dimensional (1-D) signals and for two or higher dimensional signals [Da92, AnEtAl04]. Its capability of describing the behavior of 1-D signals both in time and frequency/scale domains with time resolution intrinsically adjusted by the scales used makes this a superior tool to analyze nonstationary processes.

Extending the use of CWT, Rioul and Flandrin [RiFl92] defined the *Wavelet Scalogram* to estimate the wavelet spectrum of such processes. Hudgins et al. [HuEtAl93] introduced the definitions of *Wavelet Power Spectrum* and *Wavelet Cross Spectrum* to analyze climatological data. These definitions do not include the time varying aspect, since they are integrated over the time variable. Liu [Li94] proposed later the definitions of *Wavelet Spectrum*, *Cross Wavelet Spectrum* (here named Wavelet Cross Spectrum, WCS) and *Wavelet Coherence* (WCO) to study the behavior of ocean wind waves. These definitions included time and scale as independent variables and they constitute the so-called Cross Wavelet Analysis (CWA), being the most used ones in the subsequent works. Torrence and Compo [ToCo98] presented some applications of *Wavelet Power Spectrum* (WPS), *Cross Wavelet Spectrum*, and *Cross Wavelet Power*, and were the first establishing their confidence levels. Furthermore, they pointed out that Liu's definition for the Wavelet Coherence estimator can be identically unitary at all times and scales and

A. Schuck Jr.
Electrical Engineering Department (DELET), UFRGS, Porto Alegre, RS, Brazil

B. E. J. Bodmann (✉)
Mechanical Engineering Department (DEMEC), (UFRGS), Porto Alegre, RS, Brazil
e-mail: bardo.bodmann@ufrgs.br

© Springer Nature Switzerland AG 2019   409
C. Constanda, P. Harris (eds.), *Integral Methods in Science and Engineering*,
https://doi.org/10.1007/978-3-030-16077-7_32

recommended some sort of smoothing in time or scale dimension to circumvent this problem. Maraun and Kurths [MaKu04] pointed out other drawbacks with the CWA and improved the significance tests presented in [ToCo98] by means of Monte Carlo simulations. These significance tests were further developed in a posterior work [MaEtAl07]. Thereafter, the CWA definitions have been largely used to analyze geophysical data [GrEtAl04, Nt10], electroencephalographic (EEG) and other biomedical signals [LaEtAl02, HeEtAl05, KlEtAl06] as well as in transient detection [Pl07] among others. Today, these definitions make part of the Wavelet Toolbox of Matlab®R2012a.

Recalling that the definitions of CWA are obtained by point-by-point multiplication of the CWT of the time series, then, if transients appear at different times for different realizations of processes, the estimators of WPS and WCS will be inconclusive, with as many local maxima as the total number of transients in the time series. Therefore they cannot be properly used for detection of these non-synchronized transients. This issue is exemplified in Sect. 32.3. Furthermore, the estimator of WCO has the inherent pitfall of being identically unitary as mentioned in [ToCo98, MaKu04].

In [ScEtAl13], the authors presented alternative definitions for WPS, WCS, and WCO to detect multi-channel synchronized transients in noise and estimate the delay (or lag) of the target transients present in two processes. The preliminary assessment of the technique was made by a very simple signal test. Therefore, the objective of this work is to present and discuss in a deeper way those alternative definitions for WPS, WCS, and WCO that solve the aforementioned issues and improve the assessment of the technique with alternative signals.

32.2 Developments

32.2.1 Classic Definitions

The auto-correlation functions of stationary processes x and y and the cross-correlation function of the same stationary processes are respectively [BePi00]

$$R_{xx}(\tau) = E\{x(t+\tau)\overline{x}(t)\}, \tag{32.1}$$

$$R_{yy}(\tau) = E\{y(t+\tau)\overline{y}(t)\} \tag{32.2}$$

$$R_{xy}(\tau) = E\{x(t+\tau)\overline{y}(t)\} \tag{32.3}$$

where τ is the lag or time interval between samples of time series and the "overline" indicates the complex conjugate of the function. The Fourier transform of (32.1), (32.2), and (32.3) are the auto and cross spectral density functions, respectively. These functions are the traditional way to analyze and describe the behavior of stationary stochastic processes in time or frequency domain. Also the estimation errors of them are well established. But, as said before, they were developed for stationary processes. Thus, it seems that the definitions of CWA derived from CWT revealed to be a more appropriate tool to analyze non-stationary stochastic processes.

Let $x(t)$ be a square integrable time series from a stochastic process x, the CWT of $x(t)$ is given by [Da92, AnEtAl04]

$$W_x(b, a) = \frac{1}{\sqrt{a}} \int_{-\infty}^{\infty} x(t) \overline{\psi} \left(\frac{t-b}{a} \right) dt \qquad (32.4)$$

where $\psi(t)$ is the mother wavelet function used, the "overline" indicates the complex conjugate of the function, $a \in \mathbb{R}, a > 0$ is the scale variable and $b \in \mathbb{R}$ is the time translation variable. The $W_x(b, a)$ can be performed also in the frequency domain,

$$W_x(b, a) = \sqrt{a} \int_{-\infty}^{\infty} X(\omega) \overline{\Psi}(a\omega) e^{j\omega b} \, d\omega , \qquad (32.5)$$

where $X(\omega)$ and $\Psi(\omega)$ are the Fourier transforms of $x(t)$ and $\psi(t)$, respectively.

The $W_y(b, a)$ is obtained upon substituting $x(t)$ by $y(t)$ in (32.4). The usual mother wavelet used for CWA is the Morlet wavelet, described by [AnEtAl04, ToCo98, MaKu04]

$$\psi_{Mor}(t) = \pi^{-1/4} e^{-t^2/2} e^{j\omega_0 t} , \qquad (32.6)$$

where t and ω_0 are dimensionless.

Its Fourier transform is given by

$$\Psi_{Mor}(\omega) = \pi^{-1/4} e^{-(\omega-\omega_0)^2/2} . \qquad (32.7)$$

This wavelet is not admissible (admissibility means $\int_{\mathbb{R}} |\Psi(\omega)|/|\omega| \, d\omega < \infty$) but for practical purposes, for $\omega_0 > 5.5$ it can be considered [AnEtAl04]. The Morlet wavelet is usually chosen for CWA because being a complex function, it provides complex functions for its CWTs, as in Fourier analysis. Also, its scale a is easily related with the Fourier frequency f [ToCo98] by

$$1/f = 4\pi a/(\omega_0 + \sqrt{2 + \omega_0^2}) . \qquad (32.8)$$

The scalogram of the CWT is defined as [RiFl92]

$$|W(b, a)|^2 = W(b, a) \overline{W}(b, a) \qquad (32.9)$$

and describes the energy distribution of the signal at a certain time t with a scale a.
Further, the Wavelet Power Spectrum (WPS) of a process x is defined by [ToCo98,
GrEtAl04, MaKu04]

$$WPS_x(b, a) = E\left\{W_x(b, a)\overline{W}_x(b, a)\right\}, \qquad (32.10)$$

where $E\{\cdot\}$ denotes the expectation value operator and the overline indicates
the complex conjugate. The same definition can be applied to process y (named
$WPS_y(b, a)$), substituting x by y in (32.10). Notice that WPS can be interpreted
as the expectation value of scalograms. The Wavelet Cross Spectrum (WCS) is an
extension of expression (32.10) and is defined by

$$WCS_{xy}(b, a) = E\left\{W_x(b, a)\overline{W}_y(b, a)\right\}. \qquad (32.11)$$

Finally, the Wavelet Coherence (WCO) is defined by

$$WCO_{xy}(b, a) = \frac{|WCS_{xy}(b, a)|^2}{|WPS_x(b, a)||WPS_y(b, a)|}. \qquad (32.12)$$

These definitions can appear in a slightly different form according to several
authors, for instance, in [HuEtAl93, LaEtAl02, KlEtAl06] the complex conjugate
is applied to the first element of the expression. Also, in [HuEtAl93, Nt10], an
integration over b is performed on expressions (32.10) and (32.11), resulting in
functions depending only on scale, instead of scale and time. However, they
essentially express the same ideas. If there are N realizations of processes, (32.11)
can be estimated by

$$W\tilde{C}S_{xy}(b, a) = \frac{1}{N}\sum_{k=1}^{N} W_x^k(b, a)\overline{W}_y^k(b, a), \qquad (32.13)$$

where $W_x^k(b, a)$ and $W_y^k(b, a)$ are the CWT of realizations $x_k[n]$ and $y_k[n]$,
respectively. $W\tilde{P}S_x(b, a) = \frac{1}{N}\sum_{k=1}^{N} |W_x^k(b, a)|^2$ and $W\tilde{P}S_y(b, a) = $
$\frac{1}{N}\sum_{k=1}^{N} |W_y^k(b, a)|^2$ are performed in a similar way. If there is only one realization
of each process, expression (32.11) could be estimated either by a local averaging
along a time direction (smoothing), assuming the process to be stationary over a
certain time interval, or in scale direction, assuming that neighboring scales have
similar power [MaKu04].

With respect to expression (32.12), it was mentioned that if no smoothing is
performed in time or scale domain over both numerator and denominator estimators,
then $W\tilde{C}S_{xy}(b, a)$ will result in a trivial unitary response because numerator and
denominator of (32.12) will be equal [ToCo98, MaKu04, MaEtAl07]. This was one
of the reasons to suggest a smoothing separately for numerator and denominator in
order to solve this flaw.

32.2.2 Alternative Definitions for Cross Wavelet Spectrum and Wavelet Coherence

When analyzing stationary processes, correlation functions perform a time translation of one sequence in the time domain and the spectral density functions perform
a point by point multiplication in the frequency domain [BePi00]. However, even
though the CWT has both time and frequency domain representation, WPS and
WCS are obtained just by a point by point multiplication of the CWT of x and y.
This fact, together with the averaging process defined in (32.13) for transient events
to occur at different times in x and y series, renders the WCS inconclusive, with
as many local maxima at the respective positions of each event. Therefore, if there
is some fixed delay between event appearance in an x and y process, it will be
hardly estimated by (32.13), contrarily to the traditional correlation analysis. This
scenario gets worse if noise is present in the realizations, which will be exemplified
in the next section. Moreover, the estimation of WCO can result in a trivial unitary
response. For these issues and since CWT has both scale (or related frequency)
and time as independent variables, it seems more appropriate if definitions for
WPS, WCS, and WCO had some sort of correlation or convolution operation in the
time domain along with multiplication in the scale domain. With this idea in mind
and drawing an analogy with the Wiener-Khintchine relations [Nt10], we propose
different definitions for WPS and WCS by:

$$WPS_x(b, a) = E\left\{|CWT\left\{R_{xx}(t)\right\}|\right\},$$ (32.14)

$$WPS_y(b, a) = E\left\{|CWT\left\{R_{yy}(t)\right\}|\right\},$$ (32.15)

$$WCS_{xy}(b, a) = E\left\{CWT\left\{R_{xy}(t)\right\}\right\},$$ (32.16)

where $R_{xx}(\tau)$, $R_{yy}(\tau)$, and $R_{xy}(\tau)$ are the auto-correlation functions of stationary
processes x and y and the cross-correlation function of stationary processes as
defined in (32.1), (32.2), and (32.3), respectively. Now the variable b of CWT is
related with the *lag* τ between samples of time series and for a fixed scale a,
$WPS_x(b, a)$, $WPS_y(b, a)$, and $WCS_{xy}(b, a)$ can be seen as the auto-correlation
of the signals x and y and cross-correlation filtered by a scaled mother wavelet ψ.

Taking into account that since $R_{xx}(0) \geq |R_{xx}(\tau)|$ the maximum value of
WPS will be at $b = 0$, but for WCS, this maximum value can occur at a
different time position. So, in a quite similar way as in correlation analysis, one
can normalize (32.16) in order to obtain a new WCO by:

$$WCO_{xy}(b, a) = \frac{|WCS_{xy}(b, a)|^2}{|WPS_x(0, a)||WPS_y(0, a)|}.$$ (32.17)

For K time series $x_k[n]$ and $y_k[n]$, (32.14), (32.15), and (32.16) can be estimated by:

$$W\tilde{P}S_x(b,a) = \frac{1}{K}\sum_{k=1}^{K}|CWT\left\{\hat{R}_{xx}^k[m]\right\}|, \tag{32.18}$$

$$W\tilde{P}S_y(b,a) = \frac{1}{K}\sum_{k=1}^{K}|CWT\left\{\hat{R}_{yy}^k[m]\right\}|, \tag{32.19}$$

$$W\tilde{C}S_{xy}(b,a) = \frac{1}{K}\sum_{k=1}^{K}|CWT\left\{\hat{R}_{xy}^k[m]\right\}|, \tag{32.20}$$

where

$$\hat{R}_{xx}^k[m] = \frac{1}{N-m}\sum_{n=1}^{N-m}x_k[n+m]\overline{x_k}[n], 0 \le m \le N-1 \tag{32.21}$$

$$\hat{R}_{yy}^k[m] = \frac{1}{N-m}\sum_{n=1}^{N-m}y[n+m]\overline{y}[n], 0 \le m \le N-1 \tag{32.22}$$

$$\hat{R}_{xy}^k[m] = \frac{1}{N-m}\sum_{n=1}^{N-m}x_k[n+m]\overline{y_k}[n], 0 \le m \le N-1 \tag{32.23}$$

are the unbiased auto-correlation and cross-correlation estimators [BePi00].

32.3   Signal Composition and Transient Analysis

Let us assume that one wants to find a single transient event $s(t)$ added to a zero mean, unitary variance gaussian noise $n(t)$ in realizations of $x(t) = s(t) + G_0 n_x(t)$, where G_0 is adjusted to give some desired signal-to-noise ratio (SNR) and $y(t) = s(t-\tau) + G_0 n_y(t)$. The noise sequences $n_x(t)$ and $n_y(t)$ will be independent. Furthermore, for each realization the event can appear only once and at any moment within some time interval in the x process. Then the problem here is to detect $s(t)$ in both processes and estimate the delay τ between the processes. Although the noise used here is stationary, this model can also work in nonstationary signals which can be considered stationary in a time interval, as in EEG event related potential (ERP) analysis [LaEtAl02, HeEtAl05]. The transient chosen to be detected here is a modulated Gaussian pulse $s(t) = e^{-t^2}\cos 6t$, synthesized using Matlab with a sample rate of 25 samples/s. This is shown in Fig. 32.1. The noise added in the realizations were also generated with Matlab *randn* command, with 2048 sample

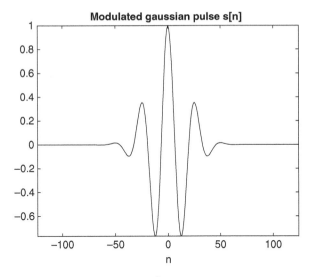

Fig. 32.1 Modulated Gaussian pulse $s(t) = e^{-t^2} \cos 6t$, sampled at a rate of 25 samples/s

points. The delay between events in x and y realizations was set to 500 samples. Three realizations of these signals were generated as shown in Fig. 32.2, for $SNR = 5$ dB, by means of averaging.

Then the CWT of those realizations were calculated using the Matlab *cwt* function with Complex Morlet mother wavelet. The Scalograms of these CWTs are shown in Fig. 32.3. Notice the events were well detected as local maxima at approximately the same scales but in different positions for each realization, as expected. Then, using the estimator of WCS given by (32.13) for these realizations, the obtained result is shown in Fig. 32.4.

To make things more challenging, two additional pulses with Gaussian shape $s_1(t) = e^{-t^2} \cos t$ and $s_2(t) = e^{-t^2} \cos 12t$ were created and these two pulses were added in each of the former realizations at random positions, creating six new realizations, as shown in Fig. 32.5. Notice that the target transient still keeps the same delay between the realizations x and y, but the other transients do not. Because of the different frequencies, the new transients will appear also in different scales at the plane time-scale. The scalograms of these CWTs are shown in Fig. 32.6 and the estimator of WCS for these new realizations is shown in Fig. 32.7.

Although the events are still detected, one cannot define the delay between transients in x and y processes. Also, the more realizations one has, the more local maxima will appear in this estimator. Thus for a good transient detection with traditional CWA techniques, the events should occur always at same time for each realization and this feature makes this technique inappropriate for an analysis of nonsynchronized transients or if one wants to estimate the delay between two processes.

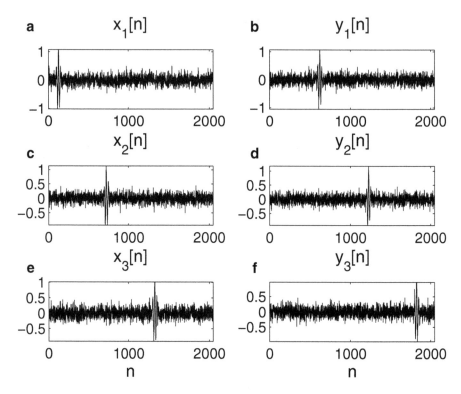

Fig. 32.2 Three realizations of processes x (**a**), (**c**), and (**e**) and y (**b**), (**d**), and (**f**) for $SNR = 5\,\text{dB}$ and delay of 500 samples between events in x and y processes

Now using the new definitions in the detection problem proposed in this work, the correlation estimators of (32.1), (32.2), and (32.3), given by

$$\tilde{R}_{xx}[m] = \frac{1}{K}\sum_{k=1}^{K}\hat{R}_{xx}^{k}[m],\tag{32.24}$$

$$\tilde{R}_{yy}[m] = \frac{1}{K}\sum_{k=1}^{K}\hat{R}_{yy}^{k}[m],\tag{32.25}$$

$$\tilde{R}_{xy}[m] = \frac{1}{K}\sum_{k=1}^{K}\hat{R}_{xy}^{k}[m]\tag{32.26}$$

where $\hat{R}_{xx}^{k}[m]$, $\hat{R}_{yy}^{k}[m]$, and $\hat{R}_{xy}^{k}[m]$ of these signals were determined by Matlab *xcorr* function, were calculated. Then, the CWT of $\tilde{R}_{xx}[m]$, $\tilde{R}_{yy}[m]$, and $\tilde{R}_{xy}[m]$ of the three first realizations (the ones without other pulses randomly added) were calculated, providing the new definitions of WPS and WCS. The result of WCS (the

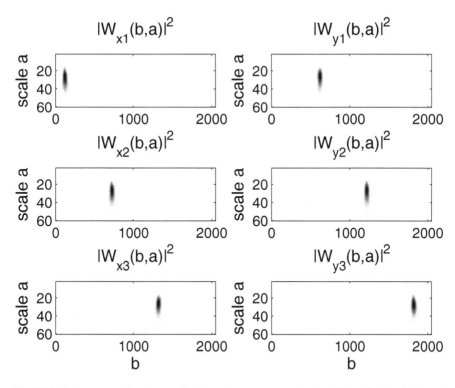

Fig. 32.3 Scalograms of the three realizations of processes x and y for $SNR = 5$ dB and delay of 500 samples between events in x and y realizations

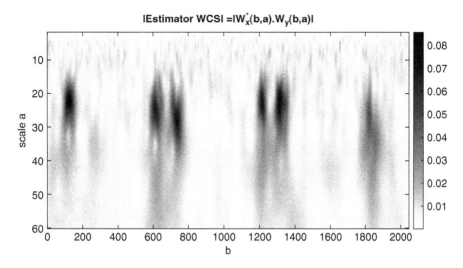

Fig. 32.4 WCS estimator for the three realizations from processes x and y, for $SNR = 5$ dB and delay of 500 samples between events in x and y processes

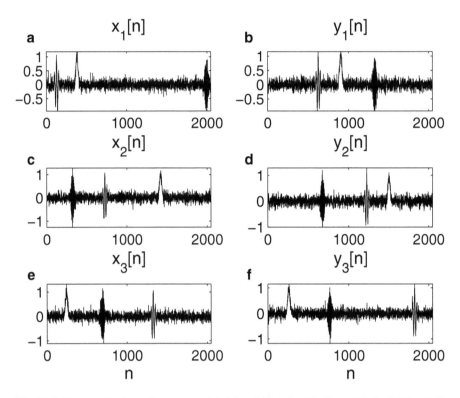

Fig. 32.5 Three realizations of processes x (**a**), (**c**), and (**e**) and y (**b**), (**d**), and (**f**) for $SNR = 5\,\mathrm{dB}$ and delay of 500 samples between events in x and y processes and some spikes randomly added

point of our interest here) is Fig. 32.8. The same procedure was repeated for the last three realizations (the ones with other pulses randomly added) and the WCS obtained is shown in Fig. 32.9.

32.4   Discussion and Conclusions

Comparing the results obtained with the traditional definition of WCS (Figs. 32.4 and 32.7) and the one proposed here (Figs. 32.8 and 32.10) for the detection problem proposed, one can see that the latter provides a significant information about the frequency/scale location of the transients and the delay time between transients in processes x and y. With the increase of realizations the traditional WCS estimator suffers a degradation which is not found in the new WCS estimation, which is perturbed only by a reduction in the SNR. The new estimators of WCO are normalized versions of WCS and the results are still significant in contrast to the traditional version which can be identically unity if no smoothing is applied. Also,

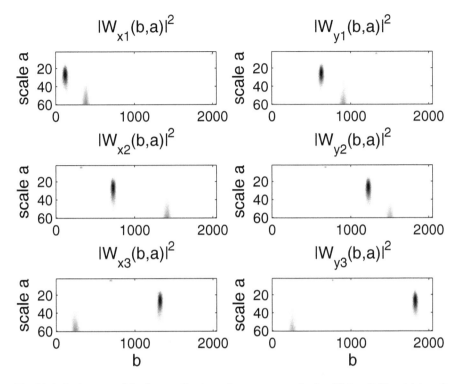

Fig. 32.6 Scalograms of the three realizations of processes x and y for $SNR = 5$ dB and delay of 500 samples between events in x and y realizations with some spikes randomly added

the new technique was found more robust to the presence of other noncorrelated spikes and transients in the realizations (Fig. 32.11).

Concluding, these new definitions proposed in the present contribution work better in a nonsynchronized transient event detection in noise as compared to the traditional definitions. Furthermore the new definitions may be seen as filtered versions of auto and cross-correlation of the signals, all estimator errors and confidence tests developed for correlation functions or spectral density functions [BePi00] can be applied to these new definitions. These ideas will be the focus of future works, where further theoretical considerations as well as experimental issues will be addressed. Finally, by means of CWT-2D use [AnEtAl04], the presented idea can easily be extended to nonstationary processes, where the definitions of correlation functions have two time axes as independent variables. This is also subject of future research.

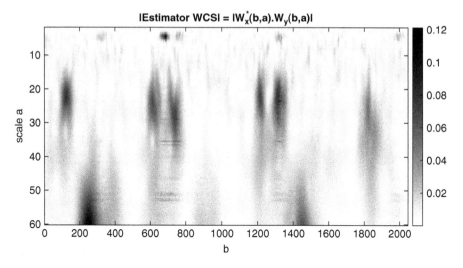

Fig. 32.7 WCS estimator for the three realizations from processes x and y, for $SNR = 5\,\mathrm{dB}$ and delay of 500 samples between events in x and y processes with some spikes randomly added

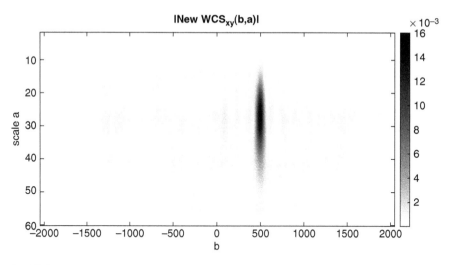

Fig. 32.8 New estimator of $WCS_{xy}(b, a)$, for the three realizations from processes x and y, for $SNR = 5\,\mathrm{dB}$ and delay of 500 samples between events in x and y processes

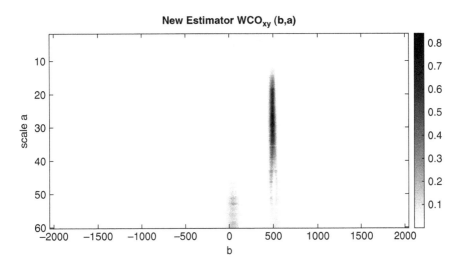

Fig. 32.9 New estimator of WCO_{xy}, for the three realizations from processes x and y, for $SNR = 5\,dB$ and delay of 500 samples between events in x and y processes with some spikes randomly added

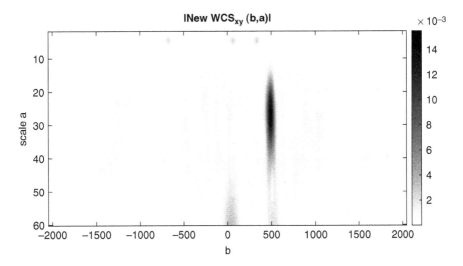

Fig. 32.10 New estimator of $WCS_{xy}(b, a)$, for the three realizations from processes x and y, for $SNR = 5\,dB$ and delay of 500 samples between events in x and y processes with some spikes randomly added

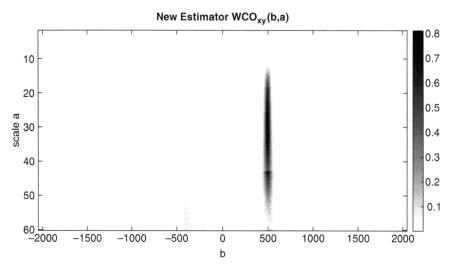

Fig. 32.11 New estimator of WCO_{xy}, for the three realizations from processes x and y, for $SNR = 5$ dB and delay of 500 samples between events in x and y processes

References

[AnEtAl04] Antoine, J-P., Murenzi, R., Vandergheynst, P., and Ali, S.T.: *Two-Dimensional Wavelets and their Relatives*. Cambridge University Press, Cambridge (UK), 2004.

[BePi00] Bendat, J.S., and Piersol, A.G.: *Random Data: analysis and measurement procedures*. John Wiley and Sons, Inc., 3rd edition, 2000.

[Da92] Daubechies, I.: *Ten Lectures on Wavelets*. SIAM, Philadelphia, 1992.

[Nt10] Ntirwihisha, E.-E.: *Corrélation des transformées en ondelettes et ondes sismiques*. PhD thesis, Université Catholique de Louvain, Louvain-la-Neuve, BE, 2010.

[GrEtAl04] Grinsted, A., Moore, J.C., and Jevrejeva, S.: Application of cross wavelet transform and wavelet coherence to geophysical time series. *Nonlinear Processes in Geophysics*, **11**, 561–566 (2004).

[HeEtAl05] Herrmann, C.S., Grigutsch, M., and Bush, N.A.: EEG oscillations and wavelet analysis. In *Event-related potentials: A methods handbook*. L.L. Schumaker (Editor), MIT Press, Cambridge, MA, USA, 2005.

[HuEtAl93] Hudgins, L., Friehe, C.A., and Mayer, M.E.: Wavelet transforms and atmospheric turbulence. *Physical Review Letters*, **71**(20), 3279–3282 (1993).

[KlEtAl06] Klein, A., Sauer, T., Jedynak, A., and Skrandies, W.: Conventional and wavelet coherence applied to sensory-evoked electrical brain activity. *IEEE Transactions on Biomedical Engineering*, **53**(2), 266–271 (2006).

[LaEtAl02] Lachaux, J.-P., Lutz, A., Rudrauf, D., Cosmelli, D., Le Van Quyen, M., Martinerie, J., and Varela, F.: Estimating the time-course of coherence between single-trial brain signals: An introduction to wavelet coherence. *Neurophysiol. Clin.*, **32**, 2002.

[Li94] Liu, P.C.: Wavelet spectrum analysis and ocean wind waves. In *Wavelets in Geophysics*. E. Foufoula and P.Kumar (Editors), Academic Press, San Diego, CA, USA, 1995.

[MaKu04] Maraum, D., and Kurths, J.: Cross wavelet analysis: significance testing and pitfalls. *Nonlinear Processes in Geophysics*, **11**, 505–514 (2004).

[MaEtAl07] Maraum, D., Kurths, J., and Holschneider, M.: Nonstationary gaussian processes in wavelet domain: Synthesis, estimation and significance testing. *Physical Review*, **E75**, 016707–1–016707–14 (2007).

[Pl07] Plett, M.I.: Transient detection with cross wavelet transforms and wavelet coherence. *IEEE Transactions on Signal Processing*, **55**(5), 1605–1611 (2007).

[RiFl92] Rioul, O., and Flandrin, P.: Time-scale energy distributions: A general class extending wavelet transforms. *IEEE Transactions on Signal Processing*, **40**(07), 1746–1757 (1992).

[ScEtAl13] Schuck Jr., A., Balbinot, A., Zabadal, J.R., and Bodmann, B.E.J.: Alternative propositions of cross wavelet analysis for use in a transient detection problem. *International Review of Chemical Engineering I.R.E.C.H.E.*, **05**(6), 1–6 (2013).

[ToCo98] Torrence, C. and Compo, G.P.: A practical guide to wavelet analysis. *Buletin of the American Meteorological Society*, **79**(1), 61–78 (1998).

Chapter 33
A Simple Non-linear Transfer Function for a Wiener-Hammerstein Model to Simulate Guitar Distortion and Overdrive Effects

Adalberto Schuck Jr., Luiz F. Ferreira, Ronaldo Husemann, and Bardo E. J. Bodmann

33.1 Introduction

In music, the holy grail for guitarists, bassist and keyboard players is the signal response by valve (or tube) amplifiers. This preference is justified by the kind of distorted sound they can provide at high volumes, which many musicians appreciate and use to define their sonic identity. This is due to the particular way valves distort the sound when they are overdriven, a clearly particular non-linear behaviour. More precisely, the input–output signals are related by $I(t) = K(v_{in}(t))^{\frac{3}{2}}$. In the beginning of effect pedal developments and in order to reproduce the same kind of distortion caused by overdriven valves, many electronic devices based on diode limiters were designed, the so-called effects of Fuzz, Distortion and Overdrive. More recently, a new philosophy emerged, namely valve amplifiers are modulated using digital signal processing techniques. One of the frequently used models is the Wiener-Hammerstein cell [Og07], an extension of the Volterra-Wiener theory for modeling non-linear systems (NLS) [Sj12, RoEtAl14, ScEtAl14, MoEtAl15, EiZo16, EiEtAl17]. This model and its relation to the real signal amplification is sketched in Fig. 33.1.

For the Static Non-Linear Transfer function (NLTF) block, usually a sigmoid like function is used, such as the hyperbolic tangent (*tanh*) or arc hyperbolic sine (*arc*

A. Schuck Jr. · L. F. Ferreira · R. Husemann
Electrical Engineering Department (DELET), Federal University of Rio Grande do Sul (UFRGS),
Porto Alegre, RS, Brazil
e-mail: rhusemann@inf.ufrgs.br

B. E. J. Bodmann (✉)
Mechanical Engineering Department (DEMEC), Federal University of Rio Grande do Sul
(UFRGS), Porto Alegre, RS, Brazil
e-mail: bardo.bodmann@ufrgs.br

© Springer Nature Switzerland AG 2019
C. Constanda, P. Harris (eds.), *Integral Methods in Science and Engineering*,
https://doi.org/10.1007/978-3-030-16077-7_33

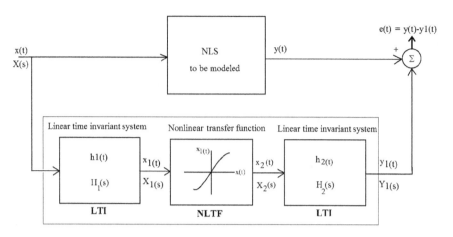

Fig. 33.1 The Wiener-Hammerstein Cell and the modeling diagram

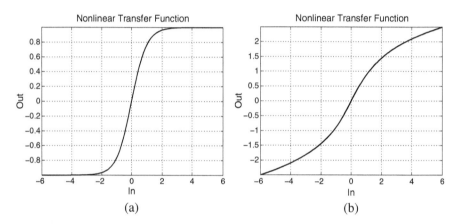

Fig. 33.2 Typical non-linear transfer functions: (**a**) hyperbolic tangent function and (**b**) arc hyperbolic sine function

sinh) functions (shown in Fig. 33.2). Also composite or piecewise functions were proposed by [Pa09, Ye08, MoEtAl15, EiZo16] to describe this block.

All these techniques have their advantages, but also have their drawbacks, like computational complexity among others. Especially, the determination of all the parameters of the entire system with compliance to a high fidelity criterion mainly with composite non-linear transfer functions is a tedious task for commonly used *brute force* adaptive algorithms. Hence, the objectives of this work are lined out in the sequel.

1. Development of a non-linear transfer function,

   • that is adequate to describe hard or soft, symmetrical or asymmetrical clipping limiters;

- that is as simple as possible, i.e., with as few parameters as possible, which are to be determined by some algorithm (in this work a non-linear least square algorithm);
- that is consistent with the physics behind the circuits used to implement these limiters.

2. Identification and evaluation of the non-linear transfer function parameter set by means of non-linear parametric inference with real data.

33.2   The Development of the NLTF

The idea behind the current NLTF developments is that analogical distortion, overdrive and fuzz pedal effects are based on semiconductor devices (diodes and LEDs) that limit the audio signal and depending on the specific choice of the electrical components they produce soft or hard clipping. Figure 33.3 shows an example for a semiconductor limiter circuit.

Considering a generic semiconductor limiter circuit, the Shockley equation for semiconductor PN junctions and the Kirchhoff laws for circuits are the starting point to develop an expression that describes the non-linear behaviour of the limiter circuit. Then, in order to open up for some degrees of freedom that allow to take influence on the shape of the NLTF some of the parameters are generalised and the equation that characterises the transfer function obtained is a generic mathematical expression for which the parameters shall be determined. Our choice so far is a non-linear least squares algorithm, but other parametric inference methods may be used, which might even improve convergence of parameter estimation.

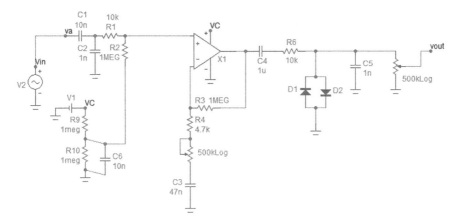

Fig. 33.3 An example for a semiconductor limiter circuit, where the "secret" of the limiter effect is marked in red

Fig. 33.4 Generic limiter
circuit with diodes to define
the NLTF parameterisation

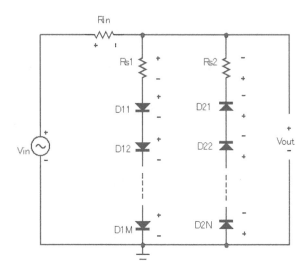

Fig. 33.4 Generic limiter circuit with diodes to define the NLTF parameterisation

The generic limiter circuit with diodes shown in Fig. 33.4 is used to define the non-linear response of the input signal and defines the parametrised form of the aforementioned generic expression to be calibrated by parametric inference.

The principal features of this circuit that define some of its characteristics are different electric resistances R_{s1} and R_{s2} that constrain the asymptotic behaviour of the non-linear transfer function curve at its limits. Further, the number of diodes in either polarisation direction in general can be different, which allows to model symmetric as well as asymmetric limiters.

The standard Kirchhoff equations for this circuits are

$$i_{in} = i_1 - i_2, \tag{33.1}$$

and

$$V_{out} = R_{s1}i_1 + \sum_{k=1}^{M} V_{D_{1k}} \tag{33.2}$$

$$= -\left(R_{s2}i_2 + \sum_{k=1}^{N} V_{D_{2k}} \right). \tag{33.3}$$

The Shockley equation is a well-established equation for PN junctions (one diode) in the literature [MiHa72].

$$i_d = I_s \left\{ \exp\left(\frac{V_d}{\eta V_t} \right) - 1 \right\}, \tag{33.4}$$

Here I_d and V_d are the current and voltage over the diode, I_s is its reverse current, η is a phenomenological constant depending on the semiconductor used and typically in the range of $[1\ldots 2]$, and $V_t = \frac{kT}{q}$ with k the Boltzmann constant, T the temperature and q is the electron charge. V_t is approximately $26\,\text{mV}$ at $25\,^\circ\text{C}$. The voltage across the diode V_d is then

$$V_d = \eta V_t \ln \left(\frac{i_d}{I_s} + 1 \right). \tag{33.5}$$

Due to the functionality of diodes if $i_{in} > 0$, only the diodes D_{1k} conduct and if $i_{in} < 0$ only the diodes D_{2k} contribute to the current. In view of constructing a composed non-linear transfer function, Eq. (33.1) can be decomposed.

$$i_{in} = \begin{cases} i_1 \text{ if } i_{in} > 0 \\ -i_2 \text{ if } i_{in} < 0 \end{cases}. \tag{33.6}$$

Considering that all diodes have equal physical properties and upon substituting Eqs. (33.6) and (33.5) in Eqs. (33.2) and (33.3) yields the voltage of the output signal as a function of the input current.

$$V_{out} = \begin{cases} M\eta V_T \ln \left(\frac{i_{in}}{I_s} + 1 \right) + R_{s1} i_{in} \text{ if } i_{in} \geq 0 \\ -N\eta V_T \ln \left(\frac{-i_{in}}{I_s} + 1 \right) + R_{s2} i_{in} \text{ if } i_{in} < 0 \end{cases}. \tag{33.7}$$

Here $i_{in} = \frac{V_{in} - V_{out}}{R_{in}}$.

Finally, obtaining the generalisation by treating the *constants* I_s, $M\eta V_T$ and $N\eta V_T$, R_{s1} and R_{s2} as independent coefficients (parameters) to be determined from parametric inference the generic parameterisation for the NLTF reads

$$V_{out} = \begin{cases} A \ln \left(\frac{i_{in}}{B} + 1 \right) + C\, i_{in} & \text{if } i_{in} \geq 0 \\ -\left\{ D \ln \left(\frac{-i_{in}}{E} + 1 \right) - F\, i_{in} \right\} & \text{if } i_{in} < 0 \end{cases}, \tag{33.8}$$

where A, B, C and D, E, F are the parameters to be estimated from the non-linear least square fit (33.8) of the experimental data set. From the decomposition it is evident that the positive part ($i_{in} > 0$) of the non-linear transfer function is modeled in a different way than the negative part ($i_{in} < 0$), i.e. allows for asymmetric transfer functions.

33.3   Model Validation

In order to validate the proposed model, six different limiters were built (see Fig. 33.5), its response recorded and the best fit for the transfer function determined. The following components were analysed:

- (1) An asymmetrical limiter with 1N4148 silicon signal diodes;
- (2) a symmetrical limiter with 1N35 germanium diodes;
- (3) a symmetrical limiter with 3mm red LEDs.
- (4) an asymmetrical limiter, composed by two MOSFET transistors and in one branch with 1N4148 silicon signal diodes.

In addition two active circuits with operational amplifiers (OpAmp) were tested because those are used by the most popular commercial overdrive pedals, and whose principles are different from the Shockley equation because they use Operational Amplifiers (OpAmp) as active gain element and the diodes in the negative feedback. In the following we present two configurations.

- (1) Symmetrical limiter with OpAmp and 1N4148 diodes;
- (2) Asymmetrical limiter with OpAmp and 1N4148 diodes.

The related circuits are shown in Fig. 33.6.

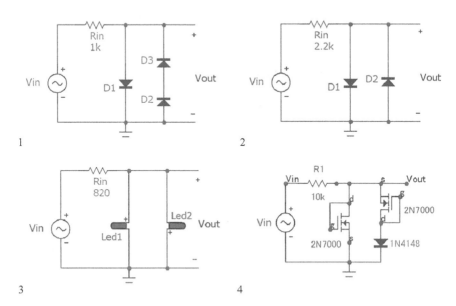

Fig. 33.5 Four different limiters: (**1**) asymmetrical limiter with 1N4148 silicon signal diodes; (**2**) symmetrical limiter with 1N35 germanium diodes; (**3**) symmetrical limiter with 3 mm red LEDs, asymmetrical limiter with two MOSFET transistors and in one branch with 1N4148 silicon signal diodes; (**4**) asymmetrical limiter with 2N7000 MOSFETs and 1N4148 diodes

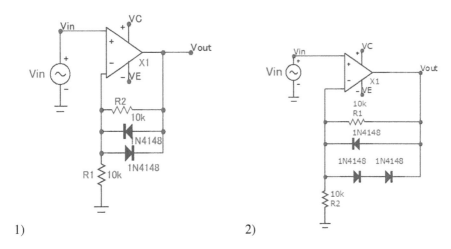

Fig. 33.6 Two active limiters: (**1**) Symmetrical limiter with OpAmp and 1N4148 diodes; (**2**) asymmetrical limiter with OpAmp and 1N4148 diodes

The signal V_{in} used as input for the circuits without OpAmps was a sine wave of 10 Hz and $10V_{pp}$ generated by a BK Precision Function Generator. For the circuits with OpAmp the voltage was changed to $5V_{pp}$ once the circuits have a voltage gain of 2. The resistor R_{in} was chosen to obtain $i_in \approx 10\,mA_{pp}$. About 10 s of V_{in} and V_{out} voltage signals were acquired with an Agilent Infinium 54833D MSO Digital Oscilloscope, with a sample rate of 100 kS/s.

For the non-linear least square fit algorithm (NLLSA) we chose the `'LevembergMarquardt'` method [Ma63] from the Matlab library using the options `'Robust''LAR'` (the Least Absolute Residual method). The fit procedure was applied separately to the positive and negative part of the signals and consequently these were also analysed separately by the aforementioned algorithm. Besides the coefficients (the parameters), the Root Mean Squared error (standard error, RMSE) between the data and the fitted curve was computed.

33.4   Results

In the following we present the results for the six models specified in the previous section. Figure 33.7 shows the obtained results for the best fit curve (V_{out} versus i_{in}) for an asymmetric Silicon limiter, a symmetric Germanium limiter, a symmetric Red LED limiter, an asymmetric MOSFETs plus 1N4148 limiter, a symmetric OpAmp limiter and an asymmetrical OpAmp limiter, respectively. The obtained coefficients by the non-linear fit together with the root mean square errors (RMSE) of the positive and the negative branch of the signal are shown in Table 33.1.

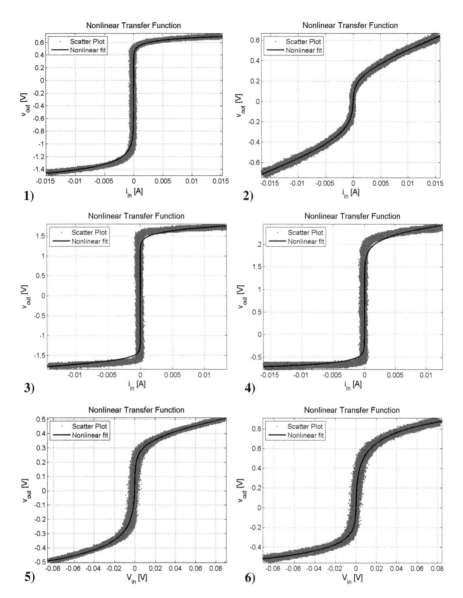

Fig. 33.7 V_{out} vs i_{in} and best fit curve for (**1**) Asymmetric Silicon limiter, (**2**) Symmetric Germanium limiter, (**3**) Symmetric Red LED limiter, (**4**) Asymmetric MOSFETs plus 1N4148 limiter, (**5**) Symmetric OpAmp limiter and (**6**) Asymmetrical OpAmp limiter

Table 33.1 Calculated coefficients and RMSE errors

Coeffs. and RMSE	ckt.1	ckt.2	ckt.3	ckt.4	ckt.5	ckt.6
A	0.06116	0.0573	0.1336	0.166	0.05309	0.1541
B	9.567e−08	1.116e−05	1.861e−08	4.27e−09	3.547e−05	0.0002953
C	0.9863	75.36	2.617e−14	1874	0.9967	0.06554
D	0.1268	0.08274	0.1387	0.08164	0.1028	0.06554
E	8.338e−08	2.449e−05	2.541e−08	1.419e−06	0.000722 E	0.0002684
F	2.337e−14	67.72	2.564e−14	2.428e−14	3.69e−14	0.2893
RMSE(+)	1.6689e−04	3.6879e−05	7.0538e−04	8.7134e−04	3.8811e−05	6.5908e−05
RMSE(−)	3.6735e−04	3.8320e−05	8.5925e−04	2.4383e−04	3.7588e−05	4.5356e−05

33.5 Discussion

Although a non-linear transfer function model was proposed and applied to six limiter circuits the RMSE standard errors were in the range of $[3.69e-05 \ldots 8.59e-04]$ (e.g. from 0.003% to 0.08%). Strictly speaking, the motivation of the parameterisation using the Shockley equation is not justifiable from the physical point of view, due to the fact that in the case of a LED diode, it does not have a PN transition, and in the case of the two active circuits with operational amplifiers the diodes control the amplification behaviour through the feedback loop and thus are not directly related to the output voltage. Nevertheless, all non-linear transfer functions for the six circuits show a fairly good agreement between the experimental findings and its parameterised representation.

Application of the Marquardt-Levenberg algorithm showed a very sensitive parameter dependence on the data. Thus in order to improve convergence it was necessary to provide reasonable *starting points* for the algorithm, which were obtained from the devices' data sheets for the constants of the Shockley equation. Although visually the Red LED circuit got the poorest fit, the RMSE is acceptably low 7.0538e−04 and 8.5925e−04 for the positive and negative branches of the signal, which indicates that the model can also be used for this circuit. As a matter of fact the LED diode has a different construction material (gallium arsenide), whereas diodes are based on germanium or silicium diodes [Sz06], which probably explains why the Shockley equation did work less for the LED in comparison with the other diodes. Since the LED among the considered circuits may be recognised as a *very hard clipping limiter circuit* one may reason that the proposed non-linear transfer function here does only work in limited cases for diodes with different physical principles, such as LED and Schottky diodes. Although active circuits based on operational amplifiers are different as compared to the passive limiter versions, as shown above also for these cases the proposed transfer function parameterisation worked surprisingly well and thus may be used as an operational model.

33.6 Conclusions

In the present work we proposed a "semi-universal" non-linear transfer function to model semiconductor limiter circuits and simulate its response. The parameterisation was based on Kirchhoff's laws together with the Shockley equation for the semiconductors with PN transition. Although three of the circuits may not be related physically to the Shockley equation the derived parameterisation reproduces in all cases with a fairly good fidelity the transfer function. For effect pedal simulations they provide an efficient starting point since from the computational point of view function calls in general work faster than iterative or schemes that use interpolation techniques.

Without further justification of the employed method the coefficients for each model were determined using parametric inference based on the Levenberg-Marquardt algorithm, where the initialisation of the algorithm was provided from an a priori knowledge of the reverse current I_s, the η parameter of the diodes from the diode datasheets. The goodness-of-fit measured by the RMSE was in the range of $[3.69e{-}05\dots 8.59e{-}04]$ (e.g. from 0.003% to 0.08%), where the poorest fit obtained (but still good enough to be used) was for the red LED circuit. Evidently, more efficient inference methods may be used, however we leave this issue for a future investigation.

Note that existing Fuzz, Distortion or Overdrive models used in digital effect pedals in general use numerical algorithms and thus need more computational power to make them work satisfactory in usage. The present proposal resulted in an analytical representation of the non-linear transfer function which can generate directly from a discretised input signal the corresponding output signal according to a desired response of a chosen limiter model. Concluding the authors consider the presented findings as a step into a direction, where analogue effects such as the typical characteristic response of valve amplification may be simulated by a digital implementation of analytical non-linear transfer functions in devices which make use of CPUs.

References

[EiEtAl17] Eixas, F., Möller, S., and Zölzer, U.: Block-oriented gray box modeling of guitar amplifier. In *Proc. of the 20th Int. Conference on Digital Audio Effects (DAFx-17)*, Edinburg, UK (2016), 184–191.

[EiZo16] Eixas, F., and Zölzer, U.: Black-box modeling of distortion circuits with block-oriented models. In *Proc. of the 19th Int. Conference on Digital Audio Effects (DAFx-16)* Brno, Czech Republic (2016) 39–45.

[Ma63] Marquandt, D.W.: An algorithm for least-squares estimation of nonlinear parameters. *J. Soc. for Industrial and Applied mathematics*, **11**(2), 431–441 (1963).

[MiHa72] Millman, J., and Halkias, C.C.: *Integrated Electronics: Analog and digital circuits and systems*. McGraw-Hill,Inc. (1972).

[MoEtAl15] Möller, S., Eichas, F., and Zölzer, U.: Block-oriented modeling of distortion audio effects using iterative minimization. In *Proc. of the 18th Int. Conference on Digital Audio Effects (DAFx-15)*, Trondheim, Norway (2015), 1–6.

[Og07] Ogunfunmi, T.: *Adaptive Nonlinear System Identification: The Volterra and Wiener Model Approaches*. Springer Science and Business, LCC (2007).

[Pa09] Pakarinen, J., and Yeh, D.T.: A review of digital techniques for modeling vacuum-tube guitar amplifiers. *Computer Music Journal*, **33**(2), 85–100 (2009).

[RoEtAl14] Rolain, Y., Schoukens, M., Vandersteen, G., and Ferranti, F.: Fast identification of Wiener-Hammerstein systems using discrete optimisation. *Electronic letters*, **50**(25), 1942–1944 (2014).

[ScEtAl14] Schoukens, M., Zhang, E., and Schoukens, J.: Structure detection of wiener-hammerstein systems with process noise. *IEEE Trans. on Instrumentation and Measurement*, **66**(3), 569–576 (2014).

[Sj12] Sjöberg, J., and Schoukens, J.: Initializing Wiener-Hammerstein models based on partitioning of the best linear approximation. *Automatica*, **48**(1), 353–359 (2012).

[Sz06] Sze, S.M., and Kwok, K.N. *Physics of semiconductor devices*. John Wiley & Sons, Ltd., third edition (2006).

[Ye08] Yeh, D.T., Abel, J.S., Vladimirescu, A., and Smith, J.O.: Numerical methods for simulation of guitar distortion circuits. *Computer Music Journal*, **32**(2), 23–42 (2008).

Chapter 34
Existence of Nonlinear Problems: An Applicative and Computational Approach

Aditya Singh, Mudasir Younis, and Deepak Singh

34.1 Introduction

In recent times, many results introduced, related to metric fixed point theory endowed with a partial order. An initiated result in this track was provided by Ran and Reurings [RaEtAl04], where they presented a fixed point result, which can be considered as a junction of two fixed point theorems: Banach contraction principle and Knaster-Tarski fixed point theorem. Moreover, the result achieved in [RaEtAl04] was extended and generalized by many researchers, some of which are in [Ag08], [Al10], [La09].

Recently, Wardowski [Wa12] introduced the notion of F-contraction. This kind of contractions generalizes the Banach contraction. Newly, Piri and Kumam [Pi14] enhanced the results of Wardowski [Wa12] by initiating the idea of an F-Suzuki contraction and obtained some interesting fixed point results.

In this article, we introduce the concept of (F, ψ)-rational type contraction in the setup of metric space and examine the existence of fixed points for such type of contraction. Some examples and applications are given to illustrate the realized improvement.

A. Singh (✉)
Department of Civil Engineering, Indian Institute of Technology (IIT) Indore, Indore, M.P., India
e-mail: ce160004003@iiti.ac.in

M. Younis
Department of Applied Mathematics, UIT-Rajiv Gandhi Technological University (State Technological University of M.P.), Bhopal, M.P., India

D. Singh
Department of Applied Sciences, NITTTR, Under Ministry of HRD, Government of India, Bhopal, M.P., India

© Springer Nature Switzerland AG 2019
C. Constanda, P. Harris (eds.), *Integral Methods in Science and Engineering*,
https://doi.org/10.1007/978-3-030-16077-7_34

34.2 Preliminaries

Throughout the article, we denote by $\mathbb{R}$ the set of all real numbers, by $\mathbb{R}^+$ the set of all positive real numbers and by $\mathbb{N}$ the set of all positive integers.

Definition 1 Let X be a nonempty set and let $\preceq$ be a binary relation on X. We say that $\preceq$ is a partial order on X if the following conditions are satisfied:

(i) For every $x \in X$, we have $x \preceq x$.
(ii) For every $x, y, z \in X$, we have $x \preceq y$, $y \preceq x \Longrightarrow x = y$.
(iii) For every $x, y, z \in X$, we have $x \preceq y$, $y \preceq z \Longrightarrow x \preceq z$.

Definition 2 ([Jl16]) Let (X, d) be a metric space and $\preceq$ be a partial order on X. we say that the partial order $\preceq$ is d-regular if the following condition is satisfied:
For every sequences $\{a_n\}, \{b_n\} \subset X$, we have

$$\lim_{n\to\infty} d(a_n, a) = \lim_{n\to\infty} d(b_n, b) = 0, \quad a_n \preceq b_n, \quad \text{for all } n \quad \Longrightarrow a \preceq b,$$

where $(a, b) \in X \times X$.

Definition 3 ([Wa12, Pi14, Se13]) Let us denote by Δ_F the set of all functions $F : \mathbb{R}^+ \to \mathbb{R}$ satisfying the following conditions:

$(\Delta_{(F1)})$. F is strictly increasing, that is, for $\alpha, \beta \in \mathbb{R}^+$ such that $\alpha < \beta$ implies $F(\alpha) < F(\beta)$;
$(\Delta_{(F2)})$.   There is a sequence $\{\alpha_n\}_{n=1}^{\infty}$ of positive real numbers such that $\lim_{n\to\infty} F(\alpha_n) = -\infty$;
$(\Delta_{(F3)})$. F is continuous on $(0, \infty)$.

Take $\Psi = \{\psi : [0, \infty) \to [0, \infty) : \psi \text{ is upper semi continuous and non-decreasing with } \psi(t) < t \text{ for each } t > 0\}$.

34.3 Fixed Point Problem Under Constraint Inequality for (F, ψ)-Rational Type Contraction

In this section, the following problem has been discussed: Find $x \in X$ such that

$$\begin{cases} x = Tx, \\ Ax = Bx. \end{cases} \tag{34.1}$$

where $T, A, B : X \to X$ are given operators and (X, d) is a metric space with a partial order $\preceq$. Now, we introduce the following definition:

Definition 4 Let (X, d) be a complete metric space including a partial order $\preceq$. Let $T, A, B : X \to X$ are given operators. We say that T is (F, ψ)-rational type contraction on a metric space X, if there exist $F \in \Delta_F$, $\tau > 0$ and $\psi \in \Psi$ such that for all $x, y \in X$ with $Tx \neq Ty$, we have

$$Ax \preceq Bx, \quad By \preceq Ay \implies F\Big(d(Tx, Ty)\Big) \leq F\Big(\psi\big(M(x, y)\big)\Big) - \tau \qquad (34.2)$$

where

$$M(x, y) = \max \left\{ d(x, y), \frac{d(y, Ty)[1 + d(x, Tx)]}{1 + d(x, y)}, \frac{d(y, Tx)[1 + d(x, Ty)]}{1 + d(x, y)} \right\},$$
$$(34.3)$$

Our main result runs as follows.

Theorem 1 *Let (X, d) be a complete metric space including a partial orders $\preceq$. Let $T, A, B : X \to X$ are given operators. Assume that the following assumptions are true, for the problem (34.1):*

1. *$\preceq$ is d-regular;*
2. *T, A, B are continuous;*
3. *there exists $x_0 \in X$ such that $Ax_0 \preceq Bx_0$;*
4. *for all $x \in X$, we have $Ax \preceq Bx \implies BTx \preceq ATx$;*
5. *for all $x \in X$, we have $Bx \preceq Ax \implies ATx \preceq BTx$;*
6. *if there exist $F \in \Delta_F$, $\tau > 0$ and $\psi \in \Psi$ such that for all $x, y \in X$ with $Tx \neq Ty$, we have*

$$Ax \preceq Bx, \quad By \preceq Ay \implies F\Big(d(Tx, Ty)\Big) \leq F\Big(\psi\big(M(x, y)\big)\Big) - \tau \qquad (34.4)$$

where $M(x, y)$ is defined as in (34.3).

Then

(i) The sequence $\{x_n\}$ converges to some $u \in X$ such that $Au = Bu$.
(ii) The point $u \in X$ is a solution of the problem (34.1).

Proof To prove (i), it follows immediately from assumption (3) that there exists a point $x_0 \in X$ such that $Ax_0 \preceq Bx_0$. We construct a sequence $\{x_n\}$ in the following way:

$$x_{n+1} = Tx_n \quad \text{for all} \ \ n \in \mathbb{N} \cup \{0\}.$$

From assumption (4), we have $Ax_0 \preceq Bx_0 \implies BTx_0 \preceq ATx_0 = Bx_1 \preceq Ax_1$.

From assumption (5), we have

$$Bx_1 \preceq Ax_1 \implies ATx_1 \preceq BTx_1 = Ax_2 \preceq Bx_2.$$

By repeating the process, we derive

$$Ax_{2n} \preceq Bx_{2n} \text{ and } Bx_{2n+1} \preceq Ax_{2n+1}, \quad n \in \mathbb{N} \cup \{0\}. \tag{34.5}$$

If there exists $n \in \mathbb{N} \cup \{0\}$ such that $x_{n_0+1} = x_{n_0}$, then x_{n_0} is the desired solution of the problem (34.1), which completes the proof.

Consequently, from now on, suppose that $x_{n+1} \neq x_n$ for all $n \in \mathbb{N} \cup \{0\}$. In view of assumption (6), it establishes that

$$F\Big(d(x_n, x_{n+1})\Big) = F\Big(d(Tx_{n-1}, Tx_n)\Big) \leq F\Big(\psi\Big(M(x_{n-1}, x_n)\Big)\Big) - \tau \tag{34.6}$$

where

$$M(x_{n-1}, x_n)$$

$$= \max\left\{ d(x_{n-1}, x_n), \frac{d(x_n, x_{n+1})[1 + d(x_{n-1}, x_n)]}{1 + d(x_{n-1}, x_n)}, \frac{d(x_n, x_n)[1 + d(x_{n-1}, x_{n+1})]}{1 + d(x_{n-1}, x_n)} \right\}$$

$$= \max\left\{ d(x_{n-1}, x_n), d(x_n, x_{n+1}) \right\}.$$

If for some n, $M(x_{n-1}, x_n) = d(x_n, x_{n+1})$, then by using the definition of function ψ, inequality (34.6), turns into the following

$$F\Big(d(x_n, x_{n+1})\Big) \leq F\Big(\psi\Big(d(x_n, x_{n+1})\Big)\Big) - \tau$$

This implies

$$F\Big(d(x_n, x_{n+1})\Big) < F\Big(d(x_n, x_{n+1})\Big),$$

which leads to a contradiction. Accordingly we deduce that $M(x_{n-1}, x_n) = d(x_{n-1}, x_n)$, for any n. Therefore for any n, by repeating the same technique as mentioned above, we speculate that

$$F\Big(d(x_n, x_{n+1})\Big) \leq F\Big(\psi\Big(d(x_{n-1}, x_n)\Big)\Big) - \tau$$
$$< F\Big(d(x_{n-1}, x_n)\Big) \tag{34.7}$$

Since F is strictly increasing, this follows

$$d(x_n, x_{n+1}) < d(x_{n-1}, x_n).$$

It shows that $\{d(x_n, x_{n+1})\}$ is a decreasing sequence of positive real numbers. Taking (34.7) into account, we acquire

$$F\Big(d(x_n, x_{n+1})\Big) \leq F\Big(\psi\Big(d(x_{n-1}, x_n)\Big)\Big) - \tau.$$

In view of the fact that $\psi(t) < t$, for all $t > 0$

$$F\Big(d(x_n, x_{n+1})\Big) < F\Big(d(x_{n-1}, x_n)\Big) - \tau. \quad \text{for } n \in \mathbb{N}. \tag{34.8}$$

Note that, by the repeated use of (34.8), it establishes that

$$F\Big(d(x_n, x_{n+1})\Big) < F\Big(d(x_{n-1}, x_n)\Big) - \tau$$
$$< F\Big(d(x_{n-2}, x_{n-1})\Big) - 2\tau$$
$$\dots$$
$$< F\Big(d(x_0, x_1)\Big) - n\tau.$$

which implies that

$$F\Big(d(x_n, x_{n+1})\Big) < F\Big(d(x_0, x_1)\Big) - n\tau.$$

Since $F \in \Delta_F$, letting the limit as $n \to \infty$, then the above inequality turns into

$$\lim_{n \to \infty} F\Big(d(x_n, x_{n+1})\Big) = -\infty \iff \lim_{n \to \infty} d(x_n, x_{n+1}) = 0. \tag{34.9}$$

Next, we prove that $\{x_n\}$ is a Cauchy sequence in (X, d). We argue it by contradiction. Assume that $\{x_n\}$ is not a Cauchy sequence. In this case, there exist $\epsilon > 0$ and two sub-sequences $\{x_{n(k)}\}$ and $\{x_{m(k)}\}$ of $\{x_n\}$ such that for all positive integer k with $n(k) > m(k) > k$, we have

$$d(x_{m(k)}, x_{n(k)}) \geq \epsilon. \tag{34.10}$$

which gives

$$d(x_{m(k)}, x_{n(k)-1}) < \epsilon.$$

Now, inequality (34.10) turns into

$$\epsilon \leq d(x_{m(k)}, x_{n(k)-1}) + d(x_{n(k)-1}, x_{n(k)})$$
$$\leq \epsilon + d(x_{n(k)-1}, x_{n(k)}).$$

By taking the limit as $k \to \infty$ in above inequality and using (34.9), we obtain

$$\lim_{k \to \infty} d(x_{m(k)}, x_{n(k)}) = \epsilon. \tag{34.11}$$

Further, from (34.11), it is easy to see that

$$\lim_{k \to \infty} d(x_{m(k)-1}, x_{n(k)-1}) = \epsilon. \tag{34.12}$$

and

$$\lim_{k \to \infty} d(x_{n(k)-1}, x_{m(k)}) = \epsilon. \tag{34.13}$$

Sequentially, from (34.9), there exists a natural number $K \in \mathbb{N}$ such that for all $k \geq K$, we have

$$d(x_{m(k)}, x_{m(k)+1}) < \frac{\epsilon}{4} \ \ and \ \ d(x_{n(k)}, x_{n(k)+1}) < \frac{\epsilon}{4}. \tag{34.14}$$

Next, we will show that

$$d(T x_{m(k)}, T x_{n(k)}) = d(x_{m(k)+1}, x_{n(k)+1}) > 0, \tag{34.15}$$

for all $k \geq K$, reasoning by contradiction. Assume that there exists $r \geq K$, such that

$$d(x_{m(r)+1}, x_{n(r)+1}) = 0. \tag{34.16}$$

On account of (34.10), (34.14) and (34.16), we arrive at

$$\epsilon \leq d(x_{m(r)}, x_{n(r)}) \leq d(x_{m(r)}, x_{n(r)+1}) + d(x_{n(r)+1}, x_{n(r)})$$
$$\leq d(x_{m(r)}, x_{m(r)+1}) + d(x_{m(r)+1}, x_{n(r)+1}) + d(x_{n(r)+1}, x_{n(r)})$$
$$\epsilon < \frac{\epsilon}{4} + 0 + \frac{\epsilon}{4} = \frac{\epsilon}{2},$$

which is impossible, which means that (34.15) is proved. From (34.2), we have

$$F\left(d(x_{m(k)+1}, x_{n(k)+1})\right) = F\left(d(T x_{m(k)}, T x_{n(k)})\right)$$
$$\leq F\left(\psi\left(M(x_{m(k)}, x_{n(k)})\right)\right) - \tau \tag{34.17}$$

in which

$$M(x_{m(k)}, x_{n(k)}) = \max \left\{ d(x_{m(k)}, x_{n(k)}), \frac{d(x_{n(k)}, x_{n(k)+1})[1 + d(x_{m(k)}, x_{m(k)+1})]}{1 + d(x_{m(k)}, x_{n(k)})}, \right.$$

$$\left. \frac{d(x_{n(k)}, x_{m(k)+1})[1 + d(x_{m(k)}, x_{n(k)+1})]}{1 + d(x_{m(k)}, x_{n(k)})} \right\}$$

Letting $k \to \infty$ and using (34.9), (34.11), (34.12) and (34.13) then the above inequality deduces to

$$\lim_{k \to \infty} M(x_{m(k)}, x_{n(k)}) = \epsilon. \tag{34.18}$$

Making the limit as $k \to \infty$ in (34.17) and using (34.11), (34.18) and upper semi-continuity of ψ, we get

$$F(\epsilon) \leq F(\psi(\epsilon)) - \tau$$

$$< F(\epsilon) - \tau,$$

which is impossible, since $\tau > 0$. Through contradiction, we conclude that $\{x_n\}$ is a Cauchy sequence in a complete metric space X. Completeness of X assures that there exists $u \in X$ such that

$$\lim_{n \to \infty} x_n = u. \tag{34.19}$$

Consequently, from (34.5), we get

$$Ax_{2n} \preceq Bx_{2n} \quad n \in \mathbb{N} \cup \{0\}.$$

Due to continuity of A and B, from (34.19), we obtain that

$$\lim_{n \to \infty} d(Ax_{2n}, Au) = \lim_{n \to \infty} d(Bx_{2n}, Bu) = 0.$$

As $\preceq$ is d-regular, we get

$$Au \preceq Bu.$$

By repeating the same technique as mentioned above, one can get

$$Bu \preceq Au.$$

The proof of (i) is completed. Moreover, by (34.19) the continuity of T asserts that

$$d(Tu, u) = \lim_{n \to \infty} d(Tx_n, x_n) = \lim_{n \to \infty} (x_{n+1}, x_n) = 0 \Longrightarrow Tu = u.$$

Hence, we conclude that $u \in X$ is a solution of the problem (34.1).
    This completes the proof.

34.4 Some Consequences

In this section, some consequences of Theorem 1 are presented.

34.4.1 Common Fixed Point Problem Under One Constraint Equality for (F, ψ)-Rational Type Contraction

Here, the following problem has been considered: Find $x \in X$ such that

$$\begin{cases} x = Tx, \\ x = Bx. \end{cases} \tag{34.20}$$

where $T, B : X \to X$ are given operators and (X, d) is a metric space with a partial order $\preceq$.

Note that problem (34.1) reduces to problem (34.20) by taking $A = I_X$. So, if we take $A = I_X$ in Corollary 1, then we get the following corollary.

Corollary 1 *Let (X, d) be a complete metric space including a partial orders $\preceq$. Let $T, B : X \to X$ are given operators. Assume that the following assumptions are true, for the problem (34.20):*

1. $\preceq$ *is d-regular;*
2. T, B *are continuous;*
3. *there exists $x_0 \in X$ such that $x_0 \preceq Bx_0$;*
4. *for all $x \in X$, we have $x \preceq Bx \Longrightarrow BTx \preceq Tx$;*
5. *for all $x \in X$, we have $Bx \preceq x \Longrightarrow Tx \preceq BTx$;*
6. *if there exist $F \in \Delta_F$, $\tau > 0$ and $\psi \in \Psi$ such that for all $x, y \in X$ with $Tx \neq Ty$, we have*

$$x \preceq Bx, \quad By \preceq y \Longrightarrow F\Big(d(Tx, Ty)\Big) \leq F\Big(\psi\big(M(x, y)\big)\Big) - \tau \tag{34.21}$$

where $M(x, y)$ is defined as in (34.3).

Then

(i) *The sequence $\{x_n\}$ converges to some $u \in X$ such that $u = Bu$.*
(ii) *The point $u \in X$ is a solution of the problem (34.20).*

The following example validates our result, obtained in Corollary 1.

Example 1 Let $X = \{(0,0), (1,0), (3,0), (5,1)\}$ be a subset of $\mathbb{R}^2$ with the order $\preceq$ defined as

$$(x_1, y_1) \preceq (x_2, y_2) \iff x_1 \leq x_2, \; y_1 \leq y_2, \quad (x_1, y_1), (x_2, y_2) \in X.$$

Let $d : X \times X \to [0, \infty)$ be given by:
$$d(x, y) = \max\{|x_1 - x_2|, |y_1 - y_2|\}, \quad for \; x = (x_1, y_1), \; y = (x_2, y_2) \in X.$$
Let $T, B : X \to X$ be defined as follows:

(x, y)	$T(x, y)$	$B(x, y)$
$(0,0)$	$(1,0)$	$(3,0)$
$(1,0)$	$(1,0)$	$(1,0)$
$(3,0)$	$(0,0)$	$(1,0)$
$(5,1)$	$(0,0)$	$(0,0)$

Take $\psi : [0, \infty) \to [0, \infty)$ by $\psi(t) = \frac{122t}{123}$ and $F(t) = \log t$. All the conditions of Corollary 1 are satisfied and $u = (1, 0)$ is a common fixed point of the mappings T and B.

By using $B = T$ in Corollary 1, we get the following one:

Corollary 2 *Let (X, d) be a complete metric space including a partial orders $\preceq$. Let $T : X \to X$ be a given operators. Assume that the following assumptions are true:*

1. *$\preceq$ is d-regular;*
2. *T is continuous;*
3. *there exists $x_0 \in X$ such that $x_0 \preceq T x_0$;*
4. *for all $x \in X$, we have $x \preceq Tx \implies T^2 x \preceq Tx$;*
5. *for all $x \in X$, we have $Tx \preceq x \implies Tx \preceq T^2 x$;*
6. *if there exist $F \in \Delta_F$, $\tau > 0$ and $\psi \in \Psi$ such that for all $x, y \in X$ with $Tx \neq Ty$, we have*

$$x \preceq Tx, \; Ty \preceq y \implies F\Big(d(Tx, Ty)\Big) \leq F\Big(\psi\big(M(x, y)\big)\Big) - \tau \qquad (34.22)$$

where $M(x, y)$ is defined as in (34.3).

Then, the sequence $\{x_n\}$ converges to a fixed point of T.

34.5 Application to Integral Equation

Consider the following integral equation:

$$u(t) = p(t) + \int_0^{\Omega} \lambda(t, s) f(s, u(s)) ds. \tag{34.23}$$

Consider the space $X = C([0, \Omega], \mathbb{R})$ of continuous functions defined on $[0, \Omega]$. Obviously, the space with the metric given by

$$d(u, v) = \sup_{t \in [0, \Omega]} |u(t) - v(t)|, \quad u, v \in C([0, \Omega], \mathbb{R})$$

is a complete metric space. Consider on $X = C([0, \Omega], \mathbb{R})$ the natural partial order relation, that is,

$$u, v \in C([0, \Omega], \mathbb{R}), \quad u \leq v \iff u(t) \leq v(t), \quad t \in [0, \Omega].$$

Theorem 2 *Consider the problem (34.23) and assume that the following conditions are satisfied:*

(i) $f : [0, \Omega] \times \mathbb{R} \to \mathbb{R}$ is continuous;
(ii) $p : [0, \Omega] \to \mathbb{R}$ is continuous;
(iii) $\lambda : [0, \Omega] \times \mathbb{R} \to [0, \infty)$ is continuous;
(iv) $\psi \in \Psi$ such that for all $u, v \in \mathbb{R}$, $u \leq v$,

$$f(s, u) - f(s, v) \geq 0 \text{ and } |f(s, u) - f(s, v)| \leq e^{-\tau} \psi(|v - u|);$$

(v) assume that

$$\sup_{t \in [0, \Omega]} \int_0^{\Omega} \lambda(t, s) ds \leq 1;$$

(vi) there exists a $x_0(t) \in X$ with $(X = C([0, \Omega], \mathbb{R}))$ such that

$$x_0(t) \leq p(t) + \int_0^{\Omega} \lambda(t, s) f(s, x_0(s)) ds.$$

Then the integral equation (34.23) has a solution in X with $(X = C([0, \Omega], \mathbb{R}))$.

Proof Consider the mapping $T : X \to X$ defined by

$$Tu(t) = p(t) + \int_0^{\Omega} \lambda(t, s) f(s, u(s)) ds,$$

for all $u \in X$ and $t \in [0, \Omega]$. We prove that all the conditions of Corollary 2 are satisfied. Clearly, $\preceq$ is d-regular and by the condition (iv) of Theorem 2, for $x(t) \in X$ with $x(t) \leq Tx(t)$, $t \in [0, \Omega]$, we have

$$Tx(t) - T^2x(t) = \int_0^\Omega \lambda(t, s)f(s, x(s))ds - \int_0^\Omega \lambda(t, s)f(s, Tx(s))ds$$

$$= \int_0^\Omega \lambda(t, s)(f(s, x(s)) - f(s, Tx(s)))ds \geq 0.$$

which yields that $Tx(t) \geq T^2x(t)$, for all $x(t) \in X$. Similarly, one can show that for all $x(t) \in X$, with $Tx(t) \leq x \implies Tx(t) \leq T^2x(t)$.

Now, for $u, v \in X$ with $u \leq v$, we obtain

$$Tu(t) - Tv(t) = \int_0^\Omega \lambda(t, s)f(s, u(s))ds - \int_0^\Omega \lambda(t, s)f(s, v(s))ds$$

$$= \int_0^\Omega \lambda(t, s)(f(s, u(s)) - f(s, v(s)))ds$$

$$\leq e^{-\tau} \int_0^\Omega \lambda(t, s)\psi(|v(s) - u(s)|)ds$$

As ψ is nondecreasing function, we have

$$\psi(|v(s) - u(s)|) \leq \psi\left(\sup_{s \in [0,\Omega]} |u(s) - v(s)|\right)$$

$$= \psi\left(d(u, v)\right).$$

Hence, from the above inequality, we arrive at

$$\sup_{t \in [0,\Omega]} |Tu(t) - Tv(t)| \leq e^{-\tau}\psi\left(d(u, v)\right) \sup_{t \in [0,\Omega]} \int_0^\Omega \lambda(t, s)ds$$

$$d\left(Tu, Tv\right) \leq \psi\left(d(u, v)\right)e^{-\tau}$$

$$d\left(Tu, Tv\right) \leq \psi\left(M(u, v)\right)e^{-\tau}.$$

in which

$$M(u, v) = \max\left\{d(u, v), \frac{d(v, Tv)[1 + d(u, Tu)]}{1 + d(u, v)}, \frac{d(v, Tu)[1 + d(u, Tv)]}{1 + d(u, v)}\right\}$$

Consequently, by passing to logarithms, one can obtain

$$\log d\left(Tu, Tv\right) \leq \log \psi\left(d(u, v)\right) - \tau.$$

This turns up to

$$F\left(d\left(Tu, Tv\right)\right) \leq F\left(\psi\left(d(u, v)\right)\right) - \tau.$$

This shows that the contractive condition in Corollary 2 is satisfied.

From condition *(vi)* of Theorem 2, we have $x_0 \leq T x_0$.

As a result of Theorem 1, T has a fixed point in X, that is, the integral equation has a solution.

The following example illustrates Theorem 2.

Example 2 Consider the following integral equation in $X = C([0, 1], \mathbb{R})$.

$$u(t) = \frac{t^2 + 1}{t^3 + 0.1} + \frac{1}{3} \int_0^1 \frac{s^2}{(t+1)} \frac{1}{(1 + u(s))} ds; \quad t \in [0, 1]. \tag{34.24}$$

Observe that this equation is a special case of (34.23), in which

$p(t) = \frac{t^2+1}{t^3+0.1}$;
$\lambda(t, s) = \frac{s^2}{(t+1)}$;
$f(s, t) = \frac{1}{3(1+t)}$.

Indeed, the functions p, λ and f are continuous. Thus the assumptions *(i)–(iii)* are satisfied. Further, for all $u, v \in \mathbb{R}$ with $u \leq v$, we get

$$0 \leq |f(s, u) - f(s, v)| \leq \left|\frac{1}{3(1 + u)} - \frac{1}{3(1 + v)}\right| \leq \frac{1}{3}|v - u|$$

$$\leq e^{-0.1}\frac{2}{3}(|v - u|)$$

$$\leq e^{-\tau}\psi(|v - u|)$$

for $\tau = 0.1$ and $\psi(t) = \frac{2t}{3}$. Hence, condition *(iv)* of Theorem 2 is fulfilled. For condition *(v)*, we have

$$\sup_{t\in[0,1]} \int_0^1 \lambda(t, s)ds = \sup_{t\in[0,1]} \int_0^1 \frac{s^2}{(t+1)} ds = \sup_{t\in[0,1]} \frac{1}{3(t+1)} \leq 1.$$

Thus, condition *(v)* is proved. Consider $x_0(t) = 1$, then we arrive at

$$p(t) + \int_0^1 \lambda(t,s) f(s, x_0(s)) ds = \frac{t^2 + 1}{t^3 + 0.1} + \int_0^1 \frac{s^2}{(t+1)} f(s, 1) ds$$

$$= \frac{t^2 + 1}{t^3 + 0.1} + \frac{1}{6} \int_0^1 \frac{s^2}{(t+1)} ds = \frac{t^2 + 1}{t^3 + 0.1} + \frac{1}{18(t+1)}$$

$$> 1$$

$$= x_0(t),$$

for all, $t \in [0, 1]$. This shows that all the conditions of Theorem 2 are satisfied. Hence, the integral equation (34.24) has a solution in $X = C([0, 1], \mathbb{R})$. Further, the approximate solution of the integral equation (34.24) is

$$u(t) = \frac{1.02733\, t^3 + t^2 + t + 1.002733}{(t^3 + 0.1)(t + 1)}. \tag{34.25}$$

For the justification of the approximate solution, from (34.24) with (34.25), we arrive at

$$u(t) = \frac{t^2 + 1}{t^3 + 0.1}$$

$$+ \frac{1}{3(t+1)} \int_0^1 \frac{s^2(s+1)(s^3 + 0.1)\, ds}{s^4 + 2.02733\, s^3 + s^2 + 1.1s + 1.102733}; \quad t \in [0, 1]. \tag{34.26}$$

From Fig. 34.1a and b, one can easily deduce that the plot of approximate solution with purple surface almost coincides with the value of $u(t)$ with dark blue surface (see Fig. 34.1b). Hence, Fig. 34.1a and b confirms the validity of the approximate solution.

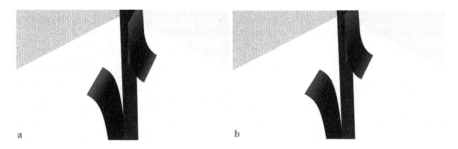

a   b

Fig. 34.1 (a) Approximate solution of (34.24) and (b) plot of inequality (34.26)

Fig. 34.2 Plot of error
function

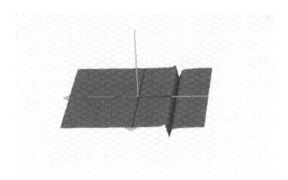

The error between the approximate solution and the value of $u(t)$ is given in Fig. 34.2.

References

[Ag08] Agarwal, R.P., EI-Gebeily, M.A. and OŔegan, D. : Generalized contractions in partially ordered metric spaces. *Appl. Anal.*, **87**, 1–8(2008).

[Al10] Altun, I. and Simsek, H. : Some fixed point theorems on ordered metric spaces and application. *Fixed Point Theory Appl.*, **2010**, Article ID 621492(2010).

[Jl16] Jleli, M. and Samet, B. : A fixed point problem under two constraint inequalities. *Fixed Point Theory Appl.*,**2016**, 18(2016).

[La09] Lakshmikantham, V. and Ciric, L. : Coupled foxed point theorems for nonlinear contractions in partially ordered metric spaces. *Nonlinear Anal.,* **70** 136, 4341–4349(2009).

[Pi14] Piri, H. and Kumam, P. : Some fixed point theorems concerning F-contraction in complete metric spaces. *Fixed Point Theory and Appl.*,**2014**, 210(2014).

[RaEtAl04] Ran, A.C.M. and Reurings, M.C.B. : A fixed point theorem in partially ordered sets and some applications to matrix equations. *Proc. Am. Math. Soc.*, **132**, 1435–1443(2004).

[Se13] Secelean, N.A. : Iterated function system consisting of F-contractions. *Fixed Point Theory and Appl.,* **2013**, https://doi.org/10.1166/1687-1812-2013-277, 277(2013).

[Wa12] Wardowski, D. : Fixed points of a new type of contractive mappings in complete metric spaces. *Fixed Point Theory and Appl.*, **2012**, 94(2012).

Chapter 35
Solving Existence Problems via F-Reich Contraction

Mudasir Younis, Deepak Singh, and Anil Goyal

35.1 Introduction and Basic Facts

Czerwik [Cz93] introduced b-metric spaces and established Banach contraction principle [Ba22] in these spaces. The notation of rectangular metric spaces were proposed by Branciari [Br00] in a different setting. Afterwards, in 2015, George et al. [GeEtAl15] launched rectangular b-metric spaces (in short $RbMS$) which are not necessarily Hausdorff and claimed these spaces to be generalization of other spaces. Acknowledging the concept of George et al. [GeEtAl15], many authors paid attention towards these spaces and published many research articles. For a further synthesis on this space, we refer the reader to [Mi17], [DiEtAl15], [Ka15] and the related references therein.

On the other side, Reich [Re71] generalized Banach fixed point theorem for single valued as well as multivalued mappings. Since then Reich type mappings have been the center of intensive research for many authors. In recent investigations, Wardowski [War12] described a new contraction, where the author proves fixed point results in a very general setting in the so-called F-contraction. Later on Secelean et al. [Sec13], Piri and Kumam [Pir14] refined the result of Wardowski [War12] by launching some weaker conditions on the self mapping regarding a complete metric space and over the mapping F (for more details on F-contraction see, e.g., [SiEtAl17], [SiEtAl18], [Na17], [NaEtAl17] and the related references

M. Younis · A. Goyal
Department of Applied Mathematics, UIT-Rajiv Gandhi Technological University (University of Technology of M.P.), Bhopal, M.P., India

D. Singh (✉)
Department of Applied Sciences, NITTTR, Under Ministry of HRD, Government of India, Bhopal, M.P., India

© Springer Nature Switzerland AG 2019   451
C. Constanda, P. Harris (eds.), *Integral Methods in Science and Engineering*,
https://doi.org/10.1007/978-3-030-16077-7_35

therein). Secelean [Sec13] established Lemma 1 by utilizing an equivalent but a more simple condition (F_2') instead of condition (F_2).

In the rest of the analysis, we denote the set of all functions satisfying $(F1)$ of [War12], $(F3')$ of [Sec13] and $(F3')$ of [Pir14] by $\varXi$.

In this paper, by extending F-contraction to Reich type mappings, we inaugurate F-Reach contraction in the context of $RbMS$. Some nontrivial illustrative examples embellish the established results along with the computer simulation. Materiality of the presented results is governed by the application part. Package of F-contraction and Reich type mappings in the framework of $RbMS$ makes our results novel and newfangled, since $RbMS$ generalizes the concepts of b-metric space, rectangular metric space. Hence our results revamp and generalize some of the existing state-of-the-art in the literature which is also discussed in the later part of the article.

For the rest of the hypothesis, $\mathbb{R}$, $\mathbb{N}$ and $\mathbb{R}^+$ denote the set of all positive real numbers, natural numbers and the set of all real non-negative numbers, respectively.

We now enunciate some primary concepts and notations which are productive for the succeeding part of the paper.

Definition 1 [GeEtAl15] A rectangular b-metric on a nonempty set Y is a mapping $r_b : Y \times Y \to [0, \infty)$ with $s \geq 1$ satisfying the following conditions:

$(r_b M1)$ $b_r(u, v) = 0 \iff u = v$ for all $u, v \in Y$;
$(r_b M2)$ $b_r(u, v) = r_b(v, u)$ for all $u, v \in Y$;
$(r_b M3)$ $b_r(u, v) \leq s[b_r(u, a) + b_r(a, b) + b_r(b, v)]$ for all $u, v \in Y$ and all distinct points $a, b \in Y \setminus \{u, v\}$.

The pair (Y, b_r) is called rectangular b-metric $(R_b MS)$ space with coefficient s on Y.

The following lemma will be productive for establishing our main results.

Lemma 1 ([Sec13]) *Let $F : \mathbb{R}^+ \to \mathbb{R}$ be an increasing map and q_n be a sequence of positive real numbers. Then the following assertions hold:*

(a) if $\lim\limits_{n \to \infty} F(q_n) = -\infty$ then $\lim\limits_{n \to \infty} q_n = 0$;

(b) if $\inf F = -\infty$ and $\lim\limits_{n \to \infty} q_n = 0$; then $\lim\limits_{n \to \infty} F(q_n) = -\infty$.

For other terminology and notations like completeness, continuity and topology along with the noteworthy remarks in the associated spaces, see [GeEtAl15].

35.2 F-Reich Contraction

Now, we introduce our main definition as follows:

Definition 2 Let (Y, b_r, s) be a $RbMS$. A mapping $J : Y \to Y$ is said to be an F-Reich contraction (in short F-RC) on a $RbMS$ if $\exists\ \wp > 0$ and $F \in \Xi$ such that

$$\wp + F\Big(b_r(Ju, Jv)\Big) \le F\Big(\kappa b_r(u, v) + \zeta b_r(u, Ju) + \xi b_r(v, Jv)\Big), \qquad (35.1)$$

$\forall u, v \in Y$, where κ, ζ, ξ are nonnegative constants with $\kappa + \zeta + \xi < 1$ and $\kappa < \frac{1}{s}$.

Example 1 Let (Y, b_r, s) be a $RbMS$. Define a mapping $F : \mathbb{R}^+ \to \mathbb{R}$ by $F(q) = q + \ln q$, then $F \in \Xi$ and each mapping $J : Y \to Y$ satisfying (35.1) is F-RC such that

$$e^{-\wp} \ge \frac{b_r(Ju, Jv)e^{b_r(Ju, Jv) - \{\kappa b_r(u,v) + \zeta b_r(u, Ju) + \xi b_r(v, Jv)\}}}{\kappa b_r(u, v) + \zeta b_r(u, Ju) + \xi b_r(v, Jv)}$$

for all $u, v \in Y$ and $Ju \ne Jv$.

Example 2 Let (Y, b_r, s) be a $RbMS$ and consider $F(q) = -(q)^{-\frac{1}{2}} \forall\ q > 0$ then $F \in \Xi$ and each mapping $J : Y \to Y$ satisfying (35.1) is F-RC such that

$$b_r(Ju, Jv) \le \frac{\kappa b_r(u, v) + \zeta b_r(u, Ju) + \xi b_r(v, Jv)}{\Big(1 + \wp\sqrt{\kappa b_r(u, v) + \zeta b_r(u, Ju) + \xi b_r(v, Jv)}\Big)^2}$$

for all $u, v \in Y$ and $Ju \ne Jv$.

Our main result runs as follows.

Theorem 1 Let (Y, b_r, s) be a complete $RbMS$ and $J : Y \to Y$ be an F-RC. Then J admits a unique fixed point in Y.

Proof Let $y_0 \in Y$ and define a Picard sequence $\{y_m\}$ with initial point y_0, that is, $y_m = J^m y_0 = J y_{m-1}$. If $y_m = y_{m+1}$, for some $m \in \mathbb{N}$, then y_m is the desired fixed point of J and we are through in this case. So suppose that $y_m \ne y_{m+1} \forall n \in \mathbb{N}$. Utilizing contractive condition (35.1), for $u = y_{m-1}$ and $v = y_m$, we have

$$\wp + F\Big(b_r(y_m, y_{m+1})\Big) = \wp + F\Big(b_r(Jy_{m-1}, Jy_m)\Big)$$

$$\le F\Big(\kappa b_r(y_{m-1}, y_m) + \zeta b_r(y_{m-1}, Jy_{m-1}) + \xi b_r(y_m, Jy_m)\Big)$$

$$\le F\Big(\kappa b_r(y_{m-1}, y_m) + \zeta b_r(y_{m-1}, y_m) + \xi b_r(y_m, y_{m+1})\Big)$$

$$= F\Big((\kappa + \zeta)b_r(y_{m-1}, y_m) + \xi b_r(y_m, y_{m+1})\Big)$$

Since $\wp > 0$ and F is strictly increasing, we deduce

$$b_r(y_m, y_{m+1}) < \left(\frac{\kappa + \zeta}{1 - \xi}\right) b_r(y_{m-1}, y_m)$$

By hypothesis, $\kappa + \zeta + \xi < 1$, it follows that

$$F\Big(b_r(y_m, y_{m+1})\Big) \le F\Big(b_r(y_{m-1}, y_m)\Big) - \wp$$

Continuing this process, we obtain

$$
\begin{aligned}
F\Big(b_r(y_m, y_{m+1})\Big) &\le F\Big(b_r(y_{m-1}, y_m)\Big) - \wp \\
&= F\Big(b_r(Jy_{m-2}, Jy_{m-1})\Big) - \wp \\
&\le F\Big(b_r(y_{m-2}, y_{m-1})\Big) - 2\wp \\
&= F\Big(b_r(Jy_{m-3}, Jy_{m-3})\Big) - 2\wp \\
&\le F\Big(b_r(y_{m-3}, y_{m-3})\Big) - 3\wp \\
&\qquad\vdots \\
&\le F\Big(b_r(y_0, y_1)\Big) - m\wp,
\end{aligned}
$$

and so $\lim\limits_{m \to \infty} F\Big(b_r(Jy_{m-1}, Jy_m)\Big) = -\infty$, which along with Lemma 1 and F_2' gives

$$\lim_{m \to \infty} b_r(y_m, y_{m+1}) = 0. \tag{35.2}$$

Furthermore, taking $u = y_{m-1}$ and $v = y_{m+1}$ in (35.1), and using rectangular inequality, we have

$$
\begin{aligned}
\wp + F\Big(b_r(y_m, y_{m+2})\Big) &= \wp + F\Big(b_r(Jy_{m-1}, Jy_{m+1})\Big) \\
&\le F\Big(\kappa b_r(y_{m-1}, y_{m+1}) + \zeta b_r(y_{m-1}, Jy_{m-1}) + \xi b_r(y_{m+1}, Jy_{m+1})\Big) \\
&= F\Big(\kappa b_r(y_{m-1}, y_{m+1}) + \zeta b_r(y_{m-1}, y_m) + \xi b_r(y_{m+1}, y_{m+2})\Big) \\
&\le F\Big(s\kappa\, b_r(y_{m-1}, y_{m+2}) + s\kappa\, b_r(y_{m+2}, y_m) + s\kappa\, b_r(y_m, y_{m+1}) \\
&\qquad + \zeta b_r(y_{m-1}, y_m) + \xi b_r(y_{m+1}, y_{m+2})\Big)
\end{aligned}
$$

$$\leq F\Big(s^2\kappa\ b_r(y_{m-1}, y_m) + s^2\kappa\ b_r(y_m, y_{m+1}) + s^2\kappa\ b_r(y_{m+1}, y_{m+2})$$

$$+ s\kappa\ b_r(y_{m+2}, y_m) + s\kappa\ b_r(y_m, y_{m+1}) + \zeta b_r(y_{m-1}, y_m)$$

$$+ \xi b_r(y_{m+1}, y_{m+2})\Big)$$

Again, since $\wp > 0$ and F is increasing, we acquire

$$(1 - s\kappa)b_r(y_m, y_{m+2}) < (s^2\kappa + \zeta)b_r(y_{m-1}, y_m) + (s^2\kappa + s\kappa)b_r(y_m, y_{m+1})$$

$$+ (s^2\kappa + \xi)b_r(y_{m+1}, y_{m+2}),$$

$$(35.3)$$

Also $\kappa < \frac{1}{s}$, passing limit $m \to \infty$ and utilizing (35.2), inequality (35.3) gives rise to

$$\lim_{n\to\infty} b_r(y_m, y_{m+2}) = 0 \tag{35.4}$$

Now, we claim that $\{y_m\}$ is a Cauchy sequence in Y. Assume to the contrary that $\exists$ $\delta > 0$ and sequences $\{y_{n_j}\}$ and $\{y_{m_j}\}$ of natural numbers, where m_j is the smallest index such that

$$m_j > n_j > j \quad b_r(y_{n_j}, y_{m_j}) \geq \delta, \quad b_r(y_{n_j}, y_{mj-1}) < \delta, \ \forall m \in \mathbb{N}. \tag{35.5}$$

Hence, we have

$$\delta \leq b_r(y_{n_j}, y_{m_j}) \leq s[b_r(y_{n_j}, y_{n_{j+1}}) + b_r(y_{n_{j+1}}, y_{m_{j+1}}) + b_r(y_{m_{j+1}}, y_{m_j})]$$

Thanks to (35.2) and (35.4), we get

$$\frac{\delta}{s} \leq \limsup_{m\to\infty} b_r(y_{n_{j+1}}, y_{m_{j+1}}) \tag{35.6}$$

Again

$$b_r(y_{n_j}, y_{m_j}) \leq s[b_r(y_{n_j}, y_{mj-1}) + b_r(y_{mj-1}, y_{m_{j+1}}) + b_r(y_{m_{j+1}}, y_{m_j})]$$

By virtue of (35.2), (35.4), and (35.5), we obtain

$$\limsup_{m\to\infty} b_r(y_{n_j}, y_{m_j}) \leq s\delta. \tag{35.7}$$

Taking $u = y_{n_j}$ and $v = y_{m_j}$ in(35.1), it follows that

$$F\left(b_r(y_{n_j+1}, y_{m_j+1})\right) = F\left(b_r(J y_{n_j}, J y_{m_j})\right)$$
$$\leq F\left(\kappa b_r(y_{n_j}, y_{m_j}) + \zeta b_r(y_{n_j}, y_{n_j+1}) + \xi b_r(y_{m_j}, y_{m_j+1})\right)$$
$$- \wp,$$

Making use of (F_2'), (35.6), and (35.7), the above inequality reduces to

$$F\left(\frac{\delta}{s}\right) < F(s\delta\kappa),$$

which is a contradiction in view of hypothesis.

Hence it follows that $\{y_m\}$ is a Cauchy sequence in (Y, b_r). By completeness of (Y, b_r) ∃ $y^* \in Y$ such that

$$\lim_{m \to \infty} y_m = y^* \tag{35.8}$$

For the existence of fixed point, we assume that $y^* \neq J y^*$

$$F\left(b_r(y^*, J y^*)\right) \leq F\left(s[b_r(y^*, y_m) + b_r(y_m, y_{m+1}) + b_r(y_{m+1}, J y^*)]\right)$$
$$= F\left(s[b_r(y^*, y_m) + b_r(y_m, y_{m+1}) + b_r(J y_m, J y^*)]\right)$$
$$\leq F\left(s[b_r(y^*, y_m) + b_r(y_m, y_{m+1}) + \kappa b_r(y_m, y^*)\right.$$
$$\left. + \zeta b_r(y_m, y_{m+1}) + \xi b_r(y^*, J y^*) - \wp]\right)$$

Since F is continuous, passing limit $m \to \infty$; and on account of (F_1), (35.2), and (35.8), we get

$$b_r(y^*, J y^*) < \xi b_r(y^*, J y^*),$$

which is absurd, since $\kappa + \zeta + \xi < 1$. Hence we must have $J y^* = y^*$, which guarantees that y^* is the fixed point of J.

For the uniqueness, suppose y be another fixed point of J, then

$$F\left(b_r(y^*, y)\right) = F\left(b_r(J y^*, J y)\right)$$
$$\leq F\left(\kappa b_r(y^*, y) + \zeta b_r(y^*, J y) + \xi b_r(y, J y) - \wp\right)$$
$$= F\left(\kappa b_r(y^*, y) - \wp\right),$$

which amounts to say that $b_r(y^*, y) < \kappa b_r(y^*, y)$, which is a contradiction in view of hypothesis. Hence $b_r(y^*, y) = 0$ and consequently, J admits a unique fixed point.

To make Theorem 1 explicit, we expound the following example.

Example 3 Let $Y = [0, 2]$ and consider a rectangular b-metric $b_r : Y \times Y \to [0, \infty)$ defined by

$$b_r(u, v) = (u - v)^2 ; \quad \forall\ u, v \in Y.$$

Then (Y, b_r) is a complete $RbMS$ with $s = 3$. Let the mapping $J : Y \to Y$ be defined by

$$Jy = \frac{e^{\log(5+y^2)} + \sin(1 + y)}{\sqrt{4 + y^2}}; \quad \forall\ y \in Y.$$

In order to check the validation of inequality (35.1) with $F(q) = -\frac{1}{q}$, $\wp = \frac{12}{13}$ and $\kappa = \frac{1}{5}, \zeta = \frac{3}{5}, \xi = \frac{4}{25}$. Clearly $F \in \Xi$ and $\kappa + \zeta + \xi < 1$ with $\kappa < \frac{1}{s}$.
Now, consider the left-hand side of the inequality (35.1), we have

$$\wp + F\left(b_r(Ju, Jv)\right) = \frac{12}{13} - \frac{1}{\left(\frac{e^{\log(5+u^2)}+\sin(1+u)}{\sqrt{4+u^2}} - \frac{e^{\log(5+v^2)}+\sin(1+v)}{\sqrt{4+v^2}}\right)^2},$$

Calculating various terms appearing in the inequality (35.1), right-hand side comes out to be

$$F\left(\kappa b_r(u, v) + \zeta b_r(u, Ju) + \xi b_r(v, Jv)\right)$$

$$= -\frac{1}{0.2(u - v)^2 + 0.6\left(u - \frac{e^{\log(5+u^2)}+\sin(1+u)}{\sqrt{4+u^2}}\right)^2 + 0.16\left(v - \frac{e^{\log(5+v^2)}+\sin(1+v)}{\sqrt{4+v^2}}\right)^2}$$

Figure 35.1 exemplifies that the surface representing the left-hand side of the inequality (35.1) is dominated by the surface representing the right-hand side, thereby authenticating the validity of inequality (35.1). This proves that all the assertions of Theorem 1 are verified. Hence J is an F-RC and has a unique fixed point $u = 1.2784$ in Y.

Remark 1 Since every metric space is $RbMS$ but the converse is not necessarily true. Thus Theorem 1 is a real generalization of Reich contraction [Re71] in the context of $RbMS$ in the sense of F-contraction.

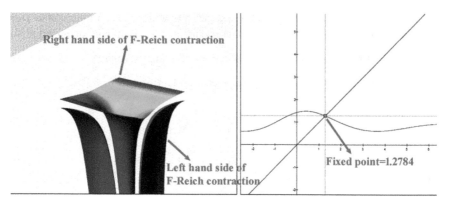

Fig. 35.1 Validation of F-Reich contraction

Taking $\kappa = 0$ in Theorem 1, we obtain the F-contraction version of Kannan type result [Ka68] in the framework of $RbMS$ as follows:

Corollary 1 *Let (Y, b_r, s) be a RbMS and $J : Y \to Y$ be such that $\exists\ \wp > 0$ and $F \in \Xi$ with*

$$\wp + F\left(b_r(Ju, Jv)\right) \leq F\left(\zeta b_r(u, Ju) + \xi b_r(v, Jv)\right),$$

$\forall u, v \in Y$, where ζ, ξ are nonnegative constants and $\zeta + \xi < 1$. Then J admits a unique fixed point.

Taking $\zeta = \xi = 0$ in Theorem 1, we obtain the F-contraction version of Banach contraction principle [Ba22] in the framework of $RbMS$ as follows:

Corollary 2 *Let (Y, b_r, s) be a RbMS and $J : Y \to Y$ be such that $\exists\ \wp > 0$ and $F \in \Xi$ with*

$$\wp + F\left(b_r(Ju, Jv)\right) \leq F\left(\kappa b_r(u, v)\right),$$

$\forall u, v \in Y$, where $0 < \kappa < 1$. Then J admits a unique fixed point.

Remark 2 In view of the established results in this article, we extend and generalize some pioneer results given in Cosentino and Vetro [Cos14], Vetro [Ve16] and [MiEtAl18] in the context of $RbMS$.

35.3 Applications

This section is devoted to signify the materiality of the obtained results.

35.3.1 Application to Concentration of a Diffusing Substance

Consider a diffusing substance placed in an absorbing medium between parallel walls such that β_1, β_2 are the stipulated concentrations at walls . Moreover, let $\Omega(r)$ be the given source density and $\Gamma(r)$ be the known absorption coefficient. Then the concentration $\chi(r)$ of the substance under the aforementioned hypothesis governs the following boundary value problem

$$\begin{cases} -\chi'' + \Gamma(r)\chi = \Omega(r) \quad ; \; r \in [0, 1] = I \\ \qquad\quad \chi(0) = \beta_1, \chi(0) = \beta_2, \end{cases} \tag{35.9}$$

Problem (35.9) is equivalent to the succeeding integral equation

$$\chi(r) = \beta_1 + (\beta_2 - \beta_1)r + \int_0^1 \Theta(r, \tau) \left(\Omega(\tau) - \Gamma(\tau)\chi(\tau)\right), \quad r \in [0, 1], \tag{35.10}$$

where $\Theta(r, \tau) : [0, 1] \times \mathbb{R} \to \mathbb{R}$ is the Green's function which is continuous and is given by

$$\Upsilon(r, \tau) = \begin{cases} r(1 - \tau) & 0 \leq r \leq \tau \leq 1, \\ \tau(1 - r) & 0 \leq \tau \leq r \leq 1. \end{cases} \tag{35.11}$$

Suppose that $C(I, \mathbb{R}) = Y$ is the space of all real valued continuous functions defined on I and let Y be endowed with the rectangular b-metric b_r defined by

$$b_r(\chi, \chi^*) = \|(\chi - \chi^*)^2\|,$$

where $\|\chi\| = \sup\{|\chi(r)| : r \in I\}$. Obviously (Y, b_r) is a complete $RbMS$ with $s = 2$.

Let the operator $J : Y \to Y$ be defined by

$$J\chi(r) = \chi(r) = \beta_1 + (\beta_2 - \beta_1)r + \int_0^1 \Theta(r, \tau) \left(\Omega(\tau) - \Gamma(\tau)\chi(\tau)\right).$$

Then χ^* is a unique solution of (35.10) $\Longleftrightarrow$ it is a fixed point of J. Subsequent Theorem is furnished for the assertion of the existence of fixed point of J

Theorem 2 *Consider the problem (35.10) and suppose that there exists $\wp > 0$ such that for all $\tau \in I$, the following assertion holds:*

$$0 \leq |\Gamma(\tau)\chi(\tau) - \Gamma(\tau)\chi^*(\tau)| \leq \frac{1}{\sqrt{e^\wp}} \left(\chi^*(\tau) - \chi(\tau)\right),$$

Then the integral equation (35.10) has a unique solution in Y.

Proof Clearly for $\chi \in Y$ and $r \in I$ the mapping $J : Y \to Y$ is well defined.

$$|J\chi(r) - J\chi^*(r)|$$

$$= \left| \int_0^1 \Theta(r,\tau)\left(\Omega(\tau) - \Gamma(\tau)\chi(\tau)\right) d\tau - h(r) \right.$$

$$\left. - \int_0^1 \Theta(r,\tau)\left(\Omega(\tau) - \Gamma(\tau)\chi^*(\tau)\right) d\tau \right|$$

$$\leq \int_0^1 \Theta(r,\tau)\left|\left(\Omega(\tau) - \Gamma(\tau)\chi(\tau)\right) - \left(\Omega(\tau) - \Gamma(\tau)\chi^*(\tau)\right)\right| d\tau$$

$$= \int_0^1 \Theta(r,\tau)\left|\Gamma(\tau)\chi(\tau) - \Gamma(\tau)\chi^*(\tau)\right| d\tau$$

$$\leq \frac{1}{\sqrt{e^\wp}} \int_0^1 \Theta(r,\tau)\left|\chi(\tau) - \chi^*(\tau)\right| d\tau$$

$$= \frac{1}{\sqrt{e^\wp}} \int_0^1 \Theta(r,\tau)\sqrt{(\chi(\tau) - \chi^*(\tau))^2} d\tau$$

$$\leq \frac{1}{\sqrt{e^\wp}} \int_0^1 \Theta(r,\tau)\sqrt{\|(\chi(\tau) - \chi^*(\tau))^2\|} d\tau$$

$$\leq \frac{1}{\sqrt{e^\wp}} \sqrt{\|(\chi - \chi^*)^2\|} \sup_{r\in[0,1]} \int_0^1 \Theta(r,\tau) d\tau.$$

Since $\int_0^1 \Theta(r,\tau) d\tau = \frac{r - r^2}{2}$ and so $\sup_{r\in[0,1]} \int_0^1 \Theta(r,\tau) d\tau = \frac{1}{8}$. Hence for all $\chi, \chi^* \in Y$, we obtain

$$b_r(J\chi, J\chi^*) \leq e^{-\wp} \frac{b_r(\chi, \chi^*)}{64}.$$

Passing logarithm on both sides, we acquire

$$\wp + \log\left(b_r(J\chi, J\chi^*)\right) \leq \log\left(\frac{b_r(\chi, \chi^*)}{64}\right).$$

Taking $\frac{1}{64} = \kappa$, and noting that $F \in \Xi$ with $F(q) = log(q)$, for every $q > 0$, it follows that

$$\wp + F\left(b_r(J\chi, J\chi^*)\right) \leq F\left(\kappa b_r(\chi, \chi^*)\right)$$

$$\leq F\left(\kappa b_r(\chi, \chi^*) + \zeta b_r(\chi, J\chi) + \xi b_r(\chi^*, J\chi^*)\right),$$

where κ, ζ, ξ are nonnegative constants with $\kappa + \zeta + \xi < 1$ and $\kappa < \frac{1}{s} = \frac{1}{2}$.

Hence all the hypotheses of Theorem 1 are contented. We conclude that J has a unique fixed point χ in Y, which guarantees that the integral equation (35.10) has a unique solution and consequently the boundary value problem (35.9) has a unique solution.

35.3.2 Application to Integral Equation

As a consequence of our results, we furnish an existence theorem for the unique solution of the following integral equation

$$z(r) = h(r) + \int_0^1 \mathfrak{F}(r, t, z(t))dt \; ; \quad r \in [0, 1] = I, \tag{35.12}$$

where $h : I \to \mathbb{R}$ and $\mathfrak{F} : I^2 \times \mathbb{R} \to \mathbb{R}$ are continuous functions.

Suppose $C(I, \mathbb{R}) = Y$ is the space of all real valued continuous functions defined on I and the rectangular b-metric b_r be defined as in the Sect. 35.3.1.

Consider the operator $J : Y \to Y$ defined by

$$Jz(r) = h(r) + \int_0^1 \mathfrak{F}(r, t, z(t))dt.$$

Then z^* is a unique solution of (35.12) $\iff$ it is a fixed point of J.

Theorem 3 *Consider the problem (35.12) and suppose that $\exists \; 1 < a < 2$ and $\wp > 0$ such that for every $t \in I$ and for all $r \in I$, following assertion holds:*

$$0 \le \left| \mathfrak{F}(r, t, z(t)) - \mathfrak{F}(r, t, z^*(t)) \right| \le \Lambda(r, t) \left(z^*(t) - z(t) \right),$$

and

$$\sup_{0 \le r \le 1} \int_0^1 \Lambda(r, t)dt \le \frac{1}{\sqrt{a(e^\wp)}}$$

Then the integral equation (35.12) has a unique solution in Y.

Proof In order to prove the theorem, it amounts to show that the operator $J : Y \to Y$ is an F-RC.

$$\left| Jz(r) - Jz^*(r) \right| = \left| h(r) + \int_0^1 \mathfrak{F}(r, t, z(t))dt - h(r) - \int_0^1 \mathfrak{F}(r, t, z^*(t))dt \right|$$

$$= \left| \int_0^1 \mathfrak{F}(r, t, z(t))dt - \int_0^1 \mathfrak{F}(r, t, z^*(t))dt \right|$$

$$\leq \int_0^1 \left| \mathfrak{F}(r, t, z(t)) - \mathfrak{F}(r, t, z^*(t)) \right| dt$$

$$\leq \int_0^1 \Lambda(r, t) \left| z(t) - z^*(t) \right| dt$$

$$= \int_0^1 \Lambda(r, t) \sqrt{(z(t) - z^*(t))^2} dt$$

$$\leq \int_0^1 \Lambda(r, t) \sqrt{\left\| (z(t) - z^*(t))^2 \right\|} dt$$

$$\leq \frac{1}{\sqrt{a(e^\wp)}} \sqrt{\left\| (z - z^*)^2 \right\|}$$

Equivalently, for all $z, z^* \in Y$, we have

$$b_r(Jz, Jz^*) \leq \frac{e^{-\wp}}{a} b_r(z, z^*).$$

Passing logarithm on both sides, we acquire

$$\wp + \log\left(b_r(Jz, Jz^*)\right) \leq \log\left(\frac{1}{a} b_r(z, z^*)\right).$$

Since $1 < a < 2$, taking $\frac{1}{a} = \kappa$, and noting that $F \in \Xi$ with $F(q) = log(q)$, for every $q > 0$, it follows that

$$\wp + F\left(b_r(Jz, Jz^*)\right) \leq F\left(\kappa b_r(z, z^*)\right)$$

$$\leq F\left(\kappa b_r(z, z^*) + \zeta b_r(z, Jz) + \xi b_r(z^*, Jz^*)\right),$$

where κ, ζ, ξ are nonnegative constants with $\kappa + \zeta + \xi < 1$ and $\kappa < \frac{1}{s} = \frac{1}{2}$.

Hence all the hypotheses of Theorem 1 are contented. We conclude that J has a unique fixed point z in Y, which is the desired unique solution of integral equation (35.12).

Open Problems

- Establish analogous results of Edelstein, Meir Keelar, Hardy-Roger type contractions in the underlying space in the sense of F-contraction with applications.
- Can the results manifested in this article or their variants be applied to establish the existence of solution of the following Bessel function of first kind of order m

$$u^2 v'' + uv' + (\lambda^2 u - \frac{m^2}{u})v = 0$$

for some non-negative constant λ?

References

[Ba22]   Banach, S.: Sur les operations dans les ensembles abstraits et leur application aux equations integrales. *Fundam. Math.*, **3**, 133–181 (1922).

[Br00]   Branciari, A.: A fixed point theorem of Banach-Caccippoli type on a class of generalised metric spaces. *Publ. Math. Debrecen*, **57**, 31–37 (2000).

[Cos14]   Cosentino, M. and Vetro, P.: Fixed Point Results for F-Contractive Mappings of Hardy-Rogers-Type. *Filomat*, **28**(4), 715–722 (2014).

[Cz93]   Czerwik, S.: Contraction mappings in b-metric spaces. *Acta Math. Inform. Univ. Ostraviensis*, **1**(1), 5–11 (1993).

[DiEtAl15] Ding, H.S., Ozturk, V. and Radenović, S.: On some new fixed point results in b-rectangular metric spaces. *J. Nonlinear Sci.*, **8**, 378–86 (2015).

[GeEtAl15] George, R., Radenovic, S., Reshma, K.P., and Shukla, S.: Rectangular b-metric spaces and contraction principle. *J. Nonlinear Sci. Appl.*, **8**(6), 1005–1013 (2015).

[Ka15]   Kadelburg, Z. and Radenović, S.: Pata-type common fixed point results in b-metric and b-rectangular metric spaces. *J. Nonlinear Sci. Appl.*, **8**(6), 944–954 (2015).

[Ka68]   Kannan, R: Some results on fixed points. *Bull. Calcutta Math. Soc.*, **60**, 71–76 (1968).

[Mi17]   Mitrović, Z.D.: On An open problem in rectangular b-metric space. *J. Anal.*, **25**(1), 135–137 (2017).

[MiEtAl18] Mitrović, Z.D., George, R. and Hussain, N.: Some remarks on contraction mappings in rectangular b-metric spaces. *Bol. Soc. Paran. Mat.*, **22**(2), 1–9 (2018).

[NaEtAl17] Nashine, H.K., Agarwal, R.P., Shukla, S. and Gupta, A.: Some Fixed Point Theorems for Almost $(GF; \delta_b)$-Contractions and Application. *Fasciculi Mathematici*, **58**(1):123–43 (2017).

[Na17]   Nashine, H.K., Kadelburg, Z.: Existence of solutions of cantilever beam problem via $(\alpha - \beta - FG)$-contractions in b-metric-like spaces. *Filomat*, **31**(11):3057–74 (2017).

[Pir14]   Piri, H. and Kumam, P.: Some fixed point theorems concerning F-contraction in complete metric spaces. *Fixed Point Theory Appl.*,**2014**, 210(2014).

[Re71]   Reich, S.: Some remarks concerning contraction mappings., *Canad. Math. Bull.*, **14**, 121–124 (1971).

[Sec13]   Secelean, N.A.: Iterated function system consisting of F-contractions. *Fixed Point Theory Appl.*, **2013**, https://doi.org/10.1166/1687-1812-2013-277, 277(2013).

[SiEtAl17] Singh, D., Chauhan, V. and Altun I.: Common fixed point of a power graphic $(F - \psi)$-contraction pair on partial b-metric spaces with application. *Nonlinear analysis: Modelling and control*, **22**(5), 662–678 (2017).

[SiEtAl18] Singh, D., Chauhan, V., Kumam, P. and Joshi. V.: Some applications of fixed point results for generalized two classes of Boyd-Wong's F-contraction in partial b-metric spaces. *Math. Sc.*, **12**(2), 111–127 (2018).

[Ve16]   Vetro, F.: F-contractions of Hardy-Rogers type and application to multistage decision processes. *Nonlinear Anal. Model. Control*, **21**(4), 531–46 (2016).

[War12]   Wardowski, D.: Fixed points of a new type of contractive mappings in complete metric spaces. *Fixed Point Theory Appl.*, **2012**, 94(2012).

Chapter 36
On the Convergence of Dynamic Iterations in Terms of Model Parameters

Barbara Zubik-Kowal

36.1 Introduction

The goal of the paper is to investigate the convergence of the dynamic iterations $\{x^{(k)}(t)\}_{k=0}^{\infty}$, which satisfy the system

$$\dot{x}^{(k)}(t) = Ax^{(k)}(t) + Bx^{(k-1)}(t) + g(t) \tag{36.1}$$

constructed for

$$\dot{x}(t) = Mx(t) + g(t), \tag{36.2}$$

so that $\lim_{k\to\infty} x^{(k)}(t) = x(t)$, where the split matrix $M = A + B$ is given and $g(t)$ is a given vector function, and to address the question of how to optimally reorder the equations in (36.2) so that the convergence of the resulting modified iterations $\{y^{(k)}(t)\}_{k=0}^{\infty}$ is faster than the convergence of the original iterations $\{x^{(k)}(t)\}_{k=0}^{\infty}$. The sequence $\{y^{(k)}(t)\}_{k=0}^{\infty}$ satisfies the condition $\lim_{k\to\infty} y^{(k)}(t) = y(t) = \sigma(x(t))$, where σ is a corresponding permutation on the components $x_i(t)$ of the solution $x(t)$, and the iterates $y^{(k)}(t)$ are determined via

$$\dot{y}^{(k)}(t) = \tilde{A}y^{(k)}(t) + \tilde{B}y^{(k-1)}(t) + \tilde{g}(t). \tag{36.3}$$

That is, the iterates $y^{(k)}(t)$ are determined similarly to how the iterates $x^{(k)}(t)$ are determined from (36.1), except that (36.3) is applied to the alternative system that consists of the following reordered equations,

B. Zubik-Kowal (✉)
Department of Mathematics, Boise State University, Boise, ID, USA
e-mail: zubik@math.boisestate.edu

© Springer Nature Switzerland AG 2019   465
C. Constanda, P. Harris (eds.), *Integral Methods in Science and Engineering*,
https://doi.org/10.1007/978-3-030-16077-7_36

$$\dot{y}(t) = \tilde{M}y(t) + \tilde{g}(t),$$

where the matrix $\tilde{M} = \tilde{A} + \tilde{B}$ is generated from M by reordering its rows and the vector function $\tilde{g}(t)$ is generated from the original $g(t)$ by reordering its components, respectively.

For example, consider the following dynamic iteration scheme, also known as Gauss-Seidel waveform relaxation,

$$\begin{cases} \dot{x}_1^{(k)}(t) = -x_1^{(k)}(t) + 10^2 x_2^{(k-1)}(t) + 10^2 x_3^{(k-1)}(t) + g_1(t) \\ \dot{x}_2^{(k)}(t) = 10^{-2} x_1^{(k)}(t) - 10^2 x_2^{(k)}(t) + 10^2 x_3^{(k-1)}(t) + g_2(t) \\ \dot{x}_3^{(k)}(t) = 10^{-2} x_1^{(k)}(t) + 10^{-2} x_2^{(k)}(t) - 10^2 x_3^{(k)}(t) + g_3(t) \end{cases} \tag{36.4}$$

constructed for the system

$$\begin{cases} \dot{x}_1(t) = -x_1(t) + 10^2 x_2(t) + 10^2 x_3(t) + g_1(t) \\ \dot{x}_2(t) = 10^{-2} x_1(t) - 10^2 x_2(t) + 10^2 x_3(t) + g_2(t) \\ \dot{x}_3(t) = 10^{-2} x_1(t) + 10^{-2} x_2(t) - 10^2 x_3(t) + g_3(t). \end{cases} \tag{36.5}$$

and note that by exchanging the first and the third equations in (36.5), we get

$$\begin{cases} \dot{y}_3(t) = 10^{-2} y_1(t) + 10^{-2} y_2(t) - 10^2 y_3(t) + g_3(t) \\ \dot{y}_2(t) = 10^{-2} y_1(t) - 10^2 y_2(t) + 10^2 y_3(t) + g_2(t) \\ \dot{y}_1(t) = -y_1(t) + 10^2 y_2(t) + 10^2 y_3(t) + g_1(t) \end{cases} \tag{36.6}$$

and the following alternative Gauss-Seidel waveform relaxation scheme

$$\begin{cases} \dot{y}_3^{(k)}(t) = 10^{-2} y_1^{(k-1)}(t) + 10^{-2} y_2^{(k-1)}(t) - 10^2 y_3^{(k)}(t) + g_3(t) \\ \dot{y}_2^{(k)}(t) = 10^{-2} y_1^{(k-1)}(t) - 10^2 y_2^{(k)}(t) + 10^2 y_3^{(k)}(t) + g_2(t) \\ \dot{y}_1^{(k)}(t) = -y_1^{(k)}(t) + 10^2 y_2^{(k)}(t) + 10^2 y_3^{(k)}(t) + g_1(t). \end{cases} \tag{36.7}$$

Note that Gauss-Seidel waveform relaxation reflects the fact that the present iterates $x_2^{(k)}(t)$ and $x_3^{(k)}(t)$ are not used in the first equation of (36.4) and since the left-hand side of this equation involves the derivative $\dot{x}_1^{(k)}(t)$ from the present iterate, only $x_1^{(k)}(t)$ is used in this equation. Similarly, the present iterates $y_1^{(k)}(t)$ and $y_2^{(k)}(t)$ are not used in the first equation of (36.7). In place of $x_2^{(k)}(t)$ and $x_3^{(k)}(t)$ the previous iterates $x_2^{(k-1)}(t)$ and $x_3^{(k-1)}(t)$ are used in the first equation of (36.4) and, similarly, the previous iterates $y_1^{(k-1)}(t)$ and $y_2^{(k-1)}(t)$ are used in the first equation of (36.7) in place of the present iterates $y_1^{(k)}(t)$ and $y_2^{(k)}(t)$.

Gauss-Seidel waveform relaxation also reflects the fact that the present iterate $x_3^{(k)}(t)$ is replaced by $x_3^{(k-1)}(t)$ in the second equation of (36.4) and, similarly,

$y_1^{(k)}(t)$ is replaced by $y_1^{(k-1)}(t)$ in the second equation of (36.7). Moreover, only present iterates are used in the third equations of both (36.4) and (36.7). In summary, schemes (36.4) and (36.7) are obtained for the same system of ordinary differential equations and the difference between them is that Gauss-Seidel waveform relaxation is applied to its equations differently, resulting in different schemes that have different rates of convergence.

Although the scheme (36.7) has been created for (36.6) similarly (by applying Gauss-Seidel waveform relaxation) to the way in which the scheme (36.4) has been created for (36.5), it is illustrated in Sect. 36.3 that scheme (36.7) is 50% faster than scheme (36.4). This example demonstrates that we can accelerate convergence of the same method by a simple reordering of the equations in a given system. It also demonstrates that the coefficients given in the system play a decisive role in the construction of the iterative processes. In the next section, we analyze the convergence of the successive iterates in terms of the given coefficients and address the question of how to construct dynamic iterations in order to speed up their convergence.

Dynamic iteration schemes, also known as waveform relaxation techniques, were introduced as numerical schemes by Lelarasmee, Ruehli, and Sangiovanni-Vincentelli [LeEtAl82] for electrical system simulation. They were then developed by many authors for solving systems of ordinary differential equations, see, for example, [Bu95] and [MiNe96] and the references therein. They were also proposed for solving delay differential equations, see, e.g., [Bj94] and [Bj95] and more general functional differential equations, see, e.g., [ZuVa99], [Zu00], [Zu04]. However, the comparison of different rates of convergence of the different sequences of dynamic iterations obtained by a reordering of the differential equations in a given system was not considered in these papers. The influence of the order of differential equations in a given system on the rates of convergence of the resulting dynamic iteration schemes was investigated in [Zu17]. However, the results provided in [Zu17] are obtained for 2-dimensional iterates. In the present paper, we provide results for 3-dimensional iterates.

36.2   Convergence Analysis

In what follows, we use the convention that the sum $\sum_{j=0}^{-1}$ denotes zero. Let $M = [a_{ij}]_{i,j=1}^3$, let L be the lower triangular matrix created from M, let D be the diagonal matrix created from M, and let U be the upper triangular matrix created from M. We investigate the error propagated by the iterates $x^{(k)}(t)$, which satisfy scheme (36.1) with $A = L + D$ and $B = U$.

For completeness, system (36.2) is supplemented by the initial condition $x(0) = x_0$. The initial value x_0 is used for the initial conditions $x^{(k)}(0) = x_0, k = 0, 1, 2, \ldots$ for (36.1).

To provide an error analysis for alternative schemes (like, for example, (36.4) and (36.7)) constructed for a given differential system by reordering its equations, we use the following theorem.

Theorem 1 *Let*

$$\omega_{ml}^{(k)} = \sum_{j=0}^{k-1} a_{mm}^{k-1-j} a_{ll}^{j}, \quad \Omega_{ml}(t) = \sum_{k=1}^{\infty} \frac{t^k}{k!} \omega_{ml}^{(k)},$$

$$\gamma^{(k)} = a_{31}\omega_{13}^{(k)} + a_{21}a_{32} \sum_{i=0}^{k-2} a_{11}^{k-2-i} \omega_{23}^{(i+1)}, \quad \Gamma(t) = \sum_{k=1}^{\infty} \frac{t^k}{k!} \gamma^{(k)},$$

for $m, l = 1, 2, 3$, $k = 1, 2, \ldots$, and $t \geq 0$.

Then, the components of the error $e^{(k)}(t) = x^{(k)}(t) - x(t)$ are given by the formulas

$$e_1^{(k+1)}(t) = \int_0^t L_1^{(k)}(s)e^{a_{11}(t-s)}ds,$$

$$e_2^{(k+1)}(t) = \int_0^t \left(a_{21}\Omega_{12}(t-s)L_1^{(k)}(s) + e^{a_{22}(t-s)}L_2^{(k)}(s) \right)ds, \qquad (36.8)$$

$$e_3^{(k+1)}(t) = \int_0^t \left(\Gamma(t-s)L_1^{(k)}(s) + a_{32}\Omega_{23}(t-s)L_2^{(k)}(s) \right)ds,$$

where

$$L_1^{(k)}(t) = a_{12}e_2^k(t) + a_{13}e_3^k(t),$$

$$L_2^{(k)}(t) = a_{23}e_3^k(t),$$

for $k = 0, 1, \ldots$ and $t \geq 0$.

Proof Subtracting (36.2) from (36.1), we get the following relation for the error

$$\dot{e}^{(k)}(t) = Ae^{(k)}(t) + Be^{(k-1)}(t),$$

which implies the recurrence relation

$$e^{(k+1)}(t) = \int_0^t \exp\left((t - s)(L + D) \right) U e^k(s)ds. \qquad (36.9)$$

Since

$$\omega_{12}^{(1)} = 1, \quad \omega_{23}^{(1)} = 1, \quad \omega_{13}^{(1)} = 1$$

and

$$\gamma^{(1)} = a_{31}\omega_{13}^{(1)} + a_{21}a_{32}\sum_{i=0}^{-1} a_{11}^{-1-i}\omega_{23}^{(i+1)} = a_{31},$$

we get

$$L + D = \begin{bmatrix} a_{11} & 0 & 0 \\ a_{21} & a_{22} & 0 \\ a_{31} & a_{32} & a_{33} \end{bmatrix} = \begin{bmatrix} a_{11} & 0 & 0 \\ a_{21}\omega_{12}^{(1)} & a_{22} & 0 \\ \gamma^{(1)} & a_{32}\omega_{23}^{(1)} & a_{33} \end{bmatrix}.$$

We now suppose that

$$\left(L + D\right)^k = \begin{bmatrix} a_{11} & 0 & 0 \\ a_{21} & a_{22} & 0 \\ a_{31} & a_{32} & a_{33} \end{bmatrix}^k = \begin{bmatrix} a_{11}^k & 0 & 0 \\ a_{21}\omega_{12}^{(k)} & a_{22}^k & 0 \\ \gamma^{(k)} & a_{32}\omega_{23}^{(k)} & a_{33}^k \end{bmatrix},$$

for a certain $k = 1, 2, \ldots$. Then,

$$\left(L + D\right)^{k+1} = \begin{bmatrix} a_{11} & 0 & 0 \\ a_{21} & a_{22} & 0 \\ a_{31} & a_{32} & a_{33} \end{bmatrix} \begin{bmatrix} a_{11} & 0 & 0 \\ a_{21} & a_{22} & 0 \\ a_{31} & a_{32} & a_{33} \end{bmatrix}^k$$

$$= \begin{bmatrix} a_{11} & 0 & 0 \\ a_{21} & a_{22} & 0 \\ a_{31} & a_{32} & a_{33} \end{bmatrix} \begin{bmatrix} a_{11}^k & 0 & 0 \\ a_{21}\omega_{12}^{(k)} & a_{22}^k & 0 \\ \gamma^{(k)} & a_{32}\omega_{23}^{(k)} & a_{33}^k \end{bmatrix}$$

$$= \begin{bmatrix} a_{11}^{k+1} & 0 & 0 \\ a_{21}a_{11}^k + a_{22}a_{21}\omega_{12}^{(k)} & a_{22}^{k+1} & 0 \\ a_{31}a_{11}^k + a_{32}a_{21}\omega_{12}^{(k)} + a_{33}\gamma^{(k)} & a_{32}a_{22}^k + a_{33}a_{32}\omega_{23}^{(k)} & a_{33}^{k+1} \end{bmatrix}.$$

Note that

$$\omega_{12}^{(k)} = \omega_{21}^{(k)}, \quad \omega_{23}^{(k)} = \omega_{32}^{(k)}, \quad \omega_{13}^{(k)} = \omega_{31}^{(k)},$$

and

$$a_{21}a_{11}^k + a_{22}a_{21}\omega_{12}^{(k)} = a_{21}a_{22}\sum_{i=0}^{k-1} a_{22}^{k-1-i}a_{11}^i + a_{21}a_{11}^k$$

$$= a_{21}\sum_{i=0}^{k-1} a_{11}^i a_{22}^{k-i} + a_{21}a_{11}^k a_{22}^0$$

$$= a_{21}\sum_{i=0}^{k} a_{11}^i a_{22}^{k-i} = a_{21}\omega_{12}^{(k+1)}$$

for all $k = 1, 2, \ldots$. Similarly, we conclude that

$$a_{32}a_{22}^k + a_{32}a_{33}\omega_{23}^{(k)} = a_{32}\omega_{23}^{(k+1)}.$$

Observe that

$$
(a_{11} - a_{33})\sum_{i=0}^{k-2} a_{11}^{k-2-i}\omega_{23}^{(i+1)} = (a_{11} - a_{33})\sum_{i=0}^{k-2} a_{11}^{k-2-i}\sum_{j=0}^{i} a_{22}^{i-j}a_{33}^j
$$

$$
= \sum_{i=0}^{k-2}\sum_{j=0}^{i}\left(a_{11}^{k-1-i}a_{22}^{i-j}a_{33}^j - a_{11}^{k-2-i}a_{22}^{i-j}a_{33}^{j+1}\right).
$$

(36.10)

Since the right-hand side of (36.10) is equal to

$$
a_{11}^{k-1} - a_{11}^{k-2}a_{33} + \sum_{j=0}^{1}\left(a_{11}^{k-2}a_{33}^j - a_{11}^{k-3}a_{33}^{j+1}\right)a_{22}^{1-j} + \ldots
$$

$$
+ \sum_{j=0}^{k-2}\left(a_{11}a_{33}^j - a_{33}^{j+1}\right)a_{22}^{k-2-j},
$$

we conclude further that

$$
(a_{11} - a_{33})\sum_{i=0}^{k-2} a_{11}^{k-2-i}\omega_{23}^{(i+1)} = \left(a_{11}^{k-1} - a_{11}^{k-2}a_{33}\right) + \left(a_{11}^{k-2} - a_{11}^{k-3}a_{33}\right)a_{22}
$$

$$
+ \left(a_{11}^{k-2}a_{33} - a_{11}^{k-3}a_{33}^2\right) + \left(a_{11}^{k-3} - a_{11}^{k-4}a_{33}\right)a_{22}^2 + \left(a_{11}^{k-3}a_{33} - a_{11}^{k-4}a_{33}^2\right)a_{22}
$$

$$
+ \left(a_{11}^{k-3}a_{33}^2 - a_{11}^{k-4}a_{33}^3\right) + \cdots + \sum_{j=0}^{k-4} a_{11}^3 a_{33}^j a_{22}^{k-4-j} - \sum_{j=1}^{k-3} a_{11}^2 a_{33}^j a_{22}^{k-3-j}
$$

$$
+ \sum_{j=0}^{k-3} a_{11}^2 a_{33}^j a_{22}^{k-3-j} - \sum_{j=1}^{k-2} a_{11}a_{33}^j a_{22}^{k-2-j} + \sum_{j=0}^{k-2} a_{11}a_{33}^j a_{22}^{k-2-j} - \sum_{j=1}^{k-1} a_{33}^j a_{22}^{k-1-j},
$$

and, by reducing components, we obtain

$$(a_{11} - a_{33}) \sum_{i=0}^{k-2} a_{11}^{k-2-i} \omega_{23}^{(i+1)} = a_{11}^{k-1} + a_{11}^{k-2} a_{22} + a_{11}^{k-3} a_{22}^2 + a_{11}^{k-4} a_{22}^3 + \cdots +$$

$$+ a_{11}^3 a_{22}^{k-4} + a_{11}^2 a_{22}^{k-3} + a_{11} a_{22}^{k-2} - \sum_{j=1}^{k-1} a_{33}^j a_{22}^{k-1-j} = \sum_{j=1}^{k-1} a_{11}^j a_{22}^{k-1-j} - \sum_{j=1}^{k-1} a_{33}^j a_{22}^{k-1-j}$$

$$= \sum_{j=1}^{k-1} \left(a_{11}^j - a_{33}^j \right) a_{22}^{k-1-j} = \sum_{j=1}^{k-1} \left(a_{11}^{k-1-j} - a_{33}^{k-1-j} \right) a_{22}^j$$

$$= \sum_{j=0}^{k-1} \left(a_{11}^{k-1-j} - a_{33}^{k-1-j} \right) a_{22}^j.$$

On the other hand, we get

$$\omega_{12}^{(k)} - \omega_{23}^{(k)} = \sum_{j=0}^{k-1} a_{11}^{k-1-j} a_{22}^j - \sum_{j=0}^{k-1} a_{33}^{k-1-j} a_{22}^j.$$

Therefore,

$$\omega_{12}^{(k)} + a_{33} \sum_{i=0}^{k-2} a_{11}^{k-2-i} \omega_{23}^{(i+1)} = \omega_{23}^{(k)} + a_{11} \sum_{i=0}^{k-2} a_{11}^{k-2-i} \omega_{23}^{(i+1)}. \qquad (36.11)$$

We now apply (36.11) to prove that

$$a_{31} a_{11}^k + a_{32} a_{21} \omega_{12}^{(k)} + a_{33} \gamma^{(k)} = \gamma^{(k+1)}.$$

Therefore, $\exp((t - s)(L + D))$ equals to

$$\begin{bmatrix} \sum_{k=0}^{\infty} \dfrac{a_{11}^k (t - s)^k}{k!} & 0 & 0 \\ a_{21} \sum_{k=1}^{\infty} \dfrac{(t - s)^k}{k!} \omega_{12}^{(k)} & \sum_{k=0}^{\infty} \dfrac{a_{22}^k (t - s)^k}{k!} & 0 \\ \sum_{k=1}^{\infty} \dfrac{(t - s)^k}{k!} \gamma^{(k)} & a_{32} \sum_{k=1}^{\infty} \dfrac{(t - s)^k}{k!} \omega_{23}^{(k)} & \sum_{k=0}^{\infty} \dfrac{a_{33}^k (t - s)^k}{k!} \end{bmatrix}$$

$$= \begin{bmatrix} e^{a_{11}(t-s)} & 0 & 0 \\ a_{21} \Omega_{12}(t - s) \, e^{a_{22}(t-s)} & & 0 \\ \Gamma(t - s) & a_{32} \Omega_{23}(t - s) \, e^{a_{33}(t-s)} \end{bmatrix}$$

and

$$\exp\left((t-s)\left(L+D\right)\right)U = \begin{bmatrix} 0 & a_{12}e^{a_{11}(t-s)} & a_{13}e^{a_{11}(t-s)} \\ 0 & a_{12}a_{21}\Omega_{12}(t-s) & a_{23}e^{a_{22}(t-s)} + a_{13}a_{21}\Omega_{12}(t-s) \\ 0 & a_{12}\Gamma(t-s) & a_{23}a_{32}\Omega_{23}(t-s) + a_{13}\Gamma(t-s) \end{bmatrix}.$$

The result (36.8) follows from this and from (36.9).
    Since

$$\omega_{13}^{(k)} = \omega_{31}^{(k)},$$
$$\gamma^{(k)} = a_{31}\omega_{13}^{(k)} + a_{21}a_{32}\sum_{i=0}^{k-2} a_{11}^{k-2-i}\omega_{23}^{(i+1)},$$
$$= a_{31}\omega_{31}^{(k)} + a_{21}a_{32}\sum_{i=0}^{k-2} a_{33}^{k-2-i}\omega_{21}^{(i+1)},$$

for $k = 1, 2, \ldots$, we have that $\omega_{13}^{(k)}$, $\omega_{31}^{(k)}$ and $\gamma^{(k)}$ are symmetric with respect to a_{11} and a_{33}. Therefore, Γ is symmetric with respect to a_{11} and a_{33} and the proof is finished.

    In the next section, we provide results of numerical experiments and illustrate that reordering equations of a given system leads to faster convergence.

36.3 Numerical Examples

We now consider systems (36.5) and (36.6) and their corresponding iterative schemes (36.4) and (36.7), respectively. Note that (36.6) is obtained from system (36.5) through a reordering of its equations. We apply BDF3 with the step size $h = 10^{-2}$ to integrate (36.4) and (36.7) with respect to t, which results in numerical approximations $x_{i,n}^{(k)}$ and $y_{i,n}^{(k)}$ to $x_i(t_n)$ and $y_i(t_n)$, respectively. Here, $t_n = nh$, $n = 0, 1, \ldots$.
    The approximations $x_{1,n}^{(k)}$ computed over the interval $[0, 10]$ from (36.4) and compared to the exact solution $x_1(t_n)$ (solid lines) are presented in Fig. 36.1a for $k = 1, 2$ and in Fig. 36.1c for $k = 3, 4$. The first two iterations of (36.4) result in numerical solutions with high amplitudes (dashed and dash-dotted lines in Fig. 36.1a) that are far from the exact solution, which as a result manifests errors of about 10^2. The third and fourth iterates (dashed and dash-dotted lines in Fig. 36.1c) are closer to the exact solution but their errors are still of order 10. The curve obtained from the fifth iterate $x_{1,n}^{(5)}$ covers the curve of the exact solution $x_1(t_n)$ and the curve obtained from the first iterate $y_{1,n}^{(1)}$ computed from scheme (36.7) also covers the curve of the exact solution. This illustrates that (36.7) is faster than (36.4).
    The function

$$r(t) = \sqrt{(x_1(t))^2 + (x_2(t))^2 + (x_3(t))^2}$$

(solid lines) and the approximations

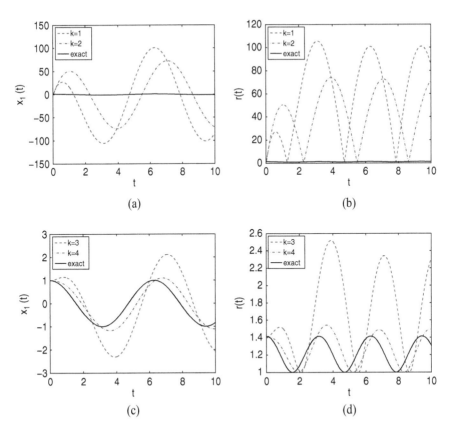

Fig. 36.1 Numerical and exact solutions

$$r^{(k)}(t) = \sqrt{(x_1^{(k)}(t))^2 + (x_2^{(k)}(t))^2 + (x_3^{(k)}(t))^2}$$

(dashed and dash-dotted lines) obtained from (36.4) are presented in Fig. 36.1b for $k = 1, 2$ and Fig. 36.1d for $k = 3, 4$. Three and four iterations of (36.4) result in approximations that are closer to the exact solution than those obtained after the first two iterations of (36.4), illustrating convergence of the scheme. Moreover, the curves obtained after five or more iterations of (36.4) cover the curve of the exact solution. On the other hand, all curves obtained after the first and higher iterations of (36.7) cover the curve of the exact solution, again confirming that scheme (36.7) is faster than (36.4).

The numerical errors

$$E_i^{(k)} = |x_i(t_n) - x_{i,n}^{(k)}| \tag{36.12}$$

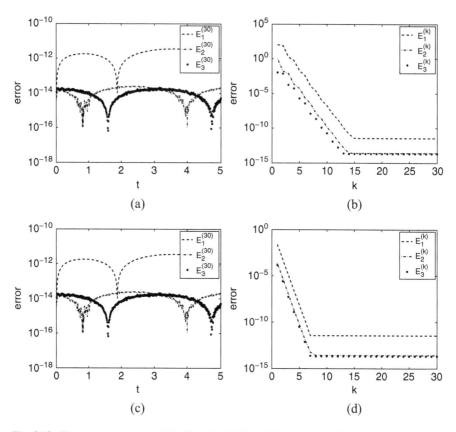

Fig. 36.2 The numerical errors (36.12) and (36.14) of the schemes (36.4) **(a)**, **(b)** and the numerical errors (36.13) and (36.15) of the scheme (36.7) **(c)**, **(d)**

of the approximations $x_{i,n}^{(k)}$ computed from (36.4) are presented in Fig. 36.2a and the errors

$$E_i^{(k)} = |y_i(t_n) - y_{i,n}^{(k)}| \tag{36.13}$$

of the approximations $y_{i,n}^{(k)}$ computed from (36.7) are presented in Fig. 36.2c on a logarithmic scale. Both errors (36.12) and (36.13) are presented for $k = 30$ as functions of time t_n over the interval $[0, 5]$.

The errors resulting from both numerical schemes (36.4) and (36.7) are also presented as functions of the iteration k at the fixed point $t_n = 5$ in Fig. 36.2b, d. Figure 36.2b presents the maximum errors

$$\tilde{E}_i^{(k)} = \max_n \{|x_i(t_n) - x_{i,n}^{(k)}|\} \tag{36.14}$$

of the numerical scheme (36.4) while Fig. 36.2d presents the errors

$$\tilde{E}_i^{(k)} = \max_n \{|y_i(t_n) - y_{i,n}^{(k)}|\} \tag{36.15}$$

of (36.7).

Figure 36.2 demonstrates convergence of both schemes, (36.4) and (36.7). However, it can be observed in Fig. 36.2b, d that scheme (36.7) is faster than (36.4). Figure 36.2a, c shows that the errors resulting from both schemes, (36.4) and (36.7), are well below the accuracy of 10^{-11}, which is in agreement with the fact that the approximations were computed by using BDF3 with the step size $h = 10^{-4}$. This accuracy is also observed in Fig. 36.2b, d. However, Figure 36.2b, d illustrates that scheme (36.4) converges in 14 iterations while scheme (36.7) converges in 7 iterations. We conclude that scheme (36.7) is twice as fast as scheme (36.7).

This numerical illustration is confirmed by Theorem 1. Since in the first equation of (36.4), the previous iterates $x_2^{(k-1)}$ and $x_3^{(k-1)}$ are multiplied by the coefficient 10^2, while in the first equation of (36.7) the previous iterates $y_1^{(k-1)}$ and $y_2^{(k-1)}$ are multiplied by the smaller coefficient 10^{-2}, and in the second equation of (36.4) the previous iterate $x_3^{(k-1)}$ is multiplied by the coefficient 10^2 while in the second equation of (36.7) the previous iterate $y_1^{(k-1)}$ is multiplied by the smaller coefficient 10^{-2}, scheme (36.7) demonstrates faster convergence than scheme (36.4). As seen in the proof of Theorem 1, the coefficients 10^2 and 10^{-2} are carried over and multiplied by the errors of the numerical schemes.

36.4   Concluding Remarks and Future Work

We have addressed the question of whether or not reordering the equations in a given system of linear ordinary differential equations written in terms of a matrix M of coefficients influences the rate of convergence of dynamic iterations based on the same splitting method applied to M and $\tilde{M}$, where $\tilde{M}$ is a matrix obtained by reordering the rows of M. We have derived formulas for the errors of the dynamic iterations and concluded that the coefficients of the given matrix M influence the convergence of the dynamic iterations, which leads to the conclusion that if the alternative matrix $\tilde{M}$ is generated in line with the formulas for the errors, then the iterations derived from $\tilde{M}$ converge faster than the iterations derived from M. Based on the derived formulas for the errors, we have concluded that the parameters of given systems should be taken into account in the generation of dynamic iterations in order to obtain faster convergence. This theoretical result confirms numerical examples, illustrating that iterative schemes generated according to the derived formulas can speed up the convergence of the resulting iterations by 50%.

Future work will address the question of how the differential equations should be ordered in higher dimensional systems so that the convergence of the applied iterative schemes is fastest.

References

[Bj94] Bjorhus, M.: On dynamic iteration for delay differential equations. *BIT*, **43**, 325–336 (1994).

[Bj95] Bjorhus, M.: A note on the convergence of discretized dynamic iteration, *BIT*, **35**, 291–296 (1995).

[Bu95] Burrage, K.: *Parallel and Sequential Methods for Ordinary Differential Equations*, Oxford University Press, Oxford (1995).

[LeEtAl82] Lelarasmee, E., Ruehli, A., and Sangiovanni-Vincentelli, A.: The waveform relaxation method for time-domain analysis of large scale integrated circuits, *IEEE Trans. CAD*, **1**, 131–145 (1982).

[MiNe96] Miekkala, U., and Nevanlinna, O.: Iterative solution of systems of linear differential equations, *Acta Numerica*, 259–307 (1996).

[ZuVa99] Zubik-Kowal, B., and Vandewalle, S.: Waveform relaxation for functional-differential equations, *SIAM J. Sci. Comput.* **21**, 207–226 (1999).

[Zu00] Zubik-Kowal, B.: Chebyshev pseudospectral method and waveform relaxation for differential and differential-functional parabolic equations, *Appl. Numer. Math.* **34**, 309–328 (2000).

[Zu04] Zubik-Kowal, B.: Error bounds for spatial discretization and waveform relaxation applied to parabolic functional differential equations, *J. Math. Anal. Appl.* **293**, 496–510 (2004).

[Zu17] Zubik-Kowal, B.: Propagation of errors in dynamic iterative schemes, Proceedings of EQUADIFF 2017, K. Mikula, D. Sevcovic and J. Urban (eds.), Published by Slovak University of Technology, SPEKTRUM STU Publishing, (2017), pp. 97–106.

Index

© Springer Nature Switzerland AG 2019
C. Constanda, P. Harris (eds.), *Integral Methods in Science and Engineering*,
https://doi.org/10.1007/978-3-030-16077-7